Annual Reports in
MEDICINAL CHEMISTRY

VOLUME **43**

Sponsored by the Division of Medicinal Chemistry
of the American Chemical Society

Editor-in-Chief

JOHN E. MACOR
Neuroscience Discovery Chemistry
Bristol-Myers Squibb
Wallingford, CT, United States

Section Editors

*ROBICHAUD • STAMFORD • BARRISH • MYLES •
PRIMEAU • LOWE • DESAI*

Amsterdam • Boston • Heidelberg • London • New York • Oxford
Paris • San Diego • San Francisco • Singapore • Sydney • Tokyo
Academic Press is an imprint of Elsevier

ELSEVIER

ACADEMIC
PRESS

Academic Press is an imprint of Elsevier
32 Jamestown Road, London NW1 7BY, UK
Radarweg 29, PO Box 211, 1000 AE Amsterdam, The Netherlands
Linacre House, Jordan Hill, Oxford OX2 8DP, UK
30 Corporate Drive, Suite 400, Burlington, MA 01803, USA
525 B Street, Suite 1900, San Diego, CA 92101-4495, USA

First edition 2008

Notice
No responsibility is assumed by the publisher for any injury and/or damage to persons or property as a matter of products liability, negligence or otherwise, or from any use or operation of any methods, products, instructions or ideas contained in the material herein. Because of rapid advances in the medical sciences, in particular, independent verification of diagnoses and drug dosages should be made

ISBN: 978-0-12-374344-2
ISSN: 0065-7743

For information on all Academic Press publications
visit our website at elsevierdirect.com

Printed and bound in USA

08 09 10 11 12   10 9 8 7 6 5 4 3 2 1

Working together to grow
libraries in developing countries
www.elsevier.com | www.bookaid.org | www.sabre.org

ELSEVIER   BOOK AID
International   Sabre Foundation

Annual Reports in
MEDICINAL CHEMISTRY

VOLUME **43**

CONTENTS

PART I: Central Nervous System Diseases

Section Editor: Albert J. Robichaud, Chemical & Screening Sciences, Wyeth Research, Princeton, New Jersey

PART II: Cardiovascular and Metabolic Diseases

Section Editor: Andy W. Stamford, Schering-Plough Research Institute, Kenilworth, New Jersey

PART III: Inflammation/Pulmonary/GI

Section Editor: Joel Barrish, Bristol-Myers Squibb R&D, Princeton, New Jersey

PART IV: Oncology

PART V: Infectious Diseases

PART VI: Topics in Biology

Section Editor: John Lowe, Pfizer Inc., Groton, Connecticut

PART VII: Topics in Drug Design and Discovery

Section Editor: Manoj C. Desai, Gilead Sciences, Inc., Foster City, California

Color Plate Section at the end of this book

CONTRIBUTORS

PREFACE

Annual Reports in Medicinal Chemistry has reached Volume 43. It continues to be a unique resource for the medicinal chemistry community, and I hope to keep it vibrant during my five-year tenure as Editor-in-Chief (2007–2011). That may sound a little strange to readers of this Preface, but, yes, I have a defined, five-year term. One of the "behind the scenes" aspects of the book today is reconnecting its governance to the Medicinal Chemistry Division of the ACS. The Division sponsors the book, and your dues to the Division pay for your copy. Thus, it makes perfect sense that YOU as Division members have a say in the content of the book, and I urge everyone to consider contributing a chapter to the book. I am serving as Editor-in-Chief at the request of the Executive Committee of the Medicinal Chemistry Division of the ACS. My five-year term is part of the plan to bring *Annual Reports* fully back under the auspices of the Executive Committee of the Medicinal Chemistry Division, and the next Editor-in-Chief will be chosen by that Committee for Volume 47 (2012). In the meantime my goal for the book is to continue its excellent tradition while looking for opportunities for improvement. Please contact me with any suggestions you may have.

Putting together an endeavor like *Annual Reports in Medicinal Chemistry* requires the assistance and dedication of many individuals, including the Section Editors, proof readers and administrative assistants. Firstly, I would like to thank the Section Editors for their hard work and dedication in creating Volume 43. John Primeau has joined a team of veteran Section Editors of Al Robichaud, Andy Stamford, Joel Barrish, John Lowe, Manoj Desai and David Myles. We have separated Infectious Diseases from Oncology with John Primeau focused on the Infectious Diseases Section and David has focused on the Oncology Section. I want to thank the whole team for another seamless operation. Secondly, I continue to encourage the Section Editors to enlist a group of proof readers to help them ensure a consistent quality to the Volume. I would like to acknowledge these proof readers by listing their names below as a demonstration of our appreciation for their time and effort.

AstraZeneca – Greg Bisacchi and Jacques Dumas
Bristol-Myers Squibb – Joanne Bronson, Percy Carter, Peter Cheng, Andrew Degnan, James Duan, Gene Dubowchik, Rick Ewing, John Kadow, George Karageorge, Nicholas Meanwell, Richard Olson, Kenneth Santone, Richard Schartman, Paul Scola, Drew Thompson, Michael Walker, David Weinstein, Mark Wittman, Stephen Wrobleski and Christopher Zusi
Gilead Sciences – Randall Halcomb, Jay Parrish and Will Watkins
Pfizer – Chris Shaffer

Schering-Plough – Hubert Josien
Wyeth – Wayne Childers, Jonathan Gross, Jean Kim and David Rotella

I would also like to acknowledge another quality effort of Shridhar Hegde and Michelle Schmidt for putting together our "To-Market-to-Market" review. I would also like to thank Ms. Catherine Hathaway, who was the key Administrative Assistant for the Volume.

Annual Reports in Medicinal Chemistry is entirely put together by volunteers starting with the authors themselves to the Section Editors to the proof readers to the Editor-in-Chief. I would like to commend my colleagues at Bristol-Myers Squibb who made significant contributions to the book this year. Five of the 25 chapters in Volume 43 were authored by scientists from Bristol-Myers Squibb. Boehringer-Ingelheim, Schering-Plough and Wyeth each provided two of the chapters in the book, and AstraZeneca, Amgen, Gilead Sciences, GlaxoSmithKline, Johnson & Johnson and Pfizer each provided one chapter. A number of smaller companies and Harvard and UCLA provided the rest of the chapters for Volume 43. I want to commend and thank all who contributed to making Volume 43 a valuable contribution to the review literature for our field. It is a testament to your commitment to being conscientious members of the Medicinal Chemistry community. At the same time, my hope for Volume 44 is engaging an even wider group of contributors, including colleagues in big and small pharma who have not recently contributed to this community effort. Please take this as a challenge to be part of the best review vehicle of medicinal chemistry, *Annual Reports in Medicinal Chemistry.*

In summary, I hope that you see Volume 43 of *Annual Reports in Medicinal Chemistry* as an integral reference for the medicinal chemist. As Editor-in-Chief, I continue to look for ways to optimize and evolve the series. Please do not hesitate to contact me with suggestions for improving the series (john.macor@bms.com).

John E. Macor, Ph. D.
Bristol-Myers Squibb
Wallingford, CT, USA

PART I:
Central Nervous System Diseases

Editor: Albert J. Robichaud
Chemical & Screening Sciences
Wyeth Research
Princeton
New Jersey

CHAPTER 1

Recent Advances in Corticotropin-Releasing Factor Receptor Antagonists

Carolyn D. Dzierba, Richard A. Hartz and **Joanne J. Bronson**

1. INTRODUCTION

Corticotropin releasing factor (CRF), a 41 amino acid peptide first isolated in 1981 [1], is considered to be one of the principal regulators of the hypothalamic–pituitary–adrenal (HPA) axis, which coordinates the endocrine, behavioral, and autonomic responses to stress. CRF mediates its action through binding to two well-characterized, class B subtype G-protein coupled receptors, CRF-R1 and CRF-R2, which are widely distributed throughout the central and peripheral nervous systems. In response to stress, CRF is released from the hypothalamus and binds to CRF receptors in the anterior pituitary, resulting in release of adrenocorticotropic hormone (ACTH). Increased ACTH levels stimulate release of

Bristol-Myers Squibb Co., 5 Research Parkway, Wallingford, CT 06492, USA

Annual Reports in Medicinal Chemistry, Volume 43
ISSN 0065-7743, DOI 10.1016/S0065-7743(08)00001-8

cortisol from the adrenal cortex. Cortisol then mediates a variety of metabolic and behavioral changes that facilitate adaptation to stressful stimuli. Compelling evidence suggests that hypersecretion of CRF, and the consequent over-stimulation of the stress response, contributes to development of stress-related disorders such as depression and anxiety (for recent reviews, see Refs 2–6). To date, CRF-R1 has been the most extensively studied CRF receptor as a potential therapeutic target, with both preclinical and clinical studies suggesting that antagonists of CRF-R1 may offer promise in treatment of stress-related disorders [7–10]. By comparison, relatively few CRF-R2 antagonists have been reported, although the biological role of CRF-R2 and the possible therapeutic utility of CRF-R2 antagonists remain subjects of great interest [11]. This review will cover advances in the discovery and development of CRF-R1 antagonists since mid-2005. Recent findings on the behavioral roles of CRF-R1 and CRF-R2, particularly with respect to new therapeutic indications, will also be presented.

2. NON-PEPTIDE SMALL MOLECULE CRF-R1 ANTAGONISTS

The design and synthesis of small molecule CRF-R1 antagonists continues to be an active area of research. Investigations of structure–activity relationships (SAR) from a variety of chemotypes led to the development of a pharmacophore model for small molecule CRF-R1 antagonists [12]. The key features found in several known classes of CRF-R1 antagonists are: an aromatic heterocyclic core (monocyclic, bicyclic, or tricyclic) that includes an sp^2 nitrogen acting as a hydrogen bond acceptor; an aryl ring in an orthogonal orientation to the core ring, which is minimally substituted in the *ortho-* and *para*-positions; a halide or small alkyl group *ortho* to the sp^2 nitrogen of the heterocyclic core; and a branched, lipophilic group, with limited tolerance for polar functional groups, *para* to the sp^2 nitrogen of the heterocyclic core.

Many potent, small molecule CRF-R1 antagonists from a variety of chemical classes have been reported since the disclosure of CRF-R1 antagonist CP-154,526 (**1**) in 1996 [13,14]. Numerous excellent review articles covering progress toward the development of additional novel, small molecule CRF-R1 antagonists with potential use for the treatment of neuropsychiatric disorders, including anxiety and depression, have appeared since then [11,15–22]. This section will focus on advances toward the identification of novel CRF-R1 antagonists.

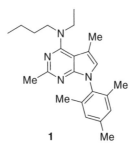

1

2.1 Monocyclic core CRF-R1 antagonists

Two compounds were recently reported from a series of pyridine-based CRF-R1 antagonists for advancement into clinical trials. Compound **2** (CP-316,311) (rat CRF-R1 $IC_{50} = 6.8$ nM) was determined to be a CRF-R1 antagonist with activity in several behavioral models, including reversal of intracerebroventricular (i.c.v.) CRF-induced excitation of locus coerulus neurons (60% inhibition at 0.3 mg/kg, intravenous (i.v.) administration), reversal of startle potentiation induced by i.c.v. CRF (100% at 32 mg/kg, oral (p.o.) administration), and activity in the defensive withdrawal model for situational anxiety at 10 mg/kg, intraperitoneal (i.p.) administration [23]. No liver toxicity was observed in five-day rat toxicology studies with **2**, and it was subsequently selected for advancement into Phase 2 clinical trials in a placebo-controlled study with depressed patients. Due to a significant food effect observed in dogs and humans with compound **2** resulting in variable oral bioavailability, compound **3** (CP-376,395) was subsequently chosen as a backup candidate [24]. Replacement of the alkyl ether group in **2** with the alkyl amino group in **3** resulted in increased basicity ($pK_a = 6.9$ for **3** vs. 3.6 for **2**) and increased solubility at low pH (5.4 mg/mL at pH 2.4 for **3** vs. 3 µg/mL at pH 2 for **2**), which reduced the food effect in dogs and humans from 10–20 fold (for compound **2**) to 2–3 fold (for compound **3**). Compound **3** was also a potent CRF-R1 antagonist (rat CRF-R1 $IC_{50} = 5.1$ nM). It was determined via *ex vivo* brain homogenate binding that **3** showed 81% occupancy of CRF-R1 in rat brain cortex after an oral dose of 3.2 mg/kg. In addition, compound **3** showed greater efficacy than **2** in several behavioral models, including reversal of i.c.v. CRF-induced excitation of locus coerulus neurons (ID_{50} of <0.01 mg/kg i.v.), complete blockade of the enhanced startle response induced by i.c.v. CRF at 17.8 mg/kg, p.o., and attenuation of fear-potentiated startle at lower doses (0.32–3.2 mg/kg, p.o.) with complete reversal at 10 mg/kg. In dog pharmacokinetic studies, **3** showed 22% and 64% oral bioavailability in fasted and fed dogs, respectively (vs. 3.6% and 37%, respectively for **2**). It was reported that **3** was advanced into the clinic for evaluation in stress-related disorders; however, no further information is available [23].

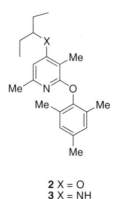

2 X = O
3 X = NH

Thiazole **4** (human CRF-R1 $K_i = 3.2$ nM) was discovered [25] as a result of further optimization of a series of previously reported thiazoles [26]. A closely related analog **5**, wherein the trifluoropropyl group was replaced with a propyl group and the *para*-chlorophenyl group was replaced with an unsubstituted phenyl group, was somewhat less potent (human CRF-R1 $K_i = 38$ nM), whereas replacement of the *para*-chlorophenyl group with a cyclohexyl to afford **6**, resulted in a significant loss of binding affinity (human CRF-R1 $K_i = 6.3$ μM). These results indicate that the phenyl group in the aminoalkyl substituent is required for good binding affinity in this chemotype.

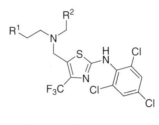

4 $R^1 = CF_3$, $R^2 = p$-chlorophenyl
5 $R^1 = CH_3$, $R^2 =$ phenyl
6 $R^1 = CH_3$, $R^2 =$ cyclohexyl

Compound **7** was reported to be the most potent analog identified in a series of novel phenylsulfonyl pyridine-based compounds (rat CRF-R1 $K_i = 17$ nM) [27]. The rat pharmacokinetic properties of this compound and two closely related analogs were assessed. It was found that these compounds were characterized by high clearance, modest half-life and low-to-moderate oral bioavailability (16% for compound **7**) in rats. More recently, a report describing the synthesis and SAR of a series of pyrazine-based compounds appeared [28]. Compound **8** (human CRF-R1 $K_i = 11$ nM) and related analogs are unique in that a *meta*-substituent is preferred on the pendant pyridyl group vs. the usually preferred *ortho*- and *para*-substituents. It was proposed that these compounds may bind *via* an alternate binding mode to the CRF-R1 receptor. Computer modeling suggested that the *meta*-substituent might occupy the region of space traditionally occupied by the *para*-substituent. Further support for this hypothesis was obtained when analog **9** (human CRF-R1 $K_i = 22$ nM), containing a 2,4-dimethylpyridyl group, was found to be slightly more potent than **10** (human CRF-R1 $K_i = 35$ nM). It is also noteworthy that the most potent analogs in this series possessed the (1R, 2S)-*cis*-aminoindanol stereochemistry (compound **10**, human CRF-R1 $K_i = 35$ nM). The corresponding (1S, 2R) enantiomer as well as the *trans*-isomers were significantly less potent (human CRF-R1 $K_i > 5$ μM).

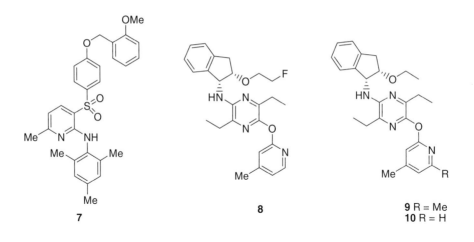

2.2 Bicyclic core CRF-R1 antagonists

Bicyclic core structures have also been studied as CRF-R1 antagonists. Compound **11** (human CRF-R1 $K_i = 42\,nM$) was one of the most potent compounds in a series of dihydroimidazoimidazoles [29]. The SAR trends revealed that the amide carbonyl was critical for good binding affinity. In contrast to the SAR surrounding compound **4** (*vide supra*), replacement of the amide in **11** with the corresponding amine resulted in a loss of binding affinity (human CRF-R1 $K_i > 10\,\mu M$). It was also found that amino groups were not tolerated within the alkyl substituents bonded to the amide nitrogen. Compound **11** was advanced into *in vivo* studies and was found to have high-to-moderate clearance (Cl = $35\,mL/min/kg$), 32% oral bioavailability in rat, and a brain-to-plasma ratio of 0.21. This compound was also evaluated in the mouse canopy stretched attend posture model to determine its anxiolytic potential and found to have a minimum effective dose of $32\,mg/kg$ (i.p.).

Recently, a report focusing primarily on the optimization of a potent series of dihydropyrrolo[2,3-*d*]pyrimidines, exemplified by **12** and **13**, appeared in the literature [30]. Compound **12** (human CRF-R1 $IC_{50} = 9.3\,nM$) was one of the most potent compounds identified in this study; however, it was determined to have poor metabolic stability in rats. Compound **13** (human CRF-R1 $IC_{50} = 48\,nM$), although less potent, was found to have good *in vivo* pharmacokinetic properties in rat (Cl = $9\,mL/min/kg$, $t_{1/2} = 6\,h$, $F = 86\%$) with a brain-to-plasma ratio of 2.3. In addition, it was reported to behave as a functional antagonist in a CRF-stimulated cAMP formation assay. Further *in vivo* evaluation demonstrated that **13** was able to decrease rat pup vocalization time by 50% when dosed at $3\,mg/kg$ (i.p.) and by 70% at $10\,mg/kg$ (i.p.).

Compounds such as **13** demonstrate that there is some tolerance for polar functional groups, such as a pyrazole, in the branched alkyl substituent, albeit with a decrease in binding affinity.

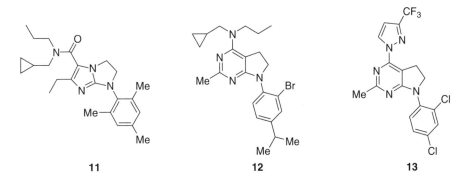

11   12   13

A variety of bicyclic series have appeared in the patent literature, represented by 1*H*-benzo[*d*]imidazoles **14** [31] and **15** [32]. It was reported that **14** showed >80% inhibition of [^{125}I]-ovine-CRF binding to human CRF-R1 at 1 μM, whereas **15** had an IC$_{50}$<10 nM for human CRF-R1. It appears that the chloro substituent on the benzo[*d*]imidazole ring system occupies the same region of space as the typical small alkyl substituent would, resulting in improved binding affinity. A series of imidazo[1,2-*b*]pyridazine derivatives, exemplified by **16** (human CRF-R1 K_i = 4.9 nM), demonstrate that the aryl or heteroaryl substituent need not be limited to six-membered rings [33]. In **16**, a substituted thiazole replaces the typical phenyl or pyridyl group. Good potency was also achieved with bicyclic aryl substituents exemplified by **17** (human CRF-R1 K_i = 19 nM) [34]. A series of pyrazolopyrimidines with a novel substitution pattern, exemplified by **18**, was disclosed in a recent patent application; however, no biological data was reported [35].

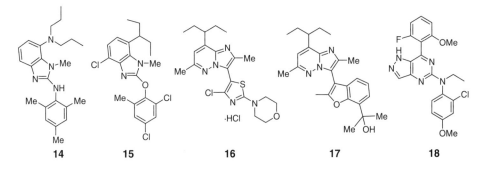

14   15   16   17   18

Two reports appeared wherein an attempt was made to replace nitrogen with oxygen as the key hydrogen bond acceptor. A report describing the synthesis and evaluation of the CRF-R1 ligand **19** indicated that this compound had moderate binding affinity (rat CRF-R1 K_i = 111 nM) [36]. Substitution of the sp^2 nitrogen with oxygen is the most likely explanation for the diminished potency of this compound relative to those in closely related chemotypes. A variety of 6,5- and 6,6-fused heterocycles exemplified by the pyrazolo[3,4-*d*]pyrimidin-4(5*H*)-one derivative **20** were evaluated [37]. The common feature in each of these analogs is the presence of a carbonyl group separated from the phenyl substituent by two atoms. Compound **20** was reported to show > 80% inhibition of [^{125}I]-ovine-CRF binding to human CRF-R1 at 10 µM.

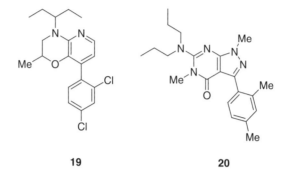

19   **20**

2.3 Tricyclic core CRF-R1 antagonists

A variety of chemotypes containing tricyclic cores have been investigated, including four related series of analogs, exemplified by compounds **21–24**. Pyrrole- and pyrazole-based compounds **21** and **22** were very potent, both in a binding assay (human CRF-R1 $K_i = 3.5$ and 2.9 nM, respectively) and in a functional assay in which ACTH release in rat pituitary cell cultures was measured (IC$_{50}$ = 14 and 6.8 nM, respectively) [38]. Rat pharmacokinetic studies indicated that **21** showed 24% oral bioavailability, but was rapidly cleared (Cl = 70 mL/min/kg). In contrast, compound **22** possessed more moderate clearance (Cl = 43 mL/min/kg), but lower oral bioavailability ($F = 7\%$) than pyrrole **21**. Both **21** and **22**, when assessed in the restraint-induced ACTH release mouse model, attenuated ACTH elevation in a dose-dependent manner when administered at oral doses ranging from 3 to 30 mg/kg; the effect was statistically significant at 10 mg/kg. Imidazolone- and pyrazole-based compounds, exemplified by **23** (human CRF-R1 $K_i = 2.0$ nM) [39] and **24** (NBI-35965) (human CRF-R1 $K_i = 3.2$ nM) [40], were also very potent. Both **23** and **24** were determined to be functional antagonists by measurement of inhibition of CRF-induced cAMP production and antagonism of CRF-stimulated ACTH release from rat anterior pituitary cell cultures. Rat pharmacokinetic studies indicated that **23** is a high clearance compound (Cl = 53 mL/min/kg) with a short half-life (1.6 h). In contrast, **24** is a low/moderate clearance compound (Cl = 17 mL/min/kg) with a longer half-life (12 h) and 34% oral bioavailability. Further *in vivo* evaluation demonstrated that **24** significantly attenuated CRF-induced ACTH release in a dose-dependent manner with statistically significant reductions observed at both

10 and 30 mg/kg doses. In addition at a dose of 20 mg/kg, **24** was able to attenuate the stress-induced increase in plasma ACTH to basal levels in a mouse-restraint stress model.

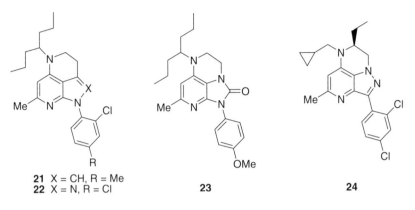

21 X = CH, R = Me
22 X = N, R = Cl

23

24

A series of imidazo[1,2-*a*]benzimidazoles, exemplified by **25**, were generally found to have binding affinities in a range similar to the closely related bicyclic series exemplified by **11** (human CRF-R1 $K_i = 7$ nM–8 µM). In contrast to compound **11** however, it was found that amine analogs are more potent than the corresponding amides. Compound **25** was reported to be one of the more potent compounds in this series (human CRF-R1 $K_i = 23$ nM) [41]. Compound **25** was advanced into rat pharmacokinetic studies and was determined to have a low plasma clearance (5 mL/min/kg) and 35% oral bioavailability; however, it had poor brain penetration (brain-to-plasma ratio = 0.03). Subsequently, an article appeared disclosing a closely related series of tetraaza-cyclopenta[*a*]indenes, exemplified by **26**, an aza analog of **25**, which had improved aqueous solubility (54 µg/mL) [42]. Compound **26** was found to be one of the most potent compounds in this series (human CRF-R1 $K_i = 7.8$ nM) and was selected for further *in vivo* characterization. Compound **26** was determined to have low oral bioavailability ($F = 4\%$) in a mouse pharmacokinetic study; nevertheless, it was found to be efficacious when dosed at both 30 and 60 mg/kg in the mouse canopy stretched attend posture model of anxiety.

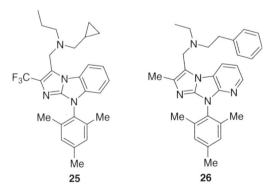

25

26

In an article describing a series of tetrahydrotriazaacenaphthylenes and related derivatives, **27** was one of the most potent compounds tested (human CRF-R1 $IC_{50} = 4.0$ nM) [43]. Evaluation in rat pharmacokinetic studies indicated that **27** is characterized by high clearance (Cl = 54 mL/min/kg) and low oral bioavailability (< 5%). A bis-trifluoromethyl analog (**28**, human CRF-R1 $IC_{50} = 100$ nM) demonstrated lower clearance (Cl = 8 mL/min/kg) and excellent oral bioavailability ($F = 71\%$) in rats. A closely related compound, **29**, was reported to be more potent (human CRF-R1 $IC_{50} = 14$ nM) and also have good rat pharmacokinetic properties [44]. A series of triaza-cyclopenta[c,d]indene derivatives, exemplified by **30**, was described in a patent application to have an $IC_{50} < 100$ nM for monkey CRF-R1 [45]. In a separate patent application a series of 2-oxo-4,5-dihydro-tetraazaacenaphthylenes and related derivatives exemplified by **31** were disclosed; however, no biological data was reported [46]. The later patent specifically disclosed various heterocyclic groups as substituents at the *para*-position of the phenyl moiety.

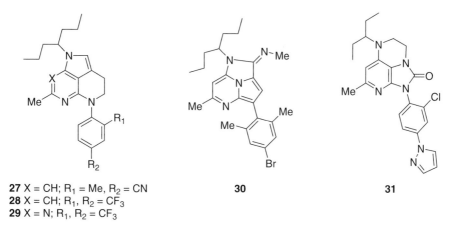

27 X = CH; R₁ = Me, R₂ = CN
28 X = CH; R₁, R₂ = CF₃
29 X = N; R₁, R₂ = CF₃

30

31

3. THERAPEUTIC INDICATIONS

The majority of research on the therapeutic utility of CRF-R1 antagonists has focused on treatment of stress-related affective disorders such as anxiety and depression [7–10]. More recently, CRF-R1 antagonists have been studied for a variety of other indications [47] including irritable bowel syndrome (IBS) [48,49], drug dependence and withdrawal [11,50], and stress-related eating disorders [51]. Recent findings in the above areas will be covered in this section. There have also been preliminary studies suggesting that CRF may play a role in other disease areas such as cardiac autonomic disorders [52], inflammatory disease [53], pain [54,55], cancer [56], and immune disorders [57]. These potential indications will not be discussed here.

3.1 Anxiety and depression

CRF is a key regulator of the HPA axis, coordinating the endocrine, behavioral, and autonomic responses to stress. It has been postulated that hypersecretion of CRF may be involved in affective disorders, including anxiety and depression [58]. Patients suffering from depression have been shown to have elevated CSF levels of CRF [59] and exhibit a blunted adrenocorticotropin (ACTH) response when injected with exogenous CRF [60,61]. Numerous reports have appeared describing the effects of various small molecule CRF-R1 antagonists in preclinical models of anxiety and depression. Recent studies on the role of CRF in anxiety and depression will be summarized in this section.

In a study of men suffering from major depression with a history of childhood trauma, it was shown that these patients exhibited higher increases in ACTH and cortisol responses when dosed with CRF after pretreatment with dexamethasone (dexamethasone/CRF test) as compared to non-depressed men, or to depressed men with no history of childhood trauma [62]. In another study, patients diagnosed with unipolar depressive disorder with melancholic features showed an attenuated ACTH and increased cortisol response to CRF administered alone (CRF test) as compared to healthy controls [63]. Interestingly, however, there was no significant difference in response between depressed patients at different stages of illness (current depressive episode vs. recovery). This may suggest that alterations to the HPA axis in depressed patients may remain after recovery.

The role of CRF-R1 in anxiety and depression has been extensively studied, whereas the role of CRF-R2 is still not fully understood. Some recent studies have attempted to further elucidate the role and interrelationship of these two receptor subtypes. In one study, mice were subjected to restraint-induced stress followed by evaluation for impairment in context- and tone-dependent fear conditioning models [64]. When evaluated immediately following the period of restraint, impairment could be blocked by injection of the CRF-R2-specific peptide antagonist anti-sauvagine-30 (1,900-fold selective for CRF-R2 over CRF-R1) into the lateral septum. Additionally, immediately following restraint, mice showed an increase in anxiety-like behaviors in the elevated plus maze model. When evaluation was delayed for 1 h after restraint, however, the mice were able to recover, no longer showing stress-induced anxiety or impairment of context-dependent fear conditioning. This recovery could be blocked by intrahippocampal administration of CRF-R1 antagonist DMP696 (32) (hCFR-R1 $K_i = 1.8$ nM [12]) or the non-selective CRF peptide antagonist astressin (human CRF-R1 $K_i = 12$ nM; human CRF-R2 $K_i = 1.5$ nM [12]). Furthermore, intraventricular administration of CRF-R2 selective peptide agonist urocortin 2 (human CRF-R1 $K_i = 4,500$ nM; human CRF-R2 $K_i = 3.3$ nM [11]) attenuated stress-induced anxiety and learning impairment without a reduction in HPA axis activation. These studies suggest that during early stress response, anxiety-like behaviors and contextual conditioned fear are mediated through septal CRF-R2, while later activation of hippocampal CRF-R1 brings the HPA system back to baseline levels.

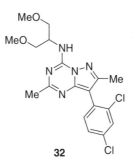

32

In another study, the selective CRF-R2 peptide agonist urocortin 3 (CRF-R1 K_i $\gg 1{,}000$ nM; CRF-R2 $K_i = 5.0$ nM) had no effect on anxiety-like behaviors in rats as demonstrated in the social interaction and shock-probe tests across a range of doses, but did decrease anxiety-like behavior in the defensive withdrawal model [65]. Urocortin 3 also did not decrease social interaction. Conversely, a selective CRF-R1 agonist, stressin$_1$-A (rat CRF-R1 $K_i = 1.7$ nM; rat CRF-R2 $K_i = 220$ nM), caused an increase in anxiety-like behaviors in the social interaction and shock-probe tests in rats, whereas the selective CRF-R1 antagonist MJL-1-109-2 (**33**) (rat CRF-R1 $K_i = 1.9$ nM) showed an attenuation of anxiety-like behaviors in the shock-probe test. These data indicate that effects of CRF-R1 and CRF-R2 on behavioral activation and anxiety-like behaviors are different, but not opposite to each other.

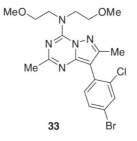

33

3.2 Irritable bowel syndrome

In addition to modulating the HPA axis, CRF is considered to be a key modulator of the gut–brain axis. Thus, CRF may play a role in mediating stress-related alterations of gut motor function associated with functional bowel disorders such as IBS [48,49]. In rats it has been shown that i.p. administration of CRF-R1 antagonist JTC-017 (**34**) (human CRF-R1 $K_i = 5.2$ nM) blocked an increase in fecal output induced by exposure to chronic colorectal distention [66]. Additionally, **34** attenuated the anxiety-related behavior seen after exposure to acute colorectal distention. CRF-stimulated colonic motility in rats was also attenuated by central administration of CRF-R1/2 peptide antagonist astressin [67]. In healthy humans, i.v. administration of CRF was shown to affect rectal hypersensitivity and mimic a stress-induced visceral response specific to IBS patients [68]. CRF-R2 may also play a role in gastric emptying. It was shown that the selective CRF-R2 peptide antagonist astressin$_2$-B (510-fold selective for CRF-R2 over CRF-R1) attenuated gastric emptying in rats induced by the selective CRF-R2 peptide agonist urocortin 2 [69]. These data suggest that CRF antagonists may be useful for the treatment of IBS.

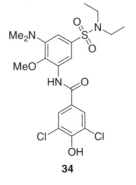

34

3.3 Drug dependence and withdrawal

Withdrawal from nicotine, cocaine, opiates, and alcohol often leads to a negative emotional state and elevated levels of anxiety. These undesirable effects can sometimes be counteracted by increasing self-administration of the substance, which leads to relapse to the addicted state. External stressors can often lead to a relapse in abuse as well. Antagonists of CRF-R1 are increasingly being examined as potential treatments for addiction and the negative aspects of drug withdrawal [11]. The role of CRF-R2 in drug withdrawal and dependence is less well understood and will not be discussed here.

3.3.1 Nicotine withdrawal
CRF receptor antagonists may be useful for treatment of the negative affective aspects of withdrawal from nicotine. Pretreatment of nicotine-dependent rats with CRF-R1/2 peptide antagonist D-Phe CRF$_{(12-41)}$ was shown to prevent the elevations in brain reward threshold associated with nicotine withdrawal [70]. D-Phe CRF$_{(12-41)}$ also caused a decrease in stress-induced reinstatement of nicotine-seeking behavior in rats [71]. Additionally, an increase in nicotine intake after a period of abstinence, often seen with nicotine dependence, could be blocked in rats by pretreatment with the CRF-R1 antagonist MPZP (**35**) (rat CRF-R1 $K_i = 5–10\,\text{nM}$ [78]) [72].

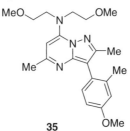

35

3.3.2 Cocaine and opiate withdrawal

Animal studies also suggest that the effects of cocaine and morphine withdrawal and relapse may be attenuated by antagonism of the CRF receptor. The CRF-R1 antagonist CP-154,526 (**1**) was shown to attenuate spiradoline-induced reinstatement of cocaine-seeking behavior in squirrel monkeys [73] as well as cue-induced reinstatement of methamphetamine-seeking behavior in rats [74]. Lorazepam-dependent rats pretreated with the CRF-R1 antagonist R121919 (**36**) (human CRF-R1 $K_i = 3.5\,nM$ [7]) before precipitation of withdrawal showed reduced HPA axis activation and reduced anxiety behaviors in the defensive withdrawal model [75]. Compound **36** was similarly able to attenuate the severity of precipitated morphine withdrawal and withdrawal-induced HPA axis activation [76]. The amount of opiate exposure during self-administration as well as the length of abstinence can affect relapse. Rats allowed to self-administer cocaine for longer periods of time (6 h daily) were more susceptible to reinstatement by cocaine, electric foot shock, or administered CRF than those allowed to self-administer for shorter periods (2 h daily) [77]. In another study, CRF-R1 antagonists **35** and antalarmin (**37**) (rat CRF-R1 $K_i = 1.0\,nM$) were shown to reduce cocaine self-administration in rats with extended daily cocaine access [78]. Interestingly, **37** did not have an effect on cocaine self-administration or cocaine discrimination in rhesus monkeys [79]. In this study, only the effects of **37** on active cocaine self-administration were studied, not on self-administration after a period of abstinence.

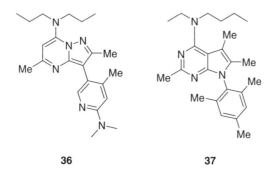

36   37

3.3.3 Ethanol withdrawal

CRF-R1 antagonists may help block the negative emotional aspects, excessive alcohol drinking, and stress-induced relapse seen in ethanol dependence [50]. Ethanol-dependent wild-type mice show an increase in ethanol self-administration during withdrawal, but only after a period of abstinence [80]. This effect was reversed by administration of the CRF-R1 antagonist **37**. CRF-R1 knockout (KO) mice do not show this tendency toward increased self-administration. When treated with CRF-R1 antagonists **33**, **36**, or **37**, ethanol-dependent rats showed a reduction in excessive ethanol self-administration during acute withdrawal [81]. Non-dependent rats treated with these CRF-R1 antagonists, however, showed no effect on ethanol self-administration. Similarly, CRF-R1 antagonist **35** selectively reduced excessive ethanol self-administration during acute withdrawal in dependent rats [82]. In another study, a novel CRF-R1 antagonist, **16**, selectively reduced excessive ethanol self-administration induced by stress in dependent rats [83]. These studies demonstrate that antagonism of CRF-R1 can selectively block excessive ethanol self-administration without affecting basal self-administration levels. This suggests that CRF-R1 antagonists could be useful for the treatment of alcohol dependence.

3.4 Eating disorders

High states of stress or anxiety may exacerbate a variety of eating disorders including binge eating, anorexia nervosa, and bulimia nervosa [51]. Alleviation of stress and anxiety by antagonism of CRF receptors may be useful for the treatment of these maladaptive eating habits. Yohimbine, an α-2 adrenoceptor antagonist commonly used to induce stress- and anxiety-like symptoms in humans and non-humans, causes stress- and anxiety-induced reinstatement of food-seeking in rats [84]. CRF-R1 antagonist **37** was shown to attenuate this food-seeking behavior as well as block the stress and anxiety effects when tested in the social interaction behavioral model. Compound **37** did not block pellet-priming-induced reinstatement of food-seeking, suggesting it selectively blocked the effects of yohimbine. While studies show CRF-R1 plays a role in food-seeking behavior, there is mounting evidence indicating that the CRF-R2 receptor plays a role in controlling food intake. CRF-R2 KO mice showed increased nocturnal food intake relative to wild-type mice, characterized by an increase in meal size rather than meal frequency [85]. In contrast, when subjected to acute restraint stress, the KO mice showed an immediate anorectic response, with an increase in latency to eat and decrease in meal size, but not prolonged restraint-induced anorexia. This suggests that CRF-R2 may play a role in controlling meal size during feeding and after acute exposure to stress. Chow-fed rats genetically prone to diet-induced obesity show an increase in feeding, faster rate of eating and faster return to feeding after meals than their obesity-resistant counterparts [86]. The selective CRF-R2 peptide agonist urocortin 2 reduced food intake and duration of meals in the obesity-prone rats but with reduced potency compared to the obesity-resistant rats. This anorectic effect was reversed with the selective CRF-R2 peptide antagonist astressin$_2$-B. Non-food-deprived rats also showed astressin$_2$-B reversible anorexia resulting from reduced meal frequency, prolonged post-meal interval, and slowed rate of eating when treated intrahypothalamically with the selective CRF-R2 peptide agonist urocortin 3 [87]. These data suggest that antagonists of CRF-R1 are likely to exert an anxiolytic effect to control stress-induced overeating, whereas CRF-R2 agonists may have an anorectic effect to control food intake during feeding and after acute exposure to stress.

4. CLINICAL FINDINGS WITH CRF-R1 ANTAGONISTS

This section will review CRF-R1 antagonists for which clinical trial results have been published, as well as compounds currently reported to be in clinical trials.

While a number of companies have actively pursued progression of CRF-R1 antagonists into the clinic, results of these trials have appeared for only a small number of compounds [7].

The first clinical efficacy results were reported in 2000 from an open-label study with R121919/NBI-30775 (**36**) in patients with major depressive disorder [88]. R121919 is a potent CRF-R1 antagonist, having a K_i of 3.5 nM (human CRF-R1) and efficacy in a number of preclinical animal models [89]. In the open-label trial, 10 patients were treated with **36** at an escalating dose of 5–40 mg/day over 30 days, while 10 patients were treated with an escalating dose of 40–80 mg/day over 30 days. R121919 was found to be safe and well-tolerated at all doses. Significant reductions in anxiety and depression scores were obtained using both patient and clinician ratings, with the group receiving the higher dose showing a better overall response. While these results were promising, progression of **36** was halted due to elevated liver enzymes in normal subjects in an extended Phase 1 study [7]. It was noted that these findings are unlikely to be CRF-R1 mediated since these receptors are absent in the liver.

The next report of the effects of a CRF-R1 antagonist in human subjects did not appear until 2007 [90]. The tricyclic pyrazolopyridine NBI-34041 (**38**), a potent CRF-R1 antagonist (human CRF-R1 $K_i = 4.0$ nM) [39,91], was evaluated in a Phase 1 proof-of-concept study in 24 healthy male volunteers for its effect on psychosocial stress and the responsiveness of the HPA axis. Subjects received 10, 50, or 100 mg of **38** or placebo for 14 days. At day 9, subjects underwent the Trier Social Stress Test (TSST), which is a public speaking task involving a mock job interview and mental arithmetic. The responsiveness of the HPA axis to exogenous CRF stimulation was not affected by treatment with NBI-34041, indicating that **38** does not impair basal regulation of the HPA system. By comparison, in subjects undergoing the psychosocial stress test, the cortisol response was significantly lower in the 100 mg/day treatment group than in the placebo group. These results suggest that CRF-R1 antagonists such as **38** may improve resistance to psychosocial stress by reducing stress hormone secretion. While there was no difference in emotional response to the acute stressful task, reduction of stress hormone secretion may predict utility in the treatment of anxiety and depression-related disorders that arise from exposure to chronic stress.

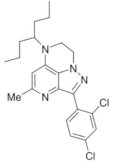

38

Clinical results have recently appeared for the pyridyl CRF-R1 antagonist CP-316,311 (**2**) (rat CRF-R1 $IC_{50} = 6.8$ nM) [23,92,93]. In a trial to evaluate the safety and efficacy of **2** in major depressive disorder, interim primary analysis showed that the CP-316,311-treated group was not significantly different in the primary efficacy endpoint from the placebo-treated group. Although the trial was terminated because of lack of efficacy, it was noted that there were no safety concerns. It has been reported that the closely related compound pyridyl compound CP-376,395 (**3**) (rat CRF-R1 $IC_{50} = 5.1$ nM) has been advanced into the clinic [24]. As noted in Section 2.1, compound **3** was reported to have better bioavailability than **2** in dogs and a lower difference in bioavailability for fasted vs. fed states [24]. Clinical results with **3** have not yet appeared.

The latest clinical trials results with a CRF-R1 antagonist appeared in the 2007 fourth quarter and year end results report from Neurocrine Biosciences [94], which described evaluation of GW-876008 (structure not disclosed) in patients with social anxiety disorder. This was the first proof-of-concept trial for a CRF-R1 antagonist in a double-blind, randomized, placebo-controlled study. While detailed information from the study was not disclosed, the report indicates that no statistically significant differences were observed in the key efficacy endpoints between patients receiving GW-876008 and those receiving placebo. GW-876008 continues to be evaluated in the clinic for treatment of IBS in a Phase 2 double-blind, randomized, placebo-controlled study [94–96]. Data is expected to be available in late 2008. Two additional compounds related to GW-876008 are also reported to be in clinical trials: GSK-561679 in a depression trial [97] and GSK-586259 in a Phase I safety study [94]. The structures of these two compounds have not been disclosed.

Several additional CRF-R1 antagonists are reported to be in Phase 1 and Phase 2 clinical studies according to sources such as www.ClinicalTrials.gov, although results have not yet been publicly disclosed. Chemical structure information is available for some of these compounds, while structures for others have not been published. The pyrazolotriazine BMS-562086 (**39**, pexacerfont) [98] is reported to be in Phase 2 trials for treatment of major depressive disorder [99], generalized anxiety disorder [100], and IBS [101]. Preclinical characterization of **39** has not been reported. The potent CRF-R1 antagonist SSR-125543 (**40**, human CRF-R1 $K_i = 1.0$ nM) [102,103] was reported to be in clinical trials as of May 2007 [104]. ONO-2333Ms (structure not available) is listed as being in a Phase 2 trial in patients with recurrent major depressive disorder [105]. A proof-of-concept study with the CRF-R1 antagonist PF-572778 (structure not available) was also initiated in which the ability of the compound to attenuate naloxone-induced corticotropin-releasing hormone increases was being evaluated in healthy subjects [106]. PF-572778 was recently reported to have been discontinued from development [107].

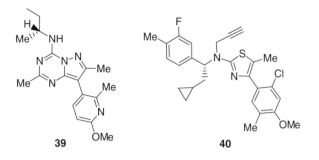

39   **40**

5. SUMMARY

Research on the role of CRF in stress-related disorders has continued at a vigorous pace over the past several years [1–10]. While much basic research supports the potential of CRF receptor antagonists for treatment of these disorders, clinical results testing this concept have been slow to emerge and are mixed in outcome. Studies such as the open-label trial with R121919/NBI-30775 suggest positive results, while the more recent trials with CP-316,311 and GW-876008 showed a lack of efficacy for depression and anxiety end points, respectively. With a number of trials ongoing for treatment of anxiety, depression, and IBS, the potential utility of these agents is expected to become clearer over the next few years, particularly as the newer compounds may have different *in vivo* profiles (e.g., pharmacokinetics and potency) than the compounds tested to date. Possible expansion of the utility of CRF-R1 antagonists into the areas of drug addiction and feeding disorders should further enhance our understanding of the role of CRF in stress-related disorders. An additional intriguing application for CRF-R1 antagonists is their use as adjunctive treatment for depression, an idea that is supported by the normalization of the HPA axis upon successful treatment with antidepressants [7]. The potential for CRF-R1 antagonists to prevent development of stress-induced disorders is also worthy of investigation. With additional compounds becoming available for clinical testing, CRF receptor antagonists continue to offer promise as novel therapeutics for the treatment of anxiety, depression, and other stress-related conditions.

REFERENCES

[1] W. Vale, J. Spiess, C. Rivier and J. Rivier, *Science*, 1981, **213**, 1394.

[2] M. E. Keck, *Amino Acids*, 2006, **31**, 241.

[3] T. L. Bale, *Hormones Behav.*, 2006, **50**, 529.

[4] J. H. Kehne, *CNS Neurol. Disord. Drug Targets*, 2007, **6**, 163.

[5] E. Arzt and F. Holsboer, *Trends Pharmacol. Sci.*, 2006, **27**, 531.

[6] V. B. Risbrough and M. B. Stein, *Hormones Behav.*, 2006, **50**, 550.

[7] F. Holsboer and M. Ising, *Eur. J. Pharmacol.*, 2008, **583**, 350.

[8] C. F. Hemley, A. McCluskey and P. A. Keller, *Curr. Drug Targets*, 2007, **8**, 105.

[9] E. Zoumakis, K. C. Rice, P. W. Gold and G. P. Chrousos, *Ann. New York Acad. Sci.*, 2006, **1083**, 239.

[10] M. Ising and F. Holsboer, *Exp. Clin. Psychopharmacol.*, 2007, **15**, 519.

[11] T. Steckler and F. M. Dautzenberg, *CNS Neurol. Disord. Drug Targets*, 2006, **5**, 147.

[12] P. J. Gilligan, D. W. Robertson and R. Zaczek, *J. Med. Chem.*, 2000, **43**, 1641.

[13] D. W. Schulz, R. S. Mansbach, J. Sprouse, J. P. Braselton, J. Collins, M. Corman, A. Dunaiskis, S. Faraci, A. W. Schmidt, T. Seeger, P. Seymour, F. D. Tingley, III, E. N. Winston, Y. L. Chen and J. Heym, *Proc. Natl. Acad. Sci., USA*, 1996, **93**, 10477.

[14] Y. L. Chen, R. S. Mansbach, S. M. Winter, E. Brooks, J. Collins, M. L. Corman, A. R. Dunaiskis, W. S. Faraci, R. J. Gallaschun, A. Schmidt and D. W. Schulz, *J. Med. Chem.*, 1997, **40**, 1749.

[15] P. J. Gilligan, P. R. Hartig, D. W. Robertson and R. Zaczek, in *Annual Reports in Medicinal Chemistry* (ed. J. A. Bristol), Vol. 32, Academic Press, San Diego, 1997, p. 41.

[16] J. R. McCarthy, S. C. Heinrichs and D. E. Grigoriadis, *Curr. Pharm. Des.*, 1999, **5**, 289.

[17] D. E. Grigoriadis, M. Haddach, N. Ling and J. Saunders, *Curr. Med. Chem.*, 2001, **1**, 63.

[18] J. Saunders and J. P. Williams, in *Annual Reports in Medicinal Chemistry* (ed. A. M. Doherty), Vol. 36, Academic Press, San Diego, 2001, p. 21.

[19] M. Lanier and J. P. Williams, *Expert Opin. Ther. Patents*, 2002, **12**, 1619.

[20] P. J. Gilligan and Y.-W. Li, *Curr. Opin. Drug Discov. Dev.*, 2004, **7**, 487.

[21] D. E. Grigoriadis, *Expert Opin. Ther. Targets*, 2005, **9**, 651.

[22] C. Chen, *Curr. Med. Chem.*, 2006, **13**, 1261.

[23] Y. L. Chen, J. Braselton, J. Forman, R. J. Gallaschun, R. Mansbach, A. W. Schmidt, T. F. Seeger, J. S. Sprouse, F. D. Tingley, III, E. Winston and D. W. Schulz, *J. Med. Chem.*, 2008, **51**, 1377.

[24] Y. L. Chen, R. S. Obach, J. Braselton, M. L. Corman, J. Forman, J. Freeman, R. J. Gallaschun, R. Mansbach, A. W. Schmidt, J. S. Sprouse, F. D. Tingley, III, E. Winston and D. W. Schulz, *J. Med. Chem.*, 2008, **51**, 1385.

[25] D. Zuev, J. A. Michne, S. S. Pin, J. Zhang, M. T. Taber and G. M. Dubowchik, *Bioorg. Med. Chem. Lett.*, 2005, **15**, 431.

[26] G. M. Dubowchik, J. A. Michne, D. Zuev, W. Schwartz, P. M. Scola, C. A. James, E. H. Ruediger, S. S. Pin, K. D. Burris, L. A. Balanda, Q. Gao, D. Wu, L. Fung, T. Fiedler, K. E. Browman, M. T. Taber and J. Zhang, *Bioorg. Med. Chem. Lett.*, 2003, **13**, 3997.

[27] R. A. Hartz, A. G. Arvanitis, C. Arnold, J. P. Rescinito, K. L. Hung, G. Zhang, H. Wong, D. R. Langley, P. J. Gilligan and G. L. Trainor, *Bioorg. Med. Chem. Lett.*, 2006, **16**, 934.

[28] J. W. Corbett, M. R. Rauckhorst, F. Qian, R. L. Hoffman, C. S. Knauer and L. W. Fitzgerald, *Bioorg. Med. Chem. Lett.*, 2007, **17**, 6250.

[29] X. Han, J. A. Michne, S. S. Pin, K. D. Burris, L. A. Balanda, L. K. Fung, T. Fiedler, K. E. Browman, M. T. Taber, J. Zhang and G. M. Dubowchik, *Bioorg. Med. Chem. Lett.*, 2005, **15**, 3870.

[30] R. Arban, R. Benedetti, G. Bonanomi, A.-M. Capelli, E. Castiglioni, S. Contini, F. Degiorgis, P. Di Felice, D. Donati, E. Fazzolari, G. Gentile, C. Marchionni, C. Marchioro, F. Messina, F. Micheli, B. Oliosi, F. Pavone, A. Pasquarello, B. Perini, M. Rinaldi, F. M. Sabbatini, G. Vitulli, P. Zarantonello, R. Di Fabio and Y. St-Denis, *Chem. Med. Chem.*, 2007, **2**, 528.

[31] A. C. Gyorkos, C. P. Corrette, S. Y. Cho, T. M. Turner, S. A. Pratt, K. Aso, M. Kori and M. Gyoten, *PCT Publication WO 2005/044793*.

[32] A. Gyorkos, C. Corrette, S. Cho, S. Pratt, C. Siedem, K. Aso and M. Gyoten, *PCT Publication WO 2006/116412*.

[33] H. J. Barbosa, E. A. Collins, C. Hamdouchi, E. J. Hembre, P. A. Hipskind, R. D. Johnston, J. Lu, M. J. Rupp, T. Takakuwa and R. C. Thompson, *PCT Publication WO 2006/102194*.

[34] E. A. Collins, P. Garcia-Losada, C. Hamdouchi, P. A. Hipskind, J. Lu and T. Takakuwa, *PCT Publication WO 2006/107784*.

[35] T. Kashiwagi, I. Takamuro, Y. Watanabe and M. Yato, *PCT Publication WO 2006/126718*.

[36] R. A. Hartz, K. K. Nanda and C. L. Ingalls, *Tetrahedron Lett.*, 2005, **46**, 1683.

[37] A. C. Gyorkos, C. P. Corrette, S. Y. Cho, T. M. Turner, K. Aso, M. Kori, M. Gyoten, K. R. Condroski, C. S. Siedem and S. A. Boyd, *PCT Publication WO* 2005/099688.

[38] B. Dyck, D. E. Grigoriadis, R. S. Gross, Z. Guo, M. Haddach, D. Marinkovic, J. R. McCarthy, M. Moorjani, C. F. Regan, J. Saunders, M. K. Schwaebe, T. Szabo, J. P. Williams, X. Zhang, H. Bozigian and T. K. Chen, *J. Med. Chem.*, 2005, **48**, 4100.

[39] Z. Guo, J. E. Tellew, R. S. Gross, B. Dyck, J. Grey, M. Haddach, M. Kiankarimi, M. Lanier, B.-F. Li, Z. Luo, J. R. McCarthy, M. Moorjani, J. Saunders, R. Sullivan, X. Zhang, S. Zamani-Kord, D. E. Grigoriadis, P. D. Crowe, T. K. Chen and J. P. Williams, *J. Med. Chem.*, 2005, **48**, 5104.

[40] R. S. Gross, Z. Guo, B. Dyck, T. Coon, C. Q. Huang, R. F. Lowe, D. Marinkovic, M. Moorjani, J. Nelson, S. Zamani-Kord, D. E. Grigoriadis, S. R. J. Hoare, P. D. Crowe, J. H. Bu, M. Haddach, J. McCarthy, J. Saunders, R. Sullivan, T. Chen and J. P. Williams, *J. Med. Chem.*, 2005, **48**, 5780.

[41] X. Han, S. S. Pin, K. Burris, L. K. Fung, S. Huang, M. T. Taber, J. Zhang and G. M. Dubowchik, *Bioorg. Med. Chem. Lett.*, 2005, **15**, 4029.

[42] X. Han, R. Civiello, S. S. Pin, K. Burris, L. A. Balanda, J. Knipe, S. Ren, T. Fiedler, K. E. Browman, R. Macci, M. T. Taber, J. Zhang and G. M. Dubowchik, *Bioorg. Med Chem. Lett.*, 2007, **17**, 2026.

[43] G. Gentile, R. Di Fabio, F. Pavone, F. M. Sabbatini, Y. St-Denis, M. G. Zampori, G. Vitulli and A. Worby, *Bioorg. Med. Chem. Lett.*, 2007, **17**, 5218.

[44] Y. St-Denis, R. Di Fabio, G. Bernasconi, E. Castiglioni, S. Contini, D. Donati, E. Fazzolari, G. Gentile, D. Ghirlanda, C. Marchionni, F. Messina, F. Micheli, F. Pavone, A. Pasquarello, F. M. Sabbatini, M. G. Zampori, R. Arban and G. Vitulli, *Bioorg. Med. Chem. Lett.*, 2005, **15**, 3713.

[45] A. Nakazato, T. Okubo, D. Nozawa, T. Tamita and L. E. J. Kennis, *PCT Publication WO* 2005/066178.

[46] Z. Luo, J. E. Tellew and J. Williams, *PCT Publication WO* 2005/063749.

[47] E. Chatzaki, V. Minas, E. Zoumakis and A. Makrigiannakis, *Curr. Med. Chem.*, 2006, **13**, 2751.

[48] S. Fukudo, *J. Gastroenterol.*, 2007, **42(Suppl. XVII)**, 48.

[49] Y. Taché and B. Bonaz, *J. Clin. Invest.*, 2007, **117**, 33.

[50] M. Heilig and G. F. Koob, *Trends Neurosci.*, 2007, **30**, 399.

[51] K. J. Steffen, J. L. Roerig, J. E. Mitchell and S. Uppala, *Expert Opin. Emerg. Drugs*, 2006, **11**, 315.

[52] S. K. Wood and J. H. Woods, *Expert Opin. Ther. Targets*, 2007, **11**, 1401.

[53] M. O'Kane, E. P. Murphy and B. Kirby, *Exp. Dermatol.*, 2006, **15**, 143.

[54] S. A. McLean, D. A. Williams, P. K. Stein, R. E. Harris, A. K. Lyden, G. Whalen, K. M. Park, I. Liberzon, A. Sen, R. H. Gracely, J. N. Baraniuk and D. J. Clauw, *Neuropsychopharmacology*, 2006, **31**, 2776.

[55] S. A. Mousa, C. P. Bopaiah, J. F. Richter, R. S. Yamdeu and M. Schäfer, *Neuropsychopharmacology*, 2007, **32**, 2530.

[56] J. Wang and S. Li, *Biochem. Biophys. Res. Commun.*, 2007, **362**, 785.

[57] A. Gravanis and A. N. Margioris, *Curr. Med. Chem.*, 2005, **12**, 1503.

[58] M. J. Owens and C. B. Nemeroff, in *Corticotropin Releasing-Factor, Ciba Foundation Symposium 172* (eds D. J. Chadwick, J. Marsh, and K. Ackrill), Wiley, Chichester, UK, 1993, p. 296.

[59] C. B. Nemeroff, E. Widerlov, G. Bisette, H. Walleus, I. Karlsson, K. Eklund, C. D. Kilts, P. T. Loosen and W. Vale, *Science*, 1984, **226**, 1342.

[60] F. Holsboer, U. Von Bardeleben, A. Gerken, G. K. Stalla and O. A. Muller, *N. Engl. J. Med.*, 1984, **311**, 1127.

[61] P. W. Gold, G. Chrousos, C. Kellner, R. Post, A. Roy, P. Augerinos, H. Schulte, E. Oldfield and D. L. Loriaux, *Am. J. Psychiat.*, 1984, **141**, 619–627.

[62] C. Heim, T. Mletzko, D. Purselle, D. L. Musselman and C. B. Nemeroff, *Biol. Psychiat.*, 2008, **63**, 398.

[63] L. Pintor, X. Torres, V. Navarro, M. Jesús Martinez de Osaba, S. Matrai and C. Gastó, *Prog. Neuro-Psychopharmacol. Biol. Psychiat.*, 2007, **31**, 1027.

[64] C. Todorovic, J. Radulovic, O. Jahn, M. Radulovic, T. Sherrin, C. Hippel and J. Spiess, *Eur. J. Neurosci.*, 2007, **25**, 3385.

[65] Y. Zhao, G. R. Valdez, É. M. Fekete, J. E. Rivier, W. W. Vale, K. C. Rice, F. Weiss and E. P. Zorrilla, *J. Pharmacol. Exp. Ther.*, 2007, **323**, 846.

[66] K. Saito, T. Kasai, Y. Nagura, H. Ito, M. Kanazawa and S. Fukudo, *Gastroenterology*, 2005, **129**, 1533.
[67] K. Tsukamoto, Y. Nakade, C. Mantyh, K. Ludwig, T. N. Pappas and T. Takahashi, *Am. J. Physiol. Regul. Integr. Comp. Physiol.*, 2006, **290**, R1537.
[68] T. Nozu and M. Kudaira, *J. Gastroenterol.*, 2006, **41**, 740.
[69] J. Czimmer, M. Million and Y. Taché, *Am. J. Physiol. Gastrointest. Liver Physiol.*, 2006, **290**, G511.
[70] A. W. Bruijnzeel, G. Zislis, C. Wilson and M. S. Gold, *Neuropsychopharmacology*, 2007, **32**, 955.
[71] G. Zislis, T. V. Desai, M. Prado, H. P. Shah and A. W. Bruijnzeel, *Neuropharmacology*, 2007, **53**, 958.
[72] O. George, S. Ghozland, M. R. Azar, P. Cottone, E. P. Zorrilla, L. H. Parsons, L. E. O'Dell, H. N. Richardson and G. F. Koob, *Proc. Natl. Acad. Sci., USA*, 2007, **104**, 17198.
[73] G. R. Valdez, D. M. Platt, J. K. Rowlett, D. Rüedi-Bettschen and R. D. Spealman, *J. Pharm. Exp. Ther.*, 2007, **323**, 525.
[74] M. C. Moffett and N. E. Goeders, *Psychopharmacology*, 2007, **190**, 171.
[75] K. H. Skelton, D. A. Gutman, K. V. Thrivikraman, C. B. Nemeroff and M. J. Owens, *Psychopharmacology*, 2007, **192**, 385.
[76] K. H. Skelton, D. Oren, D. A. Gutman, K. Easterling, S. G. Holtzman, C. B. Nemeroff and M. J. Owens, *Eur. J. Pharmacol.*, 2007, **571**, 17.
[77] J. R. Mantsch, D. A. Baker, D. M. Francis, E. S. Katz, M. A. Hoks and J. P. Serge, *Psychopharmacology*, 2008, **195**, 591.
[78] S. E. Specio, S. Wee, L. E. O'Dell, B. Boutrel, E. P. Zorrilla and G. F. Koob, *Psychopharmacology*, 2008, **196**, 473.
[79] N. K. Mello, S. S. Negus, K. C. Rice and J. H. Mendelson, *Pharmacol. Biochem. Behav.*, 2006, **85**, 744.
[80] K. Chu, G. F. Koob, M. Cole, E. P. Zorrilla and A. J. Roberts, *Pharmacol. Biochem. Behav.*, 2007, **86**, 813.
[81] C. K. Funk, E. P. Zorrilla, M.-J. Lee, K. C. Rice and G. F. Koob, *Biol. Psychiat.*, 2007, **61**, 78.
[82] H. N. Richardson, Y. Zhao, É. M. Fekete, C. K. Funk, P. Wirsching, K. D. Janda, E. P. Zorrilla and G. F. Koob, *Pharmacol. Biochem. Behav.*, 2008, **88**, 497.
[83] D. R. Gehlert, A. Cippitelli, A. Thorsell, A. D. Lê, P. A. Hipskind, C. Hamdouchi, J. Lu, E. J. Hembre, J. Cramer, M. Song, D. McKinzie, M. Morin, R. Ciccocioppo and M. Heilig, *J. Neurosci.*, 2007, **27**, 2718.
[84] U. E. Ghitza, S. M. Gray, D. H. Epstein, K. C. Rice and Y. Shaham, *Neuropsychopharmacology*, 2006, **31**, 2188.
[85] A. Tabarin, Y. Diz-Chaves, D. Consoli, M. Monsaingeon, T. L. Bale, M. D. Culler, R. Datta, F. Drago, W. W. Vale, G. F. Koob, E. P. Zorrilla and A. Contarino, *Eur. J. Neurosci.*, 2007, **26**, 2303.
[86] P. Cottone, V. Sabino, T. R. Nagy, D. V. Coscina and E. P. Zorrilla, *J. Physiol.*, 2007, **583**, 487.
[87] É. M. Fekete, K. Inoue, Y. Zhao, J. E. River, W. W. Vale, A. Szücs, G. F. Koob and E. P. Zorrilla, *Neuropsychopharmacology*, 2007, **32**, 1052.
[88] A. W. Zobel, T. Nickel, H. E. Künzel, N. Ackl, A. Sonntag, M. Ising and F. Holsboer, *J. Psych. Res.*, 2000, **34**, 171.
[89] C. Chen and D. E. Grigoriadis, *Drug Dev. Res.*, 2005, **65**, 216.
[90] M. Ising, U. S. Zimmermann, H. E. Künzel, M. Uhr, A. C. Foster, S. M. Learned-Coughlin, F. Holsboer and D. E. Grigoriadis, *Neuropsychopharmacology*, 2007, **32**, 1941.
[91] S. R. J. Hoare, B. T. Brown, M. A. Santos, S. Malany, S. F. Betz and D. E. Grigoriadis, *Biochem. Pharmacol.*, 2006, **72**, 244.
[92] www.ClinicalTrials.gov, Multicentre Trial to Evaluate the Safety and Efficacy of CP-316,311 in Major Depressive Disorder (last updated June 2007).
[93] B. Binneman, D. Feltner, S. Kolluri, Y. Shi, R. Qiu and T. Stiger, *Amer. J. Psychiat.*, 2008, **165**, 617.
[94] Neurocrine Biosciences Fourth Quarter and Year-End 2007 Results, www.neurocrine.com
[95] www.ClinicalTrials.gov, Study on the Effect of GW876008 on Cerebral Blood Flow in Irritable Bowel Syndrome (IBS) in Patients and Healthy Volunteers, verified by GlaxoSmithKline August 2007.
[96] www.ClinicalTrials.gov, Effect of GW876008 on the Bowel in Patients with Irritable Bowel Syndrome, study completed as of January 2008.

[97] www.ClinicalTrials.gov, A Study to Compare the Putative Anxiolytic Effect of 2 New Drugs in Subjects with Social Anxiety Disorder, verified by GlaxoSmithKline November 2007.

[98] USP Dictionary of USAN and International Drug Names, www.ama-assn.org

[99] www.ClinicalTrials.gov, Study of BMS-562086 in the Treatment of Outpatients with Major Depressive Disorder, study completed as of October 2007.

[100] www.ClinicalTrials.gov, Study of Pexacerfont (BMS-562086) in the Treatment of Outpatients with Generalized Anxiety Disorder, study ongoing as of January 2008.

[101] www.ClinicalTrials.gov, A Study of BMS-562086 in Patients with Irritable Bowel Syndrome, study ongoing as of January 2008.

[102] D. Gully, M. Geslin, L. Serva, E. Fontaine, P. Roger, C. Lair, V. Dare, C. Marcy, P. Rouby, J. Simiand, J. Guitard, G. Gout, R. Steinberg, D. Rodier, G. Griebel, P. Soubrie, M. Pascal, R. Pruss, B. Scatton, J. Maffrand and G. Le Fur, *J. Pharmacol. Exp. Ther.*, 2002, **301**, 322.

[103] G. Griebel, J. Simiand, R. Steinberg, M. Jung, D. Gully, P. Roger, M. Geslin and B. Scatton, *J. Pharmacol. Exp. Ther.*, 2002, **301**, 333.

[104] Sanofi-Aventis Late Stage R&D Pipeline Report, May 10, 2007, www.sanofi-aventis.com.

[105] www.ClinicalTrials.gov, Placebo-Controlled Study of ONO-2333Ms in Patients with Recurrent Major Depressive Disorder, study ongoing as of February 2008.

[106] www.ClinicalTrials.gov, Evaluation of PF-00572778 and Alprazolam on Naloxone Challenge in Healthy Subjects, study ongoing as of February 2008.

[107] Pfizer Pipeline Report as of February 28, 2008, www.pfizer.com

CHAPTER 2

Recent Advances on the 5-HT$_{5A}$, 5-HT$_6$ and 5-HT$_7$ Receptors

Brock T. Shireman, Pascal Bonaventure and
Nicholas I. Carruthers

1. INTRODUCTION

Known for almost 50 years, the neurotransmitter serotonin (**1**, 5-HT) was first isolated and identified as 5-hydroxytryptamine (5-HT) [1,2]. As a major modulatory transmitter in the brain, 5-HT is involved in elementary functions such as feeding, sleeping and sexual behavior. In addition, 5-HT also influences pain, interactive behavior (aggression, social behavior) and mood [3]. Perceptive operational studies have shown that 5-HT elicits these pharmacological and physiological responses by acting at a diversity of receptors. Molecular cloning studies have confirmed the existence of at least 14 different subtypes, each encoded by distinct genes. These receptor subtypes are divided into seven families, designated as 5-HT$_1$ through 5-HT$_7$, based on pharmacology, amino acid

Johnson & Johnson Pharmaceutical Research and Development, L.L.C., San Diego, CA 92121, USA

Annual Reports in Medicinal Chemistry, Volume 43
ISSN 0065-7743, DOI 10.1016/S0065-7743(08)00002-X

sequences, gene organization and second messenger coupling pathways [4]. With the exception of the 5-HT$_3$ receptor [5], a ligand-gated ion-channel, all of the 5-HT receptor subtypes are G-protein-coupled receptors. In the peripheral nervous system, 5-HT modulates smooth muscle cell function throughout the gastro-intestinal [6] and cardiovascular systems [7]. This diversity of functional roles makes the 5-HT receptors excellent targets for several therapeutic agents including antidepressants, antipsychotics, anxiolytics, antimigraine agents, antiemetics, gastrokinetics and hallucinogens. Recent research has focused on developing selective ligands for the 5-HT$_5$, 5-HT$_6$ and 5-HT$_7$ receptors. Special attention is directed to the reported seven candidates in the clinic for the 5-HT$_6$ receptor, two in phase I and five in phase II (see Table 2). A number of excellent reviews have been published on the 5-HT$_5$, 5-HT$_6$ and 5-HT$_7$ receptors [8–21]. The focus of this chapter will be on the most recent efforts regarding these receptors while providing a historical perspective to afford an understanding of the field.

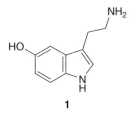

1

2. THE 5-HT$_{5A}$ RECEPTOR

2.1 Biology

The 5-HT$_5$ family consists of two receptors, 5-HT$_{5A}$ and 5-HT$_{5B}$ the identification of which defined a new family of 5-HT receptors. These 5-HT$_5$ subtypes have been isolated from both the mouse and the rat [22,30]. While the human 5-HT$_{5A}$ receptor has been cloned [31], the 5-HT$_{5B}$ receptor subtype has hitherto not been found in human tissue. Both receptors are essentially limited to distribution within the central nervous system and negatively couple to adenylate cyclase (Table 1) [27]. When exposed to novel environments, 5-HT$_{5A}$ knockout mice display increased exploratory activity with no change in anxiety-related behaviors [29]. In addition, the stimulatory effect of lysergic acid diethylamide (2, LSD) was attenuated in 5-HT$_{5A}$ knockout mice.

2.2 Medicinal chemistry

Small non-selective ligands such as LSD (2), 5-carboxamidotryptamine (3, 5-CT) and methiothepin (4) have been shown to exhibit high affinity for the 5-HT$_{5A}$ receptor [10]. Low affinity has been observed for 5-HT (1) and sumatriptan (5). The first medicinal chemistry paper identifying novel ligands for the 5-HT$_{5A}$

Table 1

Anatomical localization: periphery	Not expressed
Anatomical localization: CNS	Cortex, hippocampus, amygdala, hypothalamus (suprachiasmatic nucleus), habenula, basal ganglia, thalamus, cerebellum, dorsal raphe [22–26]
Second messenger response	Decrease cAMP [27]
Physiological response	5-HT neuronal function [28]
Phenotype of knockout mouse	Increased exploratory activity [29] Altered response to LSD (**2**) [29]
Therapeutic potential: agonist	Unknown
Therapeutic potential: antagonist	Unknown
Compounds in clinical development	None

receptor appeared in 1997 using 5-HT (**1**) as a starting point [32]. Subsequently, a series of high affinity tetrahydrocarbolines represented by **6** were described that possess modest selectivity for the 5-HT$_{5A}$ receptor versus the 5-HT$_{2A}$ receptor [33,34]. Recently, several patent applications have appeared that claim new classes of 5-HT$_{5A}$ ligands [35–43].

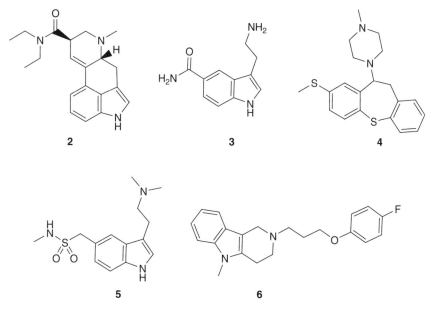

2.2.1 Agonists of the 5-HT$_{5A}$ receptor
To date no selective agonist of the 5-HT$_{5A}$ receptor has been reported.

2.2.2 Antagonists of the 5-HT$_{5A}$ receptor

Only recently has a selective human 5-HT$_{5A}$ antagonist been described [28,43,44]. SB-699551 (**7**) was shown to be at least 30-fold selective versus 5-HT$_1$, 5-HT$_2$, 5-HT$_6$ and 5-HT$_7$ receptor subtypes. In addition, 30-fold selectivity was reported versus a range of neurotransmitter receptors including dopaminergic (D$_2$–D$_4$) and adrenergic (α_1) receptors. However, species variation was observed for both the 5-HT$_{5A}$ receptor and serotonin transporter (SERT). For example, less than 10-fold selectivity was observed for the human 5-HT$_{5A}$ (5-HT$_{5A}$ pK_i = 8.2) receptor compared to SERT (SERT pK_i = 7.6). Using the respective guinea pig isoforms, **7** was reported to possess 100-fold selectivity for 5-HT$_{5A}$ (5-HT$_{5A}$ pK_i = 8.3) versus SERT (SERT pK_i = 6.3). However, in the rat isoforms, **7** was found to possess higher affinity for SERT (SERT pK_i = 8.7) than for 5-HT$_{5A}$ (5-HT$_{5A}$ pK_i = 6.3). A more recent study demonstrated enhancement of 5-HT neuronal function in guinea pig by **7**, suggesting that the 5-HT$_{5A}$ receptor serves an autoreceptor role [28].

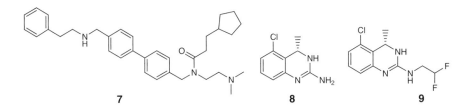

Based on an HTS hit, guanidine **8** was reported as a 5-HT$_{5A}$ antagonist (5-HT$_{5A}$ pK_i = 8.8) [45]. Selectivity versus the remaining 5-HT receptors was found to be greater than 30-fold with the exception of the 5-HT$_7$ receptor (5-HT$_7$ pK_i = 8.2). *In vivo*, brain levels of **8** >200 nM over a time course of 0.8–3 h following an oral dose of 10 mg/kg were observed. Modifications to this template led to increased brain penetration with concentrations of **9** maintained at >1 μM over 0.8–4 h after an oral dose of 6.6 mg/kg [46]. The addition of the difluoroethyl moiety was postulated to modulate the pK_a of the guanidine resulting in increased brain-to-plasma ratios. The *in vitro* selectivity profile of **9** was slightly decreased relative to **8**.

3. THE 5-HT$_6$ RECEPTOR

3.1 Biology

In 1993, the 5-HT$_6$ receptor was cloned and demonstrated to couple to adenylate cyclase (Table 2) [47]. This receptor is found exclusively in the CNS, with the highest expression levels in the striatum, and in the limbic and cortical regions. The human 5-HT$_6$ receptor gene has also been cloned and located by a chromosomal mapping study to be in close proximity to the human 5-HT$_{1D}$ receptor gene [48]. Early studies indicated that chronic administration of 5-HT$_6$

Table 2

Anatomical localization: periphery	Not expressed	*Therapeutic potential: agonist*	Anti-depressant [56] Anti-obesity [54]
Anatomical localization: CNS	Nucleus accumbens, striatum, frontal cortex, entorhinal cortex, hypothalamus, hippocampus, ventral tegmental area, anterior raphe area, cerebellum [16, 57–59]	*Therapeutic potential: antagonist*	Enhancement of cognitive function (schizophrenia, Alzheimer) [12, 13, 60] Anxiolytic [61] Anti-depressant [61] Anti-obesity [13]
Second messenger response	Increase cAMP [47, 62]	*Physiological response*	Acetylcholine/ glutamate neurotransmission
Phenotype of knockout mouse	Resistant to weight gain when exposed to high fat diet [50] Increase in anxiety-like behavior [51]		Cognition Anxiety Feeding [13, 16, 53, 54, 62–67]
Compounds in clinical development [55]	PRX-07034 (**25**, antagonist, phase I, antiobesity) BVT-74316 (**24**, antagonist, phase I, antiobesity) WAY-181187 (**12**, agonist, phase I, treatment of anxiety)		SB-742457 (structure not disclosed, antagonist, phase IIb, cognitive dysfunction associated with Alzheimer's disease) SGS-518 (structure not disclosed, antagonist, phase IIa, cognitive impairment associated with schizophrenia) SAM-315 (structure not disclosed, antagonist, phase I, cognitive dysfunction associated with Alzheimer's disease) SYN-114 (structure not disclosed, antagonist, phase I, cognitive disorders)

mRNA antisense oligonucleotides produced a significant reduction in food intake and body weight in rats [49]. In addition, 5-HT$_6$ knockout mice possess a phenotype of increased anxiety behavior and are resistant to weight gain when exposed to a high fat diet [50]. The interpretation of this phenotype is complicated by differences in the CNS distribution and pharmacology of the 5-HT$_6$ receptor in mice compared with both rats and humans [52].

Recently, several selective $5\text{-}HT_6$ receptor antagonists have been shown to suppress food intake in food-deprived animals [13,16]. Another publication demonstrated that chronic treatment with a selective $5\text{-}HT_6$ receptor partial agonist decreased food consumption and body weight in a model of diet-induced obese rats [54]. It has been suggested that $5\text{-}HT_6$ receptor ligands, either block or desensitize the serotonergic receptors resulting in a reduction of γ-aminobutyric acid (GABA) and a subsequent increase in α-melanocyte stimulating hormone (α-MSH) release, thereby suppressing food intake. This experimental data supports the hypothesis that $5\text{-}HT_6$ ligands reduce food intake by a mechanism that is consistent with an enhancement of satiety [55]. Two $5\text{-}HT_6$ compounds (PRX-07034 and BVT-74316) have entered phase I clinical trial for the treatment of obesity (Table 2).

Additional antisense oligonucleotide studies on the $5\text{-}HT_6$ receptor and behavioral studies in rats with potent and selective antagonists have also suggested an involvement of this subtype in the control of cholinergic and glutamatergic transmission (Table 2). Given the fundamental role of both acetylcholine and glutamate in cognition, $5\text{-}HT_6$ receptor antagonists have also been implicated in the modulation of cognitive function. The observation that $5\text{-}HT_6$ agonists and antagonists produce identical effects in animal models has not been restricted to their anti-obesity action. Interestingly, positive pre-clinical data in various models have also been reported with $5\text{-}HT_6$ receptor agonists [55]. Based on strong pre-clinical *in vivo* data, two $5\text{-}HT_6$ antagonists, SB-742457 and SGS-518 (Table 2), have entered phase II clinical trials for the enhancement of cognitive function in the treatment of Alzheimer's disease or schizophrenia.

3.2 Medicinal chemistry

Several reviews [8,12,14–16,66] of the $5\text{-}HT_6$ literature have been published covering research prior to 2006, consequently this section will primarily focus on more recent publications with salient points from the earlier literature. $5\text{-}HT_6$ ligands may be characterized as agonists, antagonists and inverse agonists using *in vitro* systems. However, their *in vivo* pharmacological effects in pre-clinical models of both cognitive function and body weight control can appear independent of functional behavior since both agonists and antagonists exhibit similarly favorable effects [56]. These observations pose unique challenges for the selection of compounds for pre-clinical development. Noteworthy, most of the *in vitro* work has been carried out with cloned receptors in stably transfected cell lines. Future pharmacological characterization in native tissue may shed light on this anomaly.

3.2.1 Agonists of the 5-HT$_6$ receptor

The first $5\text{-}HT_6$ receptor agonist to be described was the tryptamine analog **10** ($5\text{-}HT_6$ $pK_i = 7.3$) [67], even though the compound retained modest affinity for other serotonin receptors ($5\text{-}HT_{1A}$, $5\text{-}HT_{1D}$, $5\text{-}HT_{1E}$ and $5\text{-}HT_7$). Compounds such as **11** ($5\text{-}HT_6$ $pIC_{50} = 8.1$), in which the 2-aminoethyl side chain was replaced with a tetrahydropyridine, were obtained via modification of an HTS hit [68].

A common structural feature of 5-HT$_6$ ligands is the presence of an arylsulfonyl group, first used as an N$_1$-protecting group in the synthesis of analogs of **10**. Screening the N$_1$-phenylsulfonyl protected indoles subsequently uncovered a series of 5-HT$_6$ receptor antagonists [69]. This in turn prompted additional investigation into the arylsulfonyl fragment affording the N$_1$-arylsulfonylindole WAY-181187 (**12**, 5-HT$_6$ pK_i = 8.7), which was found to be a full agonist [70,71]. Further research around the point of attachment of the arylsulfonyl on the central indole core provided the 5-arylsulfonamidoindoles E-6801 (**13**, 5-HT$_6$ pK_i = 8.5) and E-6837 (**14**, 5-HT$_6$ pK_i = 9.1) that were determined to be partial agonists [72,73].

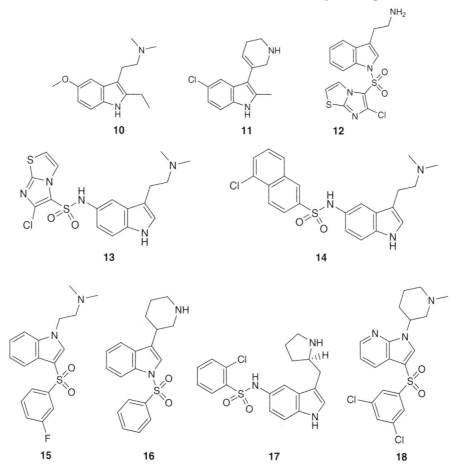

Modifications to the arylsulfonyltryptamine template include reversing the phenylsulfonyl and aminoethyl fragments resulting in the partial agonist **15** (5-HT$_6$ pK_i = 7.9) [74]. Conformational constraint of the aminoethyl side chain of the N$_1$-arylsulfonylindoles and 5-arylsulfonamidoindoles afforded agonists, **16** (5-HT$_6$ pK_i = 8.7) [75] and **17** (5-HT$_6$ pK_i = 8.7) [76]. A noteworthy feature of the latter templates was the relationship between the absolute stereochemistry of the constrained amine and intrinsic activity, with the (R)-enantiomers behaving as agonists and the (S)-enantiomers described as antagonists of the 5-HT$_6$ receptor. A further modification was the replacement of the indole nucleus with a pyrrolopyridine to give **18** (5-HT$_6$ pK$_i$ = 8.4) [77]. As observed previously, the reversal of the arylsulfonyl and aminoethyl components provided compounds exhibiting both antagonism and partial agonism.

3.2.2 Antagonists of the 5-HT$_6$ receptor

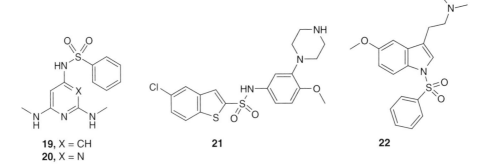

19, X = CH
20, X = N

21

22

The identification of potent and selective 5-HT$_6$ antagonists occurred shortly after the characterization of the human receptor [48]. For example, HTS screens afforded structures exemplified by the pyridinesulfonamide Ro 63-0563 (**19**, 5-HT$_6$ pK$_i$ = 7.9), pyrimidinesulfonamide Ro 04-6790 (**20**, 5-HT$_6$ pK$_i$ = 7.3), [78] and the piperazinylbenzenesulfonamide SB-271046 (**21**, 5-HT$_6$ pK$_i$ = 8.9) [79]. Modification of the endogenous ligand provided both agonists, *vide supra*, and the antagonist MS-245 (**22**, 5-HT$_6$ pK$_i$ = 8.6) [67]. The success of these complimentary approaches has yielded a range of diverse structures and led to the development of pharmacophore-based models [66,80–83]. These models, in conjunction with 5-HT$_6$ receptor homology models [84–88], have permitted medicinal chemists to discover a range of structures classified into five groups [66] with representative examples as follows: indoles, LY-483518 (**23**, 5-HT$_6$ pK$_i$ = 8.9) [89]; indole-like BVT-74316 (**24**, 5-HT$_6$ pK$_i$ = 8.9) [90]; monocyclic arylpiperazines, PRX-07034 (**25**, 5-HT$_6$ pK$_i$ = 8.4) [91]; bicyclic/tricyclic arylpiperazines (**26**, 5-HT$_6$ pK$_i$ = 8.4) [92] and other structures containing an arylsulfonyl moiety as in Ro 66-0074 (**27**, 5-HT$_6$ pK$_i$ = 9.0) [93].

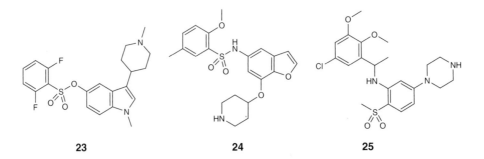

23

24

25

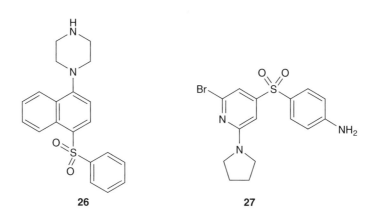

26   27

4. THE 5-HT$_7$ RECEPTOR

4.1 Biology

The 5-HT$_7$ receptor has been cloned from the rat [100,118], mouse [119], human [134] and guinea pig [135]. Coupled to adenylate cyclase, the 5-HT$_7$ receptor exhibits a high degree of interspecies homology ($\sim$95%) but a low-sequence homology with other 5-HT receptors ($<$40%). The widespread distribution of the 5-HT$_7$ receptor in the brain is suggestive of multiple central roles (Table 3). The highest densities of 5-HT$_7$ receptors have been observed in the hypothalamus, thalamus, hippocampus, cortex and brain stem. In peripheral tissue, 5-HT$_7$ receptors are present in intestinal and vascular smooth muscle. Due to the availability of selective antagonists and knockout mice, a better understanding of 5-HT$_7$ receptor function has been obtained and many important roles have been identified in circadian rhythmicity, thermoregulation, sleep, endocrine regulation, memory, pain, blood pressure, smooth muscle relaxation and control of micturition (Table 3) [17].

4.2 Medicinal chemistry

Since the cloning of the 5-HT$_7$ receptor many excellent reviews have appeared covering both the medicinal chemistry and biology of the receptor [17–21]. Pharmacophore models have been described for both 5-HT$_7$ agonists [136] and antagonists/inverse agonists [137–142]. A recent comprehensive review has been published on 5-HT$_7$ ligands and their therapeutic potential [17].

4.2.1 Agonists of the 5-HT$_7$ receptor
One of the first described 5-HT$_7$ agonists was dihydroimidazole 27 (5-HT$_7$ pK_i = 7.8) [143]. Subsequently two other classes of molecules were reported, arylpiperazines 29 and 30, and compounds 31–33 [144,145]. Through modification of compounds previously investigated as D$_4$ receptor ligands, 29 was identified and reported to be a 5-HT$_7$ agonist (5-HT$_7$ pK_i = 8.5) with good

Table 3

Anatomical localization: periphery	Intestinal and vascular smooth muscle [94–96]	Therapeutic potential: agonist	Anti-hypertension [97] Analgesic [98]
Anatomical localization: CNS	Hypothalamus (suprachiasmatic nucleus), thalamus, hippocampus, brain stem, cortex [99–105]	Therapeutic potential: antagonist	Anti-depressant [106–110] Anti-migraine [111] Analgesic [112–115] Irritable bowel syndrome [116, 117]
Second messenger response	Increase cAMP [100, 118, 119]	Compounds in clinical development	None
Physiological response	Circadian rhythmicity [100, 120–124] Thermoregulation [125, 126] Sleep [106] Endocrine regulation [127–129] Memory [99, 130–132] Pain and analgesia [112–115] Hypotension [97] Smooth muscle relaxation [94–96] Control of micturition [133]	Phenotype of knockout mouse	Resistant to 5-HT/5-CT induced hypothermia [125, 126] Anti-depressant-like behavior [106] Decrease in REM sleep [106]

selectivity over the 5-HT$_{1A}$, dopaminergic and adrenergic receptors. However, undesired 5-HT$_{2A}$ receptor affinity (5-HT$_{2A}$ pK_i = 8.1) was also observed. Replacement of the phenol with a methoxy gave **30**, a potent 5-HT$_7$ agonist (5-HT$_7$ pK_i = 9.0) that demonstrates a less favorable selectivity profile versus **29**. Subsequent to these studies, the tetrahydronaphthalenes **31** (5-HT$_7$ pK_i = 9.7), **32** (5-HT$_7$ pK_i = 7.8) and **33** (5-HT$_7$ pK_i = 9.0) were reported as 5-HT$_7$ agonists demonstrating good selectivity versus the 5-HT$_{1A}$ and 5-HT$_{2A}$ receptors with moderate D$_2$ affinity. A highly selective 5-HT$_7$ full agonist is aminotetralin **34** (5-HT$_7$ pK_i = 8.1) [146,147] exhibiting 30-fold selectivity over the 5-HT$_{1A}$ receptor and at least 100-fold selectivity over several 5-HT receptor subtypes and other neurotransmitters. Interest in the replacement of the dipropylamine with a dimethylamine afforded **35**, a selective 5-HT$_7$ receptor antagonist (5-HT$_7$ pK_i = 8.6). A recent patent application [98] has appeared claiming the use of 5-HT$_7$ agonists such as AS-19 (**36**, 5-HT$_7$ pK_i = 9.2) for the treatment of pain. This compound is reported to possess >80-fold selectivity over other 5-HT receptor subtypes including the 5-HT$_{1A}$ receptor (5-HT$_{1A}$ pK_i = 7.1).

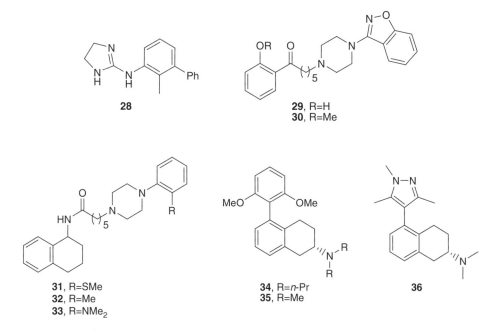

28

29, R=H
30, R=Me

31, R=SMe
32, R=Me
33, R=NMe$_2$

34, R=n-Pr
35, R=Me

36

4.2.2 Antagonists of the 5-HT$_7$ receptor

As a result of an HTS screen, the sulfonamide SB-258719 (**37**, 5-HT$_7$ pK_i = 7.5) represented the first reported selective 5-HT$_7$ antagonist [148–151]. Optimization for 5-HT$_7$ affinity and selectivity provided SB-269970 (**38**, 5-HT$_7$ pK_i = 8.9) that was shown to possess >100-fold selectivity over a range of CNS targets including the 5-HT subtypes, adrenergic and dopaminergic receptors. *In vivo* studies aimed at addressing the rapid clearance of **38** in rat (CL$_b$ = 140 mL/min/kg) resulted in indole SB-656104 (**39**, 5-HT$_7$ pK_i = 8.7) exhibiting a lower clearance (Cl$_b$ = 57 mL/min/kg) and improved oral bioavailability (F = 16%). The improvements in the pharmacokinetic profile of **39** were slightly detrimental to the *in vitro* selectivity. Identified from a back-up series, amide **40** (5-HT$_7$ pK_i = 8.7) [152] is a 5-HT$_7$ antagonist with good selectivity against the 5-HT$_1$ and 5-HT$_2$ subtypes.

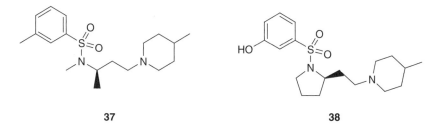

37

38

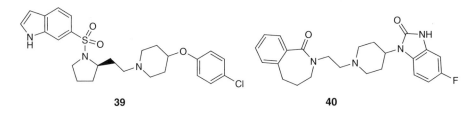

39   40

Tetrahydropyridine **41** (5-HT$_7$ pK_i = 8.7) [153] is described as a selective 5-HT$_7$ antagonist with 47-fold selectivity over the 5-HT$_2$ receptors and reported selectivity over the 5-HT$_{1A}$, 5-HT$_4$, 5-HT$_6$ and D$_2$ receptors. Tetrahydro-β-carboline **42** (5-HT$_7$ pK_i = 8.5) was found to possess improved selectivity versus other 5-HT receptor subtypes [154]. Additional heterocyclic replacements in **41** such as the tetrahydrothienopyridyl **43** (5HT$_7$ pK_i = 8.2) were also investigated [155]. The N-^{11}CH$_3$ derivative **44** (5-HT$_7$ pK_i = 8.0) was investigated as a PET tracer for the 5-HT$_7$ receptor [156]. However, specific binding to brain 5-HT$_7$ receptors was not demonstrated.

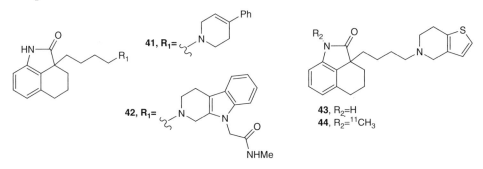

The atropisomer phenylaporphine **45** [157] was reported to possess good affinity for the 5-HT$_7$ receptor (5-HT$_7$ pK_i = 8.4) with good selectivity versus 5-HT$_{1A}$ and D$_{2A}$ receptors. A recent series of patent applications describes sulfonamides **46–50** with reported pIC$_{50}$ values ranging 7.1–7.8 [158–161].

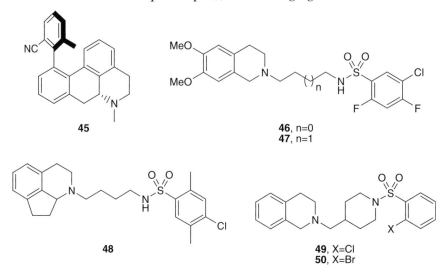

5. CONCLUSIONS

Since the cloning of the 5-HT$_5$, 5-HT$_6$ and 5-HT$_7$ receptors, research towards the development of selective agonists and antagonists has generated a large amount of excitement. For example, the identification of the selective 5-HT$_{5A}$ antagonist **1** should provide much needed insight into the role of the 5-HT$_{5A}$ receptor in CNS-mediated processes. The variety and number of 5-HT$_6$ agonists and antagonists in clinical development for obesity or cognitive disorders is representative of a large research effort from numerous pharmaceutical and academic laboratories. Finally, while there is no 5-HT$_7$ antagonist or agonist reported to be in clinical development, the diversity of structures reported for this receptor indicates continued interest in the development of selective 5-HT$_7$ receptor ligands with suitable drug-like properties for pre-clinical validation.

REFERENCES

[1] D. Hoyer, J. P. Hannon and G. R. Martin, *Pharmacol. Biochem. Behav.*, 2002, **71**, 533.
[2] D. Hoyer and G. R. Martin, *Behav. Brain Res.*, 1996, **73**, 263.
[3] N. N. Osborne and M. Hamon (eds), *Neuronal Serotonin*, Wiley, Chichester, 1988.
[4] J. R. Raymond, Y. V. Mukhin, A. Gelasco, J. Turner, G. Collinsworth, T. W. Gettys, J. S. Grewal and M. N. Garnovskaya, *Pharmacol. Ther.*, 2001, **92**, 179.
[5] A. J. Thompson and S. C. R. Lummis, *Curr. Pharm. Des.*, 2006, **12**, 3615.
[6] D. R. Brown, *Behav. Brain Res.*, 1996, **73**, 193.
[7] P. R. Saxena and C. M. Villalon, *J. Cardiovasc. Pharmacol.*, 1990, **15**, S17.
[8] R. Glennon, *J. Med. Chem.*, 2003, **46**, 2795.
[9] A. Wesolowska, *Pol. J. Pharmacol.*, 2002, **54**, 327.
[10] D. L. Nelson, *Curr. Drug Targets, CNS Neurol. Disord.*, 2004, **3**, 53.
[11] D. R. Thomas, *Pharmacol. Ther.*, 2006, **111**, 707.
[12] E. S. Mitchell and J. F. Neumaier, *Pharmacol. Ther.*, 2005, **108**, 320.
[13] J. Holenz, P. J. Pauwels, J. L. Diaz, R. Merce, X. Codony and H. Buschmann, *Drug Discov. Today*, 2006, **11**, 283.
[14] S. L. Davies, J. S. Silvestre and X. Guitart, *Drugs Fut.*, 2005, **30**, 479.
[15] W. E. Childers, Jr. and A. J. Robichaud, *Annu. Rep. Med. Chem.*, 2005, **40**, 17.
[16] M. L. Woolley, C. A. Marsden and K. C. F. Fone, *Curr. Drug Targets, CNS Neurol. Disord.*, 2004, **3**, 59.
[17] V. Pittala, L. Salerno, M. Modica, M. A. Siracusa and G. Romeo, *Mini Rev. Med. Chem.*, 2007, **7**, 945.
[18] D. R. Thomas and J. J. Hagan, *Curr. Drug Targets CNS Neurol. Disord.*, 2004, **3**, 81.
[19] M. L. Lopez-Rodriguez, B. Benhamu, M. J. Morcillo, E. Porras, J. L. Lavandera and L. Pardo, *Curr. Med. Chem.: CNS Agents*, 2004, **4**, 203.
[20] M. Leopoldo, *Curr. Med. Chem.*, 2004, **11**, 629.
[21] A. Slassi, M. Isaac and T. Xin, *Expert Opin. Ther. Pat.*, 2004, **14**, 1009.
[22] M. G. Erlander, T. W. Lovenberg, B. M. Baron, L. de Lecea, P. E. Danielson, M. Racke, A. L. Slone, B. W. Siegel, P. E. Foye, K. Cannon, J. E. Burns and J. G. Sutcliffe, *Proc. Nat. Acad. Sci., USA*, 1993, **90**, 3452.
[23] K. R. Oliver, A. M. Kinsey, A. Wainwright and D. J. S. Sirinathsinghji, *Brain Res.*, 2000, **867**, 131.
[24] M. J. Duncan, L. Jennes, J. B. Jefferson and M. S. Brownfield, *Brain Res.*, 2000, **869**, 178.
[25] A. M. Kinsey, A. Wainwright, R. Heavens, D. J. Sirinathsinghji and K. R. Oliver, *Mol. Brain Res.*, 2001, **88**, 194.
[26] M. Pasqualetti, M. Ori, D. Marazziti, M. Castagna and I. Nardi, *Ann. NY Acad. Sci.*, 1998, **861**, 245.
[27] B. J. Francken, M. Jurzak, J. F. Vanhauwe, W. H. Luyten and J. E. Leysen, *Eur. J. Pharmacol.*, 1998, **361**, 299.

[28] D. R. Thomas, E. M. Soffin, C. Roberts, J. N. C. Kew, R. M. de la Flor, L. A. Dawson, V. A. Fry, S. A. Coggon, S. Faedo, P. D. Hayes, D. F. Corbett, C. H. Davies and J. J. Hagan, *Neuropharmacology*, 2006, **51**, 566.

[29] R. Grailhe, C. Waeber, S. C. Dulawa, J. P. Hornung, X. Zhuang, D. Brunner, M. A. Geyer and R. Hen, *Neuron*, 1999, **22**, 581.

[30] J. L. Plassat, U. Boschert, N. Amlaiky and R. Hen, *EMBO J.*, 1992, **11**, 4779.

[31] S. Rees, I. den Daas, S. Foord, S. Goodson, D. Bull, G. Kilpatrick and M. Lee, *FEBS Lett.*, 1994, **355**, 242.

[32] M. Teitler, C. Scheick, P. Howard, J. E. Sullivan, T. Iwamura and R. A. Glennon, *Med. Chem. Res.*, 1997, **7**, 207.

[33] N. Khorana, A. Purohit, K. Herrick-Davis, M. Teitler and R. A. Glennon, *Bioorg. Med. Chem. Lett.*, 2003, **11**, 717.

[34] N. Khorana, C. Smith, K. Herrick-Davis, A. Purohit, M. Teitler, B. Grella, M. Dukat and R. A. Glennon, *J. Med. Chem.*, 2003, **46**, 3930.

[35] W. Amberg, A. Netz, A. Kling, M. Ochse, U. Lange, C. W. Hutchins, F. J. Garcia-Ladona and W. Wernet, *WO Patent* 07022964, 2007.

[36] W. Amberg, A. Netz, A. Kling, M. Ochse, U. Lange, A. Haupt, F. J. Garcia-Ladona and W. Wernet, *WO Patent* 07022947, 2007.

[37] W. Amberg, A. Netz, A. Kling, M. Ochse, U. Lange, C. W. Hutchins, F. J. Garcia-Ladona and W. Wernet, *WO Patent* 07022946, 2007.

[38] A. Alanine, L. C. Gobbi, S. Kolczewski, T. Luebbers, J.-U. Peters and L. Steward, *WO Patent Application* 06097391, 2006.

[39] A. Alanine, L. C. Gobbi, S. Kolczewski, T. Luebbers, J.-U. Peters and L. Steward, *US Patent Application* 0 293 349-A1, 2006.

[40] A. Alanine, L. C. Gobbi, S. Kolczewski, T. Luebbers, J.-U. Peters and L. Steward, *US Patent Application* 0 293 350-A1, 2006.

[41] A. Alanine, L. C. Gobbi, S. Kolczewski, T. Luebbers, J.-U. Peters and L. Steward, *WO Patent Application* 06117305, 2006.

[42] A. Alanine, L. C. Gobbi, S. Kolczewski, T. Luebbers, J.-U. Peters and L. Steward, *US Patent Application* 0 229 323-A1, 2006.

[43] S. M. Bromidge, D. F. Corbett, T. D. Heightman and S. F. Moss, *WO Patent* 04096771, 2004.

[44] D. F. Corbett, T. D. Heightman, S. F. Moss, S. M. Bromidge, S. A. Coggon, M. J. Longley, A. M. Roa, J. A. Williams and D. R. Thomas, *Bioorg. Med. Chem. Lett.*, 2005, **15**, 4014.

[45] J.-U. Peters, T. Luebbers, A. Alanine, S. Kolczewski, F. Blasco and L. Steward, *Bioorg. Med. Chem. Lett.*, 2008, **18**, 256.

[46] J.-U. Peters, T. Luebbers, A. Alanine, S. Kolczewski, F. Blasco and L. Steward, *Bioorg. Med. Chem. Lett.*, 2008, **18**, 262.

[47] F. J. Monsma, Jr., Y. Shen, R. P. Ward, M. W. Hamblin and D. R. Sibley, *Mol. Pharmacol.*, 1993, **43**, 320.

[48] R. Kohen, M. A. Metcalf, N. Khan, T. Druck, K. Huebner, J. E. Lachowicz, H. Y. Meltzer, D. R. Sibley, B. L. Roth and M. W. Hamblin, *J. Neurochem.*, 1996, **66**, 47.

[49] J. C. Bentley, A. J. Sleight, C. A. Mardsen and K. C. F. Fone, *Psychopharmacology*, 1997, **11**, A64.

[50] P. Caldirola, *WO Patent* 03035061-A1, 2003.

[51] L. Tecott and T. Brenman, *US Patent* 6 060 642, 2000.

[52] W. D. Hirst, B. Abrahamsen, F. E. Blaney, A. R. Calver, L. Aloj, G. W. Price and A. D. Medhurst, *Mol. Pharmacol.*, 2003, **64**, 1295.

[53] G. Perez-Garcia and A. Meneses, *Pharmacol. Biochem. Behav.*, 2005, **81**, 673.

[54] A. Fisas, X. Codony, G. Romero, A. Dordal, J. Giraldo, R. Merce, J. Holenz, D. Heal, H. Buschmann and P. J. Pauwels, *Br. J. Pharmacol.*, 2006, **148**, 973.

[55] D. J. Heal, S. L. Smith, A. Fisas, X. Codony and H. Buschmann, *Pharmacol. Ther.*, 2008, **117**, 207.

[56] P. Svenningsson, E. T. Tzavara, H. Qi, R. Carruthers, J. M. Witkin, G. G. Nomikos and P. Greengard, *J. Neurosci.*, 2007, **27**, 4201.

[57] T. A. Branchek and T. P. Blackburn, *Annu. Rev. Pharmacol. Toxicol.*, 2000, **40**, 319.

[58] C. Gerard, M.-P. Martres, K. Lefevre, M.-C. Miquel, D. Verge, L. Lanfumey, E. Doucet, M. Hamon and S. El Mestikawy, *Brain Res.*, 1997, **746**, 207.

[59] C. Gerard, S. El Mestikawy, C. Lebrand, J. Adrien, M. Ruat, E. Traiffort, M. Hamon and M. P. Martres, *Synapse*, 1996, **23**, 164.

[60] A. G. Foley, K. J. Murphy, W. D. Hirst, H. C. Gallagher, J. J. Hagan, N. Upton, F. S. Walsh and C. M. Regan, *Neuropsychopharmacology*, 2004, **29**, 93.

[61] A. Wesolowska and A. Nikiforuk, *Neuropharmacology*, 2007, **52**, 1274.

[62] M. Ruat, E. Traiffort, J. M. Arrang, J. Tardivel-Lacombe, J. Diaz, R. Leurs and J. C. Schwartz, *Biochem. Biophys. Res. Commun.*, 1993, **193**, 268.

[63] M. D. Lindner, D. B. Hodges, Jr., J. B. Hogan, A. F. Orie, J. A. Corsa, D. M. Barten, C. Polson, B. J. Robertson, V. L. Guss, K. W. Gillman, J. E. Starrett, Jr. and V. K. Gribkoff, *J. Pharmacol. Exp. Ther.*, 2003, **307**, 682.

[64] D. A. Daly and B. Moghaddam, *Neurosci. Lett.*, 1993, **152**, 61.

[65] M. L. Woolley, J. C. Bentley, A. J. Sleight, C. A. Marsden and K. C. F. Fone, *Neuropharmacology*, 2001, **41**, 210.

[66] J. Holenz, P. J. Pauwels, J. L. Diaz, R. Merce, X. Codony and H. Buschmann, *Drug Discov. Today*, 2006, **11**, 283.

[67] R. A. Glennon, M. Lee, J. B. Rangisetty, M. Dukat, B. L. Roth, J. E. Savage, A. McBride, L. Rauser, S. Hufeisen and D. K. H. Lee, *J. Med. Chem.*, 2000, **43**, 1011.

[68] C. Mattsson, C. Sonesson, A. Sandahl, H. E. Greiner, M. Gassen, J. Plaschke, J. Leibrock and H. Boettcher, *Bioorg. Med. Chem. Lett.*, 2005, **15**, 4230.

[69] Y. Tsai, M. Dukat, A. Slassi, N. MacLean, L. Demchyshyn, J. E. Savage, B. L. Roth, S. Hufesein, M. Lee and R. A. Glennon, *Bioorg. Med. Chem. Lett.*, 2000, **10**, 2295.

[70] D. C. Cole, J. R. Stock, W. J. Lennox, R. C. Bernotas, J. W. Ellingboe, L. Leung, D. Smith, G. Zhang, P. Li, L. Qian, L. A. Dawson, S. Boikess, S. Rosenzweig-Lipson, C. E. Beyer and L. E. Schechter, *230th ACS National Meeting*, Washington, DC, August, 2005, MEDI-017.

[71] D. C. Cole, J. R. Stock, W. J. Lennox, R. C. Bernotas, J. W. Ellingboe, S. Boikess, J. Coupet, D. L. Smith, L. Leung, G.-M. Zhang, X. Feng, M. F. Kelly, R. Galante, P. Huang, L. A. Dawson, K. Marquis, S. Rosenzweig-Lipson, C. E. Beyer and L. E. Schechter, *J. Med. Chem.*, 2007, **50**, 5535.

[72] J. Holenz, R. Merce, J. L. Diaz, X. Guitart, X. Codony, A. Dordal, G. Romero, A. Torrens, J. Mas, B. Andaluz, S. Hernandez, X. Monroy, E. Sanchez, E. Hernandez, R. Perez, R. Cubi, O. Sanfeliu and H. Buschmann, *J. Med. Chem.*, 2005, **48**, 1781.

[73] G. Romero, E. Sanchez, M. Pujol, P. Perez, X. Codony, J. Holenz, H. Buschmann and P. J. Pauwels, *Br. J. Pharmacol.*, 2006, **148**, 1133.

[74] R. Bernotas, S. Lenicek, S. Antane, G. M. Zhang, D. Smith, J. Coupet, B. Harrison and L. E. Schechter, *Bioorg. Med. Chem. Lett.*, 2004, **14**, 5499.

[75] D. C. Cole, W. J. Lennox, J. R. Stock, J. W. Ellingboe, H. Mazandarani, D. L. Smith, G. Zhang, G. J. Tawa and L. E. Schechter, *Bioorg. Med. Chem. Lett.*, 2005, **15**, 4780.

[76] D. C. Cole, W. J. Lennox, S. Lombardi, J. W. Ellingboe, R. C. Bernotas, G. J. Tawa, H. Mazandarani, D. L. Smith, G. Zhang, J. Coupet and L. E. Schechter, *J. Med. Chem.*, 2005, **48**, 353.

[77] H. Elokdah, D. Li, G. McFarlane, R. C. Bernotas, A. J. Robichaud, R. L. Magolda, G. M. Zhang, D. Smith and L. E. Schechter, *Bioorg. Med. Chem. Lett.*, 2007, **15**, 6208.

[78] A. J. Sleight, F. G. Boess, M. Bos, B. Levet-Trafit, C. Riemer and A. Bourson, *Br. J. Pharmacol.*, 1998, **124**, 556.

[79] S. M. Bromidge, A. M. Brown, S. E. Clarke, K. Dodgson, T. Gager, H. L. Grassam, P. M. Jeffrey, G. F. Joiner, F. D. King, D. N. Middlemiss, S. F. Moss, H. Newman, G. Riley, C. Routledge and P. Wyman, *J. Med. Chem.*, 1999, **42**, 202.

[80] A. J. Bojarski, *Curr. Top. Med. Chem. (Sharjah, UAE)*, 2006, **6**, 2005.

[81] H.-J. Kim, M. R. Doddareddy, H. Choo, Y. S. Cho, K. T. No, W.-K. Park and A. N. Pae, *J. Chem. Inf. Model.*, 2008, **48**, 197.

[82] M. R. Doddareddy, Y. S. Cho, H. Y. Koh and A. N. Pae, *Bioorg. Med. Chem.*, 2004, **12**, 3977.

[83] M. R. Doddareddy, Y. J. Lee, Y. S. Cho, K. I. Choi, H. Y. Koh and A. N. Pae, *Bioorg. Med. Chem.*, 2004, **12**, 3815.

[84] S. M. Bromidge, *Spec. Pub.-Roy. Soc. Chem.*, 2001, **264**, 101.

[85] M. L. Lopez-Rodriguez, B. Benhamu, T. de la Fuente, A. Sanz, L. Pardo and M. Campillo, *J. Med. Chem.*, 2005, **48**, 4216.

[86] W. D. Hirst, J. A. L. Minton, S. M. Bromidge, S. F. Moss, A. J. Latter, G. Riley, C. Routledge, D. N. Middlemiss and G. W. Price, *Br. J. Pharmacol.*, 2000, **130**, 1597.

[87] M. R. Pullagurla, R. B. Westkaemper and R. A. Glennon, *Bioorg. Med. Chem. Lett.*, 2004, **14**, 4569.

[88] M. Dukat, P. D. Mosier, R. Kolanos, B. L. Roth and R. A. Glennon, *J. Med. Chem.*, 2008, **51**, 603.

[89] M. M. Pineiro-Nunez, D. D. Bauzon, F. P. Bymaster, Z. Chen, E. Chernet, M. P. Clay, R. Crile, N. W. DeLapp, C. P. Denny, J. F. Falcone, M. E. Flaugh, L. J. Heinz, A. D. Kiefer, D. J. Koch, J. H. Krushinski, J. D. Leander, T. D. Lindstrom, B. Liu, D. L. McKinzie, D. L. Nelson, L. A. Phebus, V. P. Rocco, J. M. Schaus, M. C. Wolff and J. S. Ward, *229th ACS National Meeting*, San Diego, CA, March, 2005, MEDI-282.

[90] P. Caldirola, G. Johansson and L. Sutin, *WO Patent* 06134150, 2006.

[91] O. Becker, M. Lobera, R. E. Melendez, A. Sharadendu, L. Wu, X. Y. Yu, D. S. Dhanoa, S. R. Cheruku, Y. Marantz, S. Noiman, M. Fichman, H. Senderowitz, S. Shacham, A. Saha and P. Orbach, *WO Patent* 06081332, 2006.

[92] M. Lee, J. B. Rangisetty, M. R. Pullagurla, M. Dukat, V. Setola, B. L. Roth and R. A. Glennon, *Bioorg. Med. Chem. Lett.*, 2005, **15**, 1707.

[93] C. Riemer, E. Borroni, B. Levet-Trafit, J. R. Martin, S. Poli, R. H. P. Porter and M. Boes, *J. Med. Chem.*, 2003, **46**, 1273.

[94] J. A. Terron and A. Falcon-Neri, *Br. J. Pharmacol.*, 1999, **127**, 609.

[95] T. Ishine, I. Bouchelet, E. Hamel and T. J. F. Lee, *Am. J. Physiol. Heart Circ. Physiol.*, 2000, **278**, H907.

[96] B. R. Tuladhar, L. Ge and R. J. Naylor, *Br. J. Pharmacol.*, 2003, **138**, 1210.

[97] C. M. Villalon and D. Centurion, *Naunyn Schmiedebergs Arch. Pharmacol.*, 2007, **376**, 45.

[98] J. M. Vela Hernandez, A. T. Jover, H. H. Buschmann and L. Romero-Alonso, *WO Patent* 08000495, 2008.

[99] P. B. Hedlund and J. G. Sutcliffe, *Trends Pharmacol. Sci.*, 2004, **25**, 481.

[100] T. W. Lovenberg, B. M. Baron, L. de Lecea, J. D. Miller, R. A. Prosser, M. A. Rea, P. E. Foye, M. Racke, A. L. Slone, B. W. Siegel, P. E. Danielson, J. G. Sutcliffe and M. G. Erlander, *Neuron*, 1993, **11**, 449.

[101] Z. P. To, D. W. Bonhaus, R. M. Eglen and L. B. Jakeman, *Br. J. Pharmacol.*, 1995, **115**, 107.

[102] P. Vanhoenacker, G. Haegeman and J. E. Leysen, *Trends Pharmacol. Sci.*, 2000, **21**, 70.

[103] K. Varnas, D. R. Thomas, E. Tupala, J. Tiihonen and H. Hall, *Neurosci. Lett.*, 2004, **367**, 313.

[104] P. Bonaventure, D. Nepomuceno, A. Kwok, W. Chai, X. Langlois, R. Hen, K. Stark, N. I. Carruthers and T. W. Lovenberg, *J. Pharmacol. Exp. Ther.*, 2002, **302**, 240.

[105] P. Bonaventure, D. Nepomuceno, K. Miller, J. Chen, C. Kuei, F. Kamme, D.-T. Tran, T. W. Lovenberg and C. Liu, *Eur. J. Pharmacol.*, 2005, **513**, 181.

[106] P. B. Hedlund, S. Huitron-Resendiz, S. J. Henriksen and J. G. Sutcliffe, *Biol. Psychiat.*, 2005, **58**, 831.

[107] A. Wesolowska, A. Nikiforuk, K. Stachowicz and E. Tatarczynska, *Neuropharmacology*, 2006, **51**, 578.

[108] A. Wesolowska, E. Tatarczynska, A. Nikiforuk and E. Chojnacka-Wojcik, *Eur. J. Pharmacol.*, 2007, **555**, 43.

[109] P. Bonaventure, L. Kelly, L. Aluisio, J. Shelton, B. Lord, R. Galici, K. Miller, J. Atack, T. W. Lovenberg and C. Dugovic, *J. Pharmacol. Exp. Ther.*, 2007, **321**, 690.

[110] O. Mnie-Filali, L. Lambas-Senas, L. Zimmer and N. Haddjeri, *Drug News Perspect.*, 2007, **20**, 613.

[111] J. A. Terron, *Eur. J. Pharmacol.*, 2002, **439**, 1.

[112] H. I. Rocha-Gonzalez, A. Meneses, S. M. Carlton and V. Granados-Soto, *Pain*, 2005, **117**, 182.

[113] S. Doly, J. Fischer, M. J. Brisorgueil, D. Verge and M. Conrath, *J. Comp. Neurol.*, 2005, **490**, 256.

[114] J. L. Yau, J. Noble and J. R. Seckl, *Neurosci. Lett.*, 2001, **309**, 141.

[115] A. Dogrul and M. Seyrek, *Br. J. Pharmacol.*, 2006, **149**, 498.

[116] F. De Ponti, *Gut*, 2004, **53**, 1520.

[117] F. De Ponti and M. Tonini, *Drugs*, 2001, **61**, 317.

[118] M. Ruat, E. Traiffort, R. Leurs, J. Tardivel-Lacombe, J. Diaz, J. M. Arrang and J. C. Schwartz, *Proc. Natl. Acad. Sci., USA*, 1993, **90**, 8547.

[119] J. L. Plassat, N. Amlaiky and R. Hen, *Mol. Pharmacol.*, 1993, **44**, 229.

[120] R. L. Gannon, *J. Biol. Rhythms*, 2001, **16**, 19.

[121] J. D. Glass, G. H. Grossman, L. Farnbauch and L. DiNardo, *J. Neurosci.*, 2003, **23**, 7451.

[122] J. Sprouse, *Exp. Opin. Ther. Targets*, 2004, **8**, 25.

[123] J. Sprouse, X. Li, J. Stock, J. McNeish and L. Reynolds, *J. Biol. Rhythms*, 2005, **20**, 122.

[124] J. Sprouse, L. Reynolds, X. Li, J. Braselton and A. Schmidt, *Neuropharmacology*, 2004, **46**, 52.

[125] P. B. Hedlund, P. E. Danielson, E. A. Thomas, K. Slanina, M. J. Carson and J. G. Sutcliffe, *Proc. Natl. Acad. Sci., USA*, 2003, **100**, 1375.

[126] P. B. Hedlund, L. Kelly, C. Mazur, T. Lovenberg, J. G. Sutcliffe and P. Bonaventure, *Eur. J. Pharmacol.*, 2004, **487**, 125.

[127] H. Jorgensen, M. Riis, U. Knigge, A. Kjaer and J. Warberg, *J. Neuroendocrinol.*, 2003, **15**, 242.

[128] M. Hery, A. M. Francois-Bellan, F. Hery, P. Deprez and D. Becquet, *Endocrine*, 1997, **7**, 261.

[129] A. Siddiqui, M. Abu-Amara, C. Aldairy, J. J. Hagan and C. Wilson, *Eur. J. Pharmacol.*, 2004, **491**, 77.

[130] G. S. Perez-Garcia and A. Meneses, *Behav. Brain Res.*, 2005, **163**, 136.

[131] G. Perez-Garcia, C. Gonzalez-Espinosa and A. Meneses, *Behav. Brain Res.*, 2006, **169**, 83.

[132] A. Meneses, *Behav. Brain Res.*, 2004, **155**, 275.

[133] K. E. Read, G. J. Sanger and A. G. Ramage, *Br. J. Pharmacol.*, 2003, **140**, 53.

[134] J. A. Bard, J. Zgombick, N. Adham, P. Vaysse, T. A. Branchek and R. L. Weinshank, *J. Biol. Chem.*, 1993, **268**, 23422.

[135] A. P. Tsou, A. Kosaka, C. Bach, P. Zuppan, C. Yee, L. Tom, R. Alvarez, S. Ramsey, D. W. Bonhaus, E. Stefanich, L. Jakeman, R. M. Eglen and H. W. Chan, *J. Neurochem.*, 1994, **63**, 456.

[136] E. S. Vermeulen, A. W. Schmidt, J. S. Sprouse, H. V. Wikstroem and C. J. Grol, *J. Med. Chem.*, 2003, **46**, 5365.

[137] M. L. Lopez-Rodriguez, E. Porras, B. Benhamu, J. A. Ramos, M. J. Morcillo and J. L. Lavandera, *Bioorg. Med. Chem. Lett.*, 2000, **10**, 1097.

[138] M. L. Lopez-Rodriguez, E. Porras, M. J. Morcillo, B. Benhamu, L. J. Soto, J. L. Lavandera, J. A. Ramos, M. Olivella, M. Campillo and L. Pardo, *J. Med. Chem.*, 2003, **46**, 5638.

[139] E. S. Vermeulen, M. Van Smeden, A. W. Schmidt, J. S. Sprouse, H. V. Wikstroem and C. J. Grol, *J. Med. Chem.*, 2004, **47**, 5451.

[140] M. Kolaczkowski, M. Nowak, M. Pawlowski and A. J. Bojarski, *J. Med. Chem.*, 2006, **49**, 6732.

[141] R. E. Wilcox, J. E. Ragan, R. S. Pearlman, M. Y. K. Brusniak, R. M. Eglen, D. W. Bonhaus, T. E. Tenner, Jr. and J. D. Miller, *J. Comput. –Aid. Mol. Des.*, 2001, **15**, 883.

[142] A. Lepailleur, R. Bureau, S. Lemaitre, F. Dauphin, J.-C. Lancelot, V. Contesse, S. Lenglet, C. Delarue, H. Vaudry and S. Rault, *J. Chem. Info. Comput. Sci.*, 2004, **44**, 1148.

[143] V. Parikh, W. M. Welch and A. W. Schmidt, *Bioorg. Med. Chem. Lett.*, 2003, **13**, 269.

[144] R. Perrone, F. Berardi, N. A. Colabufo, E. Lacivita, M. Leopoldo and V. Tortorella, *J. Med. Chem.*, 2003, **46**, 646.

[145] M. Leopoldo, E. Lacivita, N. Contino, N. A. Colabufo, F. Berardi and R. Perrone, *J. Med. Chem.*, 2007, **50**, 4214.

[146] P. Holmberg, D. Sohn, R. Leideborg, P. Caldirola, P. Zlatoidsky, S. Hanson, N. Mohell, S. Rosqvist, G. Nordvall, A. M. Johansson and R. Johansson, *J. Med. Chem.*, 2004, **47**, 3927.

[147] P. Holmberg, L. Tedenborg, S. Rosqvist and A. M. Johansson, *Bioorg. Med. Chem. Lett.*, 2005, **15**, 747.

[148] I. T. Forbes, S. Dabbs, D. M. Duckworth, A. J. Jennings, F. D. King, P. J. Lovell, A. M. Brown, L. Collin, J. J. Hagan, D. N. Middlemiss, G. J. Riley, D. R. Thomas and N. Upton, *J. Med. Chem.*, 1998, **41**, 655.

[149] I. T. Forbes, S. Douglas, A. D. Gribble, R. J. Ife, A. P. Lightfoot, A. E. Garner, G. J. Riley, P. Jeffrey, A. J. Stevens, T. O. Stean and D. R. Thomas, *Bioorg. Med. Chem. Lett.*, 2002, **12**, 3341.

[150] P. J. Lovell, S. M. Bromidge, S. Dabbs, D. M. Duckworth, I. T. Forbes, A. J. Jennings, F. D. King, D. N. Middlemiss, S. K. Rahman, D. V. Saunders, L. L. Collin, J. J. Hagan, G. J. Riley and D. R. Thomas, *J. Med. Chem.*, 2000, **43**, 342.

[151] J. J. Hagan, G. W. Price, P. Jeffrey, N. J. Deeks, T. Stean, D. Piper, M. I. Smith, N. Upton, A. D. Medhurst, D. N. Middlemiss, G. J. Riley, P. J. Lovell, S. M. Bromidge and D. R. Thomas, *Br. J. Pharmacol.*, 2000, **130**, 539.

[152] I. T. Forbes, D. G. Cooper, E. K. Dodds, S. E. Douglas, A. D. Gribble, R. J. Ife, A. P. Lightfoot, M. Meeson, L. P. Campbell, T. Coleman, G. J. Riley and D. R. Thomas, *Bioorg. Med. Chem. Lett.*, 2003, **13**, 1055.

[153] C. Kikuchi, H. Nagaso, T. Hiranuma and M. Koyama, *J. Med. Chem.*, 1999, **42**, 533.

[154] C. Kikuchi, T. Ando, T. Watanabe, H. Nagaso, M. Okuno, T. Hiranuma and M. Koyama, *J. Med. Chem.*, 2002, **45**, 2197.

[155] C. Kikuchi, T. Hiranuma and M. Koyama, *Bioorg. Med. Chem. Lett.*, 2002, **12**, 2549.

[156] M.-R. Zhang, T. Haradahira, J. Maeda, T. Okauchi, T. Kida, S. Obayashi, K. Suzuki and T. Suhara, *J. Labelled Compd. Rad.*, 2002, **45**, 857.

[157] T. Linnanen, M. Brisander, L. Unelius, S. Rosqvist, G. Nordvall, U. Hacksell and A. M. Johansson, *J. Med. Chem.*, 2001, **44**, 1337.

[158] A. T. Jover, S. Y. Minguez, J. M. Prio, L. R. Alonso, A. D. Zueras and H. H. Buschmann, *US Patent Application* 06142332, 2006.

[159] A. T. Jover, J. M. Prio, S. Y. Minguez, M. G. Lopez, A. D. Zueras, L. R. Alonso and H. H. Buschmann, *WO Patent Application* 06018309, 2006.

[160] A. T. Jover, J. M. Prio, S. Y. Minguez, M. G. Lopez, A. D. Zueras, L. R. Alonso and H. H. Buschmann, *WO Patent Application* 06018308, 2006.

[161] A. T. Jover, S. Y. Minguez, J. M. Prio, L. R. Alonso, A. D. Zueras and H. H. Buschmann, *US Patent Application* 06142321, 2006.

CHAPTER 3

Recent Advances in Voltage-Gated Sodium Channel Blockers: Therapeutic Potential as Drug Targets in the CNS

Andreas Termin[*], **Esther Martinborough**[**] and **Dean Wilson**[***]

[*]Department of Drug Innovation, Vertex Pharmaceuticals Inc., 11010 Torreyana Road, San Diego, CA 92121, USA

[**]Apoptos Inc., 10835 Road to the Cure, Suite 205, San Diego, CA 92121, USA

[***]Department of Drug Innovation, Vertex Pharmaceuticals Inc., 130 Waverly St. Cambridge, MA 02139, USA

Annual Reports in Medicinal Chemistry, Volume 43
ISSN 0065-7743, DOI 10.1016/S0065-7743(08)00003-1

1. INTRODUCTION

Local anesthetic and anticonvulsant drugs whose mechanism of action has been attributed to inhibition of voltage-gated sodium channels (VGSCs) have been in clinical use for many decades. However, significant contemporary medicinal chemistry contributions have been limited due to low throughput assays and limited structural information. This is changing as modern high throughput assay technology becomes available and more is learned about the structure of sodium channels.

The nine VGSCs (Na$_V$1.1–1.9) discovered to date are members of a larger voltage-gated ion channel super-family which regulates the electrical activity among cells and tissues, and is involved in many physiological processes such as cognition, locomotion, and nociception [1–3].

Of the nine VGSCs, Na$_V$1.1, Na$_V$1.2, Na$_V$1.3, and Na$_V$1.6 are distributed throughout the central nervous system (CNS). Na$_V$1.4 is present in muscle tissue, and Na$_V$1.5 is in cardiac myocytes, while Na$_V$1.7, Na$_V$1.8, and Na$_V$1.9 are distributed in the peripheral nervous system (PNS) [4,5]. It is now known, however, that the expression of VGSCs can change significantly under pathological conditions such as neuropathic pain or multiple sclerosis (MS). For example, Na$_V$1.8 mRNA and protein are up-regulated within cerebellar purkinje cells in animal models of MS and in human MS conditions [6,7]. Similarly, mRNA for Na$_V$1.8 is down-regulated in the dorsal root ganglia (DRG) but up-regulated in undamaged neurons in both rodent models of neuropathic pain and in human chronic pain conditions [8,9]. *We continue to learn about the expression patterns and their functional relevance, which remain key issues for drug discovery for these targets.*

The structure of VGSCs consists of four 6-transmembrane domain alpha-subunits (I–IV) which form the channel. It has been hypothesized that pore opening and closing is controlled by a highly conserved voltage-sensing helix located on the fourth transmembrane segment (S4) of each of the four domains, comprised of four positively charged amino acids located at every third position. When stimulated by a change in transmembrane voltage, this region undergoes a large conformational shift and translocation towards the extracellular domain, thereby opening the channel. This allows the influx of sodium ions into the cell along the established concentration gradient. Within milliseconds, another conformational change occurs with the participation of the amino acids methionine, isoleucine, and phenylalanine located on domains III and IV, producing the inactivated state. A third conformational change then reprimes the channel for refiring, producing the original resting stage [10]. This cyclic sequence of events is summarized in Figure 1.

Another functional hypothesis suggests that four voltage-sensing paddles consisting of the third and fourth segments (S3b and S4) are responsible for opening the channel [11].

Each sodium channel subtype exhibits different voltage and kinetic profiles as the rapid upstroke action potential in neurons is created, ultimately leading to the initiation and propagation of electrical signals in the CNS and PNS.

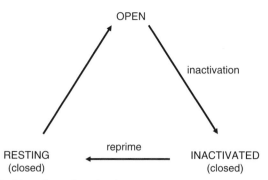

Figure 1   Voltage-gated ion channel: activation states.

With the exception of naturally occurring toxins, most small molecules acting on VGSCs are "use-dependent", i.e., preferentially binding the inactivated state of the channel outside of the pore, while non-use-dependent (tonic) blockers block the pore of the channel and are therefore frequency independent.

Under pathological conditions the channels fire at higher frequencies, resulting in channel accumulation in the inactivated state. Since use-dependent molecules preferentially block these rapidly firing neurons, it has been theorized that they could specifically target injured neurons and hence be more safe than non-use-dependent tonic blockers [10]. However, the safety and physiological implications of use-dependent block remains unclear. A recent study of the use-dependent VGSC blockers lidocaine **1** and lamotrigine **2** in two rat models of neuropathic pain suggests that the efficacy may be more closely linked to the type of injury and the resulting firing activity, rather than the firing frequency [12].

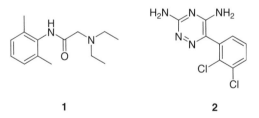

| **1** | **2** |

Small molecule VGSC blockers have been extensively reviewed in the literature, and most derive their efficacy from comparable binding affinity for all sodium channels [3,13,14]. Sodium channel modulation is surprisingly common in launched drugs. Several marketed antidepressants, for example, exhibit $Na_V1.3$ and $Na_V1.7$ activity at therapeutically relevant plasma concentrations, which might contribute to their observed efficacy in neuropathic pain [15,16].

While current sodium channel blockers may offer benefits in a number of indications, they also have been associated with CNS and cardiovascular (CV) side effects which are often dose-limiting [17–20]. New knowledge of channel distribution, structure, kinetics, and firing mechanism has bolstered the hypothesis that VGSC blockers might be developed which selectively target channels involved in disease states, while sparing those with constitutive or apathological function. For example, VGSC blockers which preferentially inhibit the sodium channels involved in neuropathic pain could lead to drugs with both improved efficacy and safety.

This review summarizes developments within the voltage-gated sodium channel field over the last two years, with particular emphasis on the structural classes recently disclosed.

2. SODIUM CHANNELS AND THERAPEUTIC OPPORTUNITIES

The therapeutic indications most frequently considered for sodium channel blockers are CNS and PNS disorders where neuronal excitability plays a key role in the manifestation of the pathology, such as neuropathic pain, epilepsy, and neuromuscular, neurodegenerative, and psychiatric disorders [21]. Neuropathic pain is a very large unmet medical need, impacting approximately 4 million people in the United States alone [22]. This disease arises from complex pathophysiological changes that develop in both the PNS and the CNS following nerve injury or disease [23]. Examples include chronic (e.g., repetitive motion disorders) or acute trauma (including surgery), diabetic neuropathy, and post-herpetic neuralgia, none of which are adequately served with current therapies. The observed CNS changes are thought to arise from abnormal signaling from the PNS, notably the electrical hyperexcitability within peripheral sensory neurons [24]. Although sodium channel blockers such as carbamazepine **3** and mexiletine **4** have shown efficacy in clinical studies, their use is limited by their side effect profile, and, as a result, they are not the standard of care [17–19].

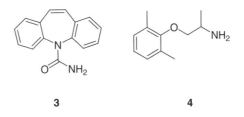

3   4

Epilepsy is another common neurological disorder, and anticonvulsant drugs known to block sodium channels are first line therapies, though ∼30% of the seizures are poorly controlled with current antiepileptic drugs (AEDs). Some AEDs have severe side effects [20], and the FDA recently reported a two-fold increase in suicidal thoughts or behavior associated with several antiseizure drugs. Taken together, these observations encourage the development of new sodium channel blockers with better safety margins and improved efficacy when compared to current antiepileptic agents.

Modulation of sodium channel activity has also been suggested for the treatment of neurodegenerative diseases such as MS, amyotrophic lateral sclerosis (Lou Gehrig's Disease or ALS), spasticity, and tremors, as well as psychiatric diseases such as schizophrenia and bipolar disorder [21,25–27]. Sodium channel blockers such as lidocaine **1** have been used with some success in the treatment of neurological diseases affecting the ear, commonly described as tinnitus, demonstrating efficacy despite dose-limiting cardiovascular side effects [28,29].

3. ASSAYS

Automated assays for ion channels have made dramatic advances in the past decade. It is now possible to utilize automated patch clamping or voltage-sensing dyes to screen large compound collections and drive medicinal chemistry and structure–activity relationship (SAR) efforts [30–33]. Recently, substantial advances with binding assays have been reported and ion flux assays with and without radiolabeled ligands have been used for screening [34–36]. Importantly, the different assay techniques make comparisons of scaffold classes and literature data difficult, and even the "gold standard" patch clamp electrophysiology experiments vary in protocol, which can influence both the observed potency and selectivity against other sodium channel subtypes. For example, the small molecule A-803467 has an activity of 79 nM at -100 mV in HEK-293 cells expressing hNa$_V$1.8 (resting state protocol) versus 8 nM at -40 mV in an inactivated state protocol [37]. Similarly, the known inactivated state protocol for Na$_V$1.7 uses a holding potential of -60 mV [37], while another group prefers -70 mV [38], confounding direct data comparisons.

4. CNS AND PNS SODIUM CHANNEL SUBTYPES AND THEIR MODULATORS

4.1 Na$_V$1.1

Of the four mammalian VGSCs that are expressed in the CNS, Na$_V$1.1 has been most frequently linked to epilepsy. In fact over 100 mutations of the SCN1A gene have been associated with inherited forms of epilepsy. Many of these mutations lead to a gain-of-function of Na$_V$1.1, producing increased persistent currents or firing rates in cerebral neurons [39]. Several non-selective use-dependent blockers of VGSCs are sold as AEDs, though an estimated 30% of patients do not respond to current standard of care [20,40]. Compounds specifically targeting Na$_V$1.1 have not been reported in the literature.

4.2 Na$_V$1.2

As with Na$_V$1.1, mutations of the SCN2A gene encoding Na$_V$1.2 are believed to produce dominant idiopathic epilepsy disorders and benign familial neonatal infantile seizures [41,42]. Recent research shows that $Na_V1.2$ is re-distributed in demyelinating disorders such as MS [43]. Given the widespread distribution in the CNS, $Na_V1.2$ is thought to give rise to the CNS side effects observed with sodium channel blockers, and hence a channel to be avoided. Recently, a spider toxin (Tx1) has been reported which is a selective, reversible inhibitor of $Na_V1.2$, providing a useful tool for elucidating the role of $Na_V1.2$ [44].

4.3 $Na_V1.3$

Peripheral nerve injury is known to increase the expression of $Na_V1.3$ (SCN3A) in a variety of primary and secondary sensory neurons, such as first-order DRG neurons, second-order dorsal horn nociceptive neurons, and even the pain-processing ventral posterolateral nucleus neurons in the thalamus [45,46]. Up-regulation of $Na_V1.3$ has also been confirmed in multiple nerve injury models designed to mimic neuropathic pain. Examples of these are the spinal cord injury model (SCI), spinal nerve ligation model (SNL), the chronic constriction injury model (CCI), as well as models of diabetic neuropathy and post-herpetic neuralgia [47–51].

Antisense experiments investigating the role of $Na_V1.3$ have, however, been confounding. In one study, lumbar intrathecal administration of $Na_V1.3$ antisense in rats attenuated both mechanical and thermal hyperalgesia after spinal cord injury (SCI model) [52], while other research reported no change in either mechanical or cold allodynia nociception [53].

Neuropathic pain develops normally in $Na_V1.3$ knock-out mice [54]. Although absent in the periphery in healthy adults, $Na_V1.3$ is significantly up-regulated in the PNS after nerve damage and has been identified in patients with spinal root avulsion injury and traumatic central axotomy [55,56].

Recently a mutation of the SCN3A gene which encodes for $Na_V1.3$ has been linked to epilepsy in man and an animal model for epilepsy [57,58].

4.4 $Na_V1.6$

Sodium channel expression in human demyelinating diseases such as MS has recently received increasing attention [25], and $Na_V1.6$ (SCN8A) has been shown to be highly expressed in degenerated axons in acute MS lesions in rodents. In addition, sodium channels in general are also believed to contribute to the activation of microglia and macrophages in MS [59].

Given the very high expression levels of $Na_V1.6$, it has been proposed that blocking these VGSCs could provide a new mechanism for neuroprotection by blocking persistent sodium current [43], a hypothesis strengthened by the observation that sodium channel blockers can have antiinflammatory properties, thus further enhancing the potential for efficacy [60]. Complete suppression of $Na_V1.6$, however, could lead to side effects. Knockout of the SCN8A gene encoding for $Na_V1.6$ in mice leads to motor failure, ataxia, tremor, muscle weakness, and dystonia [61,62]. The tetrodotoxin (TTX) analog 4,9-anhydro-TTX

5 is the only molecule which has been reported to be a specific tonic (i.e., non-use-dependent) blocker ($IC_{50} = 4\,nM$) as measured in *Xenopus laevis* oocytes [63].

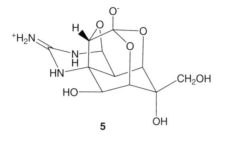

5

4.5 Na$_V$1.7

In 2006 a clinical channelopathy was described which resulted in loss-of-function of Na$_V$1.7 among a Pakistani family, leading to the complete inability to experience physical pain and providing important clinical rationale for the therapeutic relevance of this channel. *In vitro*, nonsense mutations of the SCN9A gene encoding for the alpha subunit of Na$_V$1.7 led to complete loss of function when co-expressed with the wild type or mutant human Na$_V$1.7 beta subunits in HEK-293 cells [64]. Furthermore, gain-of-function missense mutations of Na$_V$1.7 cause pain, for example primary erythromyalgia, an inherited syndrome of burning pain, flushing, and inflammation in the extremities [65–68]. The paroxysmal extreme pain disorder previously known as familial rectal pain has also been associated with Na$_V$1.7 mutation [69]. These well-defined genotype/phenotype correlations make Na$_V$1.7 one of the most convincing targets for pharmacological intervention in the treatment of neuropathic pain.

Progress has recently been made in identifying antagonists for this exciting target. A series of benzoazepinones (e.g. **6** and **7**) has been described as functionally selective blockers, targeting the inactivated state of the channel, with SAR shown in more detail in two recent publications [70,71]. The activity of the molecules was measured with voltage- sensitive fluorescence resonance energy transfer (FRET) dye in cells expressing Na$_V$1.7 [72]; the observed potencies were generally between 30 and 450 nM. When dosed orally at 10 mg/kg, compounds **6** and **7** elicited significant reversal of allodynia in the SNL model.

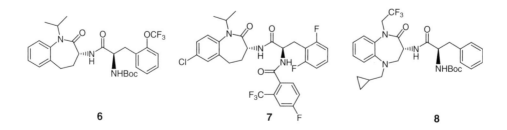

6   **7**   **8**

In a related series of 3-amino-1,5-benzodiazepinones, such as **8**, the cyclopropylmethyl moiety elicited optimal hERG K$^+$ sparing properties, while the trifluoroethyl variant limited metabolic dealkylation *in vivo*. The unsubstituted phenyl ring was retained in preferred compounds. Though selectivity was not disclosed, these molecules are the most potent state-dependent Na$_V$1.7 blockers reported to date as measured by whole cell electrophysiology ($K_i = 0.6$ nM). Despite modest bioavailability of 5%, compound **8** showed activity comparable to lamotrigine **2** and carbamazepine **3** in an epileptic mouse model at 3 mg/kg *po* 30 min post dose [38].

The same assay also led to the discovery of an alternative structural class (e.g., **9**) with an improved pharmacokinetic profile over the 1-benzazepin-2-one series **6** [73], whereby the 3-position of the imidazopyridine carried the corresponding lipophilic side chain. The *t*-butoxycarbonyl (Boc) group derived from the benzodiazepine series remained optimal and was therefore retained. Compound **9** demonstrated low *iv* clearance (15 mL/min/kg) and good exposure in rats ($F = 41\%$). Several related analogs were tested and were not efficacious when dosed orally in the SNL model, although analog **9** (Na$_V$1.7 IC$_{50}$ = 80 nM) showed significant reversal of mechanical allodynia at 3 mg/kg *po* 2 h post dose [73].

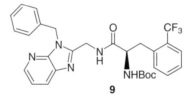

9

A third structural class, for example biaryl heterocycles exemplified by **10** (Na$_V$1.7 IC$_{50}$ = 84 nM), has been pursued. Pyrazine **11** was found to be the most active analog (IC$_{50}$ = 47 nM) in this class, and was used as a starting point for further optimization. Synthesis of a pyrazinone analog (**12**) produced lower clearance *in vivo*, and it was shown that the N-2 position tolerated a broad range of substitutions ultimately used to mitigate hERG K$^+$ and other off-target activities. The diol side chain proved optimal, providing a compound with an IC$_{50}$ on Na$_V$1.7 of 190 nM, rat clearance of 4 mL/min/kg, good bioavailability, and efficacy in the SNL model at 30 mg/kg *po* (56% reversal of mechanical allodynia at 2 h post dose) [74].

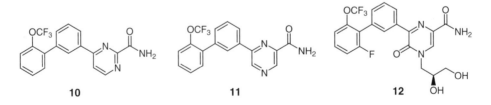

10   **11**   **12**

Additional structural classes have been published in a recent patent application [75]. The aminothiazole **13** was evaluated using a guanidine flux assay and showed modest selectivity over $Na_V1.3$. The piperazine analogs exemplified by **14** had a $Na_V1.7$ IC_{50} of 200 nM with 2–55-fold selectivity over $Na_V1.1$, $Na_V1.3$, and $Na_V1.4$ [75].

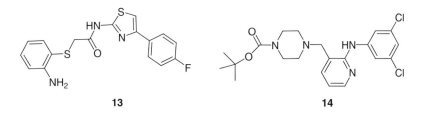

13   **14**

4.6 $Na_V1.8$

$Na_V1.8$ is distributed mostly in the PNS and appears to be an important component of pain perception [76]. Under conditions of neuropathic pain, redistribution occurs with concomitant increase in $Na_V1.8$ immunoreactivity along the uninjured unmyelinated sciatic nerve C-fiber axons, suggesting that block of the increased neuronal activity with a sodium channel blocker could produce effective analgesia [77]. Attenuating expression of $Na_V1.8$ with small interfering RNAs (siRNA) and antisense oligodeoxynucleotides (ODNs) completely reversed the development of both mechanical allodynia and thermal hyperalgesia in models of neuropathic pain in rats [78–80]. Corresponding mouse knock-out experiments show only partial analgesia, and neuropathic pain develops normally in animals lacking both $Na_V1.7$ and $Na_V1.8$ [81]. $Na_V1.8$ expression is also altered in human pain states, for example in patients with brachial plexus injury. $Na_V1.8$ channel density decreased in sensory cell bodies whose central axons had been injured while density was increased in some peripheral nerve fibers just proximal to the site of injury [82]. In addition to pain state induced redistribution, $Na_V1.8$ is also expressed in brain in an animal model for MS (experimental autoimmune encephalomyelitis, EAE) and in tissue from humans with MS [7], thus generating interest for two distinct therapeutic areas of high unmet medical need.

The arylfurans **15** and **16** represent a new class of selective $Na_V1.8$ blockers [83,84]. Initial evaluation was performed with recombinant mouse $Na_V1.8$ expressed in HEK-293 cells using an isotopic efflux assay. *Para* chloro-phenyl substitution was found to be preferred and was retained for further optimization. Interestingly, amide replacement with cyanoamidine provided the most potent analog, **15**, in the series, with a $Na_V1.8$ IC_{50} of 30 nM. Compound **16** had a $Na_V1.8$ IC_{50} of 850 nM; however this molecule showed a 100-fold improvement in the IC_{50} when measured by an assay protocol favoring the inactivated state for human $Na_V1.8$ ($IC_{50} = 8$ nM). One hundred to one thousand-fold selectivity over $Na_V1.2$, Na_V 1.3, Na_V 1.5, and $Na_V1.7$ was confirmed by electrophysiology. Furan **16** was administered *ip* and shown to be effective in rodent neuropathic

(SNL and CCI) and inflammatory pain (CFA thermal hyperalgesia) models [83]; the SNL ED_{50} was determined to be 47 mg/kg. While the molecule crossed the blood brain barrier ([brain]/[plasma] ratio = 1.1), no impairment of motor function, coordination, or balance was observed. These experiments show that selective block of $Na_V1.8$ can provide analgesia in rat animal models for both neuropathic and inflammatory pain.

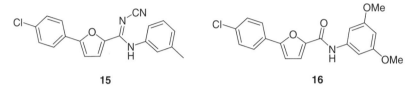

15   16

Pyridine-based $Na_V1.8$ blockers such as **17** have also been reported to be selective over $Na_V1.2$, $Na_V1.5$ and hERG K^+ [85]. This structural class had high rat clearance which was mitigated by the development of pyrazine analogs (e.g., **18**, $Na_V1.8$ IC_{50} = 30 nM). When dosed at 3 mg/kg *po*, compound **18** showed significant antinociceptive effects in rat. Paw withdrawal thresholds of 54% were recorded at the 100 mg/kg dose, leading to a calculated ED_{50} of 78 mg/kg *po* [86].

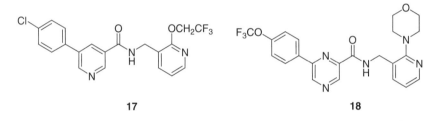

17   18

Pyrazines (e.g., **19**) and pyridines (e.g., **20**), structurally similar to the known sodium channel blocker lamotrigine **2**, have also been reported. Lamotrigine **2** (Lamical) is approved for various types of seizures and bipolar type I disorders. Its therapeutic effect is believed to be dependent upon sodium channel modulation, which may explain its off-label use in the treatment of neuropathic pain. In addition to modulation of $Na_V1.8$ (IC_{50} = 420 nM in a HEK-293 cell line expressing $Na_V1.8$), selectivity of >70 fold was observed for **19**, as measured in a SH-SY5Y cell line expressing multiple TTX-sensitive channels ($Na_V1.2$, $Na_V1.3$, $Na_V1.7$) [87].

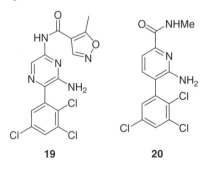

19   20

4.7 Na$_V$1.9

While the Na$_V$1.9 channel has been shown to participate in pain perception, it has proven difficult to express in recombinant systems for screening, and has thus far not appeared in the medicinal chemistry literature.

5. ADDITIONAL STRUCTURES

The majority of novel sodium channel modulators have been reported in the patent literature, where selectivity data are routinely not disclosed. The diversity of the structures shown below, all of which are reported to block sodium channels, supports the hypothesis that many different structural classes can bind to these proteins.

N-(3-sulfamoylphenyl) benzamide **21** [88], quinazolines **22** and **23** [89–91], and aryl-sulfonamides **24** and **25** [92–94] were identified using E-VIPR and membrane potential-sensitive FRET dyes measuring Na$_V$1.3 or Na$_V$1.8 activity in 293-HEK cell lines [15]. The aryl-sulfonamides **26** and **27** were identified using an ion flux assay [95,96].

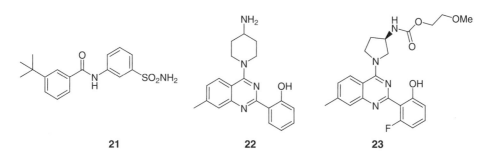

21   **22**   **23**

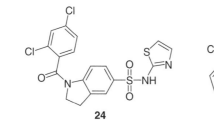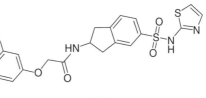

24   **25**

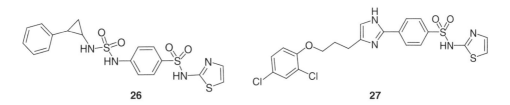

26   **27**

Prolinamide derivative **28** and related molecules [97–99] resemble the sodium channel blocker ralfinamide **29**, which is currently in phase II clinical trials for post-surgical (dental) pain.

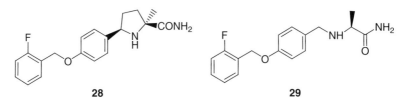

28   **29**

The ureas **30** and **31**, as well as the dihydroisoquinoline analog **32**, have been reported as Na$_V$1.8 blockers (IC$_{50}$ ∼ 100 nM), as measured in Na$_V$1.8 expressing SH-SY5Y neuroblastoma cell lines in the presence of pyrethrum to prevent channel inactivation. Compounds **30–32** [100,101] had Na$_V$1.8 IC$_{50s}$ of 440, 1,300, and 170 nM, respectively. Membrane depolarization was measured via FLIPR readout of fluorescent dyes sensitive to membrane potential changes [102].

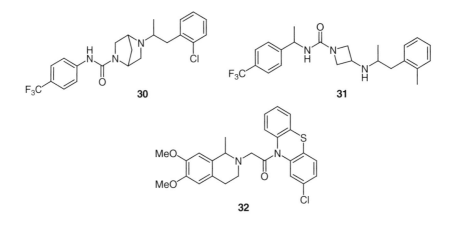

30   **31**

32

The pentyl-indolinone **33**, pyrrolo-pyrazolone **34**, spiro-indolinone **35**, and spiro-aza-indolinone **36** were discovered using a radiolabeled guanidine influx assay [103–106].

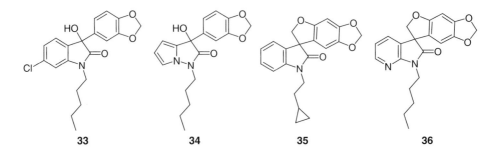

33   **34**   **35**   **36**

Compound **37** exemplifies another structural class which showed sodium channel activity, in this case demonstrated by the displacement of a radiolabled [³H]Batrachotoxin from a rat cerebral cortex sodium channel. When tested in the CCI model of neuropathic pain and the carrageenan-induced hyperalgesia model, **37** showed antihyperalgesic effects at 30 mg/kg *ip* [107].

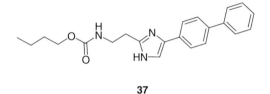

37

6. RECENT CLINICAL DATA

The sodium channel blocker ralfinamide **29** has completed phase II studies for neuropathic pain. While it did not reach its clinical endpoint in this trial, it is being re-investigated for post-operative dental pain. Lamotrigine **2** is currently in a two-year phase II clinical trial for MS in secondary progressive patients employing brain atrophy measurements via MRI. Results could be available in 2009. Lamotrigine has also been evaluated as add-on therapy to gabapentin **38**, a tricyclic antidepressant and a nonopioid analgesic. The desired clinical endpoints were not achieved at up to 400 mg daily [108]. Tetrodotoxin **39**, a potent tonic (non-use-dependent) blocker of the TTX-sensitive channels $Na_V1.1$, $Na_V1.2$, $Na_V1.3$, $Na_V1.4$, $Na_V1.6$, and $Na_V1.7$, was dosed at 15–90 µg daily *im* to treat cancer pain. The open label study showed pain relief for up to two weeks, with generally mild observed toxicity [109]. Lacosamide **40** has completed placebo-controlled clinical trials as adjunct therapy for reduction of seizures. It is believed to enhance slow inactivation of sodium channels and might be efficacious in patients with partial-onset seizures [110,111]. A phase II double-blind placebo-controlled study to treat diabetic peripheral neuropathy has also been completed. At doses up to 400 mg/day lacosamide had significantly better pain relief than placebo [112].

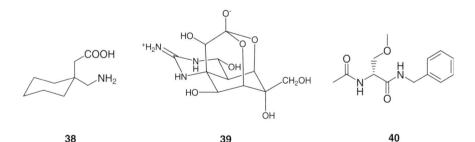

38   **39**   **40**

7. CONCLUSION AND FUTURE PERSPECTIVES

Voltage-gated sodium channels are a fascinating but challenging field of medicinal chemistry, with regular reports of new discoveries about the roles of these channels. In particular, the attention derived from the recent genetic studies of loss-of-function mutations of SCN9A (Na$_V$1.7) in man has propelled these targets to the forefront of drug discovery, resulting in a new generation of molecules which are now advancing from research into development. The pharmacological profile of these novel compounds will demonstrate whether specific channel targeting or potent, non-selective, use-dependent block will translate into better safety windows and improved efficacy in man.

REFERENCES

[1] F. H. Yu and W. A. Catterall, *Sci. STKE.*, 2004, **253**, 15.
[2] F. H. Yu, V. Yarov-Yarovoy, G. A. Gutman and W. A. Catterall, *Pharmacol. Rev.*, 2005, **57**, 387.
[3] T. Anger, D. J. Madge, M. Mulla and D. Riddall, *J. Med. Chem.*, 2001, **44**, 115.
[4] W. A. Catterall, A. L. Goldin and S. G. Waxman, *Pharmacol. Rev.*, 2005, **57**, 397.
[5] J. S. Trimmer and K. J. Rhodes, *Annu. Rev. Physiol.*, 2004, **66**, 477.
[6] J. A. Black, S. Dib-Hajj, D. Baker, J. Newcombe, M. L. Cuzner and S. G. Waxman, *Proc. Natl. Acad. Sci., USA*, 2000, **97**, 11598.
[7] S. G. Waxman, *Prog. Brain Res.*, 2005, **148**, 353.
[8] M. S. Gold, D. Weinreich, C.-S. Kim, R. Wang, J. Treanor, F. Porreca and J. Lai, *J. Neurosci.*, 2005, **25**, 10970.
[9] M. Rogers, L. Tang, D. J. Madgen and E. B. Stevens, *Semin. Cell Dev. Biol.*, 2006, **17**, 571.
[10] D. J. Kyle and V. I. Ilyin, *J. Med. Chem.*, 2007, **50**, 2583.
[11] K. Yamaoka, S. M. Vogel and I. Seyama, *Curr. Pharm. Des.*, 2006, **12**, 429.
[12] A. M. Ritter, C. Ritchie and W. J. Martin, *J. Pain*, 2007, **8**, 287.
[13] J. J. Clare, S. N. Tate, M. Nobbs and M. A. Romanos, *Drug Discov. Today*, 2000, **5**, 506.
[14] J. E. Gonzales, A. P. Termin and D. M. Wilson, in *Voltage-Gated Ion Channels as Drug Targets (Methods and Principles in Medicinal Chemistry)*, Wiley-VCH Verlag GmbH & Co. KGaA, Weinheim, 2006, p. 168.
[15] C. J. Huang, A. Harootunian, M. P. Maher, C. Quan, C. D. Raj, K. McCormack, R. Numann, P. A. Negulescu and J. E. Gonzalez, *Nat. Biotechnol.*, 2006, **24**, 439.
[16] I. E. Dick, R. M. Brochu, Y. Purohit, G. J. Kaczorowski, W. J. Martin and B. T. Priest, *J. Pain*, 2007, **8**, 315.
[17] J. Mao and L. L. Chen, *Pain*, 2000, **87**, 7.
[18] T. S. Jensen, *Eur. J. Pain*, 2002, **6**(Suppl A), 61.
[19] K. L. Petersen, A. Maloney, F. Hoke, J. B. Dahl and M. C. Rowbotham, *J. Pain*, 2003, **4**, 400.
[20] P. Yogeeswari, J. V. Ragavendran, R. Thirumurugan, A. Saxenam and D. Sriram, *Curr. Drug Targets*, 2004, **5**, 589.
[21] I. Tarnawa, H. Bolcskei and P. Kocsis, *Recent Patents CNS Drug Discov.*, 2007, **2**, 57.
[22] G. J. Benneth, in *Neuropathic Pain: New Insights, New Interventions*, Vol. 33, *Hosp. Pract.* (off edition), 1998, p. 95.
[23] B. T. Priest and G. J. Kaczorowski, *Expert. Opin. Ther. Targets*, 2007, **11**, 291.
[24] R. Amir, C. E. Argoff, G. J. Bennett, T. R. Cummins, M. E. Durieux, P. Gerner, M. S. Gold, F. Porreca and G. R. Strichartz, *J. Pain*, 2006, **7**, S1.
[25] S. G. Waxman, *Nat. Rev. Neurosci.*, 2006, **7**, 932.
[26] K. Kanai, S. Kuwabara, S. Misawa, N. Tamura, K. Ogawara, M. Nakata, S. Sawai, T. Hattori and H. Bostock, *Brain*, 2006, **129**, 953.

[27] C. G. Hahn, L. Gyulai, C. F. Baldassano and R. H. Lenox, *J. Clin. Psychiat.*, 2004, **65**, 791.
[28] T. G. Sanchez, A. P. Balbani, R. S. Bittar, R. F. Bento and J. Camara, *Auris Nasus Larynx*, 1999, **26**, 411.
[29] E. Berninger, J. Nordmark, G. Alvan, K. K. Karlsson, E. Idrizbegovic, L. Meurling and A. Al-Shurbaji, *Int. J. Audiol.*, 2006, **45**, 689.
[30] J. E. Gonzalez and Y. Tsien, *Biophys. J.*, 1995, **69**, 1272.
[31] J. E. Gonzalez and Y. Tsien, *Chem. Biol.*, 1997, **4**, 269.
[32] B. T. Priest, A. M. Swensen and O. B. McManus, *Curr. Pharm. Des.*, 2007, **13**, 2325.
[33] T. J. Dale, C. Townsend, E. C. Hollands and D. J. Trezise, *Mol. Biosyst.*, 2007, **3**, 714.
[34] R. W. Fitch and J. W. Daly, *Anal. Biochem.*, 2005, **342**, 260.
[35] S. Trivedi, K. Dekermendjian, R. Julien, J. Huang, P. E. Lund, J. Krupp, R. Kronqvist, O. Larsson and R. Bostwick, *Assay Drug Dev. Technol.*, 2007, electronic publication ahead of print, http://www.liebertonline.com
[36] B. S. Williams, J. P. Felix, B. T. Priest, R. M. Brochu, K. Dai, S. B. Hoyt, C. London, Y. S. Tang, J. L. Duffy, W. H. Parsons, G. J. Kaczorowski and M. L. Garcia, *Biochemistry*, 2007, **46**, 14693.
[37] M. F. Jarvis, P. Honore, C. C. Shieh, M. Chapman, S. Joshi, X. F. Zhang, M. Kort, W. Carroll, B. Marron, R. Atkinson, J. Thomas, D. Liu, M. Krambis, Y. Liu, S. McGaraughty, K. Chu, R. Roeloffs, C. Zhong, J. P. Mikusa, G. Hernandez, D. Gauvin, C. Wade, C. Zhu, M. Pai, M. Scanio, L. Shi, I. Drizin, R. Gregg, M. Matulenko, A. Hakeem, M. Gross, M. Johnson, K. Marsh, P. K. Wagoner, J. P. Sullivan, C. R. Faltynek and D. S. Krafte, *Proc. Natl. Acad. Sci., USA*, 2007, **104**, 8520.
[38] S. B. Hoyt, C. London, M. J. Wyvratt, M. H. Fisher, D. E. Cashen, J. P. Felix, M. L. Garcia, X. Li, K. A. Lyons, D. Euan Macintyre, W. J. Martin, B. T. Priest, M. M. Smith, V. A. Warren, B. S. Williams, G. J. Kaczorowski and W. H. Parsons, *Bioorg. Med. Chem. Lett.*, 2008, **18**, 1963.
[39] C. G. Vanoye, C. Lossin, T. H. Rhodes and A. L. George, Jr., *J. Gen. Physiol.*, 2006, **127**, 1.
[40] E. Perucca, *Br. J. Clin. Pharmacol.*, 1996, **42**, 531.
[41] R. Xu, E. A. Thomas, E. V. Gazina, K. L. Richards, M. Quick, R. H. Wallace, L. A. Harkin, S. E. Heron, S. F. Berkovic, I. E. Scheffer, J. C. Mulley and S. Petrous, *Neuroscience*, 2006, **148**, 164.
[42] R. Xu, E. A. Thomas, M. Jenkins, E. V. Gazina, C. Chiu, S. E. Heron, J. C. Mulley, I. E. Scheffer, S. F. Berkovic and S. Petrou, *Mol. Cell. Neurosci.*, 2007, **35**, 292.
[43] M. J. Craner, J. Newcombe, J. A. Black, C. Hartle, M. L. Cuzner and S. G. Waxman, *Proc. Natl. Acad. Sci., USA*, 2004, **101**, 8168.
[44] M. R. Diniz, R. D. Theakston, J. M. Crampton, M. Nascimento Cordeiro, A. M. Pimenta, M. E. De Lima and C. R. Diniz, *Protein Expr. Purif.*, 2006, **50**, 18.
[45] B. C. Hains, C. Y. Saab and S. G. Waxman, *J. Neurophysiol.*, 2006, **95**, 3343.
[46] B. C. Hains, C. Y. Saab and S. G. Waxman, *Brain*, 2005, **128**, 2359.
[47] B. C. Hains and S. G. Waxman, *Prog. Brain Res.*, 2007, **161**, 195.
[48] P. Zhao, S. G. Waxman and B. C. Hains, *Mol. Pain*, 2006, **2**, 27.
[49] M. J. Craner, J. P. Klein, M. Renganathan, J. A. Black and S. G. Waxman, *Ann. Neurol.*, 2002, **52**, 786.
[50] S. Hong, T. J. Morrow, P. E. Paulson, L. L. Isom and J. W. Wiley, *J. Biol. Chem.*, 2004, **279**, 29341.
[51] E. M. Garry, A. Delaney, H. A. Anderson, E. C. Sirinathsinghji, R. H. Clapp, W. J. Martin, P. R. Kinchington, D. L. Krah, C. Abbadie and S. M. Fleetwood-Walker, *Pain*, 2005, **118**, 97.
[52] B. C. Hains, J. P. Klein, C. Y. Saab, M. J. Craner, J. A. Black and S. G. Waxman, *J. Neurosci.*, 2003, **23**, 8881.
[53] J. A. Lindia, M. G. Kohler, W. J. Martin and C. Abbadie, *Pain*, 2005, **117**, 145.
[54] M. A. Nassar, M. D. Baker, A. Levato, R. Ingram, G. Mallucci, S. B. McMahon and J. N. Wood, *Mol. Pain*, 2006, **2**, 33.
[55] M. A. Casula, P. Facer, A. J. Powell, I. J. Kinghorn, C. Plumpton, S. N. Tate, C. Bountra, R. Birch and P. Anand, *Neuroreport*, 2004, **15**, 1629.
[56] K. Coward, A. Aitken, A. Powell, C. Plumpton, R. Birch, S. Tate, C. Bountra and P. Anand, *Neuroreport*, 2001, **12**, 495.
[57] K. D. Holland, J. A. Kearney, T. A. Glauser, G. Buck, M. Keddache, J. R. Blankston, I. W. Glaaser, R. S. Kass and M. H. Meisler, *Neurosci. Lett.*, 2008, **433**, 65, A.

[58] F. Guo, N. Yu, J. Q. Cai, T. Quinn, Z. H. Zong, Y. J. Zeng and L. Y. Hao, *Brain Res. Bull.*, 2008, **75**, 179.

[59] K. J. Smith, *Brain Pathol.*, 2007, **17**, 230.

[60] D. A. Bechtold and K. J. Smith, *J. Neurol. Sci.*, 2005, **233**, 27.

[61] M. H. Meisler, N. W. Plummer, D. L. Burgess, D. A. Buchner and L. K. Sprunger, *Genetica*, 2004, **122**, 37.

[62] J. A. Kearney, D. A. Buchner, G. De Haan, M. Adamska, S. I. Levin, A. R. Furay, R. L. Albin, J. M. Jones, M. Montal, M. J. Stevens, L. K. Sprunger and M. H. Meisler, *Hum. Mol. Genet.*, 2002, **11**, 2765.

[63] C. Rosker, B. Lohberger, D. Hofer, B. Steinecker, S. Quasthoff and W. Schreibmayer, *Am. J. Physiol. Cell Physiol.*, 2007, **293**, C783.

[64] J. J. Cox, F. Reimann, A. K. Nicholas, G. Thornton, E. Roberts, K. Springell, G. Karbani, H. Jafri, J. Mannan, Y. Raashid, L. Al-Gazali, H. Hamamy, E. M. Valente, S. Gorman, R. Williams, D. P. McHale, J. N. Wood, F. M. Gribble and C. G. Woods, *Nature*, 2006, **444**, 894.

[65] Y. Yang, Y. Wang, S. Li, Z. Xu, H. Li, L. Ma, J. Fan, D. Bu, B. Liu, Z. Fan, G. Wu, J. Jin, B. Ding, X. Zhu and Y. Shen, *J. Med. Genet.*, 2004, **41**, 171.

[66] S. D. Dib-Hajj, A. M. Rush, T. R. Cummins, F. M. Hisama, S. Novella, L. Tyrrell, L. Marshall and S. G. Waxman, *Brain*, 2005, **128**(Pt8), 1847.

[67] S. G. Waxman and S. Dib-Hajj, *Trends Mol. Med.*, 2005, **11**, 555.

[68] T. R. Cummins, S. D. Dib-Hajj and S. G. Waxman, *J. Neurosci.*, 2004, **24**, 8232.

[69] C. R. Fertleman, M. D. Baker, K. A. Parker, S. Moffatt, F. V. Elmslie, B. Abrahamsen, J. Ostman, N. Klugbauer, J. N. Wood, R. M. Gardiner and M. Rees, *Neuron*, 2006, **52**, 767.

[70] S. B. Hoyt, C. London, D. Gorin, M. J. Wyvratt, M. H. Fisher, C. Abbadie, J. P. Felix, M. L. Garcia, X. Li, K. A. Lyons, E. McGowan, D. E. MacIntyre, W. J. Martin, B. T. Priest, A. Ritter, M. M. Smith, V. A. Warren, B. S. Williams, G. J. Kaczorowski and W. H. Parsons, *Bioorg. Med. Chem. Lett.*, 2007, **17**, 4630.

[71] S. B. Hoyt, C. London, H. Ok, E. Gonzalez, J. L. Duffy, C. Abbadie, B. Dean, J. P. Felix, M. L. Garcia, N. Jochnowitz, B. V. Karanam, X. Li, K. A. Lyons, E. McGowan, D. E. Macintyre, W. J. Martin, B. T. Priest, M. M. Smith, R. Tschirret-Guth, V. A. Warren, B. S. Williams, G. J. Kaczorowski and W. H. Parsons, *Bioorg. Med. Chem. Lett.*, 2007, **17**, 6172.

[72] J. P. Felix, B. S. Williams, B. T. Priest, R. M. Brochu, I. E. Dick, V. A. Warren, L. Yan, R. S. Slaughter, G. J. Kaczorowski, M. M. Smith and M. L. Garcia, *Assay Drug Dev. Technol.*, 2004, **2**, 260.

[73] C. London, S. B. Hoyt, W. H. Parsons, B. S. Williams, V. A. Warren, R. Tschirret-Guth, M. M. Smith, B. T. Priest, E. McGowan, W. J. Martin, K. A. Lyons, X. Li, B. V. Karanam, N. Jochnowitz, M. L. Garcia, J. P. Felix, B. Dean, C. Abbadie, G. J. Kaczorowski and J. L. Duffy, *Bioorg. Med. Chem. Lett.*, 2008, **18**, 1696.

[74] F. Ye and P. Shao, MEDI-143, *234th ACS National Meeting*, Boston, MA, August 2007.

[75] R. S. Fraser, R. Sherrington, M. L. MacDonald, M. Samuels, S. Newman, J-M. Fu and R. Kamboj, *PCT Int. Appl.*, WO2007/109324, 2007.

[76] K. Zimmermann, A. Leffler, A. Babes, C. M. Cendan, R. W. Carr, J. Kobayashi, C. Nau, J. N. Wood and P. W. Reeh, *Nature*, 2007, **447**, 855.

[77] M. S. Gold, D. Weinreich, C. S. Kim, R. Wang, J. Treanor, F. Porreca and J. Lai, *J. Neurosci.*, 2003, **23**(1), 158.

[78] X.-W. Dong, S. Goregoaker, H. Engler, X. Zhou, L. Mark, J. Crona, R. Terry, J. Hunter and T. Priestley, *Neuroscience*, 2007, **146**, 812.

[79] F. Porreca, J. Lai, D. Bian, S. Wegert, M. H. Ossipov, R. M. Eglen, L. Kassotakis, S. Novakovic, D. K. Rabert, L. Sangameswaran and J. C. Hunter, *Proc. Natl. Acad. Sci., USA*, 1999, **96**, 7640.

[80] Y. Liu, S. Yao, W. Song, Y. Wang, D. Liu and L. Zen, *J. Huazhong Univ. Sci. Technol. Med. Sci.*, 2005, **25**, 696.

[81] M. A. Nassar, A. Levato, L. C. Stirling and J. N. Wood, *Mol. Pain*, 2005, **1**, 24.

[82] K. Coward, C. Plumpton, P. Facer, R. Birch, T. Carlstedt, S. Tate, C. Bountra and P. Anand, *Pain*, 2000, **85**, 41.

[83] M. F. Jarvis, P. Honore, C. C. Shieh, M. Chapman, S. Joshi, X. F. Zhang, M. Kort, W. Carroll, B. Marron, R. Atkinson, J. Thomas, D. Liu, M. Krambis, Y. Liu, S. McGaraughty, K. Chu, R. Roeloffs, C. Zhong, J. P. Mikusa, G. Hernandez, D. Gauvin, C. Wade, C. Zhu, M. Pai, M. Scanio, L. Shi, I. Drizin, R. Gregg, M. Matulenko, A. Hakeem, M. Gross, M. Johnson, K. Marsh, P. K. Wagoner, J. P. Sullivan, C. R. Faltynek and D. S. Krafte, *Proc. Natl. Acad. Sci., USA*, 2007, **104**, 8520.

[84] M. E. Kort, I. Drizin, R. J. Gregg, M. J. Scanio, L. Shi, M. F. Gross, R. N. Atkinson, M. S. Johnson, G. J. Pacofsky, J. B. Thomas, W. A. Carroll, M. J. Krambis, D. Liu, C. C. Shieh, X. Zhang, G. Hernandez, J. P. Mikusa, C. Zhong, S. Joshi, P. Honore, R. Roeloffs, K. C. Marsh, B. P. Murray, J. Liu, S. Werness, C. R. Faltynek, D. S. Krafte, M. F. Jarvis, M. L. Chapman and B. E. Marron, *J. Med. Chem.*, 2008, **51**, 407.

[85] B. E. Marron, R. N. Atkinson, J. B. Thomas, S, Johnson, G. J. Pacofsky, M. E. Secrest, L. Shi, M. E. Kort, I. Drizin, M. J. C. Scanio, R. J. Gregg, M. A. Matulenko, M. L. Chapman, D. Liu, M. J. Krambis, X. Su, C-C. Shieh, X. Zhang, G. Hernandez, S. Joshi, P. Honore, K. C. Marsh, A. Knox, R. Roeloffs, S. Werness, M. F. Jarvis, C. R. Faltynek and D. S. Krafte, MEDI-319, *233rd ACS National Meeting*, Chicago, IL, March 2007.

[86] M. J. C. Scanio, L. Shi, M. E. Kort, I. Drizin, R. J. Gregg, J. B. Thomas, R. N. Atkinson, M. S. Johnson, B. E. Marron, M. L. Chapman, D. Liu, M. J. Krambis, X. Su, C-C. Shieh, X. Zhang, G. Hernandez, S. Joshi, P. Honore, K. C. Marsh, A. Knox, S. Werness, D. S. Krafte, M. F. Jarvis and C. R. Faltynek, MEDI-321, *233rd ACS National Meeting*, Chicago, IL, 2007.

[87] K. R. Gibson, C. Poinsard, M. S. Glossop and M. I. Kemp, *PCT Int. Appl.*, WO2007/052123, 2007.

[88] P. Joshi, P. Krenitsky, J. Gonzalez, J. Wang, D. Wilson and A. Termin, *US Patent Appl.*, US2007/238733, 2007.

[89] D. M. Wilson, D. Fanning, P. Krenitsky, A. Termin, P. Joshi and U. Sheth, *PCT Int. Appl.*, WO2007/058989, 2007.

[90] J. E. Gonzalez, D. M. Wilson, A. P. Termin, P. Grootenhuis, Y. Zhang, B. Petzoldt, D. Fanning, T. Neubert, R. D. Tung, E. Martinborough and N. Zimmermann, *US Patent Appl.*, US2006/217377, 2006.

[91] D. M. Wilson, A. P. Termin, J. E. Gonzales, D. T. Fanning, T. Neubert, P. Krenitsky, P. Joshi, D. Hurley, U. Seth and J. S. Boger, *PCT Int. Appl.*, WO2006/028904, 2006.

[92] A. Kawatkar, T. Whitney, T. Neubert, N. Zimmermann, A. Termin and E. Martinborough, *PCT Int. Appl.*, WO2006/122014, 2006.

[93] T. Neubert, A. S. Kawatkar. E. Martinborough and A. Termin, *PCT Int. Appl*, WO2006/133459, 2006.

[94] J. E. Gonzalez, A. P. Termin, E. Martinborough and N. Zimmermann, *US Patent Application*, US2006/025415, 2006.

[95] A. Fulp, B. Marron, M. Suto and X. Wang, *PCT Int. Appl.*, WO2007/021941, 2007.

[96] X. Wang, A. Fulp, B. Marron, S. Beaudoin, D. Seconi and M. Suto, *PCT Int. Appl.*, WO2007/056099, 2007.

[97] G. Alvaro, D. Andreotti, M. Bergauer, R. Giovannini and A. Marasco, *PCT Int. Appl.*, WO2007/042250, 2007.

[98] G. Alvaro, D. Amantini, M. Bergauer, F. Bonetti and R. Profeta, R. Giovannini, *PCT Int. Appl.*, WO2007/042240, 2007.

[99] G. B. Alvaro, M. Bergauer, R. Giovannini and R. Profeta, *PCT Int. Appl.*, WO2007/042239, 2007.

[100] R. Hamlyn, G. Addison, C. G. Earnshaw, H. Finch, M. Huckstep, R. Lynch and S. Mellor, *PCT Int. Appl.*, WO2007/007069, 2007.

[101] R. Hamlyn, D. Callis, C. G. Earnshaw, H. Finch, M. Huckstep, R. Lynch and S. Mellor, *PCT Int. Appl.*, WO2007/007057, 2007.

[102] R. Hamlyn, M. Huckstep, C. G. Earnshaw, S. Stokes, D. Tickle, B. Allart, J. W. Boyd, L. J. Knutsen, R. Lynch and L. Patient, *PCT Int. Appl.*, WO2006/082354, 2006.

[103] M. C. Chafeev, S. Chowdhury, R. Fraser, J. Fu, D. Hou, R. Kamboj, S. Liu, S. Sun, J. Sun, S. Sviridov, M. Bagherzadeh, N. Chakka, T. Hsieh and V. Raina, *PCT Int. Appl.*, WO2006/113864, 2006.

[104] M. C. Chafeev, S. Chowdhury, R. Fraser, J. Fu, R. Kamboj, M. S. Bagherzadeh, S. Sviridov and V. Raina, *PCT Int. Appl.*, WO2006/113875, 2006.

[105] M. C. Chafeev, S. Chowdhury, R. Fraser, J. Fu, R. Kamboj, D. Hou, S. Liu, M. S. Bagherzadeh, S. Sviridov, S. Sun, J. Sun, N. Chakka, T. Hsieh and V. Raina, *PCT Int. Appl.*, WO2006/110917, 2006.

[106] M. C. Chafeev, S. Chowdhury, R. Fraser, J. Fu, R. Kamboj, S. Sviridov, M. S. Bagherzadeh, S. Liu and J. Sun, *PCT Int. Appl.*, WO2006/110654, 2006.

[107] A. M. Liberatore, J. Schulz, C. Favre-Guilmard, J. Pommier, J. Lannoy, E. Pawlowski, M. A. Barthelemy, M. Huchet, M. Auguet, P. E. Chabrier and D. Bigg, *Bioorg. Med. Chem. Lett.*, 2007, **17**, 1746.

[108] M. Silver, D. Blum, J. Grainger, A. E. Hammer and S. Quessy, *J. Pain Symptom. Manage.*, 2007, **34**, 446.

[109] N. A. Hagen, K. M. Fisher, B. Lapointe, P. du Souich, S. Chary, D. Moulin, E. Sellers and A. H. Ngoc, *J. Pain Symptom. Manage.*, 2007, **34**, 171.

[110] E. Ben-Menachem, *Drugs Today*, 2008, **44**, 35.

[111] A. C. Errington, T. Stöhr, C. Heers and G. Lees, *Mol. Pharmacol.*, 2008, **73**, 157.

[112] R. L. Rauck, A. Shaibani, V. Biton, J. Simpson and B. Koch, *Clin. J. Pain*, 2007, **23**, 150.

CHAPTER 4

Medication Development for the Treatment of Substance Abuse

Brian S. Fulton

Contents

1. INTRODUCTION

Substance abuse is a worldwide problem costing an estimated $181 billion in the USA alone [1]. A recent study [2] reported that 22.6 million persons in the USA aged 12 or older in 2006 were classified with substance dependence or abuse

Alcohol and Drug Abuse Research Center, McLean Hospital, Harvard Medical School, Department of Psychiatry, 115 Mill St. Belmont, MA 02478, USA

Annual Reports in Medicinal Chemistry, Volume 43
ISSN 0065-7743, DOI 10.1016/S0065-7743(08)00004-3

(DSM-IV criteria) in the past year (9.2% of the national population). Some pertinent data are included below.

Of the 22.6 million affected individuals, 15.6 million were dependent on or abused alcohol alone, 3.2 million were classified with dependence on or abuse of both alcohol and illicit drugs, and 3.8 million were dependent on or abused drugs but not alcohol. Alcohol is the most commonly abused drug, 50.9% of the population over the age of 12 are current drinkers (at least one drink in past month). About 23% of the total population participated in binge drinking (five or more drinks at one setting at least once in the past thirty days) and about 7% of the population were classified as heavy drinkers (five or more drinks on the same occasion on at least five different days in the past thirty days).

Regarding illicit drugs, it was estimated that 14.8 million people 12 or older were using marijuana. It was used by 76% of all current illicit drug users. The illicit drugs with the highest levels of dependence or abuse in 2006 were marijuana (4.2 million), cocaine (1.7 million), and pain relievers (1.6 million).

In terms of use/abuse potential, heroin is the most dangerous of illicit abused drugs in the sense that 58% of the heroin users were classified as dependent on or abusers of heroin whereas only 12% of those using alcohol were classified as dependent on or abusers of alcohol. Methamphetamine abuse is particularly troublesome. Its incidence of use is similar to that of crack cocaine (700,000 users within the past month). It was also reported that there are 5.2 million persons who were nonmedical users of prescription analgesics (e.g., 1.3 million persons reported the nonmedical use of OxyContin within the past year).

The DSM-IV [3] divides substance abuse into two categories: Substance Use Disorders (substance dependence and abuse) and Substance-Induced Disorders (intoxication, withdrawal, dementia, etc.). This chapter will focus on drug development for the treatment of dependence, abuse, and withdrawal.

2. THE ADDICTIVE STATE

Addiction is described as "a chronic, often relapsing brain disease that causes compulsive drug seeking and use despite harmful consequences to the individual that is addicted and those around them" [4]. The addiction process tends to follow a progression from casual social use followed by routine compulsive use then to substance dependence with the possible development of tolerance. A withdrawal response from substance dependence can occur with the degree of severity being dependent on the drug of abuse. Though it is possible to complete a withdrawal program for all drugs of abuse to the desired end point of abstinence, relapse to the dependent stage is common and is in fact, not unexpected. The most effective treatment paradigms involve a combination of psychopharmacological treatment (when available) and behavioral therapy.

The different stages of addiction have been defined as [5]:

1. Acute reinforcement/social drug taking/impulsive use
2. Escalating/compulsive use (e.g., binge drinking)

3. Dependence
4. Withdrawal
5. Protracted abstinence
6. Relapse to the compulsive stage (craving)
7. Recovery

The reasons for addiction involve a complex interaction of genetic, environmental, emotional, neurobiological, and social conditioning factors. Despite the great structural diversity of drugs of abuse they all have several similar effects. They all are acutely rewarding and result in negative emotional reinforcement upon withdrawal of the drug. Pathological and pharmacological commonalities seen among all drugs of substance abuse are long-term neurobiological and neuroanatomical changes and the effect of the drug (directly or indirectly) upon the mesolimbic dopamine system of the brain with modulation of dopamine transmission and levels [6].

The treatment of addiction is complicated due to the different stages of addiction as defined above. Future points of entry for medicinal chemistry are at the stages of dependence/withdrawal, and, most importantly, relapse.

2.1 Dependence/withdrawal stage

At this stage the patient is dependent upon the drug and can be considered to be abusing the drug with harmful effects upon themselves. Dependence has been described as a state where discontinuation of the drug causes withdrawal symptoms and the person compulsively takes the drug [7]. Physiological tolerance to the drug can develop meaning that increasing doses are required to achieve the same effect.

Chronic drug use has been shown to produce a variety of neural adaptations [8]. It increases the activity of the cAMP pathway and the transcription factors CREB and ΔFosB. Short-term abstinence (hours to days) results in increased activity of glutamatergic, noradrenergic, and corticotropin-releasing hormone (CRF) activities with a decrease in dopaminergic and serotonergic activity.

Withdrawal occurs when the drug is no longer available. Withdrawal can occur by voluntary cessation of use or can be induced (e.g., the opioid antagonist naltrexone will induce immediate withdrawal symptoms in heroin users). The negative emotional and physiological affects that can occur upon withdrawal will often lead the substance abuser to resume drug use.

Medication that can alleviate these negative effects will assist in the recovery process. A well-known example is the use of the mu opioid receptor partial agonist methadone (1) for the treatment of heroin use. A more recent example is the use of the $\alpha4\beta2$ nAChR partial agonist varenicline (2) for smoking cessation [9].

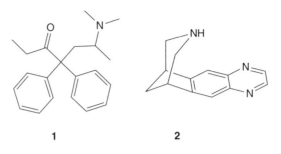

1   2

2.2 Relapse

Relapse is the resumption of drug-taking following detoxification and abstinence. A distinction of relapse is that it occurs following chronic drug use. On a molecular level, various theories for relapse involve long-term changes in ΔFosB [10] and dysregulation of glutamatergic transmission [11,12].

3. BEHAVIORAL PHARMACOLOGY MODELS OF ADDICTION

As with many other psychiatric disorders, it is difficult to determine the exact neurobiological etiology of substance abuse. A distinct molecular target that is solely, individually, responsible for the addictive state is unknown making drug design an interesting challenge. Drug abuse research therefore relies heavily on the use of animal behavioral pharmacology models. The first stage of testing involves receptor-binding studies to determine affinity and selectivity toward a given receptor of interest followed by a functional assay to determine potency. These must be closely followed by behavioral pharmacology studies to determine the effect of the medication on various stages of the addiction that can be modeled in animals [13,14].

Following the definitions above, the models will be divided into two broad categories that will be of interest to the medicinal chemist. These are briefly described.

3.1 Impulsive use: self-administration

These methods study the positive reinforcing effects of drugs. Drugs that are self-administered by animals (rodent and non-human primates) tend to have a high abuse potential in humans. These studies are performed under various "schedules" to explore different aspects of reinforcement.

3.2 Relapse (craving)

Three broad animal models of drug-seeking reinstatement are used to model relapse. They are all based on cues that the animals have been trained to associate with the reinforcing effects of a drug. These models are

1. Drug-induced reinstatement
    This model is a measure of the motivation for drug-seeking behavior following extinction (removal of the drug from a drug-dependent animal). The animal is primed with a dose of the drug (or drug substitute) and the resultant rate of self-administration is measured. An example of this in humans would be giving a single drink to a recovering alcoholic.
2. Cue-induced reinstatement
    This model is a measure of the motivation for drug-seeking behavior based on cues (sound, light) that the animal has been trained to associate with the availability of a drug. In recovering cocaine and opioid addicts, a common cue that can initiate craving and relapse is the visual sighting of drug-taking paraphernalia.
3. Stress-induced reinstatement
    This model measures the ability of stress (e.g., foot-shock) to induce an animal to resume drug-seeking behavior following extinction. Negative affective states such as stress, anger, anxiety, and depression have been identified in humans as common mechanisms that can lead to relapse.

4. PARTIAL AGONIST APPROACH TO MEDICATION DEVELOPMENT

Antagonist treatment has been used in opioid addiction (e.g., oral naltrexone) but the results have been mixed. There is a possibility that a depot form of naltrexone being developed for the treatment of alcoholism could be useful in the treatment of opioid addiction [15].

The most successful approach to date has been the development of partial agonists. These compounds mimic the activity of the drug of abuse by occupying the same binding sites on the receptor and producing a similar response, but of lower intensity, to that of the drug of abuse. Partial agonists share some of the pharmacological properties of the drug of abuse they replace. A goal of the partial agonist approach is to reduce the level of craving during and after withdrawal in order to prevent relapse while also having a low abuse liability. A partial agonist can also act as an antagonist of the drug of abuse if it has a similar or higher binding affinity than the drug of abuse [16].

The advantages and disadvantages of agonist and antagonist approaches in stimulant disorders have been discussed in detail in the literature [17]. In general, agonist treatment results in better patient compliance and reduces the risk of unknown side affects. There is however the risk of overdose.

This chapter will focus on some recent developments in the treatment of narcotic (opioids) and stimulant (cocaine and methamphetamine) abuse. The National Institute of Drug Abuse (NIDA) is performing the majority of medication development for narcotic and stimulant abuse. Several excellent reviews have been written on medication development for the treatment of drug dependence [18–24].

5. OPIOIDS

5.1 Targets

Most activity is centered on the development of agonists targeting the mu, delta, and kappa opioid receptors. It has been reported that muscarinic receptor antagonists may also be useful in treating opioid withdrawal [25]. Several muscarinic antagonists with varying levels of selectivity; for example the M1 subtype selective antagonist pirenzepine (**3**), the M1/M2/M4 antagonist metoctramine (**4**), the M2/M4 antagonist himbacine (**5**), the M1/M3/M4/M5 antagonist 4-DAMP (**6**), and the M4 antagonist PD 102807 (**7**) were all shown to inhibit the naloxone-induced contraction of isolated guinea-pig ileum in an *in-vitro* functional model.

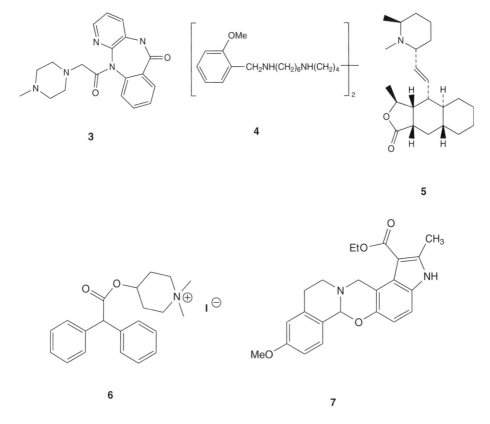

The M1, M4, and M5 receptor subtypes are predominately expressed in the CNS while the M2 and M3 subtypes are expressed both in the CNS and peripheral tissue. Mice lacking the M5 subtype (in mutant M5R$^{-/-}$) show a reduction in the rewarding effects of morphine [26].

Stress is an important cue for relapse. CP-154,526 (**8**), a potent CRF-1 receptor antagonist (IC$_{50}$ = 5.5 nM), can reduce reinstatement of morphine-conditioned place preference following foot-shock stressors [27]. Additional research indicates that the mode of action of this compound is complicated. When CP-154,526 was injected into three regions of the rat brain; the bed nucleus of the stria terminalis (BNST), amygdala, or the nucleus accumbens (NAc), it was found that the attenuated foot-shock stress-induced reinstatement of morphine-conditioned place preference occurred only when injected into the BSNT, while the attenuated morphine priming-induced reinstatement of morphine-conditioned place preference occurred only when injected into the amygdala or NAc.

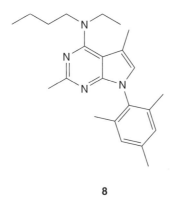

8

5.2 Medication development

Tramadol (9) is a mu opioid agonist and a reuptake inhibitor of NET and SERT. In a human clinical trial at doses of 200 and 400 mg (QD, po), it showed evidence of suppressed opioid withdrawal symptoms without any significant agonist effects [28]. Memantine (10), a noncompetitive NMDA antagonist, is approved for moderate to severe Alzheimer's disease. It produced modest reductions in subjective aspects (craving) in heroin-dependent patients [29]. Lofexidine (11), an α_{2A} agonist, reduced stress-related opioid craving and relapse [30]. AP-267 (structure unavailable) is a serotonin receptor-modulating compound in development for opioid withdrawal [31].

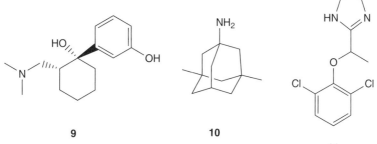

9   10

11

6. COCAINE

6.1 Targets

The primary action of cocaine (**12**) is binding to the dopamine transporter (DAT) and preventing reuptake of synaptic dopamine. Most drug strategies have focused on methods to reduce the effects of excess dopamine. These include partial antagonists at dopamine receptors, D3 agonists, or indirect modulation by interaction with the opioid, serotonin [32], norepinephrine [20], and sigma receptor systems [33].

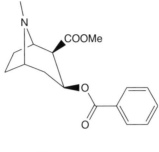

12

6.2 Medication development

To date there have been no drugs approved by the FDA for use in the treatment of cocaine abuse.

Drugs previously approved by the FDA for other indications are under study by the National Institute of Drug Abuse (NIDA) for the treatment of acute and chronic cocaine use in human clinical trials. The majority of these compounds modulate the concentration of dopamine or mechanisms involved in stress.

Modafinil (**13**) is a stimulant drug that is approved for the treatment of narcolepsy. Modafinil binds weakly to DAT ($IC_{50} = 6.4\,\mu M$) and NET ($IC_{50} = 35.6\,\mu M$) but not to SERT ($IC_{50} > 500\,\mu M$) and has been shown to occupy striatal DAT and thalamic NET sites in rhesus monkeys using positron emission tomography (PET) imaging [34]. In a small-scale clinical trial, modafinil, at 200–400 mg/day, attenuated smoked cocaine self-administration in frequent users [35]. Tiagabine (**14**) at 24 mg/day was shown to reduce cocaine-taking behavior among methadone-stabilized cocaine abusers [36]. N-acetylcysteine reduced cue-induced reinstatement [37].

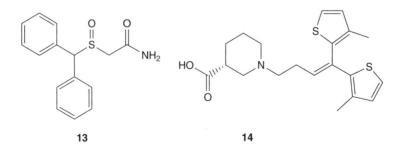

13   **14**

Combination therapy of the dopamine modulator amantadine (**15**) and the GABA$_B$ agonist baclofen (**16**) has been reported to reduce craving for cocaine [38].

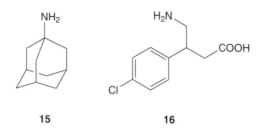

Disulfiram (**17**), an aldehyde dehydrogenase and dopamine-β-hydroxylase inhibitor, reduces the cocaine "high" and "rush" in cocaine-dependent individuals when administered prior to cocaine injection [39]. Dopamine-β-hydroxylase is a key enzyme in the conversion of dopamine into norepinephrine and the inhibition of this enzyme has been shown to modulate the levels of dopamine in the CNS. JDTic (**18**), a selective kappa opioid antagonist (kappa $K_e = 20\,\text{pM}$), has shown promise in reducing the stress-induced relapse of cocaine-seeking behavior in abstinent rats [40,41].

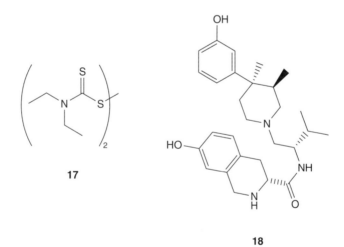

7. STIMULANTS: METHAMPHETAMINE

7.1 Targets

Methamphetamine (**19**), like cocaine, is a stimulant but it acts by a different mechanism. Whereas cocaine binds to the DAT on the cell surface, methamphetamine acts intracellularly by binding to vesicular monoamine transporters (VMAT2) causing the release of dopamine [42]. PET imaging of five methamphetamine abusers showed that chronic use of methamphetamine reduced dopaminergic tone with loss of dopamine transporters [43]. Methamphetamine also activates trace amine-associated receptor 1 (TAAR1) and reduces dopamine accumulation when coexpressed with DAT [44]. Stress is an important cue for relapse. Dysfunction of the CRF-1 receptor system has been implicated in stress-related disorders [45]. It is being shown that CRF-1 receptor antagonists for example can attenuate stress-induced physiological and behavioral changes [46].

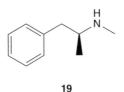

19

7.2 Medication development

No drug has been approved for use in the treatment of methamphetamine abuse.

Aripiprazole (**20**), a D2 partial agonist, attenuates behavorial effects of d-amphetamine and is proposed to be of possible use for the treatment of stimulant abuse [47]. Lobeline (**21**) has shown promise as a potential medication for treating methamphetamine abuse [48]. This compound is fairly promiscuous acting as a mu opioid receptor antagonist ($K_i = 740$ nM), a $\alpha 4 \beta 2$ and $\alpha 7$ nicotinic acetylcholine receptor agonist/antagonist ($K_i = 10$–20 nM), VMAT2 inhibitor ($IC_{50} = 1 \mu M$), and a DAT inhibitor ($K_i = 40$–$100 \mu M$) [49].

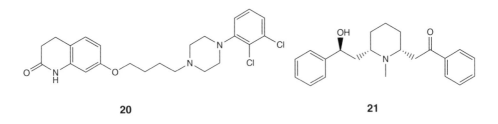

20   **21**

CP-154,526 (**8**) has shown promise in its ability to attenuate the cue-induced and methamphetamine-induced reinstatement of extinguished methamphetamine-seeking behavior in rats [50]. It has also shown promise in it's ability to attenuate the stress-induced increase of alcohol consumption in mice [51]. Bupropion (**22**), a DAT inhibitor, was tested for efficacy in increasing the weeks of abstinence in methamphetamine-dependent patients ($n = 151$). Seventy-two patients were randomized to placebo and seventy-nine to sustained-released bupropion 150 mg twice daily. It was found that bupropion was effective in increasing the number of weeks of abstinence in patients with a low-to-moderate methamphetamine dependence [52].

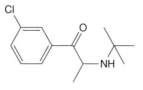

22

8. CONCLUSION

While some drugs are in use for the treatment of opiod withdrawal and maintenance, there are no effective, approved medications for the treatment of stimulant dependence and abuse. The problem of effective pharmacological treatment for relapse (a chronic condition with all drugs of abuse) remains unsolved. It must be emphasized that history has shown that an effective treatment strategy for substance abuse cannot be achieved by the use of psychopharmacological drugs alone. An important component to recovery will require concomitant behavioral and psychology therapy.

REFERENCES

[1] NIDA info facts 8/2006, http://www.nida.nih.gov
[2] 2006 National Survey on Drug Use & Health, http://www.oas.samhsa.gov
[3] American Psychiatric Association. *Diagnostic and Statistical Manual of Mental Disorders*, 4th edition, Text Revision, DSM-IV-TR. Washington, DC, 2000.
[4] NIDA Info Facts 9/2007.
[5] G. F. Koob and Michel Le Moal, in *Neurobiology of Addiction*, Academic Press, New York, 2006, p. 3.
[6] E. J. Nestler, *Nature Neuroscience*, 2005, **8**, 1445.
[7] W. A. McKim, in *Drugs and Behavior*, 4th edition, Prentice Hall, Upper Saddle River, NY, 2003, p. 42.
[8] E. J. Nestler and G. K. Aghajanian, *Science*, 1997, **278**, 58.
[9] J. W. Coe, P. R. Brooks, M. G. Vetelino, M. C. Wirtz, E. P. Arnold, J. Huang, S. B. Sands, T. I. Davis, L. A. Lebel, C. B. Fox, A. Shrikhande, J. H. Heym, E. Schaeffer, H. Rollema, Y. Lu, R. S. Mansbach, L. K. Chambers, C. C. Rovetti, D. W. Schulz, F. D. Tingley, III and B. T. O'Neill, *J. Med. Chem.*, 2005, **48**, 3472.
[10] E. J. Nestler, *Trends Pharmacol. Sci.*, 2004, **25**, 210.
[11] P. W. Kalivas, *Curr. Opin. Pharmacol.*, 2004, **4**, 23.
[12] P. W. Kalivas, *Am. J. Addict.*, 2007, **16**, 71.
[13] G. F. Koob and Michel Le Moal, in *Neurobiology of Addiction*, Academic Press, New York, 2006, Chapter 2.
[14] C. P. O'Brien and E. L. Gardner, *Pharmacol. Therapeut.*, 2005, **108**, 18.
[15] J. H. Jaffe, *Addiction*, 2006, **101**, 468.

[16] H. Rollema, J. W. Coe, L. K. Chambers, R. S. Hurst, S. M. Stahl and K. E. Williams, *Trends Pharmacol. Sci.*, 2007, **28**, 316.

[17] J. Grabowski, J. Shearer, J. Merrill and S. S. Negus, *Addict. Behav.*, 2004, **29**, 1439.

[18] F. Vocci and W. Ling, *Pharmacol. Therapeut.*, 2005, **108**, 94.

[19] M. J. Kreek, K. S. LaForge and E. Butelman, *Nat. Rev. Drug Discover.*, 2002, **1**, 710.

[20] D. Weinshenker and J. P. Schroeder, *Neuropsychopharmacology*, 2007, **32**, 1433.

[21] A. M. Thayer, *C&EN*, 2006, (September 25), 21.

[22] M. Rocio, A. Carrera, M. M. Meijler and K. D. Janda, *Bioorg. Med. Chem.*, 2004, **12**, 5019.

[23] F. J. Vocci, J. Acri and A. Elkashef, *Am. J. Psychiat.*, 2005, **162**, 1432.

[24] G. A. Kenna, D. M. Nielsen, P. Mello, A. Schiesl and R. M. Swift, *CNS Drugs*, 2007, **21**, 213.

[25] A. Capasso, *Lett. Drug Des. Discover.*, 2007, **4**, 207.

[26] J. Wess, R. M. Eglen and D. Gautam, *Nat. Rev. Drug Discover.*, 2007, **6**, 721.

[27] J. Wang, Q. Fang, Z. Liu and L. Lu, *Psychopharmacology*, 2006, **185**, 19.

[28] M. R. Lofwall, S. L. Walsh, G. E. Bigelow and E. C. Strain, *Psychopharmacology*, 2007, **194**, 381.

[29] S. D. Comer and M. A. Sullivan, *Psychopharmacology*, 2007, **193**, 235.

[30] R. Sinha, A. Kimmerling, C. Doebrick and T. R. Kosten, *Psychopharmacology*, 2007, **190**, 569.

[31] H. S. Sharma, T. Lundstedt, A. Boman, P. Lek, E. Seifert, L. Wiklund and S. F. Ali, *Ann. N.Y. Acad. Sci.*, 2006, **1074**, 482.

[32] P. W. Kalivas, *Am. J. Addict.*, 2007, **16**, 71.

[33] R. R. Matsumoto, Y. Liu, M. Lerner, E. W. Howard and D. J. Bracket, *Eur. J. Pharmacol.*, 2003, **469**, 1.

[34] B. K. Madras, Z. Xie, Z. Lin, A. Jassen, H. Panas, L. Lynch, R. Johnson, E. Livni, T. J. Spencer, A. A. Bonab, G. M. Miller and A. J. Fischman, *J. Pharmacol. Exp. Ther.*, 2006, **319**, 561.

[35] C. L. Hart, M. Haney, S. K. Vosburg, E. Rubin and R. Foltin, *Neuropsychopharmacology*, 2007, **33**, 761.

[36] G. Gonzalez, R. Desai, M. Sofuoglu, J. Poling, A. Oliveto, K. Gonsai and T. Kosten, *Drug Alcohol Dep.*, 2007, **87**, 1.

[37] S. D. LaRowe, H. Myrick, S. Hedden, P. Mardikain, M. Saladin, A. McRae, K. Brady, P. W. Kalivas and R. Malcom, *Am. J. Psychiat.*, 2007, **164**, 1115.

[38] E. Rotheram-Fuller, R. De La Garza, II, J. J. Mahoney, III, S. Shoptaw and T. F. Newton, *Psychiat. Res.*, 2007, **152**, 205.

[39] J. R. Baker, P. Jatlow and E. F. McCance-Katz, *Drug Alcohol Dep.*, 2007, **87**, 202.

[40] I. Carroll, J. B. Thomas, L. A. Dykstra, A. L. Granger, R. M. Allen, J. L. Howard, G. T. Pollard, M. D. Aceto and L. S. Harris, *Eur. J. Pharmacol.*, 2004, **501**, 111.

[41] P. M. Beardsley, J. L. Howard, K. L. Shelton and F. I. Carroll, *Psychopharmacology*, 2005, **183**, 118.

[42] J. S. Partilla, A. G. Dempsey, A. S. Nagpal, B. E. Blough, M. H. Baumann and R. B. Rothman, *J. Pharmacol. Exp. Ther.*, 2006, **319**, 237.

[43] N. D. Volkow, L. Chang, G. J. Wang, J. S. Fowler, D. Franceschi, M. Sedler, S. J. Gatley, E. Miller, R. Hitzemann, Y. S. Ding and J. Logan, *J. Neurosci.*, 2001, **21**, 9414.

[44] Z. Xie and G. M. Miller, *J. Pharmacol. Exp. Ther.*, 2007, **312**, 128.

[45] J. H. Kehne, *CNS & Neurol. Disorders – Drug Targets*, 2007, **6**, 163.

[46] P. A. Seymour, A. W. Schmidt and D. W. Schulz, *CNS Drug Rev.*, 2003, **9**, 57.

[47] W. W. Stoops, J. A. Lile, P. E. A. Glaser and C. Rush, *Drug Alcohol Dep.*, 2006, **84**, 206.

[48] D. K. Miller, in *New Research on Methamphetamine Abuse* (ed. G. H. Toolaney), Hauppauge, NY, Nova Science Publishers, 2006.

[49] D. K. Miller, J. R. Lever, K. R. Rodvelt, J. A. Basket, M. J. Will and G. R. Kracke, *Drug Alcohol Dep.*, 2007, **89**, 282.

[50] M. C. Moffett and N. E. Goeders, *Psychopharmacology*, 2007, **190**, 171.

[51] E. G. Lowery, A. M. Sparrow, G. R. Breese, D. J. Knapp and T. E. Thiele, *Alcohol.: Clin. Exp. Res.*, 2008, **32**, 240.

[52] A. M. Elkashef, R. A. Rawson, A. L. Anderson, S. H. Li, T. Holmes, E. V. Smith, N. Chiang, R. Kahn, F. Vocci, W. Ling, V. J. Pearce, M. McCann, J. Campbell, C. Gorodetzky, W. Haning, B. Carlton, J. Mawhinney and D. Weis, *Neuropsychopharmacology*, 2008, **33**, 1162.

PART II:
Cardiovascular and Metabolic Diseases

Editor: Andy W. Stamford
Schering-Plough Research Institute
Kenilworth
New Jersey

CHAPTER 5

GPR40 (FFAR1) Modulators

Julio C. Medina and **Jonathan B. Houze**

<table>
<tr><td>Contents</td><td>1.</td><td>Introduction</td><td>75</td></tr>
<tr><td></td><td>2.</td><td>Biology of GPR40</td><td>76</td></tr>
<tr><td></td><td>3.</td><td>Arylpropionic Acids</td><td>79</td></tr>
<tr><td></td><td></td><td>3.1 Hydroxyphenylpropionic acid derivatives and related compounds</td><td>79</td></tr>
<tr><td></td><td></td><td>3.2 Aminophenylpropionic acid derivatives and related compounds</td><td>81</td></tr>
<tr><td></td><td>4.</td><td>Long Chain Alkyl Acid Derivatives</td><td>82</td></tr>
<tr><td></td><td>5.</td><td>Acid Isosteres</td><td>82</td></tr>
<tr><td></td><td>6.</td><td>Conclusion</td><td>83</td></tr>
<tr><td></td><td></td><td>References</td><td>84</td></tr>
</table>

1. INTRODUCTION

The consequences of diabetes are severe and far-reaching. Despite the development of multiple therapeutic options in the decades since the introduction of isolated insulin first revolutionized the treatment of diabetes, complications from inadequately controlled diabetes still inflict blindness, kidney failure, amputations, and death upon thousands of patients every year. Alongside the staggering human costs of diabetes stands the burden on overstretched healthcare systems worldwide. As of 2002, 11% of all direct healthcare costs were due to diabetes and its associated co-morbidities. This burden is only expected to increase with the spread of sedentary lifestyles and the "western diet" leading to an estimated 366 million diabetics worldwide by the year 2030 [1].

Diabetes is classically characterized by an inability of the patient to maintain glucose homeostasis. This defect arises due to both increased resistance of bodily tissues to the effects of insulin as well as a marked decrease in secretion of insulin

Amgen Inc., 1120 Veterans Boulevard, South San Francisco, CA 94080, USA

Annual Reports in Medicinal Chemistry, Volume 43
ISSN 0065-7743, DOI 10.1016/S0065-7743(08)00005-5

by the pancreatic islets in response to glucose. Therapeutic agents targeting the latter of these two defects, insulin secretagogues, have been in use for nearly 50 years. Still in wide use, sulfonylurea insulin secretagogues target the ATP-sensitive potassium (K_{ATP}) channel present in pancreatic beta cells and elicit a sustained elevation in insulin release [2]. However, a drawback associated with the use of sulfonylurea insulin secretagogues is that their activity is not modulated by the plasma glucose level in the patient. Continued insulin secretion during periods of relatively low blood sugar such as an extended break between meals or vigorous exercise leads to hypoglycemia with its attendant symptoms of sweating, nervousness, dizziness, and confusion. Without intervention, patients can lose consciousness or even fall comatose. Newer generation therapeutics targeting the K_{ATP} channel such as the glinides seek to avoid the risk of hypoglycemia by acting for a short period of time. When taken shortly before or with a meal, the shorter acting insulin secretagogues provide a surge of insulin that allows glucose and other nutrients to be absorbed by body tissues but then the relatively rapid clearance of these compounds leads to a cessation of secretagogue activity [3].

An elegant solution to the risk of hypoglycemia would be for the activity of the insulin secretagogue to be dependent upon elevated plasma glucose levels. Glucagon-like peptide 1 (GLP-1) receptor-based therapeutics have shown just such activity [4]. While GLP-1 itself is unsuitable as a therapeutic agent due to its short half-life, a variety of peptide analogs have been developed clinically. One in particular, exenatide, won approval by the United States Food and Drug Administration (FDA) in 2005 [4]. The peptidic nature of GLP-1 and its analogs necessitates dosing by injection. An alternative approach that allows oral dosing of a small molecule therapeutic agent is inhibition of dipeptidyl peptidase IV (DPP-IV), the protease primarily responsible for inactivation of endogenous GLP-1. DPP-IV inhibitors have been widely sought, and the first one to win FDA approval, sitagliptin, was the subject of a recent case study [5]. While these agents are highly successful, there is still room for improvement in the class of glucose-dependent insulin secretagogues due to the need for additional hemoglobin A1c (Hb_{A1c}) lowering beyond the 0.5–1% obtained with DPP-IV inhibitors [4] and the need for injection of peptidic GLP-1 analogs.

Fatty acids have been repeatedly shown to increase the responsiveness of pancreatic islets to glucose both *in vitro* and *in vivo* [6]. The recent identification of GPR40 as a receptor for free fatty acids that is localized to the pancreatic islet cells has therefore stimulated interest in obtaining small molecule modulators of its function as glucose-dependent insulin secretagogues for the treatment of Type II Diabetes [7–9].

2. BIOLOGY OF GPR40

GPR40 is a member of a family G-protein-coupled receptors that sense free fatty acids [7–10]. The other members of the family, GPR41 and GPR43, respond to short chain fatty acids while GPR40 prefers long chain fatty acids (LCFA, $>C_6$).

The potency of LCFA effects on GPR40 increases in parallel with increasing chain length and degree of unsaturation [8]. Among the common naturally occurring fatty acids, docosahexaenoic acid (DHA) shows the most potent effects displaying an EC50 of approximately 1 μM in CHO cells expressing GPR40 [8].

Expression of GPR40 was initially reported in the pancreas and small intestine of rats [8], and pancreas and the brain in humans [7]. Expression in both humans and rodents is highest in the pancreas, and is particularly concentrated in the pancreatic islets [11]. The high expression of GPR40 in pancreatic islets in humans was verified from samples taken from patients undergoing pancreatectomy and was found to be comparable to that of targets of clinically used antidiabetic agents such as the K_{ATP} channel and the GLP-1 receptor [12]. Within the islet, expression of GPR40 mRNA is highest in insulin-secreting beta cell lines such as the mouse insulinoma (MIN6) line [8]. Reports conflict as to whether GPR40 is present in other islet cell types such as glucagon-secreting alpha cells [8,13].

When pancreatic beta cell lines such as MIN6 are treated with LCFA, insulin secretion is stimulated in a glucose-dependent manner. The effects of LCFA are evident when the medium contains 5.5 mM glucose, and is near maximal at 11 mM glucose [8]. The effects of LCFA on MIN6 cells are GPR40-mediated since pre-treatment of MIN6 cells with GPR40-specific small interfering RNA (siRNA) nearly abolishes the effect of LCFA on glucose-dependent insulin secretion while leaving GLP-1-stimulated insulin secretion unchanged [8].

Signal transduction from the GPR40 receptor occurs primarily through $G\alpha_q$. When treated with long chain fatty acids MIN6, Chinese hamster ovary (CHO), and human embryonic kidney (HEK293) cells overexpressing GPR40 show an increase in intracellular calcium ions ($[Ca^{2+}]_i$) [7,8,14]. This increase in $[Ca^{2+}]_i$ is attenuated by a phospholipase C (PLC) inhibitor, consistent with a $G\alpha_q$-coupled pathway [14]. The products of PLC, inositol triphosphate (IP3) and diacylglycerols (DAG), can each further propagate signaling through GPR40. IP3 increases $[Ca^{2+}]_i$ by triggering release of calcium ions from stores in the endoplasmic reticulum. DAG, either alone or in combination with increased $[Ca^{2+}]_i$ can activate certain isoforms of protein kinase C (PKC). Furthermore, it has been reported that GPR40 can additionally signal through $G\alpha_s$ and thus leading to an increase in intracellular cAMP levels and activation of protein kinase A (PKA) [15] (Figure 1).

To understand how activation of GPR40 can amplify glucose-stimulated insulin secretion, the signaling pathway of GPR40 must be considered in the context of overall control of insulin release from the beta cell. Once taken into the beta cell by the glucose transporter GLUT2, glucose triggers insulin release through a complex series of events starting with initiation of glycolysis and oxidative phosphorylation leading to an increase of the ATP/ADP ratio [16]. This leads to the triggering of an action potential through the closure of K_{ATP} channels. The initial depolarization then leads to opening of L-type voltage-dependent calcium channels (VDCC) that propagate the action potential. Calcium flux though the L-type VDCC has been shown to be tightly coupled to insulin

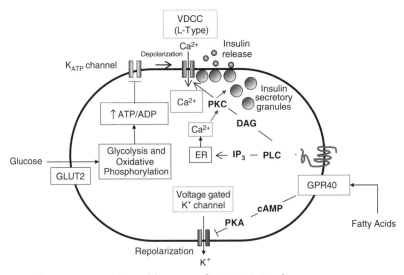

Figure 1 Insulin secretion and possible targets for GPR40 signaling.

secretion [17]. Insulin release ceases when the action potential is dissipated by opening of voltage-gated potassium channels allowing potassium ions to exit the cell [16]. PKC and PKA, both of which are potentially activated by GPR40 signaling, can phosphorylate the ion channels described above that regulate insulin release from the beta cell. However, the exact mechanism by which GPR40 exerts its effects *in a glucose-dependent manner* remains to be elucidated.

To study the effects of GPR40 at the whole-animal level, GPR40$^{-/-}$ mice have been generated in two different genetic backgrounds [18,19]. GPR40$^{-/-}$ mice from both backgrounds were apparently healthy with no overt signs of metabolic or other abnormalities. As expected, pancreatic islets taken from GPR40$^{-/-}$ mice of both backgrounds showed greatly reduced LCFA amplification of glucose-stimulated insulin secretion compared to their wild-type littermates. Here, however, the similarities ended. One group of investigators have reported little difference between islets taken from wild-type and KO mice in their ability to maintain insulin secretion after exposure to palmitate for 72 h [19]. In contrast, a separate group reported that islets taken from the KO mice were apparently protected from the lipotoxic effects of 48-hour exposure to palmitate [18]. In addition, they also reported that GPR40$^{-/-}$ mice resisted the development of hyperinsulinemia, hepatic steatosis, hypertriglyceridemia, and glucose intolerance after being fed a high-fat diet for 8 weeks as compared to wild-type mice. Based on these results, they have postulated that GPR40 antagonists may represent a viable therapeutic strategy for treating diabetes. Given the contradiction in reported results, the viability of either GPR40 agonists or antagonists will likely be settled in the clinic. As will be shown in the next sections, the preponderance of GPR40 modulators reported to date are agonists and thus they are more likely to be the first tested in the clinic.

3. ARYLPROPIONIC ACIDS

3.1 Hydroxyphenylpropionic acid derivatives and related compounds

4-Hydroxyphenylpropionic acid derivatives, typified by **1** and **2**, have been reported as GPR40 agonists [20]. The ability of these compounds to flux Ca^{2+} was evaluated in a FLIPR assay using CHO cells that stably express the human GPR40 receptor. Using this assay, compounds **1** and **2** were reported to have EC_{50} values of 6 and 11 nM, respectively.

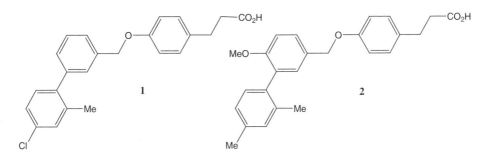

Several research groups have reported on the utility of phenylpropionic acid derivatives with substituents at the 3-position of the propionic acid moiety as GPR40 agonists [21–23]. A series of 3-aryl-3-(4-phenoxy)-propionic acid derivatives typified by compounds **3** and **4** were discovered by high-throughput screening and optimized for GPR40 agonistic activity using a Ca^{2+} flux assay in HEK293 cells expressing human GPR40 [21]. In this assay, compounds **3** and **4** afforded ED_{50} values of 300 and 640 nM, respectively, and intrinsic efficacy similar to that observed with linoleic acid. Furthermore, the ability of **4** to stimulate insulin release in a MIN6 cell line in the presence of 5 and 25 mM glucose was evaluated. MIN6 cells, in the presence of 25 mM glucose, treated with 30 µM of **4** afforded a statistically significant increase in insulin release similar to that observed with linoleic acid. As with linoleic acid, no significant increase in insulin release was observed with MIN6 cells treated in the presence of 5 mM glucose. The EC_{50} for **4** in MIN6 cells treated with 25 mM glucose was 2.2 µM. The pharmacokinetic properties for one of the enantiomers of **3** in the rat have also been reported.

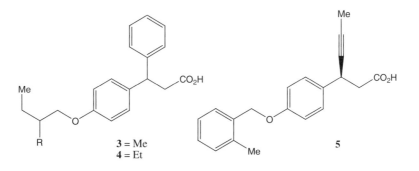

Propynyl-substituted phenylpropanoic acid derivatives such as compound **5** show an excellent combination of potency on the GPR40 receptor combined with favorable pharmacokinetic properties in rodents [22,23]. In a Ca^{2+} flux assay in CHO cells co-expressing human GPR40 and aequorin, compound **5** showed an EC_{50} value of 50 nM. The same compound displays good oral bioavailability (66%) along with low clearance (0.02 L/kg/h) when dosed in Sprague-Dawley (SD) rats [22]. This favorable profile allowed for compound **5** to be evaluated *in vivo* in SD rats undergoing an *i.p.* glucose tolerance test (ipGTT). Thirty minutes following a *p.o.* dose of compound **5**, the rats were given 1 g/kg glucose *i.p.* and plasma glucose levels were monitored for the following 2 h [22]. Compound **5** was reported to lower the peak plasma glucose levels following the *i.p.* glucose bolus comparably to the known insulin secretagogue nateglinide. Researchers reported a drop in plasma glucose levels prior to the glucose bolus in the nateglinide-treated group but not in the GPR40 agonist-treated animals, thus highlighting one of the potential advantages that GPR40 agonists may display compared to non-glucose-dependent secretagogues.

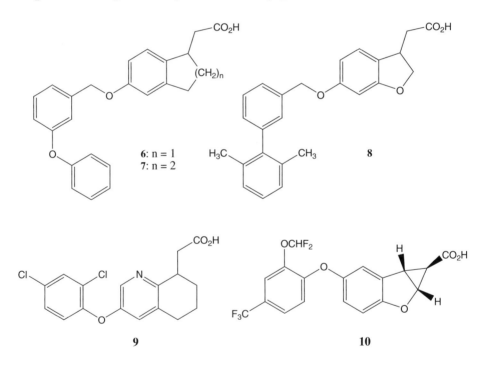

Several bicyclic acetic acid derivatives typified by **6**, **7**, **8** and **9** have been reported as GPR40 agonists [24,25]. Compounds **6**, **7** and **8** were evaluated using CHO cells expressing the human GPR40 receptor in a FLIPR assay. In this assay the EC_{50} values of these compounds were reported to be <100 nM.

Researchers have also reported on a series of conformationally rigid dihydrobenzofuran acids derivatives exemplified by **10** as GPR40 agonists

[26,27]. Compounds were evaluated using an NFAT (nuclear factor of activated T cells)-BLA(beta-lactamase) assay with CHO cells stably transfected with GPR40. Compounds were also evaluated in a FLIPR assay and in an inositol phosphate turnover assay using HEK cells stably expressing GPR40 receptors.

3.2 Aminophenylpropionic acid derivatives and related compounds

A series of 3-(4-(amino)phenyl)propanoic acid derivatives was discovered after screening of a chemical library using a FLIPR high-throughput assay in HEK293 cells expressing human GPR40 [28–30]. Compounds in this series are exemplified by **11** which was reported to have a pEC_{50} of 6.03 and efficacy similar to linoleic acid in a Gal4/Elk1 luciferase reporter assay with CHO cells expressing GPR40. Researchers evaluated the effect of optimizing the tether between the carboxylic acid and the phenyl moiety by evaluating linear and branched alkylene moieties of different lengths. The effects of introducing heteroatoms on the tether were also evaluated. Compound **12** (GW9508) was reported to afford a pEC_{50} value of 7.19 ± 0.54 and an efficacy of $82 \pm 24\%$ compared to **11** [31]. Similarly, compound **13**, featuring a cyclopropanyl tether was reported to afford a pEC_{50} of 8.31 ± 0.27 and an efficacy of $89 \pm 16\%$ [30]. Compound **12** was reported to afford a clearance in the rat of 24.9 mL/min/kg, a V_{dss} of 4.7 L/kg and a $T_{1/2}$ of 5.3 h after a 1 mg/kg i.v. dose, and an oral bioavailability of 65% after a dose of 3.9 mg/kg [28]. In addition, **12** was found to be selective for GPR40 over other lipid receptors such as GPR41, GPR43, S1P1, HM74A, PPARα, PPARδ, PPARγ, EP$_1$, EP$_3$ and EP$_4$ [28]. Regulation of insulin release in MIN6 cells in a dose and glucose-dependent manner by **12** was also reported. It has also been demonstrated that a GPR40 selective antagonist, **14** (GW1100), inhibited stimulation induced by **12** of GPR40 receptors expressed in HEK293 and MIN6 cells [31].

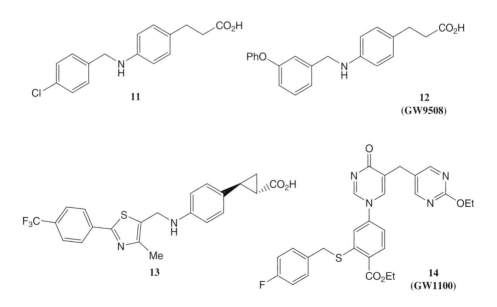

A series of aminobenzoxazole derivatives, exemplified by **15**, has been reported as GPR40 agonists [32]. Compounds of this series were evaluated in an FLIPR assay using recombinant HEK293 cells expressing GPR40 receptors. Compound **15** was reported to display 111% efficacy at 10 μM compared to linoleic acid.

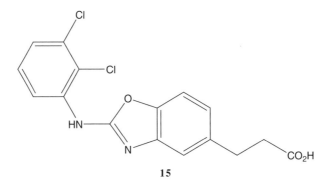

15

4. LONG CHAIN ALKYL ACID DERIVATIVES

Several diphenylpyrimidine derivatives typified by **16** [33] and **17** [34] have been reported as GPR40 agonists. These compounds feature a long alkyl tether between the pyrimidine moiety and the carboxylic acid moiety. Compounds **16** and **17** were reported to display 101% and 104% activation, respectively, at 10 μM compared to linoleic acid in a FLIPR assay using recombinant HEK293 cells expressing GPR40 receptors.

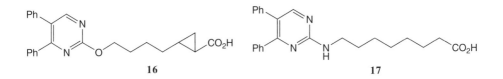

16   **17**

5. ACID ISOSTERES

One of the original reports that identified fatty acids as ligands for GPR40 also reported that the thiazolidinedione-containing antidiabetic agent rosiglitazone is also a potent activator of GPR40 [9]. Several research groups have reported GPR40 agonists that feature related carboxylic acid isosteres. Structures **18–22** typify several series of thiazolidinedione, oxazolidinedione and oxadiazolidinedione derivatives that have been reported to activate

GPR40 receptors [25,35,36].

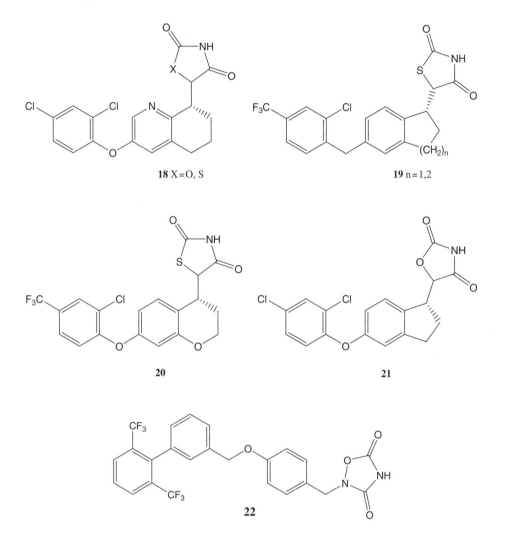

18 X = O, S

19 n = 1,2

20

21

22

6. CONCLUSION

GPR40 is a G-protein couple receptor expressed primarily in the pancreas, but also found in the small intestine and brain. Within the pancreas, GPR40 mRNA levels suggest that GPR40 is expressed mainly in the islet β-cells. GPR40 is activated by fatty acids. Experiments with siRNA specific for GPR40 inhibits insulin secretion mediated by fatty acids *in vitro*. Furthermore, GPR40 agonists potentiate glucose-stimulated insulin secretion in MIN6 cells, a mouse pancreatic β-cell line, suggesting that GPR40 agonists can function as glucose-dependent insulin secretagogues. A GPR40 agonist has also been reported to lower glucose levels in a rat *i.p.* glucose tolerance test (ipGTT). Taken together, this data suggests that GPR40 may function as a lipid sensor that mediates insulin secretion from islet β-cells in a glucose-dependent manner and that GPR40 agonists will be useful in the treatment of Type II diabetes.

REFERENCES

[1] S. Wild, G. Roglic, A. Green, R. Sicree and H. King, *Diabetes Care*, 2004, **27**, 1047.

[2] P. Proks, F. Reimann, N. Green, F. Gribble and F. Ashcroft, *Diabetes*, 2002, **51**, S368.

[3] J. B. Kalbag, Y. H. Walter, J. R. Nedelman and J. F. McLeod, *Diabetes Care*, 2001, **24**, 73.

[4] D. J. Drucker and M. A. Nauck, *Lancet*, 2006, **368**, 1696.

[5] A. E. Weber and N. Thornberry, *Annu. Rep. Med. Chem.*, 2007, **42**, 95.

[6] R. L. Dobbins, M. W. Chester, M. B. Daniels, J. D. McGarry and D. T. Stein, *Diabetes*, 1998, **47**, 1613.

[7] C. P. Briscoe, M. Tadayyon, J. L. Andrews, W. G. Benson, J. K. Chambers, M. M. Eilert, C. Ellis, N. A. Elshourbagy, A. S. Goetz, D. T. Minnick, P. R. Murdock, H. R. Sauls, Jr., U. Shabon, L. D. Spinage, J. C. Strum, P. G. Szekeres, K. B. Tan, J. M. Way, D. M. Ignar, S. Wilson and A. I. Muir, *J. Biol. Chem.*, 2003, **278**, 11303.

[8] Y. Itoh, Y. Kawamata, M. Harada, M. Kobayashi, R. Fujii, S. Fukusumi, K. Ogi, M. Hosoya, Y. Tanaka, H. Uejima, H. Tanaka, M. Maruyama, R. Satoh, S. Okubo, H. Kizawa, H. Komatsu, F. Matsumura, Y. Noguchi, T. Shinohara, S. Hinuma, Y. Fujisawa and M. Fujino, *Nature*, 2003, **422**, 173.

[9] K. Kotarsky, N. E. Nilsson, E. Flodgren, C. Owman and B. Olde, *Biochem. Biophys. Res. Commun.*, 2003, **301**, 406.

[10] A. J. Brown, S. Jupe and C. P. Briscoe, *DNA Cell Biol.*, 2005, **24**, 54.

[11] T. Tomita, H. Masuzaki, M. Noguchi, H. Iwakura, J. Fujikura, T. Tanaka, K. Ebihara, J. Kawamura, I. Komoto, Y. Kawaguchi, K. Fujimoto, R. Doi, Y. Shimada, K. Hosoda, M. Imamura and K. Nakao, *Biochem. Biophys. Res. Commun.*, 2005, **338**, 1788.

[12] T. Tomita, H. Masuzaki, H. Iwakura, J. Fujikura, M. Noguchi, T. Tanaka, K. Ebihara, J. Kawamura, I. Komoto, Y. Kawaguchi, K. Fujimoto, R. Doi, Y. Shimada, K. Hosoda, M. Imamura and K. Nakao, *Diabetologia*, 2006, **49**, 962.

[13] E. Flodgren, B. Olde, S. Meidute-Abaraviciene, M. S. Winzell, B. Ahren and A. Salehi, *Biochem. Biophys. Res. Commun.*, 2007, **354**, 240.

[14] K. Fujiwara, F. Maekawa and T. Yada, *Am. J. Physiol. Endocrinol. Metab.*, 2005, **289**, E670.

[15] D. D. Feng, Z. Luo, S.-g. Roh, M. Hernandez, N. Tawadros, D. J. Keating and C. Chen, *Endocrinology*, 2006, **147**, 674.

[16] P. E. MacDonald, J. W. Joseph and P. Rorsman, *Phil. Trans. R. Soc. B*, 2005, **360**, 2211.

[17] S.-N. Yang and P.-O. Berggren, *Endocrinol. Rev.*, 2006, **27**, 621.

[18] P. Steneberg, N. Rubins, R. Bartoov-Shifman, M. D. Walker and H. Edlund, *Cell Metab.*, 2005, **1**, 245.

[19] M. G. Latour, T. Alquier, E. Oseid, C. Tremblay, T. L. Jetton, J. Luo, D. C. H. Lin and V. Poitout, *Diabetes*, 2007, **56**, 1087.

[20] T. Yasuma, S. Kitamura and N. Negoro, *PCT International Publication No. WO* 2005/063729.

[21] F. Song, S. Lu, J. Gunnet, J. Z. Xu, P. Wines, J. Proost, Y. Liang, C. Baumann, J. Lenhard, W. V. Murray, K. T. Demarest and G.-H. Kuo, *J. Med. Chem.*, 2007, **50**, 2807.

[22] J. Houze, W. Qiu, A. Zhang, R. Sharma, L. Zhu, Y. Sun, M. Akerman, M. Schmitt, Y. Wang, J. Liu, J. Liu, J. Medina, J. Reagan, J. Luo, G. Tonn, J. Zhang, J. Lu, M. Chen, E. Lopez, K. Nguyen, L. Yang, L. Tang, H. Tian, S. Shuttleworth and D. Lin, in *234th ACS National Meeting*, Boston, MA, August 19–23, 2007, MEDI-251.

[23] M. Akerman, J. Houze, D. C. H. Lin, J. Liu, J. Luo, J. C. Medina, W. Qiu, J. D. Reagan, R. Sharma, S. J. Shuttleworth, Y. Sun, J. Zhang and L. Zhu, *PCT International Publication No. WO* 2005/086661.

[24] T. Yasuma, N. Negoro and K. Fukatsu, *European Patent Application Publication No. EP* 1,630,152.

[25] M. Ge, L. Yang, C. Zhou, S. Lin and E. D. Cline, *PCT International Publication No. WO* 2006/083612.

[26] M. Ge, J. He, F. W. Y. Lau, G.-B. Liang, S. Lin, W. Liu, S. P. Walsh and L. Yang, *PCT International Publication No. WO* 2007/136572, 2007.

[27] M. Ge, J. He, F. W. Y. Lau, G.-B. Liang, S. Lin, W. Liu, S. P. Walsh and L. Yang, U.S. *Patent Application Publication No.* 2007/0265332.

[28] S. C. McKeown, D. F. Corbett, A. S. Goetz, T. R. Littleton, E. Bigham, C. P. Briscoe, A. J. Peat, S. P. Watson and D. M. B. Hickey, *Bioorg. Med. Chem. Lett.*, 2007, **17**, 1584.

[29] D. F. Corbett, K. A. Dwornik, D. M. Garrido, S. C. McKeown, W. Y. Mills, A. J. Peat and T. L. Smalley, Jr., *PCT International Publication No. WO* 2005/051890.

[30] D. M. Garrido, D. F. Corbett, K. A. Dwornik, A. S. Goetz, T. R. Littleton, S. C. McKeown, W. Y. Mills, T. L. Smalley, C. P. Briscoe and A. J. Peat, *Bioorg. Med. Chem. Lett.*, 2006, **16**, 1840.

[31] C. P. Briscoe, A. J. Peat, S. C. McKeown, D. F. Corbett, A. S. Goetz, T. R. Littleton, D. C. McCoy, T. P. Kenakin, J. L. Andrews, C. Ammala, J. A. Fornwald, D. M. Ignar and S. Jenkinson, *Br. J. Pharm.*, 2006, **148**, 619.

[32] E. Defossa, M. Follmann, T. Klabunde, V. Drosou, G. Hessler, S. Stengelin, G. Haschke, A. Herling and S. Bartoschek, *PCT International Publication No. WO* 2007/131622.

[33] E. Defossa, J. Goerlitzer, T. Klabunde, V. Drosou, S. Stengelin, G. Haschke, A. Herling and S. Bartoschek, *PCT International Publication No. WO* 2007/131619.

[34] E. Defossa, J. Goerlitzer, T. Klabunde, V. Drosou, S. Stengelin, G. Haschke, A. Herling and S. Bartoschek, *PCT International Publication No. WO* 2007/131620.

[35] M. Ge, L. Yang, C. Zhou, S. Lin, H. Tang, E.D. Cline, S. Malkani, U.S. *Patent Application Publication No.* 2006/003255.

[36] K. Negoro, F. Iwasaki, K. Ohnuki, T. Kurosaki, Y. Yonetoku, N. Asai, S. Yoshida and T. Soga, *PCT International Publication No. WO* 2007/123225.

CHAPTER 6

Rho-Kinase Inhibitors for Cardiovascular Disease

Robert A. Stavenger

Contents

1. INTRODUCTION

Rho kinases (also known as Rho-associated kinases, Rho-associated coiled-coil forming protein kinases, ROCK) are two highly similar (92% homology in their catalytic domains, 65% homology overall) Ser/Thr kinases belonging to the AGC kinase family [1]. ROCK1 (ROCKI, ROKβ) and ROCK2 (ROCKII, ROKα) are both expressed ubiquitously, though ROCK1 has been shown to be more abundant in the kidney, liver, lung, spleen and testes, while ROCK2 is more abundant in brain and skeletal muscle [2]. Genetic ablation of either ROCK1 or ROCK2 is developmentally lethal showing that their functions are not redundant [3–5].

GlaxoSmithKline, Antibacterial Chemistry, Infectious Diseases CEDD, 1250 South Collegeville Rd., Collegeville P.A. 19426, USA

Annual Reports in Medicinal Chemistry, Volume 43
ISSN 0065-7743, DOI 10.1016/S0065-7743(08)00006-7

In addition, until recently small molecule inhibitors lacked selectivity and thus had not been helpful in elucidating the individual function of the two isoforms. Further details of the cellular and molecular biology of ROCK and its potential role in human disease can be found in several recent reviews [6–12].

2. PRECLINICAL STUDIES

The majority of the known pharmacology of ROCK has been determined with two inhibitors, neither of which was originally designed to inhibit rho kinase. Fasudil (HA-1077, **1**) and Y-27632 (**2**) have been used extensively to investigate the function of ROCK in cellular and whole animal environments. Compound **1** was approved in Japan in 1995 as a drug for the treatment of cerebral vasospasm, thus becoming the first protein kinase inhibitor drug approved for human treatment [13–14]. Compound **1** inhibits ROCK1 and ROCK2 with reported $IC_{50}s$ between 300 and 1,900 nM, and shows modest selectivity over a panel of protein kinases, with a notable exception being PKA [11,13,15]. It has also been shown that **1** undergoes oxidation *in vivo* to the isoquinolinone **3** (often referred to as hydroxyfasudil) which retains potency at ROCK [8]. In humans, **3** has superior pharmacokinetic properties with a much longer plasma half-life ($T_{1/2}>5$ h) in comparison to **1** ($T_{1/2} = 0.5$ h).

The pyridine-based ROCK inhibitor **2** ($IC_{50} \sim 800$ nM) is slightly more potent than **1** and shows a different selectivity profile, inhibiting PRK2 with potency that is similar to its potency at ROCK, and inhibiting MSK1 $\sim$10-fold less potently [15]. No pharmacokinetic parameters for **1** or **2** in preclinical species have been published.

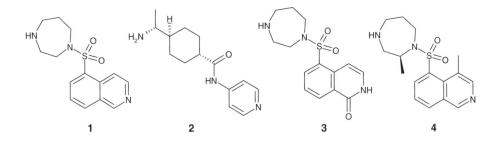

Compounds **1** and **2** have been shown to relax arteries and bronchi that had been preconstricted with a variety of agonists [13,16]. In addition, further studies in three rat models of hypertension (spontaneously hypertensive rat (SHR), deoxycorticosterone (DOCA) salt rats and renal-obstruction hypertensive rats) showed that a single oral dose of 30 mg/kg of **2** led to a robust (max $\sim$60–80 mmHg) and long lived (changes statistically significant up to 7 h post dosing) drop in blood pressure (bp) in all three groups of rats [17]. Interestingly, when normotensive (Wistar-Kyoto) rats were similarly treated, only a modest reduction (max. ~20 mmHg) in bp was observed. These results have led to suggestions that not only is ROCK critically involved in bp control, but also that it is more important in maintaining tone in hypertensive animals relative to normotensive animals.

Building on the anti-hypertensive effects of ROCK inhibition, several groups have published reports that either **1** or **2** has a beneficial effect in various models of heart failure. Compound **2** was shown to reduce bp, left ventricular hypertrophy and cardiac remodeling upon chronic administration to N-nitro-L-arginine methyl ester (L-NAME) hypertensive rats [18]. Similarly, **2** was shown to decrease myocardial hypertrophy and improve contractility in Dahl salt-sensitive rats [19–20]. Compounds **2** and **3** have also been shown to have protective effects in rat and dog (respectively) reperfusion models [21–22]. Compound **2** has been found to be efficacious in a rat balloon injury model [23,24]. Furthermore, **2** provided a 35% reduction in lesion size in the LDLr-KO mouse model of atherosclerosis [25].

ROCK inhibitors have been studied in a variety of preclinical models to assess their potential in non-cardiovascular indications. The use of ROCK inhibitors in a range of diseases including cancer, erectile dysfunction, glaucoma, stroke, pulmonary diseases, and Alzheimer's disease can be found in a number of reviews [8,10–12].

3. CLINICAL STUDIES

Compound **1** is currently the only ROCK inhibitor approved for human use (in Japan) and the only ROCK inhibitor for which clinical data is available. It is used in Japan to treat cerebral vasospasm as a two-week (iv) treatment, and has shown considerable success in this arena [13]. Due to its availability in the clinical setting, **1** has also been tested in several other areas. In one study, directly relating to the preclinical work on hypertension, **1** was found to increase forearm blood flow in hypertensive patients to a larger degree than in normotensive patients [26]. Notably one of the few side effects of **1** in humans has been mild hypotension [27].

Compound **1** has also been tested as an oral treatment for stable angina, and showed positive outcomes in a Phase II study reported in 2002 with minimal side effects [28]. Smaller trials have also shown that intravenously administered **1** had positive effects in pulmonary hypertension and acute stroke [29,30].

4. STRUCTURE

X-ray structures of the ROCK1 dimer complexed to inhibitors **1–4** have been solved [31]. In addition, the X-ray structures of these inhibitors in the related kinase PKA have been reported and an analysis of the two sets of structures has led to some gross conclusions regarding the selectivity of this compound set

[31–33]. In particular, the position of Ala-215 in ROCK in comparison with Thr-183 in PKA seems important. Since the larger Thr side chain in PKA disrupts the preferred position of the inhibitor in the ATP-binding site, selectivity for ROCK over PKA was high for compounds 2–4. In contrast, the position of Thr-183 does not significantly displace 1 in PKA, relative to its position in the ROCK active site, leading to roughly equal inhibitory potency at the two enzymes.

5. MEDICINAL CHEMISTRY PROGRESS

The large body of preclinical and clinical data with several ROCK inhibitors has led to a significant effort in the industry to identify new, more potent and/or selective inhibitors for a wide variety of indications. Although somewhat arbitrary, these newer developments are divided below based on the presumed hinge-binding element present in the inhibitors. Note that since several preferred hinge binders (notably pyridine, 7-azaindole and indazole) occur in a variety of classes, their placement below was sometimes chosen to best illustrate SAR trends, rather than to necessarily keep all similar hinge binders together.

5.1 Pyridines and pyrimidines

Several classes of pyridine derivatives, more or less related to 2, have been reported. Pyridine 5 was identified by a combination of analog synthesis around a related pyridine high-throughput screening (HTS) hit and a computational docking approach [34]. Although 5 was only a modest ROCK2 inhibitor ($IC_{50} = 200\,nM$), this research group progressed the lead by probing other pyridine replacements, identifying indazoles (see below) which led to larger research efforts. Another simple acylaminopyridine 6 was shown to be a potent inhibitor of ROCK1 ($IC_{50} = 13\,nM$) though no data for other analogs was presented [35].

The pyridyl indole 7 was identified from a novel phenotypic screen designed to detect molecules with an effect on wound healing [36]. Further experiments provided evidence that ROCK was the likely molecular target of 7, and this compound was eventually shown to have an IC_{50} of $\sim 25\,\mu M$ against the purified kinase domain of ROCK.

4-Phenylpyridines and -pyrimidines bearing a pendant benzyl amide have also been shown to be ROCK1 inhibitors. The pyridine analog 8 is somewhat more potent ($IC_{50} = 14\,nM$) than the corresponding pyrimidine analog 9 ($IC_{50} = 42\,nM$) [37]. Extension to a phenethyl amide, inversion of the amino stereocenter, and incorporation of the aniline nitrogen into an indoline ring was also shown to provide a potent ROCK1 inhibitor (**10**, $IC_{50} = 53$ nM).

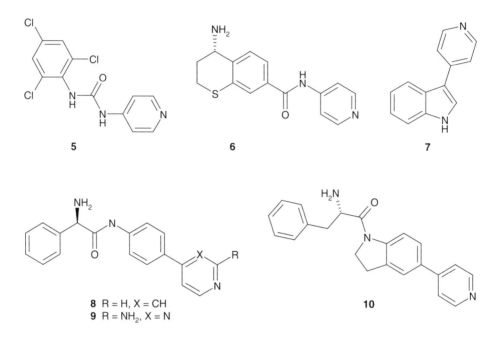

5   **6**   **7**

8 R = H, X = CH
9 R = NH_2, X = N

10

A series of inhibitors based on an anilinopyrimidine scaffold on which various potential hinge-binding motifs are tethered by a heteroatom have been reported to be potent ROCK inhibitors and have been shown to relax precontracted rabbit saphenous arteries. Example **11** bearing a pendent quinoline ring has potent ROCK2 inhibition ($IC_{50} = 7$ nM) and was found to be active ($IC_{50} = 760$ nM) in a rabbit artery relaxation assay [38]. Replacement of the quinoline with a saturated pyrrolopyridine (**12**) resulted in a slight loss of potency ($IC_{50} = 11$ nM, arterial relaxation $IC_{50} = 1,700$ nM), but when a second fluorine substituent (**13**) was added to the phenyl ring, potency in both assays was improved ($IC_{50} = 5$ nM, arterial relaxation $IC_{50} = 360$ nM).

Replacement of the pyridothio group with an indazole-ether (**14**, ROCK2 $IC_{50} = 6$ nM) resulted an equipotent ROCK inhibitor, but activity in the functional assay was improved to ($IC_{50} = 350$ nM) relative to **12** [39]. When the saturated pyrrolopyridine substituent of **14** was replaced by the structurally simpler 4-pyridyl group (**15**) only a small decrease in potency was observed (ROCK2 $IC_{50} = 20$ nM; rabbit tissue $IC_{50} = 1,020$ nM). Replacement of the head group with a 4-linked azaindole (**16**) afforded improved potency in both assays (ROCK2 $IC_{50} = 2$ nM; rabbit tissue $IC_{50} = 350$ nM). Unfortunately, no data on kinase selectivity or additional pharmacology of these interesting and potent inhibitors has been reported.

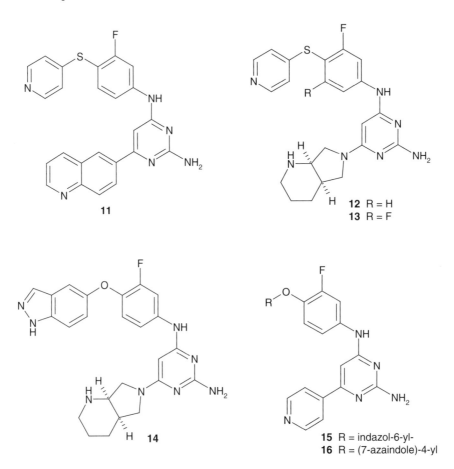

The only ROCK inhibitors reported thus far that show significant selectivity between the ROCK1 and ROCK2 isoforms are exemplified by the indazolaminoquinazolines **18–19** [40]. The compound **17**, containing a primary amide was found to have modest potency at ROCK2 (IC$_{50}$ = 520 nM) with nearly equipotent activity for ROCK1 (IC$_{50}$ = 2,600 nM). However, through elaboration of the amide selectivity was achieved. The pyridin-3-yl analog **18** provided a dramatic boost in ROCK2 potency (IC$_{50}$ = 90 nM) and a loss in ROCK1 potency (IC$_{50}$ > 10,000 nM), giving > 100-fold selectivity between the isoforms. Changing the oxygen-linked amide to a simple propylamide, with the addition of solubilizing groups on the quinazoline core (**19**), also lead to potent ROCK2 inhibition with good selectivity over ROCK1 (IC$_{50}$s of 50 nM and > 3,000 nM, respectively). Compounds such as these should prove to be valuable tools to help elucidate the functions of the two isoforms (see Recent Developments section).

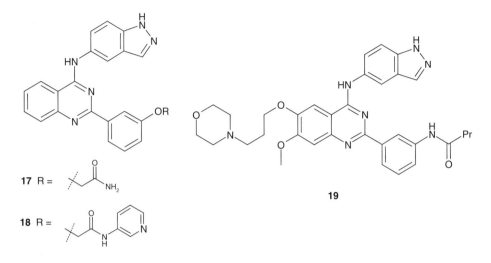

17 R =

18 R =

19

A series of aminopyrazines with various head groups and a hexahydrodiazepine substituent reminiscent of **1** has been reported [41]. Several of the common head groups are tolerated (**20a–d**, ROCK2 IC_{50}s = 3.0, 2.8, 0.53 and 0.95 nM, respectively), with the azaindole and indazole groups being slightly favored over the other hinge elements presented. A variety of other mono- or diamines are also tolerated in place of the hexahydrodiazepine (data not shown). The most potent compound in the series was the azaindole **21** (IC_{50} = 0.021 nM) in which the pyrazine primary amino substituent was incorporated into a pyrrolidine ring. No further data was presented on these exceedingly potent inhibitors.

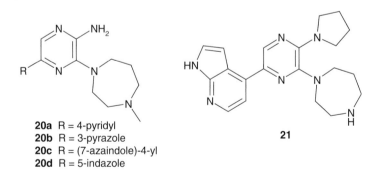

20a R = 4-pyridyl
20b R = 3-pyrazole
20c R = (7-azaindole)-4-yl
20d R = 5-indazole

21

5.2 Isoquinolines and isoquinolones

Given the clinical success of **1**, there have been a large number of analogs of **1** reported, of which **4** is the most extensively studied [42]. Relative to **1**, methylation of the isoquinoline ring provided a small improvement in potency (**22**; R6 = Me, R1–R5 = H; IC$_{50}$ from 160 to 96 nM) [43]. Other substituents tested at R6 (Et, CN, OH) led to a loss of potency, while halogen (Cl, Br) and ethynyl substituents led to an increase in potency, but with less selectivity over several key kinases. Substituting the 7-membered ring with a methyl group had little effect on potency at the R2, R3 and R4 positions, whereas when R5 is a methyl group, the ROCK IC$_{50}$ is improved to 33 nM. The combination of R1 = (*S*)-Me and R6 = Me provided **4** (R2–R5 = H) with an IC$_{50}$ of 1.6 nM (note that the enantiomer of **4** is a ∼3-fold less potent ROCK inhibitor). Larger groups (ethyl and isopropyl) were not favored at the R1-position giving a 10–80-fold reduction in potency. In addition to being more potent, **4** also showed a higher degree of selectivity versus other kinase family members, such as PKA, PKC and MLCK (selectivity for **1** = 3-, 28- and 160-fold, respectively, and for **4** = 400-, >1,000- and >1,000-fold, respectively). Compound **4** was shown to relax phenylephrine-contracted rat penile cavernosal segments with an IC$_{50}$ of 100 nM [44]. In addition, an intraperitoneal dose of **4** (100 nmol/kg) to anaesthetized rats increased the intracavernosal pressure response and erectile response after electrical stimulus.

The simple 6-aryl substituted isoquinoline **23** is a potent ROCK2 inhibitor (K_i = 77 nM). No further information about kinase selectivity or functional activity was provided. Addition of a pyrazole group to the 5-position of the isoquinoline afforded a modest boost in potency (**24**, K_i = 15 nM) [45]. Another research group has reported isoquinolinones as ROCK1 inhibitors, the most potent example being the piperidinyloxy analog **25**, IC$_{50}$ = 40 nM [46].

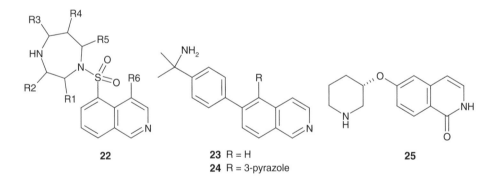

22   **23** R = H   **25**
          24 R = 3-pyrazole

5.3 7-Azaindoles

Compound **26** is an azaindole analog of **2** that was reported to be a 30-fold more potent inhibitor of ROCK [47]. In addition, **26** was found to promote neurite outgrowth more effectively than **2** [48]. Another azaindole lacking the amide spacer (**27**) has been reported to be a potent (IC$_{50}$ < 10 nM) ROCK2 inhibitor [49]. Additional azaindoles with an ether rather than an amino or amido linkage at the 4-position have also been reported. The benzamides **28** and **29** have been shown to inhibit ROCK2 with IC$_{50}$s of 49 and 12 nM, respectively [50].

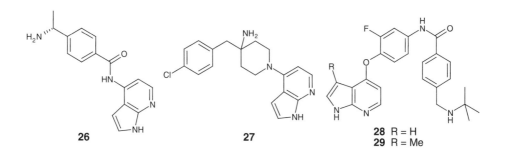

26   27   **28** R = H
                                                                           29 R = Me

Azaindole **30** was discovered from optimization of an HTS hit and is a potent inhibitor of both ROCK1 and ROCK2 (IC$_{50}$ = 0.6 and 1.1 nM, respectively), and displayed excellent selectivity for ROCK over other kinases. When screened against a panel of 112 kinases, **30** was shown to have IC$_{50}$s of $\geq$1 μM against all but two kinases, the exceptions being TRK and FLT3 (IC$_{50}$s of 252 and 303 nM, respectively) [51]. Inhibitor **30** was shown to be ATP-competitive and preincubation experiments on a 30-minute timescale suggested that its binding to ROCK1 was slowly reversible. Compound **30** has a modest half-life in mice, rats and dogs (1.5, 1.2 and 2.5 h, respectively) and good oral bioavailability in rats (48%) and dogs (73%). Compound **30** lowered bp when dosed orally to SHR (up to 40 mmHg at 10 mg/kg) with little effect seen at the same dose in normotensive rats. Upon repeat dosing (10 mg/kg po once daily for four days), the maximal bp effect appeared to lessen over time, although more detailed studies are required to make a definitive conclusion. In addition, an iv bolus (0.1 and 0.3 mg/kg) of **30** administered to anaesthetized normotensive dogs decreased bp by 10–25 mmHg, with a concomitant modest increase in heart rate. No additional analogs have been reported.

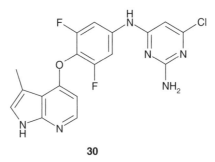

30

5.4 Indazoles

A large number of indazole-based ROCK inhibitors have been reported. Further structure–activity relationship studies related to compound **5** led to a series of 5-substituted indazoles such as **31** (IC$_{50}$ = 20 nM) [52]. Methylation at either the 3-position or the N1-position led to a decrease in activity (IC$_{50}$ = 150 and >10,000 nM, respectively). In addition, as suggested by docking studies, related

6-substituted indazoles showed >10-fold less potency at ROCK. Although a number of side chains at the 5-position were tolerated, cyclic amines were generally favored at this position. Compound **31** and related analogs were then studied both for their activity at ROCK and for their ability to inhibit the chemotaxis of a CCR2-overexpressing lymphoma cell line in response to MCP-1. There was no clear relationship between the inhibition of ROCK and the activity in the chemotaxis assay.

A related series of 5-aminoindazoles was reported as ROCK inhibitors in a study that also included assessment of pharmacokinetic and cytochrome P450 (CYP) inhibition data for selected compounds. It was shown that replacement of the hydroxyl group in **32** (ROCK2 IC_{50} = 100 nM; CYP3A4 IC_{50} = 4 µM) with an amino group significantly increased inhibitory activity against CYP3A4, while retaining ROCK inhibitory activity [53]. The hydroxyl group of **32** also imparted better selectivity for ROCK over PKA than displayed by the corresponding amino analog (27-fold vs. 5-fold selective). The *para*-bromide was about 2-fold more potent that the corresponding chloride and 8-fold more potent than the unsubstituted phenyl analog. In the rat, inhibitor **32** was shown to have high clearance (109 mL/min/kg), a modest volume of distribution (3.7 L/kg), and 28% oral bioavailability.

A number of simple 5-aminoindazole amides (**33–35**) were reported to inhibit ROCK (K_i <1 µM) [54]. Although little detail was provided, a variety of substitutions appear to be generally tolerated in this overall framework. Specifically, both 3- and 4-substitution of the aryl ring is allowed, though 2-substitution seems to be less favorable. Substitution of the benzylic position with either a hydroxyl or an amino group is tolerated, as is an alkyl group with an amine terminus. In addition, the 3-position of the indazole is at least somewhat tolerant of substituents as large as phenethyl or benzamido.

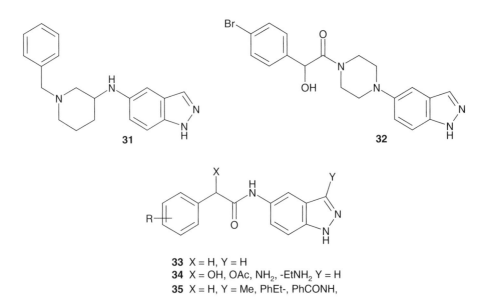

33 X = H, Y = H
34 X = OH, OAc, NH_2, -$EtNH_2$ Y = H
35 X = H, Y = Me, PhEt-, PhCONH,

The indazole amide **36** was identified as a potent ROCK1 inhibitor ($IC_{50} = 14$ nM) with good selectivity (>30-fold) over a panel of kinases, although it possessed only modest functional activity in a rat aortic ring relaxation assay ($IC_{50} = 760$ nM) and showed no oral bioavailability in the rat [55]. Replacement of one NH-group in the dihydropyrimidinone ring by a methylene to form dihydropyridinone **37** led to a small drop in ROCK1 potency ($IC_{50} = 50$ nM), but improved functional activity ($IC_{50} = 200$ nM) and rat oral bioavailability (14%). Incubation of an analog of **37** with rat hepatocytes led to hydroxylation on the indazole ring, although the hydroxylation regiochemistry was not reported. In a follow up study designed to block possible positions of metabolism, substitutions on the indazole ring were investigated. Methyl substitution at C-3 and chloro substitution at C-6 of the indazole had little effect on potency, while the C-4-fluoro analog had much lower potency. The 5-fluoroindazole **39** ($IC_{50} = 14$ nM) was slightly less potent than the corresponding desfluoro analog **38** ($IC_{50} = 5$ nM), but showed a significant increase in oral bioavailability in the rat (61% and 22%, respectively). Inhibitor **39** was also shown to be active *in vivo*, lowering bp in SHR by up to 50 mmHg after a single oral dose of 30 mg/kg.

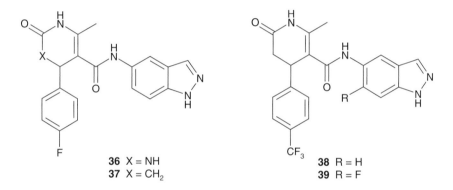

36 X = NH
37 X = CH$_2$

38 R = H
39 R = F

5.5 Aminofurazans

Several reports of aminofurazan hinge binders have appeared in the literature [56]. Compound **40** was identified as a potent ($IC_{50} = 19$ nM) ROCK1 inhibitor from a cross-screening effort, and was originally prepared as an MSK1 inhibitor [57,58]. The lead showed good overall kinase selectivity, but low selectivity over several related kinases, notably RSK1 (0.6-fold), p70S6K (3-fold) and, not surprisingly, MSK1 (0.2-fold). In addition, CDK2 and GSK3 were inhibited to different degrees (20- and 50-fold selectivity for ROCK, respectively). Addition of a phenoxy group at the C6 position of the azabenzimidazole (e.g., **41**) provided some improvement in potency ($IC_{50} = 1.8$ nM), and increased kinase selectivity, especially against RSK1, MSK1, CDK2 and GSK3. Further elaboration of the phenoxy tail, generally with solublizing substituents, provided higher levels of selectivity. For example, **42** (ROCK1 $IC_{50} = 1.8$ nM) was 83- and 30-fold selective for ROCK over RSK1 and MSK1, respectively. This compound was shown to be active in a rat aortic relaxation assay ($IC_{50} = 35\,nM$) and had a $T_{1/2}$ in the rat of 2.2 h and ~75% oral bioavailability.

Alternatively, addition of an aminopyrrolidinoamide to the 7-position of the azabenzimidazole of **40** provided **43**, which was also a potent ($IC_{50} = 6\,nM$) ROCK1 inhibitor [59]. This change had a smaller effect on selectivity, with RSK1 and MSK1 both being potently inhibited ($IC_{50} = 35$ and $14\,nM$, respectively), although selectivity against CDK2 and GSK3 was dramatically increased ($>1{,}000$-fold). Compound **43** relaxed rat aorta *in vitro* with an IC_{50} of $29\,nM$. Both **42** and **43** showed a dose-dependent drop in bp when dosed orally (0.3, 1 and 3 mg/kg) to SHR (maximum responses 45 and 40 mmHg, respectively). In addition, **43** was active in DOCA-salt hypertensive rats at the oral dose of 1 mg/kg (~45 mmHg maximum response) and had only a very modest effect on bp in normotensive (Wistar-Kyoto) rats at the same dose (12 mmHg maximum response).

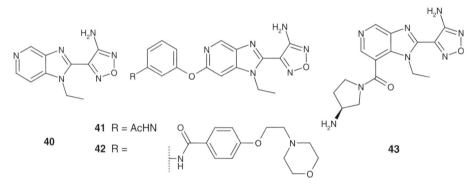

40

41 R = AcHN

42 R =

43

5.6 Other hinge-binding elements

An interesting series of benzamides has been reported to have ROCK inhibitory activity. The benzylic amine **44** showed only modest activity with an $IC_{50} = 620\,ng/mL$ [60]. Modification of the amino group from a benzylpiperidine to an aminocyclohexane, and addition of substituents at the 2- and 5-positions of the benzamide ring provided a dramatic improvement in potency (e.g., **45**, $IC_{50} = 6.5\,ng/mL$). In addition, the methoxy group could be replaced with fluoro and the chloro replaced with either fluoro or bromo with little change in potency. Interestingly, the amino linker was not required, with the ether **46** being essentially equipotent with its nitrogen analog.

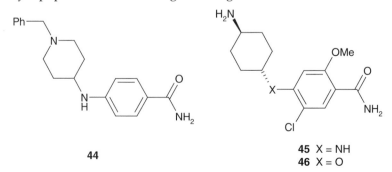

44

45 X = NH
46 X = O

6. CONCLUSION

Rho-kinase inhibitors continue to be studied for a number of potential cardiovascular and other indications. Reports on detailed pharmacology have appeared for only a small number of inhibitors. However, the volume of recent reports, many detailing inhibitors with high potency and/or selectivity, should pave the way for more in-depth studies in the coming years. Although much exciting data exists, key unanswered questions, notably the precise role of the different isoforms (see Recent Developments below) and the long-term tolerability of ROCK inhibitors await answers.

7. RECENT DEVELOPMENTS

After the final draft of this manuscript was completed, a further report appeared on the ROCK2-selective series exemplified by compounds **17–19** (see Section 5.1). Inhibitor **47** was shown to be highly selective for ROCK2 (ROCK2 $IC_{50} =$ 105 nM; ROCK1 $IC_{50} = 24,000$ nM) and was shown to reduce the plaque burden in the ApoE$^{(-/-)}$ mouse model of atherosclerosis when administered once daily at doses of 30 or 100 mg/kg [61]. Interestingly however, **47** had no effect on bp in rats (strain not reported) at doses up to 300 mg/kg, nor did it block the bp reducing effect of **1** in rats. Moreover, while the unselective inhibitors **1** and **2** were shown to relax pre-contracted aortic tissue strips, **47** did not induce such relaxation at concentrations up to 25 µM. Additional work is needed, but these data suggest that ROCK1 may be the isoform primarily responsible for maintaining vascular tone.

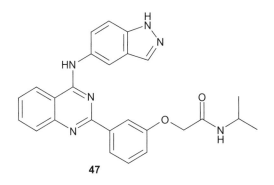

47

REFERENCES

[1] K. Riento and A. J. Ridley, *Nat. Rev. Mol. Cell Biol.*, 2003, **4**, 446.
[2] O. Nakagawa, K. Fujisawa, T. Ishizaki, Y. Saito, K. Nakao and S. Narumiya, *FEBS Lett.*, 1996, **392**, 189.
[3] D. Thumkeo, J. Keel, T. Ishizaki, M. Hirose, K. Nonomura, H. Oshima, M. Oshima, M. M. Taketo and S. Narumiya, *Mol. Cell. Biol.*, 2003, **23**, 5043.

[4] Y. Shimizu, D. Thumkeo, J. Keel, T. Ishizaki, H. Oshima, M. Oshima, Y. Noda, F. Matsumura, M. M. Taketo and S. Narumiya, *J. Cell. Biol.*, 2005, **168**, 941.

[5] Y.-M. Zhang, J. Bo, G. E. Taffet, J. Chang, J. Shi, A. K. Reddy, L. H. Michael, M. D. Schneider, M. L. Entman, R. J. Schwartz and L. Wei, *FASEB J.*, 2006, **20**, 916.

[6] K. Budzyn and C. G. Sobey, *Curr. Opin. Drug Disc. Dev.*, 2007, **10**, 590.

[7] H. Shimokawa and M. Rashid, *TRENDS Pharmacol. Sci.*, 2007, **28**, 296.

[8] J. K. Liao, M. Seto and K. Noma, *J. Cardiovasc. Pharmacol.*, 2007, **50**, 17.

[9] E. Hu, *Rec. Pat. Cardiovasc. Drug Disc.*, 2006, **1**, 249.

[10] H. Shimokawa and A. Takeshita, *Arterioscler. Thromb. Vasc. Biol.*, 2005, **25**, 1767.

[11] E. Hu and D. Lee, *Expert Opin. Ther. Targets*, 2005, **9**, 715.

[12] E. Hu and D. Lee, *Curr. Opin. Invest. Drugs*, 2003, **4**, 1065.

[13] H. Hidaka, Y. Suzuki, M. Shibuya and Y. Sasaki, in *Handbook of Experimental Pharmacology* (eds L. A. Pinna and P. T. W. Cohen). Inhibitors of Protein Kinases and Protein Phosphates, Springer GmbH, 2005, Vol. 167, p. 411.

[14] M. Shibua, Y. Suzuki, K. Sugita, I. Saito, T. Sasaki, K. Takakura, I. Nagata, H. Kikuchi, T. Takemae, H. Hidaka and M. Nakashima, *J. Neurosurg.*, 1992, **76**, 571.

[15] S. P. Davies, H. Reddy, M. Caivano and P. Cohen, *Biochem. J.*, 2000, **351**, 95.

[16] M. Asano and Y. Nomura, *Hypertens. Res.*, 2003, **26**, 97.

[17] M. Uehata, T. Ishizaki, H. Satoh, T. Ono, T. Kawahara, T. Morishita, H. Tamakawa, K. Yamagami, J. Inui, M. Maekawa and S. Narumiya, *Nature*, 1997, **389**, 990.

[18] C. Katoka, K. Egashira, S. Inoue, M. Takemoto, W. Ni, M. Koyanagi, S. Kitamoto, M. Usui, K. Kaibuchi, H. Shimokawa and A. Takeshita, *Hypertension*, 2002, **39**, 245.

[19] T. Yamakawa, S.-i. Tanaka, K. Numaguchi, Y. Yamakawa, E. D. Motley, S. Ichihara and T. Inagami, *Hypertension*, 2000, **35**, 313.

[20] N. Kobayashi, S. Horinaka, S.-i. Mita, S. Nakano, T. Honda, K. Yoshida, T. Kobayashi and H. Matsuoka, *Cardiovasc. Res.*, 2002, **55**, 757.

[21] W. Bao, E. Hu, L. Tao, R. Boyce, R. Mirabile, D. T. Thudium, X.-l. Ma, R. N. Willette and T.-l. Yue, *Cardiovasc. Res.*, 2004, **61**, 548.

[22] T. Yada, H. Shimokawa, O. Hiramatsu, T. Kajiya, F. Shigeto, E. Tanaka, Y. Shinozaki, H. Mori, T. Kiyooka, M. Katsura, S. Okhuma, M. Goto, Y. Ogasawara and F. Kajiya, *J. Am. Coll. Cardiol.*, 2005, **45**, 599.

[23] N. Sawada, H. Itoh, K. Ueyama, J. Yamashita, K. Doi, T.-H. Chun, M. Inoue, K. Masatsugu, T. Saito, Y. Fukunaga, S. Sakaguchi, H. Arai, N. Ohno, M. Komeda and K. Nakao, *Circulation*, 2000, **101**, 2030.

[24] A dominant negative approach has also been used, see: Y. Eto, H. Shimokawa, J. Hiroki, K. Morishige, T. Kandabashi, Y. Matsumoto, M. Amano, M. Hoshijima, K. Kaibuchi and A. Takeshita, *Am. J. Physiol. Heart Circ. Physiol.*, 2000, **278**, H1744.

[25] Z. Mallat, A. Gojova, V. Sauzeau, V. Brun, J.-S. Silvestre, B. Esposito, R. Merval, H. Groux, G. Loirand and A. Tedgui, *Circ. Res.*, 2003, **93**, 884.

[26] A. Masumoto, Y. Hirooka, H. Shimokawa, K. Hironaga, S. Setoguchi and A. Takeshita, *Hypertension*, 2001, **38**, 1307.

[27] M. Shibuya, Y. Suzuki, K. Sugita, I. Saito, T. Sasaki, K. Takakura, S. Okamoto, H. Kikuchi, T. Takeme and H. Hidaka, *Acta Neurochir. (Wien)*, 1990, **107**, 11.

[28] H. Shimokawa, K. Hiramori, H. Iinuma, S. Hosada, H. Kishida, H. Osada, T. Katagiri, K. Yamauchi, Y. Yui, T. Minamino, M. Nakashima and K. Kato, *J. Cardiovasc. Pharmacol.*, 2002, **40**, 751.

[29] Y. Fukumoto, T. Matoba, A. Ito, H. Tanaka, T. Kishi, S. Hayashidani, K. Abe, A. Takeshita and H. Shimokawa, *Heart*, 2005, **91**, 391.

[30] M. Shibuya, S. Hirai, M. Seto, S.-i. Satoh and E. Ohtomo, *J. Neurol. Sci.*, 2005, **238**, 31.

[31] M. Jacobs, K. Hayakawa, L. Swenson, S. Bellon, M. Fleming, P. Taslimi and J. Doran, *J. Biol. Chem.*, 2006, **281**, 260.

[32] C. Breitenlecher, M. Gaßel, H. Hidaka, V. Kinzel, R. Huber, R. A. Engh and D. Bossemeyer, *Structure*, 2003, **11**, 1595.

[33] S. Bonn, S. Herrero, C. B. Breitenlecher, A. Erlbruch, W. Lehmann, R. A. Engh, M. Gassel and D. Bossemeyer, *J. Biol. Chem.*, 2006, **281**, 24818.

[34] A. Takami, M. Iwakubo, Y. Okada, T. Kawata, H. Odai, N. Takahashi, K. Shindo, K. Kimura, Y. Tagami, M. Miyake, K. Fukushima, M. Inagaki, M. Amano, K. Kaibuchi and H. Iijima, *Bioorg. Med. Chem.*, 2004, **12**, 2115.

[35] S. Takanashi, H. Tanaka, Y. Naito, M. Uehata and K. Katayama, *WO Patent* 0168607-A1, 2001.

[36] J. C. Yarrow, G. Totsukawa, G. T. Charras and T. J. Mitchison, *Chem. Biol.*, 2005, **12**, 385.

[37] K. Sawada, T. Zenkoh, T. Terasawa, Y. Imamura, H. Fukudome, S. Kuroda, J. Maeda, J. Watanabe, H. Inami and N. Takeshita, *WO Patent* 07026920-A2, 2007.

[38] A. Feurer, S. Bennabi, H. Heckroth, J. Ergüden, T. Schenke, M. Bauser, R. Kast, J.-P. Stasch, E. Stahl, K. Münter, D. Lang and H. Ehmke, *WO Patent* 03106450-A1, 2003.

[39] A. Feurer, S. Bennabi, H. Heckroth, H. Shirock, J. Mittendorf, R. Kast, J.-P. Stasch, M. J. Gnoth, K. Münter, D. Lang, S. Figueroa Perez and H. Ehmko, *WO Patent* 04039796-A1, 2004.

[40] A. Bartolozzi, P. Sweetnam, S. Campbell, H. Foudoulakis, B. Kirk and S. Ram, *WO Patent* 06105081-A2, 2006.

[41] M. R. Hellberg and A. Rusinko, *US Patent* 060142307-A1, 2006.

[42] Y. Sasaki, M. Suzuki and H. Hidaka, *Pharmacol. Ther.*, 2002, **93**, 225.

[43] M. Tamura, H. Nakao, H. Yoshizaki, M. Shiratsuchi, H. Shigyo, H. Yamada, T. Ozawa, J. Totsuka and H. Hidaka, *Biochim. Biophys. Acta*, 2005, **1754**, 245.

[44] C. E. Teixeira, Z. Ying and R. C. Webb, *J. Pharmacol. Exp. Ther.*, 2005, **315**, 155.

[45] M. Hagihara, H. Nishida, Y. Tsuzaki, K. Yoshimura and K.-i. Komori, *WO Patent* 05035503-A1, 2005.

[46] P. C. Ray, *WO Patent* 07065916-A1, 2007.

[47] H. Tokushige, M. Inatani, S. Nemoto, H. Sakaki, K. Katayama, M. Uehata and H. Tanihara, *Invest. Opthalmol. Vis. Sci.*, 2007, **48**, 3216.

[48] H. Sagawa, H. Terasaki, M. Nakamura, M. Ichikawa, T. Yata, Y. Tokita and M. Watanabe, *Exp. Neurology*, 2007, **205**, 230.

[49] T. G. Davies, M. D. Garrett, R. G. Boyle and I. Collins, *WO Patent* 07125321-A2, 2007.

[50] S. Bennabi, H. Heckroth, H. Shirock, J. Mittendorf, R. Kast, J.-P. Stasch, M. J. Gnoth, K. Münter, D. Lang, S. F. Perez, M. Bauser, A. Feurer and H. Ehmke, *WO Patent* 05058891-A1, 2005.

[51] R. Kast, H. Schirok, S. Figueroa-Pérez, J. Mittendorf, M. J. Gnoth, H. Apeler, J. Lenz, J. K. Franz, A. Knorr, J. Hütter, M. Lobell, K. Zimmermann, K. Münter, K. H. Augstein, H. Ehmke and J. P. Stasch, *Br. J. Pharmacol.*, 2007, **152**, 1070.

[52] M. Iwakubo, A. Takami, Y. Okada, T. Kawata, Y. Tagami, H. Ohashi, M. Sato, T. Sugiyama, K. Fukushima and H. Iijima, *Bioorg. Med. Chem.*, 2007, **15**, 350.

[53] Y. Feng, M. D. Cameron, B. Frackowiak, E. Griffen, L. Lin, C. Ruiz, T. Schröter and P. LoGrasso, *Bioorg. Med. Chem. Lett.*, 2007, **17**, 2355.

[54] H. Binch, G. Brenchley, J. M. C. Golec, R. Knegtel, M. Mortimore, S. Patel and A. Rutherford, *WO Patent* 03064397-A1, 2003.

[55] K. B. Goodman, H. Cui, S. E. Dowdell, D. E. Gaitanopoulos, R. L. Ivy, C. A. Sehon, R. A. Stavenger, G. Z. Wang, A. Q. Viet, W. Xu, G. Ye, S. F. Semus, C. Evans, H. E. Fries, L. J. Jolivette, R. B. Kirkpatrick, E. Dul, S. S. Khandekar, T. Yi, D. K. Jung, L. L. Wright, G. K. Smith, D. J. Behm, R. Bentley, C. P. Doe, E. Hu and D. Lee, *J. Med. Chem.*, 2007, **50**, 6.

[56] Aminofurazans linked to 5-ring heterocycles have been reported to inhibit ROCK, but no enzyme data has been presented: J. H. Come, J. Green, C. Marhefka, S. L. Harbeson and L. Pham, *WO Patent* 05019190-A2, 2005.

[57] M. J. Bamford, M. J. Alberti, N. Bailey, S. Davies, D. K. Dean, A. Gaiba, S. Garland, J. D. Harling, D. K. Jung, T. A. Panchel, C. A. Parr, J. G. Steadman, A. K. Takle, J. T. Townsend, D. M. Wilson and J. Witherington, *Bioorg. Med. Chem. Lett.*, 2005, **15**, 3402.

[58] R. A. Stavenger, H. Cui, S. E. Dowdell, R. G. Franz, D. E. Gaitanopoulos, K. B. Goodman, M. A. Hilfiker, R. L. Ivy, J. D. Leber, J. P. Marino, Jr., H.-J. Oh, A. Q. Viet, W. Xu, G. Ye, D. Zhang, Y. Zhao, L. J. Jolivette, M. S. Head, S. F. Semus, P. A. Elkins, R. B. Kirkpatrick, E. Dul, S. S. Khandekar, T. Yi, D. K. Jung, L. L. Wright, G. K. Smith, D. J. Behm, C. P. Doe, R. Bentley, Z. X. Chen, E. Hu and D. Lee, *J. Med. Chem.*, 2007, **50**, 2.

[59] C. Doe, R. Bentley, D. J. Behm, R. Lafferty, R. Stavenger, D. Jung, M. Bamford, T. Panchel, E. Grygielko, L. L. Wright, G. K. Smith, Z. Chen, C. Webb, S. Khandekar, T. Yi, R. Kirkpatrick, E. Dul, L. Jolivette, J. P. Marino, Jr., R. Willette, D. Lee and E. Hu., *J. Pharmacol. Exp. Ther.*, 2007, **320**, 89.
[60] N. Imazaki, M. Kitano, T. Fujibayashi and S. Asano, *US Patent* 050182040-A1, 2005.
[61] A. Bartolozzi, S. Campbell, B. Cole, J. Ellis, H. Foudoulakis, B. Kirk, S. Ram, P. Sweetnam, M. Hauer-Jensen and M. Boerma, *WO Patent* 08054599-A2, 2008.

CHAPTER 7

LXR Agonists for the Treatment of Atherosclerosis: Recent Highlights

Jonathan Bennett, Andrew Cooke, Heather J. McKinnon and **Olaf Nimz**

1. INTRODUCTION

Atherosclerosis is the pathological process of arterial wall thickening that decreases blood flow, precipitates acute thrombotic occlusion and leads to life threatening clinical events such as myocardial infarction and stroke. In the United States alone, 70 million adults are thought to have one or more types of cardiovascular disease (CVD), with atherosclerosis being the primary cause and responsible for more than half of the deaths associated with this disease [1]. The economic burden of CVD in 2006 is estimated at $400 billion, and is projected to rise to $2 trillion by 2050 [2]. Atherosclerosis may develop undetected for decades, but it is usually assumed to be present in patients with dyslipidemia, with the degree of atherosclerotic progression dependent upon the patient's age and severity of the dyslipidemia. Following a clinical event, atherosclerosis can

Schering Plough Corporation, Newhouse, Motherwell ML1 5SH, UK

Annual Reports in Medicinal Chemistry, Volume 43
ISSN 0065-7743, DOI 10.1016/S0065-7743(08)00007-9

be diagnosed by imaging. However, all current treatment options are targeted at regulation of dyslipidemia, with the pharmacological reduction of LDL cholesterol via statin and/or cholesterol absorption inhibitor therapy being the standard treatment.

The pathology of atherosclerosis is complex and clinically silent lesions can exist for decades. Thus the main interest in therapy is to prevent the cause of acute thrombosis associated with disruption of an atherosclerotic plaque, notably plaque rupture. The primary cell type implicated in rupture of atherosclerotic plaques is the macrophage. Monocytes derived from bone marrow and circulating in the blood infiltrate the intimal layer of the arterial vessel wall in response to endothelial damage and the presence of extracellular lipid. Once in the vessel wall they differentiate into macrophages. Over many decades, this dynamic "response to injury" traps macrophages within the vessel wall in a state where they are overburdened with lipid, and destined to a necrotic or apoptotic fate releasing proteolytic enzymes which degrade the fibrous cap and cause mechanical instability and rupture [3].

2. LXR AND ATHEROSCLEROSIS

Liver X Receptor (LXR) is a nuclear receptor of the RXR heterodimer family, with two subtypes α and β. Nuclear receptors are transcriptional factors that regulate gene expression upon ligand binding. LXR controls cholesterol homeostasis and lipogenesis, and in particular is the key inducer of the ABCA1 (ATP-binding cassette) cholesterol transporter gene [4]. In the human population, mutations of the ABCA1 gene lead to highly atherogenic lipoprotein profiles which in the most severe form cause Tangier Disease and associated premature atherosclerosis [5–7]. LXR agonists are being developed for the treatment of atherosclerosis. They are expected to confer efficacy primarily by direct action on the vessel wall, lowering the cholesterol burden of arteries (via upregulation of ABCA1) and thus generating more stable lesions. Additionally, LXR agonists may increase circulating high-density lipoprotein (HDL) levels due to the role of ABCA1 in generation of nascent HDL by the liver and small intestine [8,9]. LXR agonists are also expected to increase cholesterol excretion via upregulation of ABCG5 and G8 in the liver and small intestine [10]. Finally, there is further potential for anti-atherosclerotic effects of LXR agonists due to suppression of inflammation and regulation of glucose metabolism [11,12].

A series of investigations in mice, including bone marrow transplantion studies, have revealed the critical role LXR plays in the development of atherosclerosis [13,14]. Importantly, LXR-induced upregulation of the ABCA1 cholesterol transporter in the arterial macrophages of the atherosclerotic vessel wall is sufficient to attenuate lesion development in the mouse [15,16]. This direct effect on the vessel wall may be critical to the clinical effectiveness of LXR agonists developed for the treatment of atherosclerosis, since other metabolic effects of LXR activation in the liver may differ substantially between species. The discovery process for LXR agonists has been clouded by the worrying observations of hepatic steatosis and elevated atherogenic plasma lipids in animal models. Successful development of LXR agonists will require confidence that compounds will be efficacious in patients and clear demonstration of safety.

3. LXR SUBTYPE SELECTIVITY

3.1 Rationale for targeting LXRβ selectivity

LXRα and β are closely related subtypes and share 77% amino acid identity in their DNA-binding and ligand-binding domains. The LXR subtypes are also highly conserved between humans and rodents [17]. The endogenous ligands for LXR are a specific group of oxysterols that include 22(R)-hydroxycholesterol, 24(S)-hydroxycholesterol, 24(S),25-epoxycholesterol and 27-hydroxycholesterol [18,19]. However, none of these ligands has been shown to differentially bind to either LXRα or β. The LXR subtypes are differentiated by their pattern of distribution. Whilst LXRα is most highly expressed in liver, adipose tissue and macrophages, LXRβ is ubiquitously distributed. The significance of this differential distribution became apparent in studies with LXR knockout mice which demonstrated that LXRα has the predominant role on lipid regulation in the liver. LXRα knockout mice are unable to regulate lipid balance when mice are given a high fat diet, with the liver becoming fatty and enlarged. Conversely, LXRβ knockout mice manage the high fat diet comparably to wild-type animals [20]. Further studies demonstrated that the induction of lipogenesis caused by administration of an LXR agonist was ablated in LXRα knockout mice [21].

These knockout studies have fuelled the interest in developing LXRβ selective compounds that are expected to have the same effect on ABCA1 upregulation in the atherosclerotic vessel wall as a pan agonist, without the adverse lipid effects seen by LXRα activation in the liver. True validation of this approach has been restricted by the lack of compounds with suitable LXRβ selectivity. However, administration of pan agonists to LXRα or β knockout mice, or macrophages derived from these mice, has supported this approach. In LXRα knockout mice a pan LXR agonist was shown to increase plasma HDL, without the associated increase in plasma triglyceride, VLDL or hepatic steatosis [21,22]. More recently, in a study with LXRα/apoE double knockout mice, an LXR pan agonist (GW3965 **2**) was shown to reduce the development of atherosclerosis by 39% as measured by en face analysis of aortic lesions or Oil Red O staining of aortic root sections [23]. This critical study provides the most direct support to date that an LXRβ selective agonist confers efficacy in the treatment of atherosclerosis.

3.2 LXRα/β X-Ray crystallography data

Currently, from a bio-structural standpoint, the analysis of LXR subtype selectivity and input to medicinal chemistry remains focused on the initial ligand-binding event. After alignment of sequences for human LXRα (Q13133) and LXRβ (P55055) with the program clustalW [24], the ligand-binding domain

Table 1 Published LXR/ligand co-crystal structures

Pdb ID[a]	Isoform-dimer	Ligand	Resolution	Reference
1upv	hLXRβ	1	2.1Å	[30]
1upw	hLXRβ	1	2.4Å	[30]
1pq9	hLXRβ/hLXRβ	1	2.1Å	[31]
1pq6	hLXRβ/hLXRβ	2	2.4Å	[31]
1pqc	hLXRβ/hLXRβ	1	2.8Å	[31]
1p8d	hLXRβ/hLXRβ	3	2.8Å	[32]
2acl	mLXRα/hRXRα	4	2.8Å	[27]
1uhl	hLXRα/hRXRβ	1	2.9Å	[33]

[a]Pdb ID as used by the protein data bank.

(LBD) shows 15 differences in amino acids that are classified as non-conservative. The majority of those are solvent-exposed residues and remote from the ligand-binding pocket. Several co-crystal X-ray structures of ligands **1–4** bound to both LXRα and β at different resolutions have been published (Table 1). Reviews on "first generation" LXR agonists have been published [25–29].

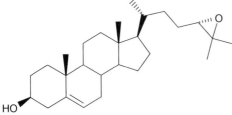

1

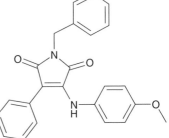

2

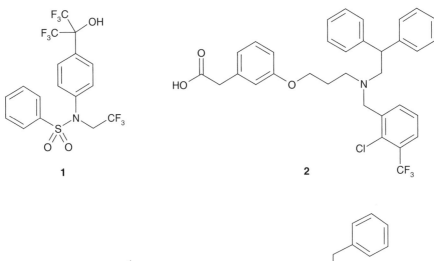

3

4

The ligand-binding pocket of LXR is predominantly hydrophobic in nature with His435 (LXRβ) mediating the agonist interaction to stabilize Trp457 (LXRβ) of Helix 12 which forms the co-activator binding groove. The pocket shows a high degree of flexibility in its side chain conformations and is able to accommodate agonists of very different sizes by an induced fit mechanism. Comparison of the amino acids forming the ligand-binding pocket shows that only a single amino acid is changed (Val295 (LXRα)→Ile311 (LXRβ)). This part of the pocket is not occupied by the published ligands in Table 1 but has recently been targeted to provide selectivity within a quinoline series of LXR agonists (see Section 3.3.1). The additional methyl group of Ile311 increases the volume of the side chain (LXRβ) which is oriented towards Helix 8. This structural change has the potential to increase the distance between Helix 5 and Helix 8 [27,30].

A secondary amino acid change (Ala292 (LXRα)→Thr308 (LXRβ)) close to the Trp/His pocket is located behind a key methionine residue (Met296 (LXRα)/Met312 (LXRβ)). This change impacts the volume of the pocket as the methyl group of the methionine residue can either orient towards or away from the pocket (see Figure 1).

The direction of the dipole moment of the methionine side chain is dependent on the location of the methyl group. The effect of the methionine dipole is underestimated in current force field parameterizations and remains unexploited in docking or molecular mechanics calculations [34].

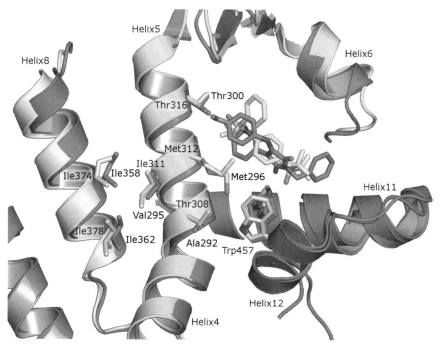

Figure 1 Different conformations of a key methionine residue in mLXRα [27] (magenta) and hLXRβ [30] (coloured from blue to red along the sequence). (See Color Plate 7.1 in Color Plate Section.)

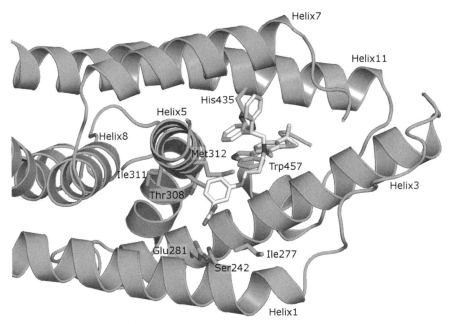

Figure 2 GW3965 **2** bound in Trp/His pocket of LXRβ extending the phenylacetic acid group into the solvent-exposed area. (See Color Plate 7.2 in Color Plate Section.)

A third amino acid change (Val263 (LXRα)→Ile277 (LXRβ)) stabilizes the usually very dynamic extension of Helix 1 with potential impact on the Glu265-Arg232 [27] or Glu281-Ser242 [31] hydrogen bonding interactions which may influence the nature of ligand-solubilizing groups tolerated in this region (see Figure 2).

From a review of the available bio-structural information, the potential to attain isoform selectivity for ligand binding appears to be limited due to the nearly complete conservation of the ligand-binding environment. The only direct change of the protein cavity is at the Val295 (LXRα)→Ile311 (LXRβ) location. Indirect effects on subtype selectivity via the second shell amino acid differences can be expected from changes to the volume of the Trp/His pocket. Other changes distal from the ligand-binding pocket may also have an effect by influencing the dynamic behaviour of the receptor, although modeling and utilizing these for rational drug design is clearly considerably more challenging than focusing on the changes proximal to the ligand-binding environment.

3.3 Progress towards LXRβ selective ligands

3.3.1 Quinolines

Recently a series of quinoline derivatives arising from initial lead **5** has been reported as LXR agonists [35,36]. Superimposition of **5** with GW3965 **2**, complexed with the LXRβ ligand-binding domain, highlighted the possibility to increase affinity through targeting an interaction with Arg319.

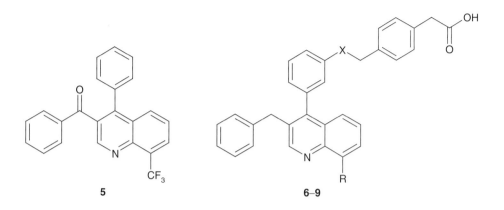

5   6–9

Analogue synthesis around **5**, targeting interaction with Arg 319, followed by subsequent optimization driven by characterization of metabolites led to the identification of analogues **6 (WAY-254011)** and **7** which possessed enhanced LXR α/β binding affinity as well as enhanced efficacy in an hLXRβ transactivation assay (Table 2).

Cell-based activity was further demonstrated for **6** and **7** through ABCA1 upregulation and concentration-dependent cholesterol efflux from THP-1 cells. No gene selectivity was evident for **6** and **7** with the EC$_{50}$'s for upregulation of SREBP-1c in Huh7 cells (45 and 15 nM, respectively) being comparable to those for ABCA1 upregulation in THP-1 cells (55 and 33 nM, respectively). Oral administration of **7** at the dose of 10 mg/kg/day for 8 weeks led to a significant 45 ± 22% ($n = 8$) reduction in lesion burden compared to the control group. However, in addition to activity at LXRα/β, compounds **6** and **7** were also shown to possess activity in human PPAR α/γ/δ transactivation assays with EC$_{50}$ values ranging from 0.3 to 1.3 μM (see Table 3) [36]. It should be noted that this PPAR activity as with any off target effects could ultimately complicate the interpretation of *in vivo* data generated with such compounds.

Table 2 Comparison of LXR binding and transactivation data for **5–9**

Compound	R	X	hLXRβ binding IC$_{50}$ (nM)	hLXRα binding IC$_{50}$ (nM)	hLXRβ EC$_{50}$ (nM/%eff)
5	CF$_3$	—	107	260	1,400 (25)
6	CF$_3$	O	2.1	9.5	71 (91)
7	CF$_3$	NH	1.9	7.6	33 (85)
8	CH$_3$	NH	5.0	21.0	90 (97)
9	Cl	NH	1.4	5.0	23 (107)

Note: Efficacy is presented as a percentage relative to the T0901317 **1** response.

Table 3 Comparison of gene regulation and PPAR transactivation data for **5–9**

Compound	R	X	ABCA1 EC$_{50}$ (nM/%eff)	SREBP-1c EC$_{50}$ (nM/%eff)	PPAR $\alpha/\gamma/\delta$ EC$_{50}$ (nM/%eff)
6	CF$_3$	O	55 (172)	45 (106)	1,312 (19), 680 (47), 627 (24)
7	CF$_3$	NH	33 (151)	15 (73)	1,300 (71), 320 (20), 1,320 (37)
8	CH$_3$	NH	438 (139)	51 (114)	1,772 (7.5), 1,030 (47), 1,975 (47)
9	Cl	NH	81 (162)	75 (137)	2,053 (12), 506 (51), 631 (41)

Note: ABCA1 and SREBP-1c induction is presented as a percentage relative to the T0901317 response.

The X-ray structure of **6** in complex with the ligand-binding domain of hLXRβ showed that, as predicted, the carboxylic acid function formed a hydrogen bond network with Arg319 and Leu330. This analysis also showed that the quinoline nitrogen was involved in a hydrogen bond interaction with the key His435 residue and that the 3-benzylic substituent occupied a hydrophobic pocket formed by Phe271, Phe340 and Phe349.

Further limited SAR data around C3 and C8 modifications of the phenylacetic acid-based quinolines exemplified by **7** have since been published, resulting in the identification of compounds **8** and **9** which possessed comparable activity to **7** in LXR α/β binding and transactivation assays [36]. In common with **6** and **7**, compounds **8** and **9** did not possess any gene selectivity (SREBP1c vs. ABCA1) and were shown to possess PPAR $\alpha/\gamma/\delta$ agonist activity (see Table 3).

More recently, optimization of the LXR α/β selectivity ratio within the quinoline series has been reported utilizing the X-ray structure of **6** bound to hLXRβ as a starting point [37]. The single amino acid difference between hLXRα and hLXRβ appeared to be accessible via modification of the phenylacetic acid moiety of **6**. To this end, a series of 2,3- and 2,5-disubstituted phenylacetic and α-substituted acetic acids were synthesized to test whether this small difference between hLXRα and hLXRβ could be targeted to increase the selectivity ratio [37].

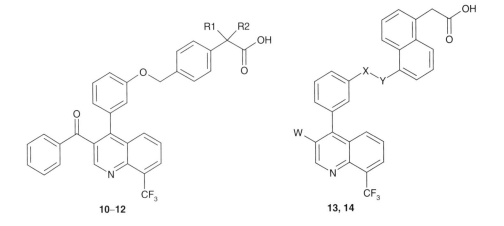

10–12   **13, 14**

Table 4 Comparison of hLXRβ/α binding and selectivity for **6** and **10–12**

Compound	R1	R2	hLXRβ binding IC$_{50}$ (nM)	hLXRα binding IC$_{50}$ (nM)	Ratio α/β
6	H	H	2.1	9.5	4.5
10	H	H	5	17	3.4
11	H	Allyl	5	16	3.2
12	Propargyl	Propargyl	26	2	0.075

Table 5 Comparison of hLXRβ/α binding and selectivity for **6**, **13** and **14**

Compound	W	X-Y	hLXRβ binding IC$_{50}$ (nM)	hLXRα binding IC$_{50}$ (nM)	Ratio α/β
6	Bn	OCH$_2$	2.1	9.5	4.5
13	Bn	CH$_2$NH	5	123	25
14	H	CH$_2$NH	15	745	50

No increase in hLXRβ-binding selectivity was obtained by α-substitution of the phenylacetic acid moiety, e.g. **11**, despite initial docking studies suggesting that this strategy would perturb the LXRβIle/LXRαVal pocket (see Table 4). However, it should be noted that an increase in LXRα selectivity was observed for compound **12**, indicating that targeting the pocket could result in selectivity differences.

An increase in binding selectivity for hLXRβ was obtained via the synthesis of 2-(naphthalene-1-yl)acetic acid analogues, e.g. **13**. Further analogue synthesis examining substitution at the C3 position led to the identification of **14** which possessed potent hLXRβ-binding affinity (IC$_{50}$ = 15 nM) and 50-fold selectivity over hLXRα (Table 5).

Whereas **13** and **14** possessed 25- and 50-fold hLXRβ-binding selectivity, respectively, this decreased to ~2-fold upon testing in functional transactivation assays. Despite this reduction in functional selectivity, compound **14** showed reduced levels of agonist activity in hLXRα transactivation (13%) as well as SREBP-1c (38%) and TG accumulation (22%) in HepG2 cells, when compared to reference compound **1** and starting point **6**, which could be suggestive of a less lipogenic profile. However, it should be noted that **13** and **14** also had reduced potency and efficacy in hLXRβ transactivation and ABCA1 upregulation in a human macrophage cell line (THP-1). As with previous analogues from this series, **13** and **14** also had PPAR agonist activity, e.g. PPARγ (EC$_{50}$ < 100 nM, >50% efficacy), which precluded further development of the series [37].

3.3.2 N-Acyl thiadiazolines

A cell-based HTS screen for LXRβ ligands (utilizing a chimeric GAL-hLXRβ LBD construct) identified N-acylthiadiazolines which activated LXRβ at micromolar

Table 6 *In vitro* data for *N*-Acyl thiadiazolines **15** and **16**

Assay	Compound **15** (racemate)	Compound **16** (R enantiomer)
IC_{50} SPA-LXRα (%E[a])	9.8 µM (52)	5 µM (80)
IC_{50} SPA-LXRβ (%E[a])	0.3 µM (100)	0.066 µM (100)
Ratio α/β	33	76
IC_{50} FRET-LXRα (%E[a])	2.3 µM (53)	1.6 µM (64)
IC_{50} FRET-LXRβ (%E[a])	0.25 µM (100)	0.067 µM (94)
Ratio α/β	9.2	24
EC_{50} hLXRα (%E[a])	0.608 µM (19)	0.346 µM (41)
EC_{50} hLXRβ (%E[a])	0.098 µM (82)	0.063 µM (98)
Ratio α/β	6.2	5.5
ABCA1 induction THP1[a] (1 µM)	34%	39%
ABCA1 induction THP1[a] (10 µM)	59%	60%
FAS induction HepG2[a] (1 µM)	10%	No data
FAS induction HepG2[a] (10 µM)	33%	No data

[a]Efficacy, ABCA1 and FAS induction is presented as a percentage relative to the T0901317 response.

levels but not LXRα [38]. The initial optimization of these HTS hits was aimed at improving potency, and led to the racemic compound **15** and its active (*R*) enantiomer **16** (Table 6, (*S*) enantiomer inactive).

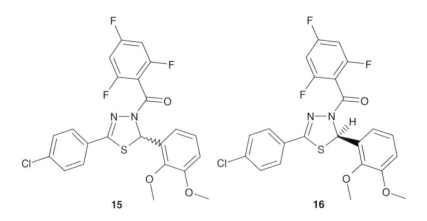

The *in vitro* LXRβ selectivity of **15** and **16** was evaluated in both cell-based and non-cell-based assays (Table 6). The (*R*) enantiomer **16** displayed greater LXRβ selectivity than its parent racemate **15** when compared by a non-cell-based fluorescence resonance energy transfer assay (76- and 33-fold, respectively) and a scintillation proximity-binding assay (24- and 9-fold, respectively). These levels of LXRβ selectivity were reduced when a cell-based assay in cells expressing the natural receptor was employed (~6-fold for both compounds). Further evidence to support a degree of functional selectivity was obtained by treating macrophages derived from LXRα and β knockout mice with **15**. A greater induction of ABCA1 was observed in macrophages expressing only the LXRβ isoform when compared to cells lacking the LXRβ subtype [38].

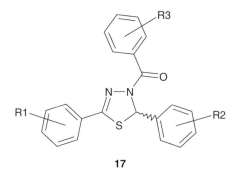

17

To further explore the LXR activity and subtype selectivity of this chemical series, a range of additional analogues was prepared by varying R1, R2 and R3 on generic structure **17** (Table 7). Data is given for LXRα and β activity for these racemic compounds in the FRET assay. Removal of the three fluorine substituents (R3) present in compound **15** gave the LXRα/β inactive analogue **18**. Replacement of the R1 chloro substituent of **15** with fluoro gave **19** which had improved LXRβ activity (~3-fold) and improved LXRα/β selectivity (~18-fold). Changing the 2,3-dimethoxy R2 substitution pattern present in **19** to a 2-methoxy, 4-fluoro substitution gave compound **20**, the most active LXRβ analogue presented. When compared to the lead compound **15**, the LXRβ potency of **20** improved by ~42-fold while maintaining comparable selectivity. Compound **19** was also shown to upregulate ABCA1 mRNA in THP1 cells (42% at 1 μM and 53% at 10 μM). Replacement of the three fluorine substituents (R3) of compound

Table 7   Comparison of hLXRα/β binding and selectivity for **18–21**

Compound	R1	R2	R3	IC$_{50}$ FRET LXRα (%E[a])	IC$_{50}$ FRET LXRβ (%E[a])	Ratio α/β
18	4-Cl	2,3-di-OMe	H	Inactive	Inactive	N/A
19	4-F	2,3-di-OMe	2,4,6-tri-F	1.4 μM (62)	0.08 μM (98)	18
20	4-F	2-OMe-4-F	2,4,6-tri-F	0.07 μM (92)	0.006 μM (100)	12
21	4-F	2,3-di-OMe	3-Cl	Inactive	0.25 μM (82)	>40

[a]Efficacy of compounds is presented as a percentage relative to the T0901317 response.

19 with 3-chloro gave compound **21**, the most selective analogue that was reported. This compound possessed LXRβ binding affinity comparable to that of the lead compound **15**, but was inactive at LXRα (<15% at 20 μM).

Compounds **15** and **19** were unable to upregulate ABCA1 and SREBP-1c in the liver or other tissues of C57BL/6 mice dosed 20–200 mg/kg for 3 days despite pharmacokinetic studies with both compounds that showed acceptable oral exposure. The authors suggest that the lack of *in vivo* activity may be related to the high plasma protein binding (>99.9%) of these analogues. However, this conclusion should be treated with caution as other NR modulators, e.g. PPAR agonists, have been shown to possess potent *in vivo* activity despite having similarly high levels of plasma protein binding.

4. CLINICAL CANDIDATE WAY-252623

WAY-252623 **22**, the first LXR agonist clinical candidate to be publicly reported, entered Phase 1 evaluation in October 2006 and its development was subsequently terminated by September 2007. To date, no preclinical data has become available in peer-reviewed journals, although limited information was recently presented [39].

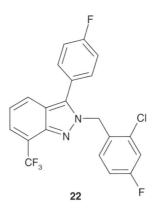

22

Compound **22** possessed potent hLXRα (179 nM IC$_{50}$) and hLXRβ (24 nM IC$_{50}$) binding affinity with a modest ~7-fold selectivity for hLXRβ. In hLXRα (6.8 μM EC$_{50}$) and hLXRβ (3.7 μM EC$_{50}$) transactivation assays, **22** showed a marked reduction in potency compared to binding affinity and also a decrease in the selectivity between hLXRβ and hLXRα to ~2-fold. Activity in ABCA1 upregulation, cholesterol efflux and SREBP-1c upregulation was also reported with EC$_{50}$ values of 639 nM, 500 nM and 2.5 μM, respectively. Data was also presented to show good selectivity over the hERG channel (inactive at 10 μM) and acceptable oral bioavailability of 63%, 28%, 35% in mice, rat and monkey, respectively, although relatively poor human microsomal stability was also reported ($T_{1/2} = 47$ min). The reason for the cessation of clinical development is currently unknown with the only information coming in a press release citing "an unfavourable profile for further development" [40].

5. CONCLUSION

The last 1–2 years has seen continued interest in the field of LXR agonists for the treatment of atherosclerosis as evidenced through recent patent activity and continued disclosures in the scientific literature. It remains a challenge to identify high quality novel compounds which are suitable for *in vivo* testing and, ultimately, progression into clinical development. The last 2 years has, however, seen public disclosure of the first LXR agonist clinical candidate. Full preclinical and clinical data disclosure for this candidate is eagerly awaited to see whether any information can be gleaned on the current assumptions relating to LXR agonist effects in humans. Of particular interest will be data on the predictivity with respect to efficacy and the potential lipogenic liability of preclinical cell-based and animal models.

The issue of potential lipogenic liability of LXR agonists remains at the front of scientists' minds and has fuelled interest in tackling the major challenge of identifying LXRβ selective agonists. The availability of X-ray crystallography data for both LXRα and β has and will continue to have an important part to play in the structure-based drug design of subtype selective chemotypes, although the difficulty of the challenge should not be underestimated. Whilst early progress has been made in identifying LXRβ selective ligands, these initial compounds are still some way short of the desired functional selectivity profile. Only once truly selective agonists have been identified will we be in a position to test the hypotheses put forward via murine knockout studies, and in animal models more predictive of human lipoprotein balance. The strong link between ABCA1 activity and cardiovascular disease will continue to provide impetus for the identification of further LXR agonist clinical candidates in an attempt to address the major unmet medical need.

REFERENCES

[1] A. G. Turpie, *Am. J. Manag. Care*, 2006, **12**, S430.
[2] D. L. Brown, B. Boden-Abala, K. M. Langa, L. D. Lisabeth, M. Fair, M. A. Smith, R. L. Sacco and L. B. Morgenstern, *Neurology*, 2006, **67**, 1390.
[3] R. Ross, *Nature*, 1993, **362**, 801.
[4] A. Venkateswaran, B. A. Laffitte, S. B. Joseph, P. A. Mak, D. C. Wilpitz, P. A. Edwards and P. Tontonoz, *Proc. Natl. Acad. Sci., USA*, 2000, **97**, 12097.
[5] R. R. Singaraja, L. R. Brunham, H. Visscher, J. J. Kastelein and M. R. Hayden, *Arterioscler. Thromb. Vasc. Biol.*, 2003, **23**, 1322.
[6] S. Rust, M. Rosier, H. Funke, J. Real, Z. Amoura, J. C. Piette, J. F. Deleuze, H. B. Brewer, N. Duverger, P. Denefle and G. Assmann, *Nat. Genet.*, 1999, **22**, 352.

[7] M. Bodzioch, E. Orso, J. Klucken, T. Langmann, A. Bottcher, W. Diederich, W. Drobnik, S. Barlage, C. Buchler, M. Porsch-Ozcurumez, W. E. Kaminski, H. W. Hahmann, K. Oette, G. Rothe, C. Aslanidis, K. J. Lackner and G. Schmitz, *Nat. Genet.*, 1999, **22**, 347.

[8] L. R. Brunham, J. K. Kruit, J. Iqbal, C. Fievet, J. M. Timmins, T. D. Pape, B. A. Coburn, N. Bissada, B. Staels, A. K. Groen, M. M. Hussain, J. S. Parks, F. Kuipers and M. R. Hayden, *J. Clin. Invest.*, 2006, **116**, 1052.

[9] J. F. Oram and A. M. Vaughan, *Curr. Opin. Lipidol.*, 2000, **11**, 253.

[10] J. J. Repa, K. E. Berge, C. Pomajzl, J. A. Richardson, H. Hobbs and D. J. Mangelsdorf, *J. Biol. Chem.*, 2002, **277**, 18793.

[11] S. B. Joseph, A. Castrillo, B. A. Laffitte, D. J. Mangelsdorf and P. Tontonoz, *Nat. Med.*, 2003, **9**, 213.

[12] B. A. Laffitte, L. C. Chao, J. Li, R. Walczak, S. Hummasti, S. B. Joseph, A. Castrillo, D. C. Wilpitz, D. J. Mangelsdorf, J. L. Collins, E. Saez and P. Tontonoz, *Proc. Natl. Acad. Sci., USA*, 2003, **100**, 5419.

[13] G. U. Schuster, P. Parini, L. Wang, S. Alberti, K. R. Steffensen, G. K. Hansson, B. Angelin and J. A. Gustafsson, *Circulation*, 2002, **106**, 1147.

[14] N. Levin, E. D. Bischoff, C. L. Daige, D. Thomas, C. T. Vu, R. A. Heyman, R. K. Tangirala and I. G. Schulman, *Arterioscler. Thromb. Vasc. Biol.*, 2005, **25**, 135.

[15] M. van Eck, I. S. Bos, W. E. Kaminski, E. Orsó, G. Rothe, J. Twisk, A. Böttcher, E. S. Van Amersfoort, T. A. Christiansen-Weber, W. P. Fung-Leung, T. J. Van Berkel and G. Schmitz, *Proc. Natl. Acad. Sci., USA*, 2002, **99**, 6298.

[16] R. K. Tangirala, E. D. Bischoff, S. B. Joseph, B. L. Wagner, R. Walczak, B. A. Laffitte, C. L. Daige, D. Thomas, R. A. Heyman, D. J. Mangelsdorf, X. Wang, A. J. Lusis, P. Tontonoz and I. G. Schulman, *Proc. Natl. Acad. Sci., USA*, 2002, **99**, 11896.

[17] D. J. Peet, B. A. Janowski and D. J. Mangelsdorf, *Curr. Opin. Genet. Dev.*, 1998, **8**, 571.

[18] B. A. Janowski, P. J. Willy, T. R. Devi, J. R. Falck and D. J. Mangelsdorf, *Nature*, 1996, **383**, 728.

[19] C. Song and S. Liao, *Endocrinology*, 2000, **141**, 4180s.

[20] S. Alberti, G. Schuster, P. Parini, D. Feltkamp, U. Diczfalusy, M. Rudling, B. Angelin, I. Bjorkhem, S. Pettersson and J. A. Gustafsson, *J. Clin. Invest.*, 2001, **107**, 565.

[21] E. M. Quinet, D. A. Savio, A. R. Halpern, L. Chen, G. U. Schuster, J. A. Gustafsson, M. D. Basso and P. Nambi, *Mol. Pharmacol.*, 2006, **70**, 1340.

[22] E. G. Lund, L. B. Peterson, A. D. Adams, M. H. Lam, C. A. Burton, J. Chin, Q. Guo, S. Huang, M. Latham, J. C. Lopez, J. G. Menke, D. P. Milot, L. J. Mitnaul, S. E. Rex-Rabe, R. L. Rosa, J. Y. Tian, S. D. Wright and C. P. Sparrow, *Biochem. Pharmacol.*, 2006, **71**, 453.

[23] M. N. Bradley, C. Hong, M. Chen, S. B. Joseph, D. C. Wilpitz, X. Wang, A. J. Lusis, A. Collins, W. A. Hseuh, J. L. Collins, R. K. Tangirala and P. Tontonoz, *J. Clin. Invest.*, 2007, **117**, 2337.

[24] R. Chenna, H. Sugawara, T. Koike, R. Lopez, T. J. Gibson, D. G. Higgins and J. D. Thompson, *Nucleic Acids Res.*, 2003, **31**, 3497.

[25] J. R. Schulz, H. Tu, A. Luk, J. J. Repa, J. C. Medina, L. Li, S. Schwendner, S. Wang, M. Thoolen, D. J. Mangelsdorf, K. D. Lustig and B. Shan, *Gene. Dev.*, 2000, **14**, 2831.

[26] J. L. Collins, A. M. Fivush, M. A. Watson, C. M. Galardi, M. C. Lewis, L. B. Moore, D. J. Parks, J. G. Wilson, T. K. Tippin, J. G. Binz, K. D. Plunket, D. G. Morgan, E. J. Beaudet, K. D. Whitney, S. A. Kliewer and T. A. Willson, *J. Med. Chem.*, 2002, **45**, 1963.

[27] M. C. Jaye, J. A. Krawiec, N. Campobasso, A. Smallwood, C. Qiu, Q. Lu, J. J. Kerrigan, M. De Los Frailes Alvaro, B. Laffitte, W-S. Liu, J. P. Marino, C. R. Meyer, J. A. Nichols, D. J. Parks, P. Perez, L. Sarov-Blat, S. D. Seepersaud, K. M. Steplewski, S. K. Thompson, P. Wang, M. A. Watson, C. L. Webb, D. Haigh, J. A. Caravella, C. H. Macphee, T. M. Willson and J. L. Collins, *J. Med. Chem.*, 2005, **48**, 5419.

[28] D. J. Bennett, L. D. Brown, A. J. Cooke and A. S. Edwards, *Expert Opin. Ther. Pat.*, 2006, **16**, 1673.

[29] D. J. Bennett, E. L. Carswell, A. J. Cooke, A. S. Edwards and O. Nimz, *Curr. Med. Chem.*, 2008, **15**, 195.

[30] S. Hoerer, A. Schmid, A. Heckel, R.-M. Budzinski and H. Nar, *J. Mol. Biol.*, 2003, **334**, 853–861.

[31] M. Färnegårdh, T. Bonn, S. Sun, J. Ljunggren, H. Ahola, A. Wilhelmsson, J.-Å. Gustafsson and M. Carlquist, *J. Biol. Chem.*, 2003, **278**, 38821–38828.

[32] S. Williams, R. K. Bledsoe, J. L. Collins, S. Boggs, M. H. Lambert, A. B. Miller, J. Moore, D. D. McKee, L. Moore, J. Nichols, D. Parks, M. Watson, B. Wisely and T. M. Willson, *J. Biol. Chem.*, 2003, **278**, 27138–27143.

[33] S. Svensson, T. Östberg, M. Jacobsson, C. Norström, K. Stefansson, D. Hallén, I. C. Johansson, K. Zachrisson, D. Ogg and L. Jendeberg, *EMBO J.*, 2003, **22**, 4625–4633.

[34] F. Wennmohs and M. Schindler, *J. Comput. Chem.*, 2005, **26**, 283–293.

[35] B. Hu, M. Collini, R. Unwalla, C. Miller, R. Singhaus, E. Quinet, D. Savio, A. Halpern, M. Basso, J. Keith, V. Clerin, L. Chen, C. Resmini, Q-Y. Liu, I. Feingold, C. Huselton, F. Azam, M. Farnegardh, C. Enroth, T. Bonn, A. Goos-Nilsson, A. Wilhelmsson, P. Nambi and J. Wrobel, *J. Med. Chem.*, 2006, **49**, 6151.

[36] B. Hu, J. Jetter, D. Kaufman, R. Singhaus, R. Bernotas, R. Unwalla, E. Quinet, D. Savio, A. Halpern, M. Basso, J. Keith, V. Clerin, L. Chen, Q-Y. Liu, I. Feingold, C. Huselton, F. Azam, A. Goos-Nilsson, A. Wilhelmsson, P. Nambi and J. Wrobel, *Bioorg. Med. Chem. Lett.*, 2007, **15**, 3321.

[37] B. Hu, E. Quinet, R. Unwalla, M. Collini, J. Jetter, R. Dooley, D. Andraka, L. Nogle, D. Savio, A. Halpern, A. Goos-Nilsson, A. Wilhelmsson, P. Nambi and J. Wrobel, *Bioorg. Med. Chem. Lett.*, 2008, **18**, 54.

[38] V. Molteni, X. Li, J. Nabakka, F. Liang, J. Wityak, A. Koder, L. Vargas, R. Romeo, N. Mitro, P. A. Mak, H. M. Seidel, J. A. Haslam, D. Chow, T. Tuntland, T. A. Spalding, A. Brock, M. Bradley, A. Castrillo, P. Tontonoz and E. Saez, *J. Med. Chem.*, 2007, **50**, 4255–4259.

[39] J. Wrobel, R. J. Steffan, E. Matelan, S. M. Bowen, B. Hu, M. Collini, C. P. Miller, R. J. Unwalla, P. Nambi, E. Quinet, L. Chen, A. Halpern, Q.-Y. Liu, D. Savio, E. Zamaratsky, L. Kruger, A. Wilhelmsson, A. G. Nilsson, C. Ursu, E. Arnelof, J. Sandberg, C. Enroth, T. Bonn and M. Farnegardh, *Abstracts of Papers, 234th ACS National Meeting*, Boston, MA, United States, August 19–23, 2007, MEDI-468.

[40] Karo Bio Press, Release 20th September 2007, http://www.karobio.com/en/IR–Media/Press-Releases/572/

CHAPTER 8

Glucagon Receptor Antagonists for Type II Diabetes

Duane E. DeMong, Michael W. Miller and **Jean E. Lachowicz**

1. INTRODUCTION

While the incidence of Type 2 Diabetes Mellitus (T2DM) in the United States continues to rise, the worldwide incidence is growing at an even higher rate, particularly in developing nations. Estimates project rise from 171 million diagnosed cases of diabetes in 2000 to 366 million diagnosed cases in 2030 [1]. The majority of these cases is T2DM, reflecting unfavorable lifestyle parameters such as increased caloric intake and reduced physical activity. Numerous acute and chronic complications such as cardiovascular disease, renal failure, neuropathy, and retinal damage arise from excessively high plasma glucose levels. Chronic elevation of plasma glucose results in a concomitant increase in glycosylated hemoglobin (HbA1c) levels, which are easily measured and routinely used to diagnose T2DM.

Drug discovery efforts addressing targets to treat T2DM have been focused largely on augmenting insulin secretion and insulin sensitivity. As goals for controlling HbA1c levels are often unmet with monotherapy T2DM treatments,

Schering-Plough Research Institute, 2015 Galloping Hill Rd, Kenilworth, NJ 07033, USA

Annual Reports in Medicinal Chemistry, Volume 43
ISSN 0065-7743, DOI 10.1016/S0065-7743(08)00008-0

new molecular entities (NMEs) that interact with different pathways and can be combined with insulin-mimetics or sensitizers are highly desired. Another pathway involved in glucose homeostasis is that of hepatic glucose output. A key regulator of both glucose production, i.e. gluconeogenesis, and glycogen catabolism to glucose, i.e. glycogenolysis, is the hormone glucagon. In the fasting state, hypoglycemia signals the pancreas to release glucagon, resulting in glucose production and delivery to plasma.

Glucagon is a 29 amino acid peptide derived from cleavage of proglucagon, a reaction that occurs in alpha cells of pancreatic islets. The peptide is used therapeutically to increase blood glucose levels in situations where they are dangerously low. The effects of glucagon are mediated through the glucagon receptor, a G-protein coupled receptor in the Class B subfamily, with a large N-terminus that is required for endogenous agonist binding. Stimulation of the receptor leads to increased cyclic adenosine monophosphate (cAMP) production in tissues including liver and kidney [2].

In healthy individuals, hypoglycemia during fasting is prevented by increasing glucagon levels, an effect which is reversed following a glucose load. However, in T2DM patients, the glucagon increase is not suppressed by a glucose load, as insulin resistance reduces hepatic glucose uptake and hepatic recognition of the hyperglycemic state. As a result, glycogenolysis continues despite elevated plasma glucose levels. This effect is not limited to advanced disease. Recently, increased fasting and glucose-stimulated glucagon levels were shown to be elevated in humans classified as IFG (impaired fasting glucose) and IGT (impaired glucose tolerance), both prediabetic states [3]. This observation suggests that unchecked glucagon may play a role in progression to T2DM. While not the primary mechanism for glycemic control, glucagon regulation is a result of glucagon-like-peptide-1 (GLP-1) receptor stimulation. Patients dosed with the GLP-1 agonist Exendin-4 (BYETTA™) demonstrated a 20% reduction in preprandial plasma glucagon concentration and a 54% decrease in endogenous glucose production [4].

A more direct means of regulating unchecked glucagon levels in T2DM would be to administer an antagonist of the glucagon receptor. This effort has been challenging, although a number of successful NMEs with glucagon receptor antagonist activity have been identified [5]. One example in particular, BAY 27-9955 (+)-**1**, has been tested in healthy, non-diabetic humans using a pharmacodynamic assay of glucagon receptor antagonism [6]. A hypergluconemic state was induced by delivering glucagon (3 ng/kg/min) for 3 h. Somatostatin was also infused to inhibit release of endogenous glucagon and insulin, the latter being replaced by an insulin infusion. This treatment resulted in a doubling of glucagon levels. Plasma glucose levels following glucagon infusion rose from 5 mg/dL to 10, 9.3 and 7.6 mg/dL in placebo-, low dose- and high dose-treated subjects, respectively. Glucose production increased by over 100% in placebo-treated subjects following glucagon infusion. This increasing phase was followed by a diminution phase of glucose production so that, at 3 h, glucose production was approximately half of the baseline value. This second phase is due to the healthy individual response to hyperglycemia by decreasing glucose output. A 70 mg dose of (+)-**1** attenuated the rise in plasma glucose output to a 72% increase, but the hyperglycemic phase was still adequate to trigger the second phase lowering of glucose output to well below basal levels. In contrast, the increase in glucose output following the 200 mg dose (25%) was sufficiently flattened so as to obliterate the phase of declining glucose production. These results showed that glucagon receptor antagonist treatment reduces plasma glucose concentration by inhibiting the increase in hepatic glucose production mediated by glucagon. As T2DM patients have an impaired response to hyperglycemia, this treatment is expected to enable improved glycemic control in T2DM.

2. GLUCAGON RECEPTOR ANTAGONISTS

2.1 Biaryl antagonists

Hindered biaryl derivatives have been investigated as glucagon receptor antagonists as exemplified by (+)-1 [7,8]. It has been reported that this compound competitively inhibits the binding of glucagon to the human glucagon receptor (hGCGR IC_{50} = 110 nM) [6]. Receptor occupancy studies in a transgenic mouse model with (+)-1 indicate that a minimum of 50–60% coverage is needed to effectively block hyperglycemia induced by exogenous glucagon [9].

Several reports in this area focused on structure–activity relationships around the analogous pyridyl analogs. The pyridyl analog 2 was identified through high-throughput screening (HTS) and showed modest binding and functional activity (hGCGR binding IC_{50} = 7 µM; hGCGR cAMP IC_{50} = 2 µM). Further optimization of the pyridyl ring substituents provided the alcohol 3 (hGCGR binding IC_{50} = 110 nM; hGCGR cAMP IC_{50} = 65 nM) [10]. The configuration of the hydroxyl stereocenter present in 3 was found to be important as the (R) isomer 3 was approximately two-fold more potent than the (S) isomer. Introduction of a 2'-hydroxyl substituent on the phenyl ring provided analogs with high affinity for the glucagon receptor (4, hGCGR IC_{50} = 16 nM) [11,12]. It should be noted that the introduction of the 2'-hydroxy substituent in this system gave rise to atropisomers due to restricted rotation about the pyridyl-phenyl ring bond.

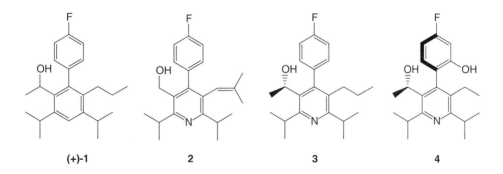

(+)-1   2   3   4

2.2 β-Alanine- and aminotetrazole-terminated antagonists

Recently, efforts to identify small molecule antagonists for the glucagon receptor have focused on compounds containing a C-terminal β-alanine group, or isosteres thereof. Compound 5 was initially prepared as part of a directed compound library aimed at mimicking the binding interactions of GLP-1 with its receptor [13]. While **5** was found to be a poor ligand for the GLP-1 receptor ($IC_{50} = 100 \, \mu M$), a high-throughput screening campaign identified it as a weak antagonist of the human glucagon receptor (hGCGR $IC_{50} = 7 \, \mu M$). A follow-up compound library identified **6** as a potent glucagon receptor antagonist (hGCGR $IC_{50} = 55 \, nM$, ratGCGR $IC_{50} = 56 \, nM$) with a modest pharmacokinetic (PK) profile (dog: $F_{po} = 10$–20%; i.v. $t_{1/2} = 101 \, min$, rat: $F_{po} = 17\%$; i.v. $t_{1/2} = 90 \, min$). In a glucagon challenge assay in Sprague-Dawley rats, urea **6** inhibited the glucagon-stimulated rise in blood glucose at a dose of 3 mg/kg i.v. Optimization of structure **6** resulted in the discovery of cyclohexenyl aniline urea **7** (hGCGR binding $IC_{50} = 27 \, nM$, ratGCGR binding $IC_{50} = 122 \, nM$). The oral bioavailability of **7** was significantly improved over that of **6** (dog: $F_{po} = 65\%$; $t_{1/2} = 92 \, min$, rat: $F_{po} = 69\%$; i.v. $t_{1/2} = 82 \, min$). In the glucagon challenge assay, compound **7** exhibited a statistically significant reduction of blood glucose at the dose of 10 mg/kg i.v., but not at 3 mg/kg i.v. Compound **7** was also profiled in the leptin deficient *ob/ob* mouse model. At the dose of 100 mg/kg *po*, a significant decrease in blood glucose concentration was observed.

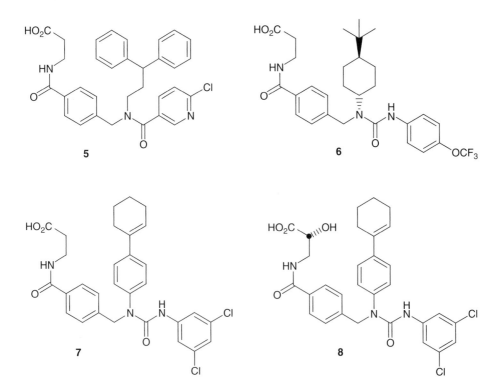

5   6

7   8

Further optimization of **7** afforded the hydroxy-β-alanine analog **8** (NNC 25-0926; hGCGR binding $IC_{50} = 12 \, nM$) [14]. After chronic administration of **8** at the dose of 300 mg/kg *via* gavage to mice with high fat diet-induced insulin resistance, both an intravenous glucose tolerance test (IVGTT) and oral glucose tolerance test (OGTT) were performed. The animals treated with compound **8** demonstrated a statistically significant reduction in fasting plasma glucose increase versus the vehicle-treated group. The antagonist-treated group also showed an increase in glucose-stimulated insulin secretion (GSIS) in the IVGTT study. Treatment with **8** also resulted in an acute increase in glucagon secretion upon administration of arginine. An increase in islet alpha cell mass was also observed.

Compound **8** was also shown to inhibit hepatic glycogenolysis at the dose of 20 mg/kg (intragastric infusion) in a conscious dog hepatic glucose production model [15]. These results correlated with the plasma levels of **8** when sampled from the hepatic portal vein.

Upon further investigation of the importance of the β-alanine moiety for *in vitro* potency, the amidotetrazole group has been confirmed as a viable β-alanine bio-isostere [16,17]. For example, the benzoylaminotetrazole **9**, and the phenethyl tetrazole **10** which lacks the benzamide moiety have been described as glucagon receptor antagonists. Fluorination of the β-alanine residue, as exemplified by compounds such as **11**, has also been shown to afford potent glucagon receptor antagonists [18].

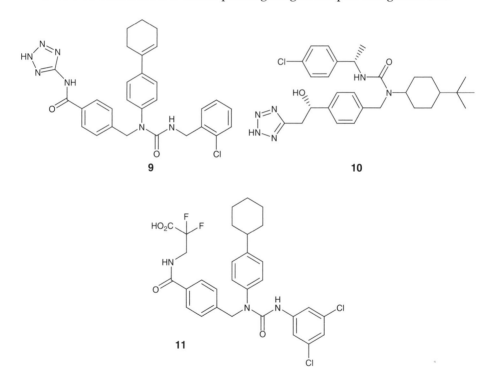

As illustrated above, the utilization of urea and related functionality as a linchpin for appending key structural motifs is a common theme in the design of glucagon receptor antagonists. This has been further exemplified by the indanyl urea analogs **12** and **13**, which displayed potent binding affinity and robust functional activity (**12**: hGCGR $IC_{50} = 4.1$ nM; hGCGR cAMP $IC_{50} = 33$ nM.

13: hGCGR IC$_{50}$ = 2.3 nM; hGCGR cAMP IC$_{50}$ = 15 nM) [19,20]. In this study, the binding affinities of the more potent enantiomers were 0.5–5 nM while the less potent enantiomers were 50–800-fold less potent. The absolute stereochemistry of the indanyl ring substitution was not disclosed. The β-alanine analog **12** provided blood glucose correction in transgenic mice that displayed moderate hyperglycemia when dosed at 3 mg/kg as an admixture with chow. The closely related spiro-urea analogs **14** and **15** displayed lower binding affinity and were less potent in a functional assay (**14**: hGCGR IC$_{50}$ = 182 nM; hGCGR cAMP IC$_{50}$ = 282 nM. **15**: hGCGR IC$_{50}$ = 34 nM; hGCGR cAMP IC$_{50}$ = 92 nM) [21,22]. Nevertheless, the aminotetrazole **15** effectively blunted blood glucose levels when orally administered at the dose of 10 mg/kg to transgenic mice expressing the human glucagon receptor. It was also noted that the *in vivo* glucagon receptor occupancy at 1 h post dose in the 10 mg/kg arm was 52 %.

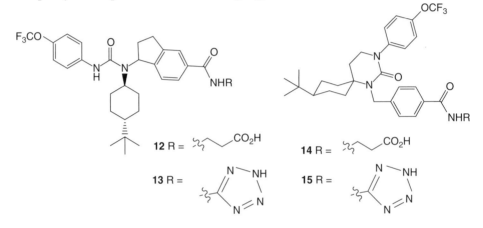

Related analogs that possess a constrained central core are exemplified by the imidazolone **16**, cyclic urea **17** and iminooxazolidine **18** [23,24]. While no specific *in vitro* data was provided for **16**, preferred compounds possess an hGCGR binding IC$_{50}$ of 100 nM or less. Compounds **17** and **18** were reported to inhibit glucagon-stimulated cAMP production at levels between 50% and 100% at 20 μM.

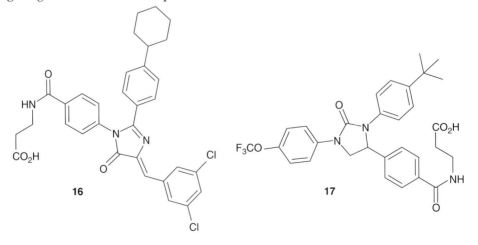

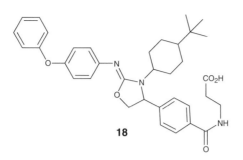

18

Related amino-heterocycles have also been utilised as core structural motifs. The aminothiazole **19** has recently been detailed as β-alanine-based glucagon receptor antagonist [25]. The benzimidazole analog **20** exhibited robust activity in a functional cAMP assay (hGCGR cAMP $IC_{50} = 58\,nM$) [26,27]. Although the corresponding IC_{50} of **20** in CHO cells expressing the dog glucagon receptor was decreased by 8-fold, the compound was effective in suppressing hyperglycemia in dogs elicited by exogenously administered glucagon. Related guanidine analogs such as **21** have appeared in the patent literature [28].

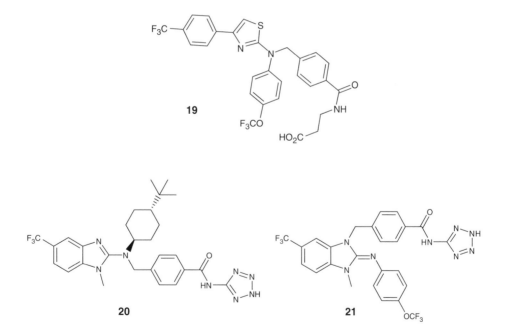

Butanedione benzamides **22**, dibenzylamino benzamides **23**, N-benzyl carbamoylbenzamides **24**, and N-benzyl ureidobenzamides **25** have all been described as viable core structures for the effective display of the β-alanine residue and its bioisosteres [29]. The compounds from these series were reported to possess a wide range of hGCGR binding affinities, with the most potent compounds displaying hGCGR binding IC$_{50}$'s of <100 nM. Additionally, a guanidine core exemplified by compound **26** has been reported [30].

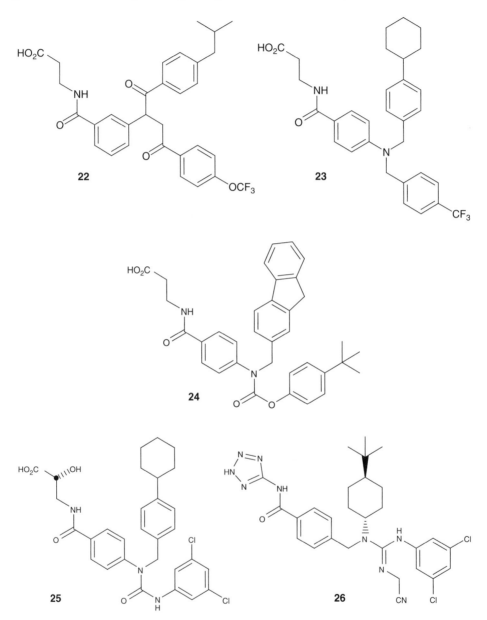

Patent disclosures of related analogs possessing an alkyl and/or aminoalkyl linker that append the benzoyl β-alanine moiety with aryl and alkyl functionality have been published. The alkyl- and amino- tethered analogs **27** and **28** have been reported to display good binding affinity with that of **28** being 6-7 fold more potent (**27**: hGCGR binding $K_i = 266$ nM; **28** hGCGR binding $K_i = 40$ nM) [31,32]. The absolute stereochemistry of these analogs was not defined. Bicyclic aryl and heteroaryl analogs such as **29** and **30** that are linked *via* a methine to the benzoyl β-alanine group have also been disclosed with moderate to high binding affinities (hGCGR binding IC$_{50}$ = 1–500 nM) [33]. Other bicyclic groups that have been incorporated into glucagon receptor antagonists related to **29** and **30** include the benzimidazole and benzoxazole moieties.

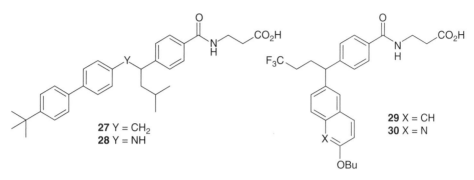

27 Y = CH$_2$
28 Y = NH

29 X = CH
30 X = N

Various ether-linked analogs possessing a β-alanine- or an aminotetrazole-terminated benzoyl group have been reported in the patent literature. The (pyridylmethoxy)benzamide derivative **31** displayed moderate binding affinity (hGCGR K_i = 801 nM) [34]. In contrast, the closely related (pyridyloxy)methylbenzamide **32** displayed more potent binding affinity (hGCGR K_i = 254 nM) [35]. Related thiophene-amido analogs have been disclosed that display similar binding affinity (**33**: hGCGR K_i = 390 nM) [36]. Further extension of the linker provided **34** (hGCGR K_i = 61 nM) with improved binding affinity [37]. The absolute stereochemistry of these analogs was not defined.

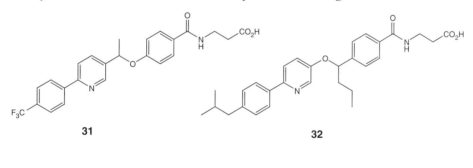

31

32

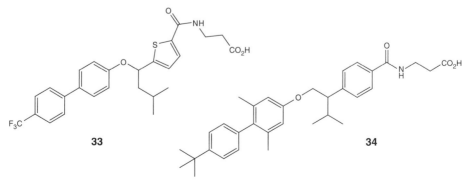

33

34

Acylindole analogs such as **35** have been described as glucagon receptor antagonists [38]. Analogs within this series display moderate to good oral bioavailability in rat and dog ($F\% = 40$–60), and they exhibit high affinity for the human glucagon receptor (hGCGR $IC_{50} = 10$–30 nM [39]. Closely related analogs exemplified by **36** have been described that display a range of binding affinities (hGCGR $IC_{50} = 1$–500 nM) [40]. All possible isomers are claimed in this application, but no preferred relative or absolute stereochemistry was disclosed. Amido analogs, exemplified by **(+)-37** and **(+)-38**, have been reported to exhibit high binding affinities with moderate potency in a functional assay ((+)-37: hGCGR binding $K_i = 11$ nM, hGCGR cAMP $K_i = 300$ nM; **(+)-38**: hGCGR binding $K_i = 9$ nM, hGCGR cAMP $K_i = 144$ nM) [41].

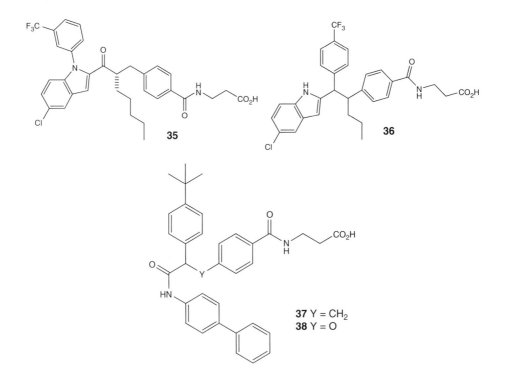

A series of pyridazine glucagon receptor antagonists such as **39** has been disclosed in the patent literature [23]. While no specific *in vitro* potency data was disclosed for any of the compounds described, the most preferred compounds have an hGCGR binding IC_{50} of 100 nM or less. Compounds containing a carbon-based core ring system for appendage of the benzoyl β-alanine group and other aromatic residues have also appeared in the patent literature. Aryl core analogs, such as **40** and **41**, have been described in which the linkage is *via* a methylene or an ether unit [42]. Compounds in which the methylene linker is constrained in a spirocycloalkyl moiety are exemplified by the spirocyclohexane **42** and the norbornane **43** [43]. The binding potencies (IC_{50}s) for these two classes of compounds range from 1 to 500 nM.

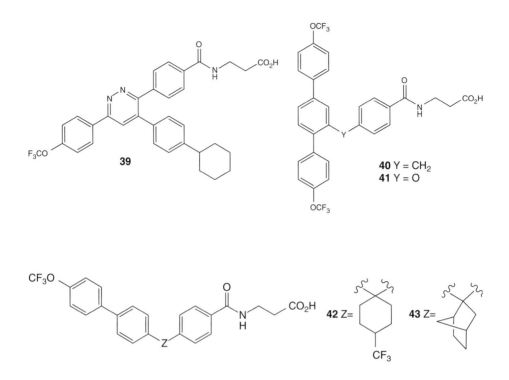

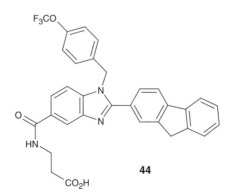

Heterocyclic ring systems utilizing a benzimidazole core motif have also been disclosed. Initial reports described substituted benzimidazoles such as **44** in which a β-alanine moiety is connected *via* an amido group to the 5-position of the benzimidazole core [44]. An *N*-dehydroabietylbenzimidazole-based glucagon receptor antagonist **45**, which incorporates the β-alanine moiety *via* a benzoyl linking group at the benzimidazole 2-position, has been reported [45]. While no *in vivo* or *in vitro* potency values were available, a variety of benzoyl β-alanine bioisosteres were disclosed including a 3-(carboxymethyloxy)-phenyl derivative **46**, a 3-(1H-tetrazol-5-ylmethylaminocarbonyl)phenyl deriva-tive **47** and a 4-(3-carboxypropyloxy-)-2-methoxyphenyl derivative **48**.

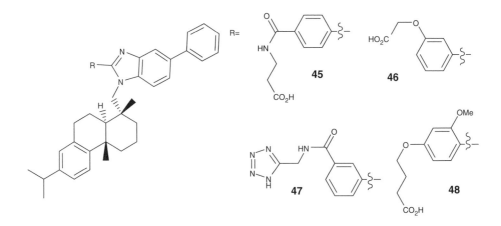

Over the last several years, numerous reports of pyrazole-based analogs have appeared in the patent literature. These analogs are characterized by the incorporation of three substituents around the pyrazole nucleus one of which is a β-alanine- or aminotetrazole-derived benzoyl group. Several applications center on pyrazoles **49** and **50** in which the pyrazole ring is substituted by an aryl and a fused bicyclic aryl ring system [46]. Phenyl pyrazoles possessing alkoxy- and trifluoromethyl-phenyl substitution exemplified by **49** were reported to exhibit potent binding and functional activity (hGCGR cAMP IC_{50} = 5–50 nM) [47]. The corresponding chloro-subsituted derivatives such as **50** were reported to exhibit moderate to high binding affinities (hGCGR IC_{50} = 1–500 nM) [48]. Related pyrazoles substituted by a bicyclic heteroaryl group, such as **51**, have been disclosed [49]. In a recent patent application, diphenyl-substituted pyrazoles, exemplified by **52**, and cycloheteroalkyl-substituted pyrazoles such as **53** were described [50]. No specific activity was disclosed in the applications describing analogs encompassing **51**, **52**, and **53**. Lastly, pyrazole carboxamides such as **54** were disclosed, which were reported to exhibit moderate to high binding affinity for the human glucagon receptor (hGCGR IC_{50} = 1–500 nM) [51].

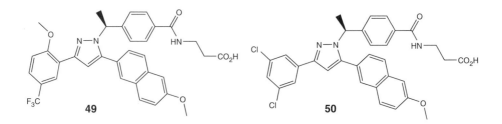

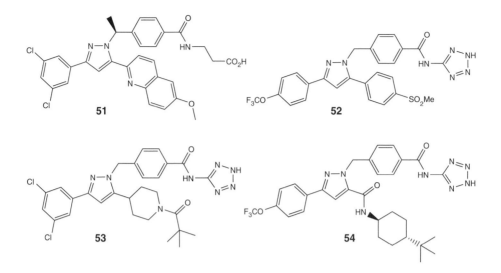

2.3 Non-carboxylic acid-based antagonists

Quinoxaline **55** is a glucagon receptor antagonist with binding affinity in the micromolar range (hGCGR $IC_{50} = 1\,\mu M$) and was the first non-peptidic glucagon receptor antagonist reported in the literature [52]. The development of a series of pyrrolo[1,2-a]quinoxaline glucagon receptor antagonists exemplified by compound **56** followed this initial report [53]. Like quinoxaline **55**, compound **56** exhibited micromolar binding affinity for the receptor (hGCGR $IC_{50} = 5\,\mu M$).

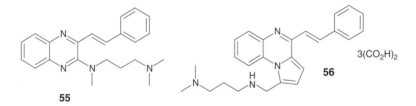

The fungal metabolite skyrin **57** and its close analog oxyskyrin **58** have been shown to block the glucagon-stimulated production of cAMP in rat hepatocytes (for **57**, 50% @ 30 μM), primary cultures of human hepatocytes (>50% @ 10 μM), and CHO cells transfected with hGCGR (for **57**, 30% @ 30 μM) [54,55]. Interestingly, it was observed that **57** (at 50 μM) had no effect on the ability of unlabelled glucagon to displace [125I]-glucagon from membranes prepared from CHO cells transfected with hGCGR. Additionally, **57** (at 30 μM) had no effect on the ability of [125I]-glucagon to bind to these same membranes. While the effects of **57** on cAMP production occur by a mechanism that is not understood, it does not appear to indiscriminately inhibit cAMP production in all cell lines as evidenced by the fact that **57** has no effect on cAMP production in CHO cells transfected with the GLP-1 receptor.

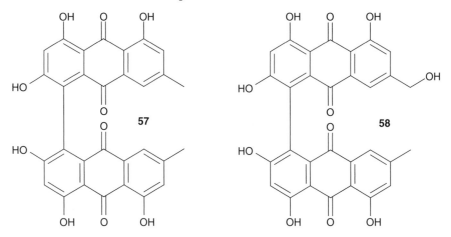

A series of unique mercaptobenzimidazoles exemplified by compound **59** has also been described as glucagon receptor antagonists [56]. Additionally, a series of alkylidene hydrazides such as compound **60** has been prepared [57]. This compound was determined to be a true non-competitive antagonist of the rat glucagon receptor, but a mixed competitive and non-competitive antagonist of the human glucagon receptor with an IC_{50} in the single-digit nanomolar range. Further characterization of **60** demonstrated its potency in a rat glucagon challenge model when administered intravenously. A structurally similar series of furanyl hydrazides such as **61** has been described [58]. Bioisosteric replacement of the hydrazide with an aminooxadiazole such as **62** (GW4123X) has also been reported [59]. This compound has been described as a functional antagonist with an IC_{50} in the 2–30 nanomolar range. Additionally, in rats, compound **62** has been shown to decrease fasting plasma glucose levels and to inhibit glucagon-stimulated increase of plasma glucose concentration at the dose of 3 mg/kg *i.v.*

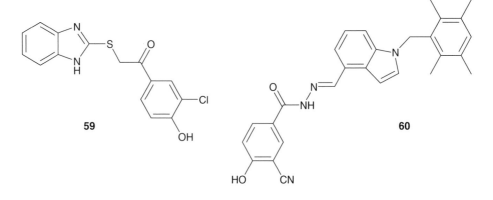

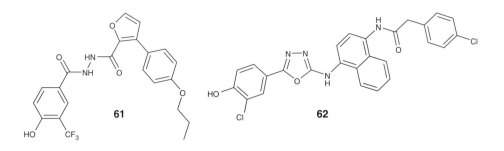

61   **62**

Disubstituted pyridyl imidazoles and disubstituted pyridyl pyrroles (known p38 MAP kinase inhibitors) have been identified as glucagon receptor antagonists. Pyridyl imidazole **63**, was optimized for improved glucagon receptor affinity (hGCGR $IC_{50} = 6.5$ nM) and reduced p38 inhibition [60]. Compound **64** has been reported to be a potent glucagon receptor antagonist (hGCGR $IC_{50} = 7$ nM) with oral bioavailability in both rats and mice [61,62]. In the course of these investigations, it was discovered that there was a significant decrease in glucagon receptor binding affinity of compounds **63** and **64** when the receptor binding assay was performed in the presence of physiological concentrations of Mg^{2+} (hGCGR IC_{50} (5 mM $MgCl_2$ present) **63** = 53 nM, **64** = 170 nM), although the mechanism by which Mg^{2+} affects potency in these assays was not elucidated. Additional functional and competitive binding studies suggested that **64** is a non-competitive antagonist of the glucagon receptor [63]. Site-directed mutagenesis of selected residues in the transmembrane region of hGCGR yielded two mutants with reduced affinity for compound **64** while maintaining their affinity for glucagon, suggesting that **64** binds in the transmembrane domain.

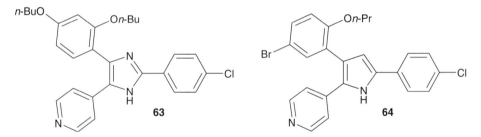

63   **64**

Bicyclic cyanothiophene **65**, of undefined stereochemistry, has been reported as a competitive antagonist of the glucagon receptor (hGCGR binding $IC_{50} = 181$ nM, hGCGR cAMP $IC_{50} = 129$ nM) [64]. Further characterization of **65** demonstrated its effectiveness when administered at 50 mg/kg intraperitone- ally in blunting a glucagon challenge in an hGCGR-containing transgenic mouse model. Lead optimization studies performed on **65** probing the necessity of the fused cyclohexylthiophene resulted in the discovery of the oxadiazole **66** [65]. The *in vitro* binding and functional potency of **66** (hGCGR binding $IC_{50} = 89$ nM, hGCGR cAMP $IC_{50} = 34$ nM) was somewhat improved over **65** while the mouse oral PK profile was significantly improved. Compounds **65** and **66** were both profiled in the hGCGR-containing transgenic mouse glucagon challenge model at

100 mg/kg *po*. No reduction in glycogenolysis was observed with oral dosing of these compounds. The observation that there was a 15-fold decrease in the potency of **66** in the hGCGR cAMP assay when performed in the presence of 5% mouse plasma has been offered as a possible explanation for the lack of *in vivo* efficacy. Structurally similar compounds **67** and **68** have also been disclosed as glucagon receptor antagonists [66,67].

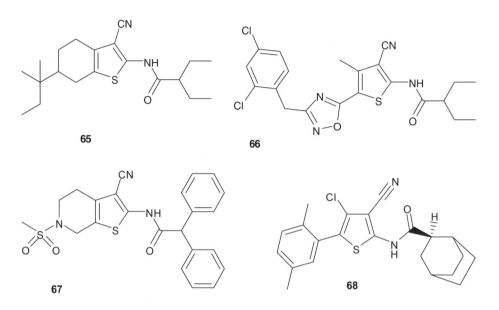

3. CONCLUSION

As the prevalence of T2DM increases at an exponential rate, new agents for the control of elevated glucose and/or increased glucose tolerance are needed. Glucagon receptor antagonists have shown promise in this regard as they act to reduce hepatic glucose output and breakdown of glycogen. Over the last several years, different chemotypes have appeared in the peer-reviewed and patent literature, with most of the recent efforts focused on structures containing a β-alanine amide and bioisosteres thereof. The ability of these agents to modulate hepatic glucose output in diabetic individuals remains untested, but hopefully, efforts in this area will identify one or more candidates to test this hypothesis.

REFERENCES

[1] S. Wild, G. Roglic, A. Green, R. Sicree and H. King, *Diabetes Care*, 2004, **27**, 1047.
[2] For a recent review see F. Authier and B. Desbuquois, *Cell. Mol. Life Sci.*, 2008, **65**, 1880.

[3] K. Faerch, A. Vaag, J. J. Holst, C. Glümer, O. Pederson and K. Borch-Johnsen, *Diabetologia*, 2008, **51**, 853.

[4] A. Cervera, E. Wajcberg, A. Sriwijitkamol, M. Fernandez, P. Zuo, C. Triplitt, N. Musi, R. A. DeFronzo and E. Cersosimo, *Am. J. Physiol. Endocrinol. Metab.*, 2008, **294**, E846.

[5] For recent reviews targeting the incretin system for the treatment of T2DM see: (a) A. H. Stonehouse and D. G. Maggs, *Curr. Drug Ther.*, 2007, **2**, 151. (b) J. L. Estall and D. J. Drucker, *Curr. Pharm. Des.*, 2006, **12**, 1731. (c) S. W. Djuric, N. Grihalde and C. W. Lin, *Curr. Opin. Invest. Drugs*, 2002, **3**, 1617.

[6] K. F. Peterson and J. T. Sullivan, *Diabetologia*, 2001, **44**, 2018.

[7] G. Schmidt, R. Angerbauer, A. Brandes, M. Muller-Gliemann, H. Bischoff, D. Schmidt, S. Wohlfeil, W. R. Schoen, G. Ladoucheur, J. H. Cook, T. G. Lease and D. J. Wolanin, *WO Patent* 04528, 1998.

[8] W. R. Schoen, G. H. Ladouceur, J. H. Cook, T. G. Lease, D. J. Wolanin, R. H. Kramss, D. L. Hertzog and M. H. Osterhout, *US Patent* 6,218, 431, 2001.

[9] Q. Dallas-Yang, X. Shen, M. Strowski, E. Brady, R. Saperstein, R. E. Gibson, D. Szalkowski, S. A. Qureshi, M. R. Candelore, J. E. Fenyk-Melody, E. R. Parmee, B. B. Zhang and G. Jiang, *Eur. J. Pharm.*, 2004, **501**, 225.

[10] G. H. Ladouceur, J. H. Cook, E. M. Doherty, W. R. Shoen, M. L. MacDougall and J. N. Livingston, *Bioorg. Med. Chem. Lett.*, 2002, **12**, 461.

[11] R. A. Smith, D. L. Hertzog, M. H. Osterhout, G. H. Ladouceur, M. Korpusik, M. A. Bobko, J. H. Jones, K. Phelan, R. H. Romero, T. Hundertmark, M. L. MacDougall, J. N. Livingston and W. R. Schoen, *Bioorg. Med. Chem. Lett.*, 2002, **12**, 1303.

[12] G. H. Ladouceur, J. H. Cook, D. L. Hertzog, J. H. Jones, T. Hundertmark, M. Korpusik, T. G. Lease, J. N. Livingston, M. L. MacDougall, M. H. Osterhout, K. Phelan, R. H. Romero, W. R. Schoen, C. Shao and R. A. Smith, *Bioorg. Med. Chem. Lett.*, 2002, **12**, 3421.

[13] J. Lau, C. Behrens, U. G. Sidelmann, L. B. Knudsen, B. Lundt, C. Sams, L. Ynddal, C. L. Brand, L. Pridal, A. Ling, D. Kiel, M. Plewe, S. Shi and P. Madsen, *J. Med. Chem.*, 2007, **50**, 113.

[14] M. S. Winzell, C. L. Brand, N. Wierup, U. G. Sidelmann, F. Sundler, E. Nishimura and B. Ahren, *Diabetologia*, 2007, **50**, 1453.

[15] N. Rivera, C. A. Everett-Grueter, D. S. Edgerton, T. Rodewald, D. W. Neal, E. Nishimura, M. O. Larsen, L. O. Jacobsen, K. Kristensen, C. L. Brand and A. D. Cherrington, *J. Pharm. Exp. Ther.*, 2007, **321**, 743.

[16] C. Behrens, J. Lau and P. Madsen, *US Patent App.* 0203946 A1, 2003.

[17] A. Ling, M. B. Plewe, L. K. Truesdale, J. Lau, P. Madsen, C. Sams, C. Behrens, J. Vagner, I. T. Christensen, B. F. Lundt, U. G. Sidelmann and H. Thøgersen, *WO Patent* 069810 A1, 2000.

[18] A. S. Jørgensen and P. Madsen, *WO Patent* 040446 A1, 2002.

[19] R. Liang, L. Abrardo, E. J. Brady, M. R. Candelore, V. Ding, R. Saperstein, L. M. Tota, M. Wright, S. Mock, C. Tamvakopolous, S. Tong, S. Zheng, B. B. Zhang, J. R. Tata and E. R. Parmee, *Bioorg. Med. Chem. Lett.*, 2007, **17**, 587.

[20] R. Liang and E. R. Parmee, *WO Patent* 104826, 2006.

[21] D.-M. Shen, F. Zhang, E. J. Brady, M. R. Candelore, Q. Dallas-Yang, V. D.-H. Ding, J. Dragovic, W. P. Feeney, G. Jiang, P. E. McCann, S. Mock, S. A. Qureshi, R. Saperstein, X. Shen, C. Tamvakopoulos, X. Tong, L. M. Tota, M. J. Wright, X. Yang, S. Zheng, K. T. Chapman, B. B. Zhang, J. R. Tata and E. R. Parmee, *Bioorg. Med. Chem. Lett.*, 2005, **15**, 4564.

[22] E.M. Parmee, F. Zhang, D.-M. Shen and J. Stelmach, *WO Patent* 050039, 2004.

[23] P. Madsen, J. Lau, J. T. Kodra and I. T. Christensen, *WO Patent* 058845 A2, 2005.

[24] R. Kurukulasuriya, J. T, Link, J. R. Patel and B. K. Sorensen, *US Patent App.* 0209928, 2004.

[25] J. Lau, I. T. Christensen, P. Madsen, P. Bloch, C. Behrens, J. K. Kodra and P. E. Nielsen, *WO Patent* 002480 A1, 2004.

[26] X. Yang, M. L. Yates, M. R. Candelore, W. Feeney, D. Hora, R. M. Kim, E. R. Parmee, J. P. Berger, B. B. Zhang and S. A. Qureshi, *Eur. J. Pharm.*, 2007, **555**, 8.

[27] (a) E. R. Parmee, R. M. Kim, R. Liang, J. Chang, E. A. Rouse and K. T. Chapman, *WO Patent* 100875, 2004. (b) R. M. Kim, J. Chiang, A. R. Lins, E. Brady, M. R. Candelore, Q. Dallas-Yang, V. Ding, J. Dragovic, S. Iliff, G. Jiang, S. Mock, S. Qureshi, R. Saperstein, D. Szalkowski,

C. Tamvakopoulos, L. Tota, M. Wright, X. Yang, J. R. Tata, K. Chapman, B. B. Zhang, E. R. Parmee, *Bioorg. Med. Chem. Lett.*, 2008, **18**, 3701.

[28] E. R. Parmee, R. M. Kim, E. A. Rouse, D. R. Schimdt, C. J. Sinz and J. Chang, *WO Patent* 065680, 2005.

[29] J. T. Kodra, C. Behrens, P. Madsen, A. S. Jørgensen and I. T. Christensen, *WO Patent* 056763 A2, 2004.

[30] P. Madsen and C. Behrens, *WO Patent* 051357 A1, 2003.

[31] J. Li and G. Zhu, *WO Patent* 123581, 2007.

[32] S. E. Conner, J. Li and G. Zhu, *WO Patent* 106181, 2007.

[33] R. M. Kim, E. R. Parmee, Q. Tan, C. Yang and A. R. Lins, *WO Patent* 136577, 2007.

[34] M. D. Chappell, S. E. Conner, I. C. Gonzalez Valcercel, J. E. Lamar, J. Li, J. S. Moyers, R. A. Owens, A. E. Tripp and G. Zhu, *WO Patent* 118542, 2005.

[35] (a) S. E. Conner, G. Zhu and J. Li, *WO Patent* 123668, 2005. (b) S. E. Conner, P. A. Hipskind, J. Li, G. Zhu, *WO Patent* 120270, 2007.

[36] M. D. Chappell, S. E. Conner, A. E. Tripp and G. Zhu, *WO Patent* 086488, 2006.

[37] M. D. Chappell, S. E. Conner, P. A. Hipskind, J. E. Lamar and G. Zhu, *WO Patent* 114855, 2007.

[38] R. M. Kim, A. R. Bittner, C. J. Sinz and E. R. Parmee, *US Patent* App. 0088071, 2007.

[39] C. J. Sinz, A. Bittner, R. M. Kim, E. Brady, M. R. Candelore, V. D.-H. Ding, G. Jiang, Z. Lin, A. R. Lins, P. McCann, C. Miller, K. Nam, S. A. Qureshi, F. Salituro, R. Saperstein, J. Shang, D. Szalkowski, L. Tota, M. Wright, R. Wang, S. Xu, X. Yang, B. Zhang, M. Hammond, J. Tata and E. Parmee, *Abstracts of Papers, 235th ACS National Meeting*, New Orleans, LA, April, 2008, MEDI 16.

[40] J. E. Stelmach, E. R. Parmee, J. R. Tata, K. G. Rosauer, R. M. Kim and A. R. Bittner, *WO Patent* 042223, 2008.

[41] R. Kurukulasuriya, B. K. Sorensen, J. T. Link, J. R. Patel, H.-S. Jae, M. X. Winn, J. R. Rohde, N. D. Grihalde, C. W. Lin, C. A. Ogiela, A. L. Adler and C. A. Collins, *Bioorg. Med. Chem. Lett.*, 2004, **14**, 2047.

[42] J. E. Stelmach, K. G. Rosauer, E. R. Parmee and J. R. Tata, *WO Patent* 102067, 2006.

[43] R. M. Kim, E. R. Parmee, Q. Tang, A. R. Lins, J. Chang and C. Yang, *WO Patent* 111864, 2007.

[44] J. Lau, I. T. Christensen, P. Madsen and C. Behrens, *WO Patent* 053938 A1, 2003.

[45] R. Streicher, J. Mack, R. Walter, I. Konetzki, T. Trieselmann and V. Austel, *US Patent* 7,151,114 B2, 2006.

[46] E. R. Parmee, Y. Xiong, J. Guo, R. Liang and Lo. Brockunier, *WO Patent* 121097, 2005.

[47] E. R. Parmee, Y. Xiaong, J. Guo and L. Brockunier, *US Patent App.* 0088070, 2007.

[48] E. R. Parmee, Y. Xiong, J. Guo, R. Liang and L. Brockunier, *US Patent App.* 0272794, 2005.

[49] L. Brockunier, J. Guo, R. Liang, E. R. Parmee, S. Raghavan, G. Tria and Y. Xiong, *WO Patent* 014618, 2006.

[50] E. R. Parmee, S. Raghavan, T. Beeson and D.-M. Shen, *WO Patent* 069158, 2004.

[51] T. Beeson, L. Brockunier, E. R. Parmee and S. Raghavan, *WO Patent* 017055, 2006.

[52] J. L. Collins, P. J. Dambek, S. W. Goldstein and W. S. Faraci, *Bioorg. Med. Chem. Lett.*, 1992, **2**, 915.

[53] J. Guillon, P. Dallemagne, B. Pfeiffer, P. Renard, D. Manechez, A. Kervran and S. Rault, *Eur. J. Med. Chem.*, 1998, **33**, 293.

[54] R. R. West, V. Labroo, J. R. Piggott, R. A. Smith and P. A. McKernan, *WO Patent* 14427 A2, 1994.

[55] J. C. Parker, R. K. McPherson, K. M. Andrews, C. B. Levy, J. S. Dubins, J. E. Chin, P. V. Perry, B. Hulin, D. A. Perry, T. Inagaki, K. A. Dekker, K. Tachikawa, Y. Sugie and J. L. Treadway, *Diabetes*, 2000, **49**, 2079.

[56] P. Madsen, L. B. Knudsen, F. C. Wiberg and R. D. Carr, *J. Med. Chem.*, 1998, **41**, 5150.

[57] P. Madsen, A. Ling, M. Plewe, C. K. Sams, L. B. Knudsen, U. G. Sidelmann, L. Ynddal, C. L. Brand, B. Andersen, D. Murphy, M. Teng, L. Truesdale, D. Kiel, J. May, A. Kuki, S. Shi, M. D. Johnson, K. A. Teston, J. Feng, J. Lakis, K. Anderes, V. Gregor and J. Lau, *J. Med. Chem.*, 2002, **45**, 5755.

[58] A. Fujii, T. Negoro, C. Migihashi, M. Murata, K. Nakamura, T. Nukuda, T. Matsumoto and K. Konno, *WO Patent* 064404, 2003.

[59] A. L. Handlon, A. Akwabi-ameyaw, K. Brown, F. De Anda, D. Drewry, J. Fang, O. Irsula, G. Li, J. A. Linn, N. O. Milliken, J. Ramanjulu, Glucagon receptor antagonists for the treatment of type 2

diabetes. *Abstracts of Papers, 226th ACS National Meeting*, New York, United States, September 7–11, 2003, MEDI-164.

[60] L. L. Chang, K. L. Sidler, M. A. Cascieri, S. de Laszlo, G. Koch, B. Li, M. MacCoss, N. Mantlo, S. O'Keefe, M. Pang, A. Rolando and W. K. Hagmann, *Bioorg. Med. Chem. Lett.*, 2001, **11**, 2549.

[61] S. E. De Laszlo, L. L. Chang, D. Kim and N. B. Mantlo, *US Patent* 5776954, 1998.

[62] S. E. De Laszlo, C. Hacker, B. Li, D. Kim, M. MacCoss, N. Mantlo, J. V. Pivnichny, L. Colwell, G. E. Koch, M. A. Cascieri and W. K. Hagmann, *Bioorg. Med. Chem. Lett.*, 1999, **9**, 641.

[63] M. A. Cascieri, G. E. Koch, E. Ber, S. J. Sadowski, D. Louizides, S. E. de Laszlo, C. Hacker, W. K. Hagmann, M. MacCoss, G. G. Chicchi and P. P. Vicario, *J. Biol. Chem.*, 1999, **274**, 8694.

[64] S. A. Qureshi, M. R. Candelore, D. Xie, X. Yang, L. M. Tota, V. D.-H. Ding, Z. Li, A. Bansal, C. Miller, S. M. Cohen, G. Jiang, E. Brady, R. Saperstein, J. L. Duffy, J. R. Tata, K. T. Chapman, D. E. Moller and B. B. Zhang, *Diabetes*, 2004, **53**, 3267.

[65] J. L. Duffy, B. A. Kirk, Z. Konteatis, E. L. Campbell, R. Liang, E. J. Brady, M. R. Candelore, V. D. H. Ding, G. Jiang, F. Liu, S. A. Qureshi, R. Saperstein, D. Szalkowski, S. Tong, L. M. Tota, D. Xie, X. Yang, P. Zafian, S. Zheng, K. T. Chapman, B. B. Zhang and J. R. Tata, *Bioorg. Med. Chem. Lett.*, 2005, **15**, 1401.

[66] S. D. Erickson, P. Gillespie and K. R. Guertin, *US Patent App.* 209943, 2004.

[67] R. Anderskewitz, G. Morschhäuser, R. Streicher, T. Trieselmann and R. Walter, *WO Patent* 042850, 2006.

PART III:
Inflammation/Pulmonary/GI

Editor: Joel Barrish
Bristol-Myers Squibb R&D
Princeton
New Jersey

CHAPTER 9

Advances Toward Dissociated Non-Steroidal Glucocorticoid Receptor Agonists

John Regan, Hossein Razavi and **David Thomson**

Contents			

1. INTRODUCTION

Glucocorticoids (GCs) are endogenous hormones exhibiting profound effects on multiple tissues and organs [1]. Synthetic GCs, such as prednisolone (**1**) and dexamethasone (**2**) are indispensable to modern clinical medicine for treating a wide variety of inflammatory, immune and allergic conditions [2,3]. However, their use causes dose- and duration-dependent side effects [4]. GCs mediate their activity via binding to the cytoplasmic glucocorticoid receptor (GR) and subsequent translocation of the receptor–ligand complex into the nucleus resulting in altered gene expression patterns [5]. The receptor–ligand complex regulates gene transcription through two broad mechanisms: transcriptional

Medicinal Chemistry Department, Boehringer Ingelheim Pharmaceuticals, Ridgefield, CT 06877, USA

Annual Reports in Medicinal Chemistry, Volume 43
ISSN 0065-7743, DOI 10.1016/S0065-7743(08)00009-2

repression (TR) that, in large part, accounts for the anti-inflammatory actions of GR agonists and transcriptional activation (TA), which is associated with steroid-induced side effects [6–9]. The complexities of GR signaling are clearly illustrated by the observations that ligands possessing very similar structures may exhibit, most likely via altered interactions with glucocorticoid response elements (GRE) and components of the transcriptional machinery, differential gene expression patterns [10]. Since the GRE and cofactor composition of various tissues differ, the ultimate effects on *in vivo* gene expression are highly ligand- and context-specific. The crystal structure of the GR ligand-binding domain (LBD) bound to **2** reveals that, although only 65% of the available LBD volume is occupied, the extensive array of H-bonding and lipophilic interactions between ligand and receptor contribute to the high-affinity binding [11,12]. Another X-ray co-crystal structure shows that the conformational flexibility of the GR LBD allows it to double its size to accommodate fluorocortivazol analog **3a** [13]. The search for agents capable of selectively binding to the GR, discriminating between the TR and TA pathways (dissociation), resulting in high anti-inflammatory efficacy and a reduced clinical side effect profile remains an ongoing activity [14–17]. This chapter will discuss recent advances toward the identification of dissociated non-steroidal GR receptor agonists from 2002 [18] to present.

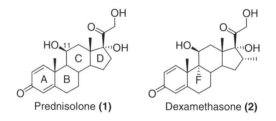

Prednisolone (**1**)   Dexamethasone (**2**)

2. FLUOROCORTIVAZOL ANALOGS

The C- and D-rings of the steroidal GR agonist fluorocortivazol (**3b**) can be substantially modified without losing activity. For example, compound **4** is a potent and selective GR ligand (binding $IC_{50} = 0.8$ nM) and shows activity in both TR (interleukin 6 (IL-6) inhibition in human A549 lung carcinoma cells $EC_{50} = 1.0$ nM, 97% efficacy vs. **2** at 100%) and TA (tyrosine amino transferase (TAT) induction in human HepG2 cells $EC_{50} = 36$ nM, 69% efficacy vs. **2** at 100%) functional assays. In a lipopolysaccharide-stimulated mouse model of tumor necrosis factor — alpha production (LPS-TNFα) **4** gives an ED_{50} of 4 mg/kg (p.o.) [19]. Conversion of the thiophene ring to benzothiophene and removal of the tertiary methyl group (**5**) negligibly affects GR binding affinity ($IC_{50} = 1.5$ nM) or dissociative activity in TR and TA assays [20]. Switching to the other hydroxyl epimer (**6**) does not influence GR binding, yet confers dissociation when examined in TR (IL-6 inhibition $EC_{50} = 18$ nM, 86% efficacy) and TA (TAT efficacy

21%) assays. A single 30 mg/kg oral dose of **6** in a LPS-TNFα model inhibits cytokine production similarly to a 3 mg/kg dose of prednisolone. Contracting the B-ring of **5** and **6** from a 6- to a 5-membered ring (**7** and **8**) has biological and pharmacokinetic (PK) consequences [21]. Compound **7** is efficacious in the TR assay (IL-6 inhibition $EC_{50} = 10$ nM, 92% efficacy) and less efficacious in the TA assay (TAT induction $EC_{50} = 675$ nM, 62% efficacy). In addition, **7** has a good PK profile and significantly reduces *in vivo* TNFα production. On the other hand, **8**, the hydroxyl epimer, is less efficacious when examined in TR and TA assays (IL-6 inhibition $EC_{50} = 5$ nM, 71% efficacy; TAT efficacy 29%). Despite modest oral bioavailability in mouse (34%), compound **8** diminishes murine TNFα production when administered in an LPS-challenged model. A metabolite, believed to be the benzothiophene dioxide, is a potent, but not dissociated, GR ligand and postulated to be responsible for the *in vivo* activity seen with **8**. Other fluorocortivazol analogs contain truncated C- and D-rings, 5- and 6-membered B-rings and 5- and 6-membered ketals [22]. As an example, **9** is a potent and selective GR ligand (GR $IC_{50} = 8.0$ nM) with a partial agonism profile in TR and TA assays (IL-6 inhibition $EC_{50} = 2.6$ nM, 58% efficacy; TAT efficacy 11%). A good murine PK profile for **9** translates into effective TNFα inhibition *in vivo* ($ED_{50} = 14$ mg/kg, p.o.).

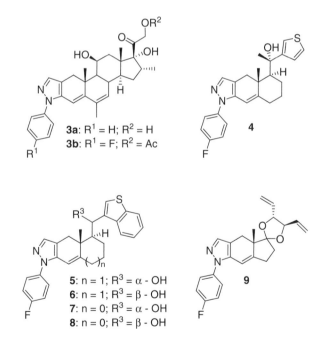

Other C- and D-ring modified fluorocortivazol analogs were systematically evaluated for TA and TR in osteosarcoma cells [23]. Ligands **10** and **11** have equal affinity for GR ($IC_{50} = 5$–8 nM) and full agonism activities in TR assays (nuclear factor κB (NFκB) $EC_{50} = 3$ nM, 63–68% efficacy; activating protein 1 (AP-1)

$EC_{50} = 3\,nM$, 73–112% efficacy) and a TA assay (mouse mammary tumor virus (MMTV) $EC_{50} = 3\,nM$, 99–133% efficacy). However, analog **12** (GR $IC_{50} = 2\,nM$) is more potent in the AP-1 assay ($EC_{50} = 3\,nM$, 71% efficacy) than the NFκB ($EC_{50} = 74\,nM$, 54% efficacy) and the MMTV assays ($EC_{50} = 22\,nM$, 127% efficacy). These ligands were further profiled for their ability to affect 17 GR target genes using quantitative real-time PCR in A549 human lung adenocarcinoma cells [10]. For example, compounds **11** and **12** (i) activate all the genes that are up-regulated by dexamethasone, (ii) have little effect on the genes induced by TNFα and (iii) fail to repress target genes that are glucocorticoid-repressed, but not TNFα-induced. Alternatively, ligand **10** (i) activates only a subset of the genes that are up-regulated by dexamethasone, (ii) inhibits a subset of the genes induced by TNFα and (iii) is similar to dexamethasone in repressing GC-repressed, but not TNFα-induced genes. This study clearly demonstrates that the transcriptional regulatory activities of GR, and the endogenous genes bearing natural GREs, are markedly influenced by subtle differences in the shape of the ligands and their ability to affect the topology of the GR–ligand complex.

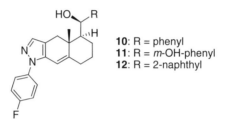

10: R = phenyl
11: R = *m*-OH-phenyl
12: R = 2-naphthyl

3. HYDROXY-TRIFLUOROMETHYL-PHENYL-PENTANOIC (HTPP) ARYL AMIDES

Evidence that a dissociated GR agonist can convincingly demonstrate good *in vivo* anti-inflammatory activity while achieving a better safety profile compared to a steroid is supplied by ZK-216348 ((+)-enantiomer, **13**) [24,25]. This compound, although partially selective over other nuclear receptors, shows good activity in TR assays (TNFα inhibition in hPBMC $IC_{50} = 90\,nM$, 63% efficacy; interleukin 8 (IL-8) inhibition in THP-1 cells $IC_{50} = 35\,nM$, 52% efficacy) and is less active in a TA assay (TAT induction in liver hepatoma cells $EC_{50} = 95\,nM$, 88% efficacy). When administered topically in the croton oil model of ear inflammation in mice, compound **13** produces an ED_{50} of $0.02\,\mu g/cm^2$. A superior safety profile (e.g., glucose levels, spleen involution and skin atrophy), compared to prednisolone, is also observed. However, similar adrenocorticotropic hormone suppression as prednisolone is seen implying that some steroid-induced side effects may be mediated through the TR pathway. The disclosure of **13** spurred extensive SAR and scaffold modifications to the 2-hydroxy-2-trifluoromethyl-4-methyl-4-phenyl-pentanoic (HTPP) aryl amides scaffold. One key finding is the identification of a function-regulating pharmacophore (FRP) that affects the biological profile of a ligand. For example, in the HTPP amide series the group at C-2 serves as a FRP. If CF_3 is present (**14**), then the ligand behaves as an agonist (IL-6 inhibition in HFF cells $EC_{50} = 10\,nM$, 87% efficacy). However, switching the group to benzyl (**15**) imparts full antagonist activity. Further support for the importance of the CF_3 group is the observation that its removal results in abolished GR binding affinity [26]. Another important result is that the reduction of the amide carbonyl does not negatively affect GR binding in this series [27]. Close analogs of **13** and **14**, such as the protected catechol (**16**) and the cyclobutyl derivative (**17**), are claimed to be more dissociated, and yet retain *in vivo* activity (e.g., **16**, 55% inhibition at 3 mg/kg in a croton oil-induced ear edema mouse model) [28].

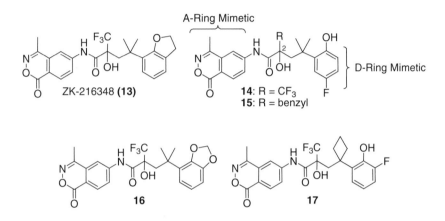

3.1 HTPP D-rings

Using an "agreement docking method" a binding conformation of HTPP aryl amide **14** was proposed [29]. Key H-bond interactions between the benzox-azinone heteroatoms and the Arg611 and Gln570 pair in the GR LBD, as well as, the tertiary hydroxyl group and Asn564, are reported. Thus, the benzoxazinone nucleus serves as a steroidal A-ring mimetic while the tertiary alcohol at C-2 (*R*-configuration) of the HTPP acts as a proxy for the C-11 hydroxyl of a steroidal GR agonist. In this model the 4-fluorophenol ring resides near the steroid D-ring binding domain. Aided by this analysis, cyclic D-ring mimetics were constructed. In a functional assay for TR (NFκB agonism in human A549 lung epithelial cells), tetrahydronaphthalene (THN) analog **18** (single enantiomer) shows weak agonist activity (NFκB agonism 51% at 10 μM) despite good GR binding affinity (pIC$_{50}$ = 8.08). This compound is active in an MMTV antagonist reporter assay (pIC$_{50}$ = 6.25). However, suberan **19** (single enantiomer) is a partial agonist of TR (NFκB agonism pIC$_{50}$ = 6.84, 60% efficacy) while also displaying weak antagonism of TA in an MMTV reporter assay (pIC$_{50}$ = 6.15). SAR of substituents on the THN nucleus reveals that a large group (i.e., 3-pentyl) at C-1 serves as a FRP by converting **20** to a full agonist in the TR assay (NFκB pIC$_{50}$ = 8.92, 105% efficacy) with diminished activity in the TA assay. In a mouse delayed type hypersensitivity (DTH) model, compound **20** gives an EC$_{50}$ of 0.35 μg when administered topically [30]. In addition to THN-based analogs, chroman is claimed as a D-ring mimetic [31,32].

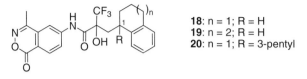

18: n = 1; R = H
19: n = 2; R = H
20: n = 1; R = 3-pentyl

3.2 HTPP A-rings

The effect of eliminating the amide bond, while retaining other space filling A-ring mimetics, in the HTPP series has been systematically explored [33]. Benzyl analog **21** is a better ligand for GR than phenyl (**22**) or phenethyl (**23**) derivatives. To recapture the H-bond network seen with the benzoxazinone nucleus and the Arg-Gln pair, polar functionality is introduced onto the benzyl ring. For example, the 2-chloro-4-cyanobenzyl analog (**24**) shows a 30-fold increase in GR binding affinity compared to **21** but only modest nuclear receptor binding selectivity. The compound displays partial agonist activity in a TR assay (IL-6 EC_{50} = 20 nM, 60% efficacy) and diminished activity in a TA assay (aromatase induction in HFF cells EC_{50} = 30 nM, 20% efficacy). The inclusion of an azaindole as an A-ring pharmacophore with H-bonding potential also has been described. Compound **25** is a potent and selective GR ligand that exhibits strong TR activity (IL-6 inhibition EC_{50} = 5 nM, 90% efficacy) and weakened TA (MMTV efficacy 30%). When dosed orally at 10 mg/kg in a 5- week mouse collagen-induced arthritis (CIA) study, ligand **25** reduces paw swelling by 62%. In addition, when compared to an equally efficacious dose of prednisolone (3 mg/kg) in the same study, **25** shows no increase in triglyceride levels and reduced free fatty acid production, insulin secretion, body weight gain and percent body fat [34]. Other A-ring mimetic pharmacophores on a modified HTPP scaffold that profitably interact with the Arg-Gln pair in the GR LBD are outlined. Of these, quinolone **26** exhibits good GR binding affinity (GR IC_{50} = 4 nM), TR activity (IL-6 inhibition EC_{50} = 11 nM, 80% efficacy) and inhibition of TNFα production *in vivo* (ED_{50} = 10 mg/kg, p.o.). This compound also displays a dissociated profile in a TA counter-screen. FRPs affecting the TR/TA profiles were highlighted on both the A-ring and D-ring mimetics in this series [35].

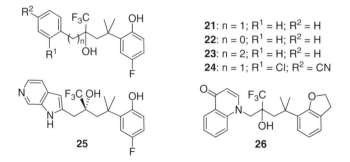

21: n = 1; R^1 = H; R^2 = H
22: n = 0; R^1 = H; R^2 = H
23: n = 2; R^1 = H; R^2 = H
24: n = 1; R^1 = Cl; R^2 = CN

25

26

Additional heterocyclic A-ring mimetics, such as quinazoline, with good selectivity over other nuclear receptors have been reported (**27**, GR $IC_{50} = 1.8$ nM vs. progesterone receptor (PR), mineralocorticoid receptor (MR), androgen receptor (AR) all IC_{50}'s >1 µM) [36]. Notably, **28** has a dissociated *in vivo* profile (41% skin thinning vs. 65% for clobetasol in a rat model) and is suitable for topical applications [37]. Combinations of **28** with immunosuppressive agents (e.g., cyclosporine A) for the treatment of inflammatory ophthalmological diseases, such as dry eye condition, are also disclosed. In particular, **28** reduces the inflammatory effects of anterior chamber paracentesis (i.e., increase levels of prostaglandin E2 and myeloperoxidase activity) in a rabbit eye model [38].

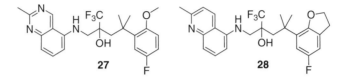

Modifications of substituted THN-based GR agonists with FRPs (e.g., **20**) are described. Incorporating an isoquinoline as an A-ring mimetic furnishes **29** (single enantiomer) having potent TR activity (NFκB $pIC_{50} = 8.66$, 90% efficacy) with diminished TA (MMTV agonism $pEC_{50}<6$, 6% efficacy) [39]. Inserting an acyl group and utilizing a fluorocortivazol-like A-ring (**30**, single enantiomer) also elicits a dissociated profile (TR: NFκB $pIC_{50} = 8.07$, 103% efficacy; TA: MMTV agonism $pEC_{50} = 7.02$, 36% efficacy). The amino group at C-5 of the pyrazole ring provides about a 10-fold improvement in TR activity compared to a methyl at this position [40].

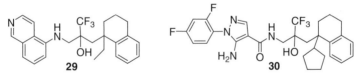

Constrained variants of the 4-acyl-5-aminopyrazole moiety of **30** include indazole and pyrazolo-pyrimidine (e.g., **31** and **32**, respectively) [41,42]. In particular, analog **33**, with a further functionalized fluorocortivazol-like A-ring, is comparable in potency to fluticasone propionate in a rat LPS-challenged neutrophilia and a mouse DTH model [43].

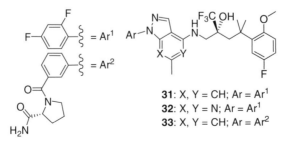

31: X, Y = CH; Ar = Ar¹
32: X, Y = N; Ar = Ar¹
33: X, Y = CH; Ar = Ar²

4. DIHYDRO-1H-BENZOPYRANO[3,4-*f*]QUINOLINES (DBQ)

The dihydro-1H-benzopyrano[3,4-*f*]quinolines (DBQ) scaffold has received considerable drug discovery attention aimed at developing versatile hormone receptor ligands. An advanced compound in this class, AL-438 (**34**), was the first dissociated GR agonist to show a spectrum of *in vivo* anti-inflammatory activity while attenuating the side effect profile [44]. Briefly, **34** is a potent and selective GR ligand (GR Ki = 2.5 nM) exhibiting cellular dissociative behavior in TR (IL-6 inhibition in human skin fibroblast) compared to TA (TAT in HepG2 cells, osteocalcin inhibition in MG63 cells and aromatase induction in HSF cells) assays. In addition, **34** displays steroid-like anti-inflammatory activity *in vivo* (rat carrageenan-paw edema (CPE) ED_{50} = 11 mg/kg; rat adjuvant-induced arthritis (AIA) ED_{50} = 9 mg/kg). Reduced side effects (bone metabolism and glucose regulation) in rats are seen with **34** when compared with prednisolone at equivalent anti-inflammatory doses. Further SAR studies of **34** reveal that substitutions at C-9 and C-10 play pivotal roles in achieving highly potent and selective GR ligands [45]. For example, **35** selectively binds to GR (IC_{50} = 0.98 nM) vs. PR (IC_{50} = 150 nM) and exhibits potent TR activity (E-selectin repression in HepG2 cells EC_{50} = 14 nM, 96% efficacy). When evaluated orally in a rat CPE model, **35** produces dose-dependant anti-inflammatory effects with an ED_{50} of 16 mg/kg. Also recognized in the DBQ scaffold is that the substituent at C-5 acts as an FRP. Aromatic groups at C-5 (e.g., **36**) impart equal TR and TA activities [46], whereas allyl (e.g., **34**) conveys diminished TA activity. Hybridizing the two groups, as in cyclohexenyl anti-isomer **37** and syn-isomer **38** ((-)-enantiomers), has a differentiating effect on transcriptional activity [47]. The profile of **37** is similar to **34**, whereas ligand **38** demonstrates a substantially reduced ability to induce GRE activation. Both compounds show a dissociated profile when gene products were measured in GR-mediated metabolic side effects (TAT) and osteoporotic (osteocalcin) assays. Introducing arylidene moieties at C-5 of the DBQ scaffold confers similar TR and TA activities as seen with **36**. For example, substituted thienylidene **39** retains potent GR binding affinity and is effective against GRE agonist (EC_{50} = 0.2 nM, 93% efficacy), as well as, GRE antagonist activations (EC_{50} = 4.1 nM, 70% efficacy) [48]. Closely related arylidines are described for the treatment of multiple myeloma [49]. Another arylidine, **40** (LGD-5552), is a potent and selective ligand for GR (K_i = 2.4 nM) vs. other nuclear receptors (PR K_i = 866 nM, MR K_i = 150 nM, AR K_i = 910 nM) and exhibits a full antagonist profile in an MMTV-LUC GRE activation assay with no agonist activity. However, the compound is a potent transrepressor of the anti-inflammatory E-selectin gene. In other gene regulation experiments **40** shows a differential pattern compared to **1**. Evaluation of **40** in a 15-day mouse (DBA/1) therapeutic CIA model produces a dose-responsive decrease in clinical score that is similar to **1**. When anti-inflammatory doses were administered orally to Swiss-Webster mice over 28 days, a decrease in bone formation rate was only observed at a high dose (30 mg/kg) of **40** compared to all doses (3, 10 and 30 mg/kg) of **1**. However, in contrast to **1**, no changes in body fat were observed with compound **40** [50].

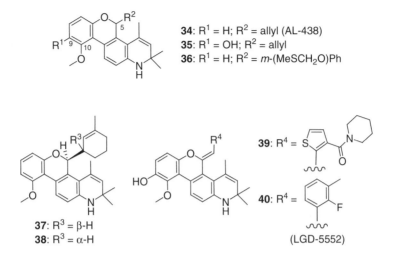

34: R^1 = H; R^2 = allyl (AL-438)
35: R^1 = OH; R^2 = allyl
36: R^1 = H; R^2 = *m*-(MeSCH$_2$O)Ph

37: R^3 = β-H
38: R^3 = α-H

39: R^4 = [structure]

40: R^4 = [structure]
(LGD-5552)

4.1 Ring-opened DBQ analogs

A series of ring-opened DBQ analogs are reported. For example, 6-phenyl-dihydroquinoline derivatives containing a phenethyl thioether at C-4 (e.g., **41**) bind selectively to GR and exhibit only antagonist activity [51]. Allyl ether analog **42** (eutomer) retains potent GR binding but is no longer selective vs. PR. The partial agonist profile of **42** (IL-6 inhibition EC$_{50}$ = 170 nM, 28% efficacy) suggests that substituents at C-4 of this modified DBQ scaffold serve as FRPs [52].

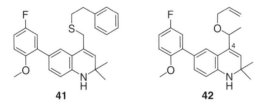

41   **42**

5. ADDITIONAL FUSED CYCLIC CORES

An octahydrophenanthrene-based scaffold with GR activity has been reported [53–55]. As an example, **43** shows good GR binding potency (IC$_{50}$ = 51 nM) and TR activity. Additionally, this compound appears dissociated in alkaline phosphatase induction (21% in primary human osteoblast vs. **2**), and does not promote the terminal differentiation (apoptosis) of osteoblasts.

Other fused cyclic GR agonists include dibenzodiazepines and dibenzoazepines [56,57]. Reported ligands have high GR affinity and TR activity.

Particularly, compound **44** exhibits efficacy in an LPS-TNFα model and anti-inflammatory activity in a CIA study (ED$_{50}$<10 mg/kg, p.o. in both murine models).

A recent report has disclosed **45**, a potent and selective GR ligand, which is active in a TR assay and dissociated in a TA counter-screen [58]. Compound **45** shows efficacy in a rat CPE model (ED$_{50}$ = 1.2 mg/kg, p.o.) and exerts a weaker effect on reducing serum octeocalcin levels than prednisolone at equivalent anti-inflammatory doses. Other distinct fused cyclic GR agonists are represented by ligands **46** (±), **47** (±) and **48** (**46**, AP-1 TR assay EC$_{50}$ = 20 nM, 79% efficacy) [59–71].

A number of tetrahydronaphthalene-based GR ligands have also been disclosed over the last several years [72–75]. Since this scaffold is derived from cyclization of HTPP precursors, typical pharmacophores, such as CF$_3$ and OH, are embedded. Examples include quinoline **49** that exhibits selective GR affinity (GR IC$_{50}$ = 20 nM, PR IC$_{50}$ > 1 µM) and TR activity (LPS-induced IL-8 secretion in THP-1 cells, IC$_{50}$ = 6.4 nM, 90% efficacy) [76]. In contrast, analog **50** is equipotent in GR binding, but less potent and efficacious in the TR assay [77].

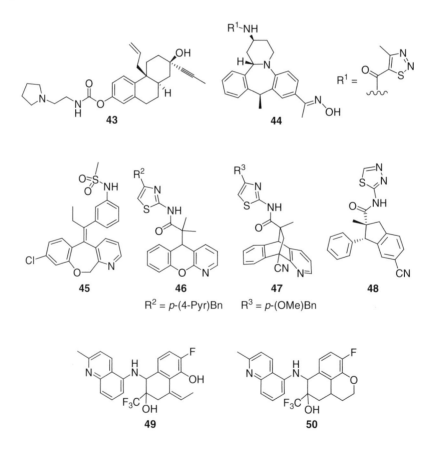

R^2 = *p*-(4-Pyr)Bn   R^3 = *p*-(OMe)Bn

6. SULFONAMIDE-LINKED SCAFFOLDS

Sulfonamide-linked α-methyltryptamines were originally identified as GR modulators during a HTS campaign [78,79]. Subsequent structural modifications led to increases in GR binding and agonistic activity [80]. As an example, **51** (eutomer) is a potent and selective ligand showing activity in a TR assay (GR $IC_{50} = 4$ nM, PR $IC_{50} > 2$ μM, IL-6 inhibition $IC_{50} = 30$ nM, 80% efficacy) and efficacy in an *in vivo* model (LPS-TNFα in mice) [81]. The SAR is expanded to include alternative heterocycles and linkers (e.g., **52** and **53**, both GR $IC_{50} = 3$ nM; **54** GR $IC_{50} = 17$ nM) [82–85].

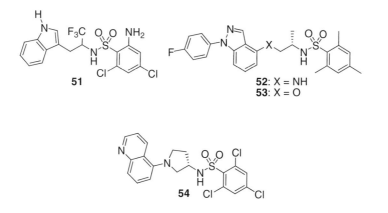

7. CONCLUSION

Notwithstanding the enormously complex biological actions of GR agonists, the promising preclinical data may hopefully be a harbinger of the long-sought-for effective clinical results in which a therapeutic advantage, vis-à-vis side effects, is clearly established.

REFERENCES

[1] N. J. Goulding and R. J. Flowers, in *Glucocorticoids: Milestones in Drug Therapy* (eds N. J. Goulding and R. J. Flowers), Birkhauser, Boston, 2001, p. 3.
[2] L. Parente, in *Glucocorticoids: Milestones in Drug Therapy* (eds N. J. Goulding and R. J. Flowers), Birkhauser, Boston, 2001, p. 35.
[3] A. R. Clark, *Mol. Cell. Endocrinol.*, 2007, **275**, 79.
[4] A. Vegiopoulos and S. Herzig, *Mol. Cell. Endocrinol.*, 2007, **275**, 43.
[5] I. Grad and D. Picard, *Mol. Cell. Endocrinol.*, 2007, **275**, 2.
[6] L. I. McKay and J. A. Cidlowski, *Mol. Endocrinol.*, 1998, **12**, 45.
[7] P. J. Barnes and I. M. Adcock, *Ann. Intern. Med.*, 2003, **139**, 359.
[8] H. Schacke, H. Rehwinkel, K. Asadullah and A. C. Cato, *Exp. Dermatol.*, 2006, **15**, 565.
[9] F. Buttgereit, G. R. Burmester and B. J. Lipworth, *Lancet*, 2005, **365**, 801.

[10] J. C. Wang, N. Shah, C. Pantoja, S. H. Meijsing, J. D. Ho, T. S. Scanlan and K. R. Yamamoto, *Genes Dev.*, 2006, **20**, 689.

[11] R. K. Bledsoe, V. G. Montana, T. B. Stanley, C. J. Delves, C. J. Apolito, D. D. McKee, T. G. Consler, D. J. Parks, E. L. Stewart, T. M. Willson, M. H. Lambert, J. T. Moore, K. H. Pearce and H. E. Xu, *Cell*, 2002, **110**, 93.

[12] R. K. Bledsoe, E. L. Stewart and K. H. Pearce, *Vitam. Horm.*, 2004, **68**, 49.

[13] K. Suino-Powell, Y. Xu, C. Zhang, Y. g. Tao, W. D. Tolbert, S. S. Simons, Jr. and H. E. Xu, *Mol. Cell. Biol.*, 2008, **28**, 1915.

[14] J. N. Miner, M. H. Hong and A. Negro-Vilar, *Expert Opin. Invest. Drugs*, 2005, **14**, 1527.

[15] M. L. Mohler, Y. He, Z. Wu, S.-S. Hong and D. D. Miller, *Expert Opin. Ther. Patents*, 2008, **17**, 37.

[16] I. H. Song, R. Gold, R. H. Straub, G. R. Burmester and F. Buttgereit, *J. Rheumatol.*, 2005, **32**, 1199.

[17] H. Takahashi, H. Razavi and D. Thomson, *Curr. Topics Med. Chem.*, 2008, **8**, 521.

[18] M. J. Coghlan, S. W. Elmore, P. R. Kym and M. E. Kort, in *Annual Reports in Medicinal Chemistry* (ed. A. M. Doherty), Vol. 37, Academic Press, 2002, p. 167.

[19] A. Ali, C. F. Thompson, J. M. Balkovec, D. W. Graham, M. L. Hammond, N. Quraishi, J. R. Tata, M. Einstein, L. Ge, G. Harris, T. M. Kelly, P. Mazur, S. Pandit, J. Santoro, A. Sitlani, C. Wang, J. Williamson, D. K. Miller, C. M. Thompson, D. M. Zaller, M. J. Forrest, E. Carballo-Jane and S. Luell, *J. Med. Chem.*, 2004, **47**, 2441.

[20] C. F. Thompson, N. Quraishi, A. Ali, J. R. Tata, M. L. Hammond, J. M. Balkovec, M. Einstein, L. Ge, G. Harris, T. M. Kelly, P. Mazur, S. Pandit, J. Santoro, A. Sitlani, C. Wang, J. Williamson, D. K. Miller, T. T. Yamin, C. M. Thompson, E. A. O'Neill, D. Zaller, M. J. Forrest, E. Carballo-Jane and S. Luell, *Bioorg. Med. Chem. Lett.*, 2005, **15**, 2163.

[21] C. F. Thompson, N. Quraishi, A. Ali, R. T. Mosley, J. R. Tata, M. L. Hammond, J. M. Balkovec, M. Einstein, L. Ge, G. Harris, T. M. Kelly, P. Mazur, S. Pandit, J. Santoro, A. Sitlani, C. Wang, J. Williamson, D. K. Miller, T. T. Yamin, C. M. Thompson, E. A. O'Neill, D. Zaller, M. J. Forrest, E. Carballo-Jane and S. Luell, *Bioorg. Med. Chem. Lett.*, 2007, **17**, 3354.

[22] C. J. Smith, A. Ali, J. M. Balkovec, D. W. Graham, M. L. Hammond, G. F. Patel, G. P. Rouen, S. K. Smith, J. R. Tata, M. Einstein, L. Ge, G. S. Harris, T. M. Kelly, P. Mazur, C. M. Thompson, C. F. Wang, J. M. Williamson, D. K. Miller, S. Pandit, J. C. Santoro, A. Sitlani, T. T. Yamin, E. A. O'Neill, D. M. Zaller, E. Carballo-Jane, M. J. Forrest and S. Luell, *Bioorg. Med. Chem. Lett.*, 2005, **15**, 2926.

[23] N. Shah and T. S. Scanlan, *Bioorg. Med. Chem. Lett.*, 2004, **14**, 5199.

[24] H. Schacke, A. Schottelius, W. D. Docke, P. Strehlke, S. Jaroch, N. Schmees, H. Rehwinkel, H. Hennekes and K. Asadullah, *Proc. Natl. Acad. Sci., USA*, 2004, **101**, 227.

[25] H. Schacke, M. Berger, H. Rehwinkel and K. Asadullah, *Mol. Cell. Endocrinol.*, 2007, **275**, 109.

[26] R. Betageri, Y. Zhang, R. M. Zindell, D. Kuzmich, T. M. Kirrane, J. Bentzien, M. Cardozo, A. J. Capolino, T. N. Fadra, R. M. Nelson, Z. Paw, D. T. Shih, C. K. Shih, L. Zuvela-Jelaska, G. Nabozny and D. S. Thomson, *Bioorg. Med. Chem. Lett.*, 2005, **15**, 4761.

[27] M. Lehmann, K. Krolikiewicz, W. Skuballa, P. Strehlke, F. Kalkbrenner, R. Ekerdt and C. Giesen, *Patent Application WO 2000/32584-A2*, 2000.

[28] N. Schmees, M. Lehmann, H. Rehwinkel, P. Strehlke, S. Jaroch, H. Schäcke and A. J. G. Schottelius, *Patent Application WO 2004/058733-A1*, 2004.

[29] M. Barker, M. Clackers, D. A. Demaine, D. Humphreys, M. J. Johnston, H. T. Jones, F. Pacquet, J. M. Pritchard, M. Salter, S. E. Shanahan, P. A. Skone, V. M. Vinader, I. Uings, I. M. McLay and S. J. Macdonald, *J. Med. Chem.*, 2005, **48**, 4507.

[30] M. Barker, M. Clackers, R. Copley, D. A. Demaine, D. Humphreys, G. G. Inglis, M. J. Johnston, H. T. Jones, M. V. Haase, D. House, R. Loiseau, L. Nisbet, F. Pacquet, P. A. Skone, S. E. Shanahan, D. Tape, V. M. Vinader, M. Washington, I. Uings, R. Upton, I. M. McLay and S. J. Macdonald, *J. Med. Chem.*, 2006, **49**, 4216.

[31] T. M. Kirrane, D. Kuzmich and J. R. Proudfoot, *Patent Application WO 2005/090336-A1*, 2005.

[32] N. Schmees, M. Berger, H. Rehwinkel and H. Schäcke, *Patent Application WO 2006/108711-A1*, 2006.

[33] D. Kuzmich, T. Kirrane, J. Proudfoot, Y. Bekkali, R. Zindell, L. Beck, R. Nelson, C. K. Shih, A. J. Kukulka, Z. Paw, P. Reilly, R. Deleon, M. Cardozo, G. Nabozny and D. Thomson, *Bioorg. Med. Chem. Lett.*, 2007, **17**, 5025.

[34] G. Nabozny, Glucocorticoids and inflammatory disease, tissue selective nuclear receptors; Keystone Symposia, Breckenridge, CO, September 19, 2005.

[35] J. Regan, T. W. Lee, R. M. Zindell, Y. Bekkali, J. Bentzien, T. Gilmore, A. Hammach, T. M. Kirrane, A. J. Kukulka, D. Kuzmich, R. M. Nelson, J. R. Proudfoot, M. Ralph, J. Pelletier, D. Souza, L. Zuvela-Jelaska, G. Nabozny and D. S. Thomson, *J. Med. Chem.*, 2006, **49**, 7887.

[36] M. Berger, S. Bäuerle, H. Rehwinkel, N. Schmees, H. Schäcke, M. Lehmann, K. Krolikiewicz, A. J. B. Schottelius, D. Nguyen, A. Mengel and S. Jaroch, *Patent Application WO 2005/003098-A1*, 2005.

[37] S. Jaroch, H. Rehwinkel, H. Schäcke, N. Schmees, W. Skuballa, M. Schneider, J. Hübner and O. Petrov, *Patent Application WO 2006/050998-A1*, 2006.

[38] E. Xia and Z. Hu, *Patent Application WO 2008/005686-A2*, 2008.

[39] K. Biggadike, M. Boudjelal, M. Clackers, D. M. Coe, D. A. Demaine, G. W. Hardy, D. Humphreys, G. G. Inglis, M. J. Johnston, H. T. Jones, D. House, R. Loiseau, D. Needham, P. A. Skone, I. Uings, G. Veitch, G. G. Weingarten, I. M. McLay and S. J. Macdonald, *J. Med. Chem.*, 2007, **50**, 6519.

[40] M. Clackers, D. M. Coe, D. A. Demaine, G. W. Hardy, D. Humphreys, G. G. A. Inglis, M. J. Johnston, H. T. Jones, D. House, R. Loiseau, D. J. Minick, P. A. Skone, I. Uings, I. M. McLay and S. J. F. Macdonald, *Bioorg. Med. Chem. Lett.*, 2007, **17**, 4737.

[41] C. D. Eldred, D. House, G. G. A. Inglis, S. J. F. MacDonald and P. A. Skone, *Patent Application WO 2006/108699-A1*, 2006.

[42] K. Biggadike, D. House, G. G. A. Inglis, S. J. F. MacDonald, I. M. McLay and P. A. Skone, *Patent Application WO 2007/054294-A1*, 2007.

[43] K. Biggadike, A. W. J. Cooper, D. House, I. M. McLay and G. R. Woollam, *Patent Application WO 2007/122165-A1*, 2007.

[44] M. J. Coghlan, P. B. Jacobson, B. Lane, M. Nakane, C. W. Lin, S. W. Elmore, P. R. Kym, J. R. Luly, G. W. Carter, R. Turner, C. M. Tyree, J. Hu, M. Elgort, J. Rosen and J. N. Miner, *Mol. Endocrinol.*, 2003, **17**, 860.

[45] P. R. Kym, M. E. Kort, M. J. Coghlan, J. L. Moore, R. Tang, J. D. Ratajczyk, D. P. Larson, S. W. Elmore, J. K. Pratt, M. A. Stashko, H. D. Falls, C. W. Lin, M. Nakane, L. Miller, C. M. Tyree, J. N. Miner, P. B. Jacobson, D. M. Wilcox, P. Nguyen and B. C. Lane, *J. Med. Chem.*, 2003, **46**, 1016.

[46] S. W. Elmore, M. J. Coghlan, D. D. Anderson, J. K. Pratt, B. E. Green, A. X. Wang, M. A. Stashko, C. W. Lin, C. M. Tyree, J. N. Miner, P. B. Jacobson, D. M. Wilcox and B. C. Lane, *J. Med. Chem.*, 2001, **44**, 4481.

[47] S. W. Elmore, J. K. Pratt, M. J. Coghlan, Y. Mao, B. E. Green, D. D. Anderson, M. A. Stashko, C. W. Lin, D. Falls, M. Nakane, L. Miller, C. M. Tyree, J. N. Miner and B. Lane, *Bioorg. Med. Chem. Lett.*, 2004, **14**, 1721.

[48] R. J. Ardecky, A. R. Hudson, D. P. Phillips, J. S. Tyhonas, C. Deckhut, T. L. Lau, Y. Li, E. A. Martinborough, S. L. Roach, R. I. Higuchi, F. J. Lopez, K. B. Marschke, J. N. Miner, D. S. Karanewsky, A. Negro-Vilar and L. Zhi, *Bioorg. Med. Chem. Lett.*, 2007, **17**, 4158.

[49] A. R. Hudson, S. L. Roach, R. I. Higuchi, D. P. Phillips, R. P. Bissonnette, W. W. Lamph, J. Yen, Y. Li, M. E. Adams, L. J. Valdez, A. Vassar, C. Cuervo, E. A. Kallel, C. J. Gharbaoui, D. E. Mais, J. N. Miner, K. B. Marschke, D. Rungta, A. Negro-Vilar and L. Zhi, *J. Med. Chem.*, 2007, **50**, 4699.

[50] J. N. Miner, B. Ardecky, K. Benbatoul, K. Griffiths, C. J. Larson, D. E. Mais, K. Marschke, J. Rosen, E. Vajda, L. Zhi and A. Negro-Vilar, *PNAS*, 2007, **104**, 19244.

[51] H. Takahashi, Y. Bekkali, A. J. Capolino, T. Gilmore, S. E. Goldrick, R. M. Nelson, D. Terenzio, J. Wang, L. Zuvela-Jelaska and J. Proudfoot, *Bioorg. Med. Chem. Lett.*, 2006, **16**, 1549.

[52] H. Takahashi, Y. Bekkali, A. J. Capolino, T. Gilmore, S. E. Goldrick, P. V. Kaplita, L. Liu, R. M. Nelson, D. Terenzio, J. Wang, L. Zuvela-Jelaska, J. Proudfoot, G. Nabozny and D. Thomson, *Bio. Med. Chem. Lett.*, 2007, **17**, 5091.

[53] L. Buckbinder, *Patent Application US 2003/0224349-A1*, 2003.

[54] A. Chantigny, E. F. Kleinman and R. P. Robinson, Jr., *Patent Application WO 2004/005229-A1*, 2004.

[55] R. P. Robinson, Jr., E. F. Kleinman and H. Cheng, *Patent Application WO 2005/047254-A1*, 2005.

[56] R. Plate, G. J. R. Zaman, P. H. H. Hermkens, C. G. J. M. Jans, R. C. Buijsman, A. P. A. Man, P. G. M. Conti, S. J. Lusher and W. H. A. Dokter, *Patent Application WO 2006/084917-A1*, 2006.

[57] R. Plate and C. G. J. M. Jans, *Patent Application WO 2007/025938-A1*, 2007.

[58] M. W. Carson and M. J. Coghlan, *Patent Application* WO 2008/008882-A2, 2008.

[59] D. S. Weinstein, H. Gong, J. Duan, T. G. M. Dhar, B. V. Yang, P. Chen and B. Jiang, *Patent Application* WO 2008/021926-A2, 2008.

[60] J. Duan, J. Sheppeck, B. Jiang and J. L. Gilmore, *Patent Application* WO 2005/072732-A1, 2005.

[61] W. Vaccaro, B. V. Yang, S.-H. Kim, T. Huynh, D. R. Tortolani, K. J. Leavitt, W. Li, A. M. Doweyko, X.-T. Chen and L. Doweyko, *Patent Application* WO 2004/009017-A2, 2004.

[62] J. Duan, B. Jiang, J. Sheppeck and J. L. Gilmore, *Patent Application* WO 2005/070207-A1, 2005.

[63] D. S. Weinstein, B. V. Yang, S.-H. Kim, W. Vaccaro, J. Sheppeck and J. Gilmore, *Patent Application* WO 2005/072132-A2, 2005.

[64] B. V. Yang, *Patent Application* WO 2005/072729-A1, 2005.

[65] B. V. Yang, *Patent Application* WO 2005/073203-A1, 2005.

[66] D. S. Weinstein, J. Sheppeck and J. L. Gilmore, *Patent Application* WO 2005/073221-A1, 2005.

[67] B. V. Yang, *Patent Application* WO 2006/076509-A1, 2006.

[68] J. E. Sheppeck, T. G. M. Dhar, A. M. P. Doweyko, L. M. Doweyko, J. L. Gilmore, D. S. Weinstein, H.-Y. Xiao and B. V. Yang, *Patent Application* WO 2006/076632-A1, 2006.

[69] B. V. Yang, L. M. Doweyko and A. M. Doweyko, *Patent Application* WO 2006/076633-A1, 2006.

[70] T. G. M. Dhar, H.-Y. Xiao and B. V. Yang, *Patent Application* WO 2006/076702-A1, 2006.

[71] J. Duan and B. Jiang, *Patent Application* WO 2007/073503-A2, 2007.

[72] H. Rehwinkel, S. Bäuerle, M. Berger, N. Schmees, H. Schäcke, K. Krolikiewicz, A. Mengel, D. Nguyen, S. Jaroch and W. Skuballa, *Patent Application* WO 2005/034939-A1, 2005.

[73] S. Bäuerle, H. Schäcke, M. Berger and A. Mengel, *Patent Application* WO 2006/108714-A2, 2006.

[74] A. Mengel, K. Krolikiewicz, S. Bäuerle and H. Schäcke, *Patent Application* WO 2006/108713-A1, 2006.

[75] C. Huwe, W. Skuballa, D. Nguyen and H. Schäcke, *Patent Application* WO 2006/108712-A1, 2006.

[76] S. Bäuerle, M. Berger, S. Jaroch, K. Krolikiewicz, D. Nguyen, H. Rehwinkel, H. Schäcke, N. Schmees and W. Skuballa, *Patent Application* WO 2006/027236-A1, 2006.

[77] M. Berger, N. Schmees, H. Schäcke, S. Bäuerle, H. Rehwinkel, A. Mengel, K. Krolikiewicz, D. Grossbach and D. Voigtländer, *Patent Application* WO 2006/066950-A2, 2006.

[78] D. R. Marshall, *Patent Application* WO 2004/019935-A1, 2004.

[79] D. R. Marshall, G. Rodriguez, D. S. Thomson, R. Nelson and A. Capolina, *Bioorg. Med. Chem. Lett.*, 2007, **17**, 315–319.

[80] R. Betageri, D. Disalvo, D. S. Thomson, D. Kuzmich, J. Regan and J. Kowalski, *Patent Application* WO 2006/071609-A2, 2006.

[81] D. S. Thomson, *233rd American Chemical Society National Meeting*, Chicago, IL, March 2007.

[82] H. Bladh, K. Henriksson, V. Hulikal and M. Lepistö, *Patent Application* WO 2006/046914-A1, 2006.

[83] H. Bladh, K. Henriksson, V. Hulikal and M. Lepistö, *Patent Application* WO 2006/046916-A1, 2006.

[84] H. Bladh, J. Dahmen, T. Hansson, K. Henriksson, M. Lepistö and S. Nilsson, *Patent Application* WO 2007/046747-A1, 2007.

[85] M. Bengtsson, H. Bladh, T. Hansson and E. Kinchin, *Patent Application* WO 2007/114763-A1, 2007.

Advances in the Discovery of IκB Kinase Inhibitors

William J. Pitts and **James Kempson**

1. INTRODUCTION

There has been a significant effort in the pharmaceutical industry focused on the discovery of small molecule kinase inhibitors for inflammatory/immunological diseases [1] and oncology [2]. IκB kinase inhibitors are capable of blocking a number of signals mediated through the nuclear transcription factor NF-κB

Bristol-Myers Squibb

Annual Reports in Medicinal Chemistry, Volume 43
ISSN 0065-7743, DOI 10.1016/S0065-7743(08)00010-9

thereby blocking the downstream production of pro-inflammatory mediators. As a result of the pleotropic effects of NF-κB modulation, there are many potential therapeutic applications for an IKK-2 inhibitor. These include inflammatory/immunological diseases such as rheumatoid arthritis, osteoarthritis, inflammatory bowel disease, chronic obstructive pulmonary disease (COPD), psoriasis, solid organ transplantation, ischemia reperfusion injury, diabetes, heart failure, and cancer [3]. As a result, there has been significant interest on the part of pharmaceutical companies to identify IKK-2 inhibitors as possible drug candidates [4]. However, there has also been longstanding concern over possible toxicities arising from IKK-2 knockout studies which revealed massive hepatic necrosis resulting in embryonic lethality [5]. Recent studies have suggested that effects on the liver appear to be context specific, and in some cases appear to be protective [6]. It now appears the major issue with respect to the progression of IKK-2 inhibitors is related to its immunosuppressive effects. Despite toxicity concerns, several compounds have advanced into phase 1 clinical studies. The results of these initial human studies are eagerly awaited.

2. BIOCHEMISTRY OF IκB KINASE INHIBITION

At the cellular level, a variety of stimuli funnel through IκB kinase resulting in the liberation of the nuclear transcription factor NF-κB. These include receptor-specific stimuli such as cytokines (TNF-α, IL-1β), CD40 ligand, and LPS, which signal through what is known as the canonical pathway (Figure 1). Recently, an alternative NF-κB pathway (non-canonical pathway [7]) has emerged which is triggered by signaling through TNF receptor members such as the lymphotoxin-beta receptor,

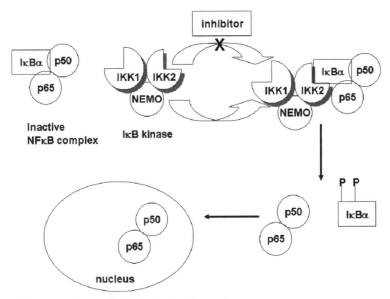

Figure 1 Schematic of canonical NF-κB signaling pathway.

CD40, the B-cell-activating factor receptor and RANK as well as oncogenic viruses [8]. This diverse biology has lead to the examination of inhibitors of IκB kinase for both immunological/inflammatory [9] and oncology indications [10]. Two enzymes (TBK-1 and IKK-ε [11]) with significant homology [12] to IκB kinase but distinct enzymology [13] and biology [14] are not discussed further in this review.

IκB kinase is a multi-subunit kinase that contains two catalytic subunits commonly designated as IKK-1 (IKKα) and IKK-2 (IKKβ), and a subunit known as NEMO (IKKγ). Despite the complexity of this system, it is generally thought that IKK-2 inhibition plays a more important role in regulating NF-κB signal transduction than IKK-1. For example, embryonic fibroblasts from IKK-2 knockout mice show a marked reduction in TNF-α induced NF-κB activation while IKK-1 knockout mice do not [5,15]. These results are consistent with the profile observed with compounds selective for inhibition of IKK-2 over IKK-1 described below. The IκB kinase complex has been shown to be phosphorylated by the upstream kinase TAK1 [16]. Although the exact activation mechanisms of IκB kinase itself are not completely understood, it has been recently shown that NEMO not only acts as a scaffolding protein for the complex but may play a role in activation of the complex [17]. Once activated IκB kinase proceeds to phosphorylate IκBα resulting in its dissociation from the p50-p65 heterodimer. This in turn permits the heterodimer to translocate to the nucleus where it promotes transcription of a number of gene products. Several of these products are pro-inflammatory including cytokines such as TNF-α, IL-1β, IL-6, IL-17, and IL-23 [18]. This results in a plethora of inflammatory responses in a number of cell types including macrophages, T cells, B cells, osteoclasts, and fibroblasts.

3. IKK-2 PHARMACOLOGY

The human genome contains over 500 protein kinases [19]. Many kinase inhibitors have varying degrees of selectivity and inhibition is frequently not restricted to a specific kinase family. The ability to screen enzyme collections using either enzyme inhibition [20] or binding [21] has significantly improved our understanding of kinase inhibitor selectivity profiles. Many of the compounds described below are from publications that disclose limited selectivity data. Whereas this caveat should be kept in mind, the pharmacologic activity reported for these compounds is generally consistent with the expectations one would have for inhibition of the NF-κB signaling pathway.

3.1 Inflammation

MLN120B (1) is reported to be highly selective (IKK-2 IC_{50} = 60 nM, IKK-1 IC_{50} >100,000 nM) [22]. The compound decreased paw swelling in a dose-dependant manner with an ED_{50} of 7 mg/kg and 12 mg/kg, respectively when dosed orally twice daily in either the prophylactic or therapeutic adjuvant induced arthritis models in rats [23]. Compound 1 also showed a similar EC_{50} in a model of LPS-induced TNF-α production in rats which suggests TNF-α may be a suitable biomarker for IKK inhibition. Histology and micro-CT (computerized tomography) imaging showed a significant protection against bone and cartilage destruction seen in the diseased controls. Notably, NF-κB regulated gene expression in the joints was significantly decreased when compared to diseased controls. mRNA levels were decreased for TNFα, IL-1β, iNOS, MMP-13, CD11b, TRAP, and RANKL, the latter two having significant inhibitory effects on osteoclast-mediated bone degradation. This profile is consistent with the decrease in cytokine production and the protective effects on bone destruction seen in this study. Compound **1** also exhibited a dose-dependent decrease in clinical symptoms compared to control animals in a collagen antibody-induced murine (BALBc) model of arthritis [24]. The maximally efficacious dose was determined to be 60 mg/kg. A significant effect was also observed on several NF-κB-related gene products including adhesion molecules (ICAM-1), cytokines (TNF-α, IL-1β, IL-6) inflammatory mediators (iNOS, COX-2), metalloproteases (MMP-3), and cathepsins (B & K). Additionally, protease-activated near-infrared fluorescence *in vivo* imaging demonstrated that IKK-2 inhibition effectively abolished protease activity in the joints.

Thienopyridine **2**, a potent IKK-2 inhibitor (IKK-2 $IC_{50} = 26$ nM), was examined in the rat collagen induced arthritis model [25]. When dosed orally at 10 mg/kg bid, a 44% decrease in paw weight was observed when compared to control animals. Pyrimidine **3** (IKK-2 $IC_{50} = 40$ nM), when dosed subcutaneously or orally at 30 mg/kg, significantly decreased serum TNF-α levels 4 h post LPS-challenge (86%, s.c., 75% p.o.). In a second experiment, Compound **3** was also efficacious (10 mg/kg s.c.) in a thioglycolate-induced peritonitis model in mouse ($\sim$50% inhibition of neutrophil extravasation) [26].

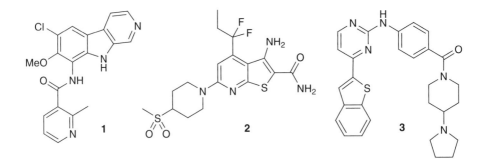

3.2 Ischemia reperfusion injury

Bay 65-1942 (**4**) (IKK-2 $K_i = 2$ nM, IKK-1 $K_i \sim 140$ nM) demonstrated activity in a mouse reperfusion injury model when dosed by the intraperitoneal route at 5 mg/kg [27]. There was a significant decrease in serum levels of TNF-α and IL-6 in the treatment group, providing additional evidence that cytokine levels can be used as a biomarker. This was the first demonstration that an IKK inhibitor could decrease infarct size. Impressively, a single dose of this compound administered two hours after reperfusion resulted in significant cardioprotection.

IMD 0354 (**5**) has been reported to inhibit NF-κB activation putatively via IKK-2 inhibition [28]. It has a positive effect in a cardiac reperfusion model with respect to remodeling events, however unlike the observations with **4**, there was no significant decrease in infarct size observed with compound **5** [29].

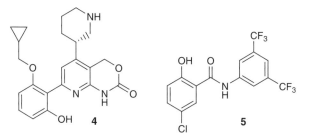

3.3 Airway disease

IKK inhibitors have been shown to decrease airway inflammation in animal models [30]. In an attempt to better understand the potential for use of an IKK inhibitor in steroid-resistant disease states, TCPA 1 (**6**) (IKK-2 $IC_{50} = 18\,nM$, IKK-1 $IC_{50} = 400\,nM$) was examined in two rat models of airway inflammation [31]. In a rat LPS-induced airway inflammation model administration of compound **6** resulted in a dose dependent decrease in these cytokine levels compared with untreated controls. In a porcine pancreatic elastase (PPE)-induced inflammation model, neither steroids nor compound **6** affected the increase in cytokine production. Whereas the relevance of the PPE model to human disease is not well understood, it is clear that IKK inhibition of cytokine production can be context selective. In a separate study compound **6** was able to decrease exhaled nitric oxide in a dose-dependent manner in the LPS-induced airway inflammation model similar to that observed in the clinic, suggesting that this may be a useful biomarker of lung inflammation [32]. Given that there is no efficacy difference between steroids and IKK inhibitors for airway disease, an understanding of the side effect profile of an IKK inhibitor would be an important differentiating factor when considering benefits over current therapy.

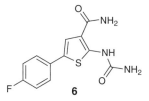

3.4 Melanoma

It has been suggested that TNF-α induced NF-κB activation in multiple myeloma cells prolongs their survival, thereby providing the expectation that IKK inhibition would decrease proliferation of multiple myeloma cells [33]. It was subsequently shown that proliferation of multiple myeloma cells can be inhibited by IKK inhibitors. Compound 1 inhibits the growth of human multiple myeloma cell lines as well as the proliferation of human IL-6-dependent myeloma cells injected into implanted human fetal bone chips in a severe combined immunodeficiency (SCID) background. The compound demonstrated a trend toward survival in the model [34]. Inhibition of NF-κB activation via IKK inhibition increased apoptotic cell death in the presence of TNF-α. Additionally, the investigators demonstrated an augmentation of the cell death when 1 was administered to multiple myeloma cells in conjunction with doxorubicin, melphalan, and dexamethasone.

BMS-345541 (7) (IKK-2 IC_{50} = 300 nM, IKK-1 IC_{50} = 4,000 nM) demonstrated dose-dependent inhibition of the growth of three different human melanoma cell lines when inoculated into nude mice [35]. Administration of this compound also resulted in a dose-dependent decrease in the incorporation of ^{32}P into GST-IκBα in one of the melanoma cell lines, consistent with IKK-2 inhibition *in vitro*, as well as a dose-dependent increase in apoptosis in melanoma cells.

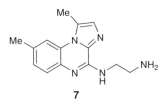

7

4. IKK TOXICOLOGY

As a result of embryonic lethality observed in both IKK-2 and NEMO knockout animals secondary to hepatocyte apoptosis, there have been long-standing concerns regarding potential toxicity associated with IKK inhibition [36]. Conditional knockout mice generated using a Cre/loxP system has emerged as a powerful tool to study protein deletion in specific tissues in adult animals [37]. Hepatocyte-specific conditional NEMO knockout mice (NEMOΔLPC) were generated to address potential developmental issues [38]. In this study there was no difference between wild-type and knockout animals with respect to survival, growth, or fertility, however upon challenge with TNF-α the NEMOΔLPC mice developed massive hepatocyte apoptosis compared to wild-type controls. In contrast the NEMOΔLPC animals were protected from ischemia reperfusion damage to the liver. A hepatocyte-specific IKK-2 conditional knockout (IKK2Δhepa) mouse has also been generated [39]. In this instance no increased sensitivity toward TNF-α-induced apoptosis was observed and these mice were protected from ischemic reperfusion injury. Additionally, treatment of wild-type animals with an IKK-2 inhibitor AS602868 (structure not disclosed) was found to attenuate liver injury after ischemia reperfusion. AS602868 also did not increase liver injury after TNF-α stimulation. Thus it appears that IKK-2 inhibition may not have the adverse effects on hepatocytes anticipated from earlier knockout data, however studies with additional inhibitors would be helpful in confirming this initial result.

Compound **1** was examined at a dose of 300 mg/kg in mice [40]. A C_{max} of 113 μM was obtained (> 100-fold higher than the IC_{50} against LPS-induced IL-6) and the predicted efficacious level was maintained for 4–6 h post dose. Significant apoptosis was noted in both the spleen and thymus and apoptotic foci were evident on histological evaluation. After compound **1** was dosed twice daily for 4 days at 100 mg/kg and 300 mg/kg, a significant decrease was observed in bone marrow-derived B cells and $CD4^+CD8^+$ thymocytes. In contrast, there was a dramatic increase in granulocytes. When compound **1** was dosed twice daily at 100 mg/kg in TNF receptor-deficient mice, B-cell depletion was attenuated significantly. It is not clear that this profile would be desirable in the treatment of inflammatory diseases, depending on the therapeutic window; however it is an encouraging result for the treatment of B cell malignancies such as multiple myeloma.

Another conditional knockout mouse was created in which IKK-2 was deleted in myeloid cells [41]. After LPS challenge, survival time for these mice was significantly reduced compared to control animals. This outcome is counter-intuitive given the role of IKK-2 inhibitors in decreasing LPS-induced TNF-α production. Administration of an IL-1 receptor antagonist protected against the LPS-induced mortality suggesting a role for IL-1β in the observed mortality. The authors provided data supporting the idea that IKK positively regulates the suppression of pro-IL-1β cleavage, and that IKK inhibition leads to enhanced processing of pro-IL-1β to biologically active IL-1β. In order to put this finding into context, LPS-induced mortality was examined with compound **1**. Administration of a single dose of **1** (300 mg/kg) resulted in no significant increase in mortality over control animals despite a significant increase in IL-1β. However, multiple dosing (300 mg/kg bid for 4 days prior to LPS challenge) recapitulated the granulocytosis and mortality observed with the conditional knockout mice. The clinical significance of this finding is not known at this time. However, when one considers the immunosuppressive nature of IKK-2 inhibition in combination with lethal outcomes on administration of bacterial degradation products like LPS reported in this study, there is an obvious cause for concern.

5. CLINICAL TRIALS

MLN-0415 (SAR 479746), an IKK-2 inhibitor of undisclosed structure, is in early development for inflammatory disorders; however, no results from clinical trials of selective IKK-2 inhibitors have been disclosed.

6. RECENT IKK-2 INHIBITORS

6.1 Benzamides

A recent publication [42] described efforts focused on a common hydrogen-bonding pharmacophore observed with several small molecule inhibitors of IKK. Based on the assumption that the constrained, unsubstituted amide group was essential for IKK inhibition, this publication described the discovery of 2-amino-3,5-diarylbenzamide inhibitors of both IKK-1 and IKK-2 emanating from an IKK-ε inhibitor series. Compound **8** represented the most potent compound from this series (IKK-2 $pIC_{50} = 7.0$) and exhibited good selectivity over IKK-ε, and across a wide variety of kinase enzyme and binding assays. The compound was also able to inhibit LPS-induced cytokine production in PBMC's (TNF-α $pIC_{50} = 6.1$, IL-1β $pIC_{50} = 6.4$, IL-6 $pIC_{50} = 5.7$).

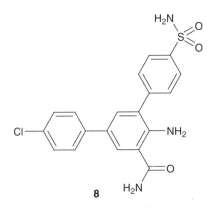

8

6.2 Thiophenes

A series of thiophene amino carboxamides has extended SAR from earlier efforts [43] focused on 5-acetylenic groups. Representative compounds **9** (IKK-2 $IC_{50} = 195\,nM$) and **10** (IKK-2 $IC_{50} = 273\,nM$) were found to inhibit IL-1β-induced IL-8 production in synovial fibroblasts derived from patients with rheumatoid arthritis ($IC_{50} = 1.1\,\mu M$ and $3.1\,\mu M$, respectively) with no cytotoxicity observed up to $30\,\mu M$.

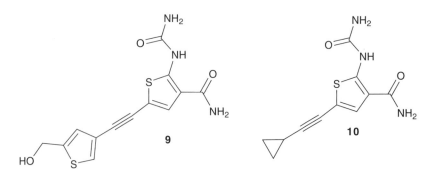

6.3 α-Carbolines

In addition to the β-carboline **1** already highlighted, a recent patent application claimed α-carbolines **11** as a new series of IKK-2 inhibitors [44]. The vast majority of examples in this disclosure were prepared by derivatization of the embedded nitrogen. Appendages included, but were not restricted to, alkyl, alkylamine, amino alcohol as well as acylated and sulfonylated derivatives. No biological data was provided.

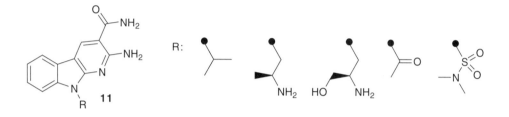

6.4 Benzothienofurans

Benzothieno[3,2-*b*]furan derivatives **12** (X = S, Y = bond) [45] have been examined as constrained versions of the structurally related thiophene-urea derivative **6**. Among the various fused furan compounds, the benzothieno[3,2-*b*]furan example **13** displayed potent inhibitory activity for IKK-2 (IC_{50} = 45 nM) due to the presence of an intramolecular non-bonded S–O interaction between the oxygen atom of the urea carbonyl and the sulfur atom of the thiophene ring. Introduction of substituents onto the benzothieno[3,2-*b*]furan to overcome the low metabolic stability of **13** (rat clearance (0.1 mg/kg i.v.) = 5,120 mL/h/kg) led to the discovery of a series of 6-alkoxy derivatives, as exemplified with compound **14** (IKK-2 IC_{50} = 37 nM, rat clearance (0.1 mg/kg i.v.) = 1,463 mL/min/kg). Furthermore, it was found that lipophilic compounds (logD > 2) exemplified by **15** showed an improved oral bioavailability while maintaining potent IKK-2 enzymatic activity (IKK-2 IC_{50} = 25 nM) and improved clearance (rat clearance (0.1 mg/kg i.v.) = 643 mL/min/kg, F% = 24).

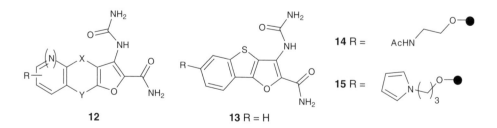

6.5 Pyrimidines and related systems

Several patent applications have appeared detailing the 2-anilinopyrimidine core structure as an IKK inhibitor scaffold (e.g. **16** [46,47]). Separate patent applications disclosed [48,49] compounds in which the *para*-fluoroanilino moiety was maintained and modifications centered on the amine portion of the sulfonamide as in **17**. The design and SAR of a series of benzothiophene substituted pyrimidines related to **18** (IKK-2 IC_{50} = 200 nM) have been disclosed [50]. Compound **3** from this series was efficacious in animal models of inflammation (vida supra). A second research group has independently identified benzothiophene-substituted pyrimidines as inhibitors of IKK-2 [51].

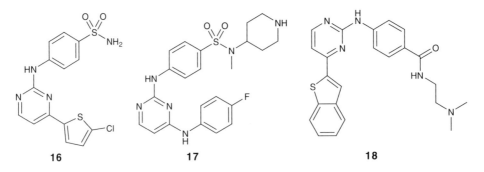

16	**17**	**18**

Two of the most recent patent applications detailed the fused thienopyrimidine core structure, **19** [52], and the related bicyclic structure typified by compound **20** [53]. No biological data was reported in either case.

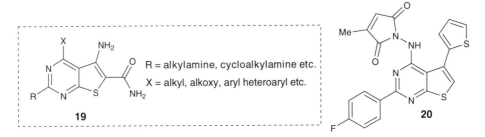

R = alkylamine, cycloalkylamine etc.

X = alkyl, alkoxy, aryl heteroaryl etc.

19	**20**

6.6 Fused pyridines

The thienopyridine IKK inhibitors represent a structural class that has received considerable attention. The evolution of this series has been reported [54] including the hit-to-lead strategy and initial structure–activity relationships. Most recently, a patent publication [55] identified compound **2** (vida supra) as a potent inhibitor of IKK-2 (IKK-2 IC_{50} = 26 nM, HeLa cell assay IC_{50} = 700 nM). Significant improvements included introduction of the halogenated propyl group which resulted in enhanced plasma exposure following a single 10 mg/kg oral dose (compound **2**, plasma concentration = 0.65 μM at 10 h vs. compound **21**, plasma concentration $= 0.08\,\mu M$ at 10 h).

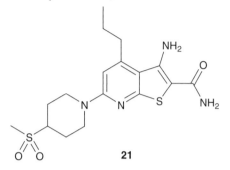

21

Constrained tricyclic analogs **22**, have also been disclosed in the recent patent literature [56], although no biological data was reported. Most examples contained the fused cyclohexyl ring, as in **23**.

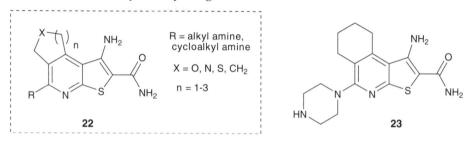

22

R = alkyl amine, cycloalkyl amine

X = O, N, S, CH₂

n = 1-3

23

6.7 Indole-7-carboxamides and related systems

After the initial disclosure of the structurally distinct indole carboxamide class of IKK-2 inhibitors **24** in 2005 [57], several patent applications on related compounds emerged. Initial exploration around the piperidine sulfonamide produced compounds such as **25** [58].

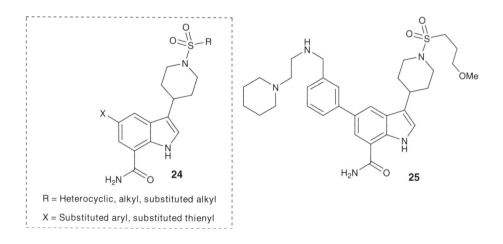

24

R = Heterocyclic, alkyl, substituted alkyl

X = Substituted aryl, substituted thienyl

25

Later, substitution was restricted to the ethyl sulfonamide **26** with significant additional SAR around the indole 5-position including the *meta*-substituted phenyl substitution pattern as in **25**, and substituted thiophene, pyrazole, and pyridine compounds [59]. An additional patent application [60] expanded the series to include examples linked with O, N, and S heteroatoms. Closely related examples which contain the 2-carboxamide moiety have also appeared in the patent literature [61].

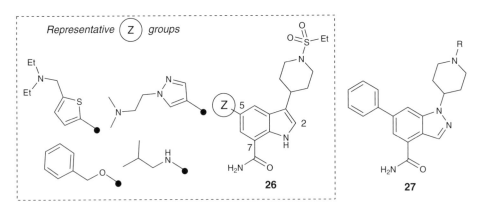

A closely related indazole system **27** has also appeared with initial examples focused on piperidine nitrogen substitution [62].

6.8 Fused imidazopyridazine and imidazopyridine systems

A series of tetracyclic structures, based on **7** as a structural lead, has been reported to display potent IKK-2 inhibition [63]. For example, compound **28** exhibited an IKK-2 IC_{50} value of 18 nM. Despite the poor physicochemical properties reported for these tetracycles, a dose proportional reduction in mouse serum TNF-α levels was observed with this compound *in vivo*. SAR studies around a tricyclic imidazothienopyrazine core identified **29** (IKK-2 IC_{50} = 13 nM) [64]. This compound had modest activity (49% reduction in serum TNF-α at 30 mg/kg) when dosed orally in a murine LPS-induced TNF-α model.

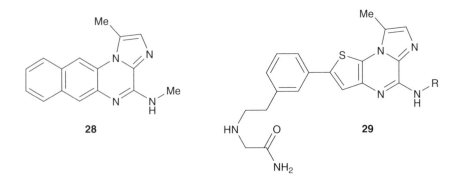

A series of patent applications closely mimicking the topological disposition of the tricyclic template in **29** describes compounds of the general structure **30** [65,66].

Structure **31** represents isomeric thiazole [67] and thiophene [68] tricyclic systems which have also been claimed in the patent literature as IKK-2 inhibitors.

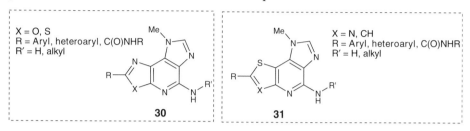

An additional application describes tricyclic compounds of general structure **32** [69]. Compound **33** represents an example where the pyrrole nitrogen has been substituted with an alkyl chain substituted with a diol functionality. This substitution pattern on the pyrrole nitrogen is reminiscent of that described for the α-carboline **11**.

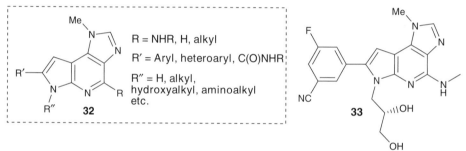

6.9 Miscellaneous

Expanding upon earlier patent examples, a series of novel, pyridyl cyanoguanidine compounds **34**, has recently been claimed as IKK inhibitors [70] for use in the treatment of hyperproliferative and neoplastic diseases. Tricyclic compounds have also been claimed in the patent literature [71] as inhibitors of a variety of kinases, including IKK-2. Compound **35** is exemplified as a compound with potent IKK-2 inhibitory activity (IKK-2 IC$_{50}$ < 100 nM).

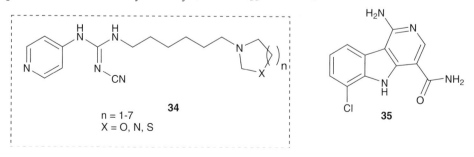

7. CONCLUSIONS

IKK-2 inhibitors have shown efficacy in a number of preclinical animal models, including arthritis and inflammation, ischemia reperfusion injury, airway disease, and melanoma. These results support the notion that IKK-2 inhibitors have the potential to ameliorate a number of human disease states. Concerns of liver toxicity may have been alleviated by studies with genetically modified animals with hepatocyte-specific deletions of IKK-2 and reports of a lack of adverse effects on the liver with administration of IKK-2 inhibitors; however additional such studies will increase the level of confidence in this positive finding. A greater toxicity concern involves the potential for IKK-2 inhibitor immunosupression to be adversely affected by bacterial degradation products (e.g. LPS) resulting in a septic shock-like lethality in mice.

In the past few years there has been an increase in the reports of new IKK-2 inhibitors in both the primary and patent literature. These represent additional diversity and a creative expansion of earlier SAR. Most importantly, several compounds have advanced to human clinical trials. Reports on the safety and efficacy of these compounds in human trials are eagerly awaited.

REFERENCES

[1]   M. Gaestel, A. Mengel, K. U. Bothe and K. Asadullah, *Curr. Med. Chem.*, 2007, **14**, 2214.
[2]   A. A. Mortlock and A. J. Backer, *Comprehensive Medicinal Chemistry II*. Elsevier LTD, Oxford, 2006, Vol. 7, p. 183.
[3]   F. C. Zusi, W. J. Pitts and J. R. Burke, *Target Validation in Drug Discovery*. Academic Press, Boston, 2007, p. 199.
[4]   P. D. G. Coish, P. L. Wickens and T. B. Lowinger, *Expert Opin. Ther. Patents*, 2006, **16**, 1.
[5]   Q. Li, D. Van Antwerp, F. Mercurio, K.-F. Lee and I. M. Verma, *Science*, 1999, **284**, 321.
[6]   T. Luedde, N. Beraza and C. Trautwein, *J. Gastroentero. Hepatol.*, 2006, **21**, S43–S46.
[7]   E. Dejardin, *Biochem. Pharmacol.*, 2006, **72**, 1161.
[8]   J. Hiscott, T.-L. A. Nguyen, M. Arguello, P. Nakhaei and S. Paz, *Oncogene*, 2006, **25**, 6844.
[9]   J. Strnad and J. R. Burke, *Trends Phamacol. Sci.*, 2007, **28**, 142.
[10] H. J. Kim, N. Hawke and A. S. Baldwin, *Cell Death and Differentiation*, 2006, **13**, 738.
[11] P. Bamborough, J. A. Christopher, G. J. Cutler, M. C. Dickson, G. W. Mellow, J. F. Morey, C. B. Patel and L. M. Shewchuk, *Bioorg. Med. Chem. Lett.*, 2006, **16**, 6236.
[12] J. Hiscott, *Cytokine Growth Factor Rev.*, 2007, **18**, 483.
[13] N. Kishore, Q. K. Huynh, S. Mathialagan, T. Hall, S. Rouw, D. Creely, G. Lange, J. Carol, B. Reitz, A. Donnelly, H. Boddupalli, R. G. Combs, K. Kretzmer and C. S. Tripp, *J. Biol. Chem.*, 2002, **277**, 13840.
[14] T.-L. Chau, R. Gioia, J.-S. Gatot, F. Patrascu, I. Carpentier, J.-P. Chapelle, L. O'Neill, R. Beyaert, J. Piette and A. Chariot, *Trends Biochem. Sci.*, 2008, **33**, 171.
[15] Z.-W. Li, W. Chu, Y. Hu, M. Delhase, T. Deerinck, M. Ellisman, R. Johnson and M. Karin, *J. Exp. Med.*, 1999, **189**, 1839.
[16] M. Windheim, M. Stafford, M. Peggie and P. Cohen, *Mol. Cell. Biol.*, 2008, **25**, 1783.
[17] L. Palkowitsch, J. Leidner, S. Ghosh and R. B. Marienfeld, *J. Biol. Chem.*, 2008, **283**, 76.
[18] F.-L. Liu, C.-H. Chen, S.-J. Chu, J.-H. Chen, J.-H. Lai, H.-K. Sytwu and D.-M. Chang, *Rheumatology*, 2007, **46**, 1266.
[19] G. Manning, D. B. Whyte, R. Martinez, T. Hunter and S. Sadarsanam, *Science*, 2002, **298**, 1912.

[20] J. Bain, L. Plater, M. Elliott, N. Shpiro, C. J. Hastie, H. McLauchlan, I. Klevernic, J. S. C. Arthur, D. R. Alessi and P. Cohen, *Biochem. J.*, 2007, **408**, 297.

[21] M. W. Karaman, S. Herrgard, D. K. Treiber, P. Gallant, C. E. Atteridge, B. T. Campbell, K. W. Chan, P. Ciceri, M. I. Davis, P. T. Edeen, R. Faraoni, M. Floyd, J. P. Hunt, D. J. Lockhart, Z. V. Milanov, M. J. Morrison, G. Pallares, H. K. Patel, S. Pritchard, L. M. Wodica and P. P. Zarrinkar, *Nat. Biotech.*, 2008, **26**, 127.

[22] D. Wen, Y. Nong, J. G. Morgan, P. Gangurde, A. Bielecki, J. DaSilva, M. Keaveney, H. Cheng, C. Fraser, L. Schopf, M. Hepperle, G. Harriman, B. D. Jaffee, T. D. Ocain and Y. Xu, *J. Pharmacol. Exp. Ther.*, 2006, **317**, 989.

[23] L. Schopf, A. Savinainen, K. Anderson, J. Kujawa, M. DuPont, M. Silva, E. Siebert, S. Chandra, J. Morgan, P. Gangurde, D. Wen, J. Lane, Y. Xu, M. Hepperle, G. Harriman, T. Ocain and B. Jaffee, *Arthritis Rheum.*, 2006, **54**, 3163.

[24] E. S. Izmailova, N. Paz, H. Alencar, M. Chun, L. Schopf, M. Hepperle, J. H. Lane, G. Harriman, Y. Xu, T. Ocain, R. Weissleder, U. Mahmood, A. M. Healy and B. Jaffee, *Arthritis Rheum.*, 2007, **56**, 117.

[25] J. D. Ginn, R. J. Sorcek, M. R. Turner and E. R. R. Young, *Patent Application US 2007293533*, 2007.

[26] R. Waelchli, B. Bollbuck, C. Bruns, T. Buhl, J. Eder, R. Feifel, R. Hersperger, P. Janser, L. Revesz, H.-G. Zerwes and A. Schlapbach, *Bioorg. Med. Chem. Lett.*, 2006, **16**, 108.

[27] N. C. Moss, W. E. Stansfield, M. S. Willis, R.-H. Tang and C. H. Selzman, *Am. J. Physiol. Heart Circ. Physiol.*, 2007, **293**, H2248.

[28] A. Tanaka, S. Muto, M. Konno, A. Itai and H. Matsuda, *Cancer Res.*, 2006, **66**, 419.

[29] Y. Onai, J.-I. Suzuki, Y. Maejima, G. Haraguchi, S. Muto, A. Itai and M. Isobe, *Am. J. Physiol Heart Circ. Physiol.*, 2006, **292**, H530.

[30] I. M. Adcock, K. F. Chung, G. Caramori and K. Ito, *Eur. J. Pharmacol.*, 2006, **533**, 118.

[31] M. A. Birrell, S. Wong, E. L. Hardaker, M. C. Catley, K. McCluskie, M. Collins, S. Haj-Yahia and M. G. Belvisi, *Mol. Pharmacol.*, 2006, **69**, 1791.

[32] M. A. Birrel, K. McCluskie, E. Hardaker, R. Knowles and M. G. Belvisi, *Eur. Respir. J.*, 2006, **28**, 1236.

[33] C. M. Annuziata, R. E. Davis, Y. Demchenko, W. Bellamy, A. Gabrea, F. Zhan, G. Lenz, I. Hanamura, G. Wright, W. Xiao, S. Dave, E. M. Hurt, B. Tan, H. Zhao, O. Stephens, M. Santra, D. R. Williams, L. Dang, B. Barlogie, J. D. Shaughnessy, W. M. Kuehl and L. M. Staudt, *Cancer Cell*, 2007, **12**, 115.

[34] T. Hideshima, P. Neri, P. Tassone, H. Yasui, K. Ishitsuka, N. Raje, D. Chauhan, K. Podar, C. Mitsiades, L. Dang, N. Mushi, P. Richardson, D. Schenkein and K. C. Anderson, *Clin. Cancer. Res.*, 2006, **12**, 5887.

[35] J. Yang, K. I. Amiri, J. R. Burke, J. A. Schmid and A. Richmond, *Clin. Cancer Res.*, 2006, **12**, 950.

[36] R. F. Schwabe and D. A. Brenner, *Gastroenterology*, 2007, **132**, 2601.

[37] V. Brault, V. Besson, L. Magnol, A. Duchon and Y. Herault, *Handbook of Experimental Pharmacology*, Springer-Verlag, New York, NY, 2007, Vol. 178, p. 121.

[38] N. Beraza, T. Ludde, U. Assmus, T. Roskams, S. V. Borght and C. Trautwein, *Gastroenterology*, 2007, **132**, 2504.

[39] T. Luedde, U. Assmus, T. Wustefeld, A. M. Vilsendorf, T. Roskams, M. Schmidt-Supprian, K. Rajewsky, D. A. Brenner, M. P. Manns, M. Pasparakis and C. Trautwein, *J. Clin. Invest.*, 2005, **115**, 849.

[40] K. Nagashima, V. G. Sasseville, D. Wen, A. Bielecki, H. Yang, C. Sompson, E. Grant, M. Hepperle, G. Harriman, B. Jaffee, T. Ocain, Y. Xu and C. C. Fraser, *Blood*, 2006, **107**, 4266.

[41] F. R. Greten, M. C. Arkan, J. Bollrath, L.-C. Hsu, J. Goode, C. Miething, S. I. Goktuna, M. Neuenhahn, J. Fierer, S. Pacian, N. Van Rooijen, Y. Xu, T. Ocain, B. B. Jaffee, D. H. Busch, J. Duyster, R. M. Schmid, L. Eckmann and M. Karin, *Cell*, 2007, **130**, 918.

[42] J. A. Christopher, B. G. Avitabile, P. Bamborough, A. C. Champigny, G. J. Cutler, S. L. Dyos, K. G. Grace, J. K. Kerns, J. D. Kitson, G. W. Mellor, J. V. Morey, M. A. Morse, C. F. O'Malley, C. B. Patel, N. Probst, W. Rumsey, C. A. Smith and M. J. Wilson, *Bioorg. Med. Chem. Lett.*, 2007, **17**, 3972.

[43] D. Bonoafoux, S. Bonar, L. Christine, M. Clare, A. Donnelly, J. Guzova, N. Kishore, P. Lennon, A. Libby, S. Mathialagan, W. McGhee, S. Rouw, C. Sommers, M. Tollefson, C. Tripp, R. Weier, S. Wolfson and Y. Min, *Bioorg. Med. Chem. Lett.*, 2005, **15**, 2870.

[44] M. E. Hepperle, J. F. Liu, R. S. Rowland and D. Vitharana, *Patent Application* WO 2007097981, 2007.

[45] H. Sugiyama, M. Yoshida, K. Mori, T. Kawamoto, S. Sogabe, T. Takagi, H. Oki, T. Tanaka, H. Kimura and Y. Ikeura, *Chem. Pharm. Bull.*, 2007, **55**, 613.

[46] F.-W. Sum, D. W. Powell, Y. Zhang, L. Chen, S. L. Kincaid, L. D. Jennings, Y. Hu, A. M. Gilbert and M. G. Bursavich, *Patent Application* US 2006079543, 2006.

[47] Y. Hu, F.-W. Sum, M. Di Grandi and E. Norton, *US Patent* 2007244140, 2007.

[48] M. Bosch, M. Bouaboula, P. Casellas, S. Jegham, J. F. Nguefack and J. Wagnon, *Patent Application* FR 2888239, 2007.

[49] J. Wagnon, J.-F. Nguefack, S. Jegham, M. Bosch, M. Bouaboula, P. Casellas, B. Tonnerre, J.-A. Olsen and S. Mignani, *Patent Application* WO 2007006926, 2007.

[50] R. Waelchli, B. Bollbuck, C. Bruns, T. Buhl, J. Eder, R. Feifel, R. Hersperger, P. Janser, L. Revesz, H.-G. Zerwes and A. Schlapbach, *Bioorg. Med. Chem. Lett.*, 2006, **16**, 108.

[51] K. R. Dahnke, H.-S. Lin, C. Shih, Q. M. Wang, B. Zhang and M. E. Richett, *Patent Application* WO 2007092095, 2007.

[52] M. Wada, N. Sueda, T. Komine, K. Katsuyama, H. Tsuchida, R. Kojima and O. Kawamura, *Patent Application* WO 2008020622, 2008.

[53] M.-H. Kim, C.-H. Park, K. Chun, B.-K. Oh, B.-Y. Joe, J.-H. Choi, H.-M. Kwon, S.-C. Huh, S. Won, K. H. Kim and S.-M. Kim, *Patent Application* WO 2007102679, 2007.

[54] T. Morwick, A. Berry, J. Brickwood, M. Cardozo, K. Catron, M. DeTuri, J. Emeigh, C. Homon, M. Hrapchak, S. Jacober, S. Jakes, P. Kaplita, T. A. Kelly, J. Ksiazek, M. Liuzzi, R. Magolda, C. Mao, D. Marshall, D. McNeil, A. Prokopowicz, C. Sarko, E. Scouten, C. Sledziona, S. Sun, J. Watrous, J. P. Wu and C. L. Cywin, *J. Med. Chem.*, 2006, **49**, 2898.

[55] J. D. Ginn, R. J. Sorcek, M. R. Turner and E. R. R. Young, *Patent Application* US 2007293533, 2007.

[56] Y. Okamoto, K. Hattori, H. Kubota, I. Sato, T. Kanayama, K. Yokoyama, Y. Terai and M. Takeuchi, *Patent Application* WO 2005123745, 2005.

[57] I. R. Baldwin, P. Bamborough, J. A. Christopher, J. K. Kerns, T. Longstaff and D. D. Miller, *WO Patent* 2005067923, 2005.

[58] J. K. Kerns, M. Lindenmuth, X. Lin, H. Nie and S. M. Thomas, *Patent Application* WO 2006034317, 2006.

[59] J. Deng, J. K. Kerns, Q. Jin, G. Lin, X. Lin, M. Lindenmuth, C. E. Neipp, H. Nie, S. M. Thomas and K. L. Widdowson, *Patent Application* WOt 2007005534, 2007.

[60] J. K. Kerns, J. Busch-Petersen, H. Li, J. C. Boehm, H. Nie and J. J. Taggart, *Patent Application* WO 2007062318, 2007.

[61] J. K. Kerns, *Patent Application* WO 2007076286, 2007.

[62] J. F. Callahan, J. K. Kerns, and X. Lin, *Patent Application* WO 2007102883, 2007.

[63] F. Beaulieu, C. Ouellet, E. H. Ruediger, M. Belema, Y. Qiu, X. Yang, J. Banville, J. R. Burke, K. R. Gregor, J. F. MacMaster, A. Martel, K. W. McIntyre, M. A. Pattoli, C. F. Zusi and D. Vyas, *Bioorg. Med. Chem. Lett.*, 2007, **17**, 1233.

[64] M. Belema, A. Bunker, V. N. Nguyen, F. Beaulieu, C. Ouellet, Y. Qiu, Y. Zhang, A. Martel, J. R. Burke, K. W. McIntyre, M. A. Pattoli, C. Daloisio, K. M. Gillooly, W. J. Clarke, P. J. Brassil, F. C. Zusi and D. M. Vyas, *Bioorg. Med. Chem. Lett.*, 2007, **17**, 4284.

[65] W. J. Pitts, M. Belema, P. Gill, J. Kempson, Y. Qiu, C. Quesnelle, S. H. Spergel and C. Zusi, *Patent Application* WO 2004106293, 2004.

[66] A. Dyckman, W. J. Pitts, M. Belema, P. Gill, J. Kempson, Y. Qiu, C. Quesnelle, S. H. Spergel and C. F. Zusi, *US Patent* 2006106051, 2006.

[67] J. Das, J. Kempson, W. J. Pitts and S. H. Spergel, *Patent Application* WO 2006053166, 2006.

[68] W. J. Pitts, J. Das, Y. Qiu and S. H. Spergel, *US Patent* 2006178393, 2006.

[69] W. J. Pitts, J. Kempson, J. Guo, J. Das, C. M. Langevine, S. H. Spergel and S. H. Watterson, *Patent Application* WO 2006122137, 2006.

[70] F. Bjorkling and H. W. Dannacher, *Patent Application* WO 2006066584, 2006.

[71] J. F. Truchon, N. Lachance, C. Lau, Y. Leblanc, C. Mellon, P. Roy, E. Isabel, R. D. Otte and J. R. Young, *WO Patent*, 2007061764, 2007.

CHAPTER 11

Protease Inhibitors for the Potential Treatment of Chronic Obstructive Pulmonary Disease and Asthma

Weimin Liu and **Eugene R. Hickey**

1. INTRODUCTION

Chronic lower respiratory diseases, including chronic obstructive pulmonary disease (COPD) and asthma, are among the leading causes of deaths and health burdens worldwide [1,2]. Despite some shared symptoms, COPD and asthma are different disease entities with distinct patterns of inflammation involving different inflammatory cells and proteins [3].

In individuals susceptible to cigarette smoke-induced lung damage, inhalational exposure to cigarette smoke triggers an inflammatory response in airways and alveoli that, when chronically present, can lead to COPD manifestation. At later stages, the disease process is thought to be mediated by an increase in protease activity along with a decrease in antiprotease activity. The disruption of

Boehringer Ingelheim Pharmaceuticals, Inc., 900 Ridgebury Road, Ridgefield, CT 06877, USA

Annual Reports in Medicinal Chemistry, Volume 43
ISSN 0065-7743, DOI 10.1016/S0065-7743(08)00011-0

the balance between proteolytic enzymes and their inhibitors has long been considered the major cause of COPD. Small molecule protease inhibitors that inhibit these proteolytic enzymes have been targeted to prevent the progression of airflow obstruction in COPD as a naturally occurring human deficiency of α-1-antitrypsin, a serine protease inhibitor, confers susceptibility to COPD [4,5].

Asthma is a chronic inflammatory disease characterized by airway hyper-responsiveness, tissue remodeling, and airflow obstruction after exposure to allergic and environmental stimuli [6]. Mast cell and leukocyte serine proteases are elevated in the airways of asthmatic patients [7,8]. In addition, patients with reduced antiprotease activity as a result of α-1-proteinase inhibitor deficiency have an increased propensity to develop asthma [9].

Serine proteases secreted by immune cells contribute to bronchial remodeling in asthma. Tryptase from mast cells stimulates the synthesis of type I collagen by human lung fibroblasts [10]. Neutrophils secrete the proteases neutrophil elastase (NE), cathepsin (Cat) G, and neutrophil proteinase 3. Differential expression of serine proteases and/or their inhibitors in asthmatic vs. healthy subjects strongly suggests their importance in asthma [11].

Cysteine proteases and matrix metalloproteinases (MMPs) have been implicated in chronic inflammation and lung disease by controlling inflammation through attenuation of the aforementioned protease–antiprotease imbalance. The cysteine protease Cat S plays a key role in regulating antigen presentation and immunity [12]. A Cat S inhibitor has been found to block the rise in IgE titers and eosinophil infiltration in the lung in a mouse model of pulmonary hypersensitivity, suggesting that Cat S may be involved in asthma [13].

A number of MMPs are implicated in the pathophysiology of COPD and asthma, yet MMP profiles in COPD and asthma differ. The current literature contains evidence supporting MMP-9 inhibitors for asthma and MMP-12 inhibitors for COPD [14], with the caveat that most studies have been done in rodent models.

Serine protease inhibitors targeting tryptase, NE, Cat G, and proteinase 3 have reached clinical trials. Inhibitors of the cysteine proteases Cats K, L, and S have not reached clinical trials for respiratory indications. MMP inhibitors targeting MMP-9 and MMP-12 are in various stages of discovery; one MMP inhibitor has recently entered a phase I trial for COPD.

2. SERINE PROTEASE INHIBITORS

A number of neutrophil serine proteases (NE, Cat G, and proteinase 3) and mast cell serine proteases (tryptase and chymase) have been implicated in airway diseases [15,16]. A recent review summarized some of the serine protease inhibitors targeting the treatment of asthma [11].

2.1 Tryptase inhibitors

The major enzymes stored and secreted by mast cells are tryptase and chymase, which play key roles in asthma. Their actions include inflammation, tissue

remodeling, bronchial hyperresponsiveness [17], and β-adrenergic desensitiza-
tion of airway smooth muscle [18]. Several mechanisms have been proposed for
the role of tryptase in allergic inflammation, including promoting histamine
release and recruiting leukocytes. In addition, tryptase has an effect on narrowing
of the bronchial lumen, promoting bronchoconstriction, and asthma-associated
airway remodeling [16]. The exploration of tryptase inhibitors, particularly
β-tryptase inhibitors, was quite active in the earlier half of the decade, with
multiple compounds entering the clinic. The state of this area prior to 2004 has
been extensively reviewed [19]. Therefore, this section will focus primarily on
developments since 2004.

AVE-8923, a β-tryptase [20] inhibitor (structure not disclosed), had advanced
into a phase I clinical trial for asthma in 2007 [21]. However, as of February 2008,
it is no longer listed in the R&D portfolio on the organization's website [22].
A number of tryptase inhibitors from the same source have been disclosed.
Compound **1** was reported to have an IC_{50} of 68 nM against β-tryptase with
54-fold selectivity against α-tryptase and at least 147-fold selectivity against factor
Xa, thrombin, chymase, and trypsin [23]. A related compound (**2**) was reported to
have a K_i of 1.3 nM against β-tryptase. The primary carboxamide was introduced
to overcome hERG inhibition [24]. Further optimization of these molecules led to
3, which was relatively weak ($K_i = 400$ nM), but had oral bioavailability (F) of
90% in rats [25].

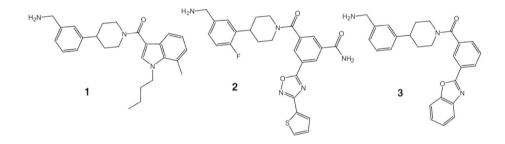

RWJ-58643 (**4**) was discontinued from an allergic rhinitis trial (phase IIa) due to
its taste [26]. In this single center, randomized and double-blind study, single
topical nasal doses of **4** (100, 300, and 600 μg) or placebo were given to 16 male
patients with grass pollen allergic rhinitis followed by a nasal allergen challenge.
Significant reductions of symptoms, as well as eosinophil and IL-5 levels were
found in the two lower dose groups, however, the efficacy was not dose responsive.
Futhermore, late eosinophilia was observed in the higher dose groups [27].

Despite the fact that some β-tryptase inhibitors have been known in the
literature for some time, their pharmacokinetic properties have only come to light
in the past few years. One example is BMS-262084 (**5**). A rat pharmacokinetic study
revealed that multiple factors appeared to impact the oral absorption of this
compound [28] leading to an F value in rat of about 4%. Trypsin binding and P-gp
efflux were suggested as potential causes. Inhibition of trypsin with aprotinin
appeared to increase the apparent oral absorption of **5** to 11–22%. In addition,

5 may also affect the opening of tight junctions in the intestinal epithelium, as evidenced by the enhancement of inulin absorption in the presence of **5**.

In efforts to improve the selectivity of an earlier series of α-ketoheterocycle-based tryptase inhibitors, a series of spirocyclic piperidine amide derivatives was identified [29]. JNJ-27390467 (**6**) is a competitive inhibitor of β-tryptase ($K_i = 3.7$ nM) and is 675-fold selective over trypsin. In an asthma model (antigen-sensitized allergic sheep) at 30 mg/kg p.o., b.i.d. dosing (0.4–1.0 μM plasma level), **6** completely blocked the late-phase response with no effect on the acute early-phase response. Similarly, in a corresponding guinea pig model with 20 mg/kg p.o., b.i.d. dosing, **6** reduced total lung resistance by 90% and increased dynamic lung compliance by 65%. Compound **6** has demonstrated good pharmacokinetic properties in multiple species.

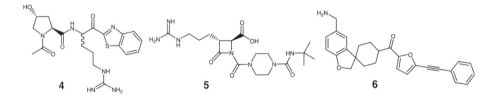

Increasing selectivity and drug-like properties of a high-throughput screening hit led another group to a 4-substituted benzylamine derivative M58539 (**7**), with an IC$_{50}$ of 5.0 nM [30]. Its potency against β-tryptase appears to be independent of Zn^{2+} ion binding. The ethylene-diamine linker offered a 300-fold boost in potency over the initial hit, while the 2-chloronaphthyl group appeared to compound this increase by another 15-fold. Compound **7** also has a selectivity factor of $>20,000$ against other serine proteases, including factor Xa, plasmin, thrombin, and elastase.

A number of guanidine-based tryptase inhibitors were identified in the past. Optimization of these compounds for oral delivery through the replacement of the guanidine moiety with primary and secondary amines led to the discovery of BMS-354326 (**8**), a 3-piperdinyl derivative [31]. It is a potent tryptase inhibitor (IC$_{50}$ = 1.8 nM) and is highly selective against trypsin ($>5,000$-fold) and other serine proteases. However, its selectivity vs. plasmin (169-fold) is modest.

A solution phase parallel synthesis yielded azepanone **9**, as a potent (IC$_{50}$ = 38 nM) inhibitor of human tryptase [32]. Its selectivity profile against other serine proteases (trypsin, plasmin, urokinase, tissue plasminogen activator, activated protein C, α-thrombin, and factor Xa) ranges from 330-fold to >870-fold.

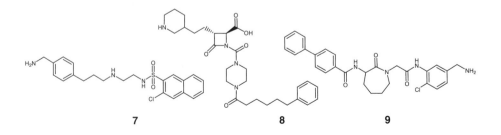

The evolution of an oxadiazole series has been chronicled by three publications, which provided some detailed SAR, pharmacokinetic data, as well as a number of X-ray co-crystal structures. Optimization of an initial hit led to **10**, with a K_i of 5.4 nM against human βII-tryptase [33]. Although the selectivity against trypsin was modest (35-fold), the selectivity against other trypsin-like serine proteases was significantly higher. However, **10** still suffered from poor oral bioavailability (8.4%) in rats. To further improve the selectivity profile, fluorinated benzamide moieties were installed in place of the carbamate (**11**). The selectivity against trypsin was improved to over 400-fold, while the tryptase activity was maintained [34]. In comparison with **10**, the introduction of halogenated phenyl rings decreased clearance and increased mean residence time in rats. Unfortunately, no improvement in absorption was observed. In an apparent parallel effort to improve kinetic solubility and to overcome hERG inhibition, the ether linker was replaced by an amide (**12**). Even though the tryptase activity was retained ($K_i = 9.9$ nM), **12** was almost equipotent against trypsin. An improvement over previous compounds was observed in the hERG inhibition profile of **12**, which demonstrated no inhibition at 1 µM compared to 40% by the initial lead at the same concentration [35].

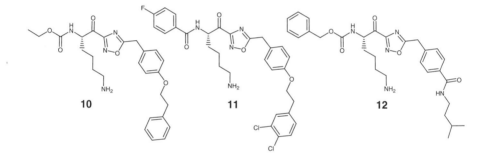

10   **11**   **12**

2.2 Chymase and cathepsin G inhibitors

Chymase is expressed and accumulated primarily in mast cells, while Cat G is expressed in neutrophils and mast cells. Chymase degrades human airway smooth muscle pericellular matrix and may affect the level of cytokines and growth factors [17]. It is capable of inducing secretion from airway submucosal gland cells at picomolar concentration [36], an effect that has also been observed with Cat G and NE [37]. Chymase has also been found to activate MMPs, which hydrolyze a broad range of matrix proteins, including collagen and elastin [38].

As of September 2007, a chymase inhibitor, TPC-806, was reported to have entered a phase II study for a cardiovascular indication [39]. However, no respiratory indication associated with this compound was reported. The structure of TPC-806 has not been confirmed, but a recent patent application from the same organization [40] describes a salt form of a single compound (**13**) with an IC_{50} against chymase at 1–10 nM.

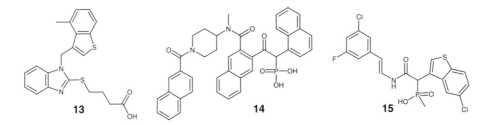

Another chymase inhibitor is reported to be in early stage development for asthma [41]. From an original hit for Cat G, a series of β-ketophosphonates has been developed, including JNJ-10311795 or RWJ-355871 (**14**), which inhibits both Cat G ($K_i = 38$ nM) and chymase ($K_i = 2.3$ nM) [42]. In a rat acute peritonitis model, **14** inhibited neutrophil influx up to 70% with i.v. dosing (total dose = 45.5 mg/kg, bolus followed by infusion). It also inhibited the increase in pro-inflammatory cytokines induced by glycogen. Similarly, in a rat acute airway inflammation model, **14** (aerosol dosing at 1 or 3 mg/kg) inhibited the LPS-induced infiltration of neutrophils and lymphocytes in the lungs.

Guided by an X-ray co-crystal structure with chymase, extensive modification of the two naphthyl groups immediately flanking the phosphonic acid moiety in **14**, as well as the phosphonic acid group itself, led to the discovery of **15** (chymase $K_i = 11$ nM) [43]. Even though the selectivity profile was not given for this compound, a related compound was reported to be selective for chymase over Cat G and trypsin. The rat pharmacokinetic data of **15** was also reported ($F = 39\%$, oral $t_{1/2} = 4.5$ h, Cmax = 6 μM).

Two new series of [1,4]diazepane-2,5-diones have been published. Compound **16**, which represents a series of sulfonamides, inhibits human chymase with an IC_{50} of 27 nM, and demonstrates 50–100-fold selectivity over Cat G [44]. The S-configuration of the chiral center is critical as the R-enantiomer is inactive. To improve the aqueous stability of **16**, the sulfonamide was replaced with a urea moiety. Compound **17** retains most of the potency and selectivity of **16** ($IC_{50} = 170$ nM against chymase and 28 μM against Cat G), while showing a 6-fold improvement in stability. In a mouse chronic dermatitis model, **17** inhibited DNFB-induced ear edema by 35.8% at 10 mg/kg p.o., q.d. dosing [45].

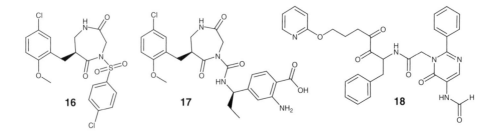

NK3201 (**18**) was reported in 2001 as an anti-tissue-remodeling agent [46]. It inhibits chymases at nanomolar potency (IC_{50} = 2.5, 1.2, and 28 nM, for human, dog, and hamster, respectively) with no significant effect on other serine proteases [47]. Its inhibitory effect on chymase prevents the development of colitis in mice via the suppression of MMP-9 activation [48]. In a bleomycin-induced pulmonary fibrosis hamster model, chymase activity from lung extract of animals treated with **18** (30 mg/kg p.o., q.d.) was less than 50% of the control and was close to the level of naïve animals. A reduced mRNA level of collagen III and a decreased ratio of fibrotic area to total pulmonary area were also observed [49].

2.3 Neutrophil elastase inhibitors

The release of NE from azurophil granules upon neutrophil activation has been suggested to play an important role in the exacerbation of asthma [50]. Cystic fibrosis and COPD are marked by neutrophil infiltration, with NE present at high concentration [51]. Therefore, targeting NE for airway diseases has been an active area for the past two decades. However, a number of compounds that advanced into the clinic for COPD eventually dropped out due to poor pharmacokinetic profiles or a narrow therapeutic window [52].

One NE inhibitor, sivelestat (**19**) has been launched in Japan [53] and South Korea [54] for acute lung injury associated with systemic inflammatory response syndrome. As of February 2008, it was reported to be in a phase II clinical study for acute respiratory failure associated with community-acquired pneumonia [55].

The elastase inhibitor BAY 71-9678 (structure not disclosed) was reported to be in a phase I trial for pulmonary hypertension in COPD [56]. In a recent PCT publication [57] from the same organization, an NE inhibitor (**20**) was described to have an IC_{50} of 5 nM.

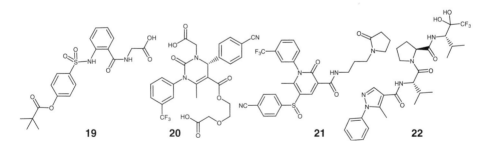

19   **20**   **21**   **22**

NE inhibitor AZD9668 was reported to be in a phase I trial for COPD [58]. Although the structure is undisclosed, a large number of patent applications focused on 2-pyridones have been published by the same organization. For example, **21** has been reported to inhibit NE with an IC_{50} of 0.5 nM [59].

In 2005, preclinical data on a peptide-based NE inhibitor was presented [60,61]. BL-5160 (**22**) is equipotent against human and rat NE (IC_{50} = 6.4 and 6.6 nM, respectively). The selectivity over pancreatic elastases and other proteases is >100-fold. Predosing of hamsters with **22** (1–10 mg/kg p.o.)

0.5–6 h before human NE instillation suppressed lung hemorrhage in a dose-dependent fashion. Studies with a rat reverse passive Arthus model revealed that dosing of **22** (1–10 mg/kg p.o.) 2 h after anti-ovalbumin serum injection into the trachea and ovalbumin injection into the jugular vein resulted in a dose-dependent inhibition of hemorrhage and leukocyte infiltration into the lungs.

2.4 Proteinase 3 inhibitors

Elafin is one of the major endogenous inhibitors of human NE and proteinase 3 [62]. A recombinant elafin for the treatment of pulmonary arterial hypertension was reported to be well tolerated in an intravenous, dose escalating phase I trial [63]. In March 2007, elafin was granted orphan drug status by the European Commission [64].

3. CYSTEINE PROTEASE INHIBITORS

Cells found to release cathepsins include, but are not limited to, macrophages (Cats B, K, L, and S), mast cells (Cats L and C (dipeptidyl-peptidase I)), and smooth muscle cells (Cats K and S) [65]. The elastinolytic cysteine proteases Cats K, L, and S have been implicated in the degradation of the main constituents of the basement membrane and elastin fibers of the pulmonary extracellular matrix, suggesting that these cathepsins may be involved in asthma and COPD [66–68].

Recently, additional mechanisms have linked cysteine proteases to lung diseases through the kinin-kallikrein system and through the regulation of antimicrobial activity [69]. The kinin-kallikrein system is a complex metabolic cascade of blood proteins that plays a role in blood pressure control, coagulation, inflammation, and pain. Cat L has been postulated to act as a kininogenase, while Cat K is suggested to be a highly potent kininase [70,71].

Cathepsin cysteine protease inhibitors have been reviewed in prior volumes of this series [72–74] and elsewhere [75–77]. Structure-based approaches toward cathepsin inhibitors have guided both the design and the rationalization of potent and selective compounds in many scaffolds. As an example, azepanone **23** is a potent inhibitor of Cat K which has been transformed into a selective Cat L inhibitor. With a $K_{i,app}$ of 0.16 nM for Cat K, **23** is >10-fold selective over Cat L ($K_{i,app} = 2.2$ nM) and Cat S ($K_{i,app} = 4.3$ nM). Compound **24**, with a $K_{i,app}$ of 0.43 nM for Cat L, is >20,000-fold and 36-fold selective for Cat L over Cats K and S, respectively [78].

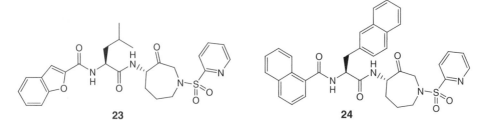

          23   **24**

The aminoethyl amide-based inhibitors and the pyrazole-based inhibitors are two classes of cathepsin inhibitors that are unique in that they do not contain an electrophilic functionality (or "warhead") common to many other cysteine protease inhibitors. Earlier analogs of the aminoethyl amide-based inhibitors were designed as Cat K inhibitors and exhibited micromolar potencies against Cat L. Compound **25** has Cat K and Cat L IC_{50}s of 12 and 7,840 nM, respectively [73]. More recent work has demonstrated how analogs such as **26** can be made selective for Cat S ($IC_{50} = 32$ nM) over Cats K and L (>30 μM each), while also possessing improvements in their pharmacokinetic profiles [79].

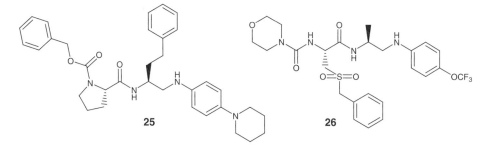

A number of cathepsin inhibitors have entered clinical studies, though not for pulmonary indications. RWJ-445380 (structure not disclosed) is in development for the treatment of plaque psoriasis and rheumatoid arthritis; two phase II studies were initiated in August 2006 [82,83]. CRA-028129 (structure not disclosed), an orally available, picomolar inhibitor of Cat S with nanomolar activity in a cellular assay, entered a phase I study for psoriasis in 2005 [84]. No current activities have been reported for this compound.

Additional progress toward Cat S selective inhibitors has been reported. The pyrazole-based inhibitor JNJ-10329670 (**27**) has Cat S potency ($IC_{50} = 100$ nM), selectivity over other cathepsins including Cats K and L ($IC_{50} >50$ μM), activity in a cellular assay measuring inhibition of cleavage of the invariant chain in JY cells ($IC_{50} = 600$–800 nM), and an acceptable pharmacokinetic profile [80]. Azaindole **28** demonstrated increased Cat S potency against the enzyme ($IC_{50} = 30$ nM) and in the cellular assay ($IC_{50} = 38$ nM) [81].

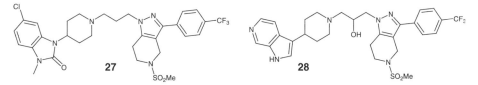

4. MATRIX METALLOPROTEINASES INHIBITORS

MMPs have been implicated in the pulmonary remodeling processes that are common to both asthma and COPD [85]. The MMP profile in asthma differs from that in COPD with MMP-9 most closely associated with asthma and MMP-12 showing a strong connection to COPD. Work is progressing toward understanding the potential application of these MMPs as possible therapeutic targets in overcoming the pathophysiology of asthma and COPD [14,86–89].

Doxycycline inhibits MMP activity at sub-antimicrobial doses [90] and is the only MMP inhibitor widely available clinically, indicated for the treatment of periodontal disease [91]. Prescribed as an antibiotic, doxycycline is also used clinically to manage COPD. In August 2007, a clinical trial for the macrolide antibiotic azithromycin commenced to study its anti-inflammatory properties in people with COPD through the inhibition of the MMP-catalyzed breakdown of collagen [92]. Broad spectrum MMP inhibitors, such as marimastat (BB-2516), have performed poorly in clinical trials. Marimastat was shown to have considerable side effects and dose-limiting musculoskeletal toxicity. The mechanism of this toxicity has not been completely elucidated, though several proposals have been advanced [93].

More selective MMP inhibitors have been the focus of intense research recently. Analyses of multiple MMP co-crystal structures have shed light on the structural differences between different MMPs. A study of phosphinic peptide libraries afforded **29** with an MMP-12 K_i of 0.19 nM and selectivities >200-fold against MMPs-3, 13, and 14 and >1,000-fold over MMPs-1, 2, 7, 8, 9, and 11. Additionally, compounds like **29** do not inhibit other proteases such as angiotensin-converting enzyme, neutral endopeptidase, and tumor necrosis factor α (TNFα) converting enzyme (TACE) (K_i >100 μM) [94]. A recent patent application claims selective inhibitors of MMPs, especially MMP-12, and presents **30** as a potent inhibitor of MMP-12 with >500-fold selectivity for MMP-12 over MMP-3 and TACE [95].

PF-00356231 (**31**) has been co-crystallized with MMP-12. Although it is not selective for MMP-12, it is an example of a non-zinc-chelating MMP inhibitor. Within this study, acetohydroxamate, added to the crystallization solutions to aid protein stability, stabilized a ternary complex with **31** and the protein through chelation with the Zn^{2+} ion [96]. The design of non-zinc-chelating MMP inhibitors is aimed at identifying more selective compounds [97,98].

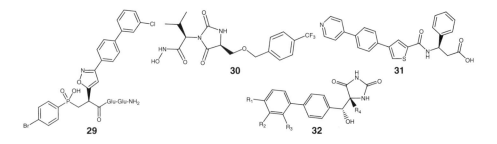

MMP inhibitors with a heterocyclic replacement of the common hydroxamate Zn^{2+} binding group have led to non-selective and selective MMP inhibitors [99,100]. Compounds containing a hydantoin (weak Zn^{2+} binding group) with a carbinol linker connecting it to a lipophilic biphenyl P1′ moiety have been discussed as selective MMP-12 inhibitors. The binding mode of these compounds was revealed by X-ray crystallography and solution state NMR. Lipophilic substituents were reported to increase the potency of these hydantoins at all MMPs, whereas direct interactions of *para-* and *meta-*substituents of the terminal ring of the P1′ domain with Lys241, a distinctive residue of MMP-12, were responsible for driving the selectivity. Compound **32** is a representative structure [101].

A recent study has shown that an orally bioavailable, non-hydroxamate, selective dual MMP-9/MMP-12 inhibitor (AZ11557272, structure not disclosed) was effective in a guinea pig model of COPD. Smoke-induced increases in inflammatory cells, desmosine (a marker of elastin breakdown), and serum TNF-α were not detected whereas increases in lung volumes and airspace size were observed. Lung function parameters were returned to control levels, and protection against small airway inflammation was found [88].

Recently, the MMP inhibitor AZD1236 was reported to be in a phase I trial for the treatment of COPD. No association to either AZ11557272 or **31** was reported, and no specific mechanism of action was disclosed [58].

5. CONCLUSIONS

COPD and asthma are two of the prevailing airway disorders that affect a large population and impose heavy financial burdens on society. Despite various treatments available today, there remains a need to discover and develop innovative therapies to battle these conditions. Among the many potential biological targets identified to date are a number of proteases implicated in the pathogenesis of these diseases. In the area of serine protease inhibitors, a number of compounds have advanced into development. However, to date most of them have been discontinued; it is yet to be seen whether the remaining compounds will emerge successfully from early stage development. Clinical trials of cysteine protease inhibitors have so far focused on inflammatory indications other than pulmonary diseases. New biological studies with selective MMP inhibitors are revealing fascinating insights into the mechanisms of action for COPD and asthma. Current research continues to support the protease–antiprotease hypothesis proposed many years ago to understand the processes that underlie both COPD and asthma.

REFERENCES

[1] Chronic Obstructive Pulmonary Disease (COPD) Fact Sheet, http://www.lungusa.org/site/pp.asp?c = dvLUK9O0E&b = 35020.
[2] N. Pearce, N. Ait-Khaled, R. Beasley, J. Mallol, U. Keil, E. Mitchell and C. Robertson, *Thorax*, 2007, **62**, 758.
[3] P. J. Barnes, *Nat. Rev. Immunol.*, 2008, **8**, 183.
[4] J. A. Elias, M. J. Kang, K. Crothers, R. Homer and C. G. Lee, *Proc. Am. Thorac. Soc.*, 2006, **3**, 494.

[5] Global Initiative for Chronic Obstructive Lung Disease, Global strategy for diagnosis, management, and prevention of COPD, Executive Summary, updated 2007, http://www.goldcopd.org/download.asp?intId = 445.

[6] B. S. Bochner, B. J. Undem and L. M. Lichtenstein, *Annu. Rev. Immunol.*, 1994, **12**, 295.

[7] S. E. Wenzel, A. A. Fowler, III and L. B. Schwartz, *Am. Rev. Resp. Dis.*, 1988, **137**, 1002.

[8] J. V. Fahy, H. Wong, J. Liu and H. A. Boushey, *Am. J. Resp. Crit. Care Med.*, 1995, **152**, 53.

[9] E. Eden, D. Mitchell, B. Mehlman, H. Khouli, M. Nejat, M. H. Grieco and G. M. Turino, *Am. J. Resp. Crit. Care Med.*, 1997, **156**, 68.

[10] J. A. Cairns and A. F. Walls, *J. Clin. Invest.*, 1997, **99**, 1313.

[11] C. Guay, M. Laviolette and G. M. Tremblay, *Curr. Top. Med. Chem.*, 2006, **6**, 393.

[12] W. Liu and D. M. Spero, *Drug News Perspect.*, 2004, **17**, 357.

[13] R. J. Riese, R. N. Mitchell, J. A. Villadangos, G. P. Shi, J. T. Palmer, E. R. Karp, G. T. De Sanctis, H. L. Ploegh and H. A. Chapman, *J. Clin. Invest.*, 1998, **101**, 2351.

[14] M. M. Gueders, J. M. Foidart, A. Noel and D. D. Cataldo, *Eur. J. Pharmacol.*, 2006, **533**, 133.

[15] F. Chua, S. E. Dunsmore, P. H. Clingen, S. E. Mutsaers, S. D. Shapiro, A. W. Segal, J. Roes and G. J. Laurent, *Am. J. Pathol.*, 2007, **170**, 65.

[16] G. H. Caughey, *Immunol. Rev.*, 2007, **217**, 141.

[17] P. Bradding, A. F. Walls and S. T. Holgate, *J. Allergy Clin. Immunol.*, 2006, **117**, 1277.

[18] M. Kobayashi, H. Kume, T. Oguma, Y. Makino, Y. Ito and K. Shimokata, *Clin. Exp. Allergy*, 2008, **38**, 135.

[19] J. A. Cairns, *Pulm. Pharmacol. Ther.*, 2005, **18**, 55.

[20] General Meeting of Sanofi-Aventis's Shareholders, Paris, May 31, 2006, http://en.sanofi-aventis.com/Images/060531_agm_en_tcm24-13038.pdf.

[21] PhRMA, 2007 Survey: Medicines in Development for Women, http://www.phrma.org/files/Women%202007.pdf.

[22] Information Meeting, Paris, February 12, 2008, http://en.sanofi-aventis.com/Images/080212_presentation_results2007_en_tcm24-20285.pdf.

[23] C. R. Hopkins, M. Czekaj, S. S. Kaye, Z. Gao, J. Pribish, H. Pauls, G. Liang, K. Sides, D. Cramer, J. Cairns, Y. Luo, H. K. Lim, R. Vaz, S. Rebello, M. Maignan, A. Dupuy, M. Mathieu and J. Levell, *Bioorg. Med. Chem. Lett.*, 2005, **15**, 2734.

[24] R. J. Vaz, Z. Gao, J. Pribish, X. Chen, J. Levell, L. Davis, E. Albert, M. Brollo, A. Ugolini, D. M. Cramer, J. Cairns, K. Sides, F. Liu, J. Kwong, J. Kang, S. Rebello, M. Elliot, H. Lim, V. Chellaraj, R. W. Singleton and Y. Li, *Bioorg. Med. Chem. Lett.*, 2004, **14**, 6053.

[25] J. Levell, P. Astles, P. Eastwood, J. Cairns, O. Houille, S. Aldous, G. Merriman, B. Whiteley, J. Pribish, M. Czekaj, G. Liang, S. Maignan, J. P. Guilloteau, A. Dupuy, J. Davidson, T. Harrison, A. Morley, S. Watson, G. Fenton, C. McCarthy, J. Romano, R. Mathew, D. Engers, M. Gardyan, K. Sides, J. Kwong, J. Tsay, S. Rebello, L. Shen, J. Wang, Y. Luo, O. Giardino, H. K. Lim, K. Smith and H. Pauls, *Bioorg. Med. Chem.*, 2005, **13**, 2859.

[26] C. Molloy, *SMi's 3rd Annual Conference on Asthma & COPD*, London, 2007.

[27] E. M. Erin, B. R. Leaker, A. Zacharasiewicz, L. A. Higgins, G. C. Nicholson, M. J. Boyce, B. P. de, R. C. Jones, S. R. Durham, P. J. Barnes and T. T. Hansel, *Clin. Exp. Allergy*, 2006, **36**, 458.

[28] A. V. Kamath, R. A. Morrison, T. W. Harper, S. J. Lan, A. M. Marino and S. Chong, *J. Pharm. Sci.*, 2005, **94**, 1115.

[29] M. J. Costanzo, S. C. Yabut, H. C. Zhang, K. B. White, G. L. de, Y. Wang, L. K. Minor, B. A. Tounge, A. N. Barnakov, F. Lewandowski, C. Milligan, J. C. Spurlino, W. M. Abraham, V. Boswell-Smith, C. P. Page and B. E. Maryanoff, *Bioorg. Med. Chem. Lett.*, 2008, **18**, 2114.

[30] Y. Miyazaki, Y. Kato, T. Manabe, H. Shimada, M. Mizuno, T. Egusa, M. Ohkouchi, I. Shiromizu, T. Matsusue and I. Yamamoto, *Bioorg. Med. Chem. Lett.*, 2006, **16**, 2986.

[31] G. S. Bisacchi, W. A. Slusarchyk, S. A. Bolton, K. S. Hartl, G. Jacobs, A. Mathur, W. Meng, M. L. Ogletree, Z. Pi, J. C. Sutton, U. Treuner, R. Zahler, G. Zhao and S. M. Seiler, *Bioorg. Med. Chem. Lett.*, 2004, **14**, 2227.

[32] G. Zhao, S. A. Bolton, C. Kwon, K. S. Hartl, S. M. Seiler, W. A. Slusarchyk, J. C. Sutton and G. S. Bisacchi, *Bioorg. Med. Chem. Lett.*, 2004, **14**, 309.

[33] D. Sperandio, V. W. Tai, J. Lohman, B. Hirschbein, R. Mendonca, C. S. Lee, J. R. Spencer, J. Janc, M. Nguyen, J. Beltman, P. Sprengeler, H. Scheerens, T. Lin, L. Liu, A. Gadre, A. Kellogg, M. J. Green and M. E. McGrath, *Bioorg. Med. Chem. Lett.*, 2006, **16**, 4085.

[34] J. T. Palmer, R. M. Rydzewski, R. V. Mendonca, D. Sperandio, J. R. Spencer, B. L. Hirschbein, J. Lohman, J. Beltman, M. Nguyen and L. Liu, *Bioorg. Med. Chem. Lett.*, 2006, **16**, 3434.

[35] C. S. Lee, W. Liu, P. A. Sprengeler, J. R. Somoza, J. W. Janc, D. Sperandio, J. R. Spencer, M. J. Green and M. E. McGrath, *Bioorg. Med. Chem. Lett.*, 2006, **16**, 4036.

[36] C. P. Sommerhoff, G. H. Caughey, W. E. Finkbeiner, S. C. Lazarus, C. B. Basbaum and J. A. Nadel, *J. Immunol.*, 1989, **142**, 2450.

[37] C. P. Sommerhoff, J. A. Nadel, C. B. Basbaum and G. H. Caughey, *J. Clin. Invest.*, 1990, **85**, 682.

[38] K. C. Fang, W. W. Raymond, S. C. Lazarus and G. H. Caughey, *J. Clin. Invest.*, 1996, **97**, 1589.

[39] Flash Report-Result of 1H & Outlook for FY07, 2007, http://www.teijin.co.jp/english/ir/doc/info071031_e.pdf.

[40] M. Teramoto, N. Tsuchiya and H. Saitoh, *Patent Applicaiton US* 2008/0015240-A1, 2006.

[41] Biotechnology, Immunology and Oncology, (http://www.jnj.com/investor/documents/presentations/pcreview07/present_peterson_siegel.pdf).

[42] L. de Garavilla, M. N. Greco, N. Sukumar, Z. W. Chen, A. O. Pineda, F. S. Mathews, C. E. Di, E. C. Giardino, G. I. Wells, B. J. Haertlein, J. A. Kauffman, T. W. Corcoran, C. K. Derian, A. J. Eckardt, B. P. Damiano, P. ndrade-Gordon and B. E. Maryanoff, *J. Biol. Chem.*, 2005, **280**, 18001.

[43] M. N. Greco, M. J. Hawkins, E. T. Powell, H. R. Almond, Jr., G. L. de, J. Hall, L. K. Minor, Y. Wang, T. W. Corcoran, C. E. Di, A. M. Cantwell, S. N. Savvides, B. P. Damiano and B. E. Maryanoff, *J. Med. Chem.*, 2007, **50**, 1727.

[44] T. Tanaka, T. Muto, H. Maruoka, S. Imajo, H. Fukami, Y. Tomimori, Y. Fukuda and T. Nakatsuka, *Bioorg. Med. Chem. Lett.*, 2007, **17**, 3431.

[45] H. Maruoka, T. Muto, T. Tanaka, S. Imajo, Y. Tomimori, Y. Fukuda and T. Nakatsuka, *Bioorg. Med. Chem. Lett.*, 2007, **17**, 3435.

[46] S. Takai, D. Jin, M. Nishimoto, A. Yuda, M. Sakaguchi, K. Kamoshita, K. Ishida, Y. Sukenaga, S. Sasaki and M. Miyazaki, *Life Sci.*, 2001, **69**, 1725.

[47] S. Takai and M. Miyazaki, *Cardiovasc. Drug Rev.*, 2003, **21**, 185.

[48] K. Ishida, S. Takai, M. Murano, T. Nishikawa, I. Inoue, N. Murano, N. Inoue, D. Jin, E. Umegaki, K. Higuchi and M. Miyazaki, *J. Pharmacol. Exp. Ther.*, 2008, **324**, 422.

[49] M. Sakaguchi, S. Takai, D. Jin, Y. Okamoto, M. Muramatsu, S. Kim and M. Miyazaki, *Eur. J. Pharmacol.*, 2004, **493**, 173.

[50] B. Chughtai and T. G. O'Riordan, *J. Aerosol Med.*, 2004, **17**, 289.

[51] A. Roghanian and J. M. Sallenave, *J. Aerosol Med.*, 2008, **21**, 1.

[52] M. F. Fitzgerald and J. C. Fox, *Drug Discov. Today*, 2007, **12**, 479.

[53] Ono Pharmaceutical Co., Ltd., press release, June 13, 2002, http://www.ono.co.jp/eng/cn/contents/sm_cn_061302.htm.

[54] Dong-A Pharma., product information, http://www.donga-pharma.com/prd/prd01_view.jsp?mitgrp=FESI.

[55] First-Third Quarter (April 1–December 31, 2007) Flash Report (unaudited) Nine months ended December 31, 2007, http://www.ono.co.jp/jp/ir_info/annual/pdf/2008/fi0803.pdf.

[56] Investor Handout Q3/2006, www.investor.bayer.com/user_upload/1418/.

[57] H. Gielen-Haertwig, B. Albrecht, J. Keldenich, V. Li, J. Pernerstorfer, K.-H. Schlemmer and L. Telan, *Patent Application WO* 2005/082864, 2005.

[58] AstraZeneca pipeline summary, http://www.astrazeneca.com/article/511390.aspx.

[59] P. Hansen, K. Lawitz, H. Lonn and A. Nikitidis, *Patent Application WO* 2006/098684, 2006.

[60] A. Kuromiya, H. Okazaki, and J. Tsuji, MEDI-332, *230th ACS National Meeting*, Washington DC, August, 2005.

[61] R. Shiratake, T. Deguchi, Y. Inoue, N. Imayoshi, T. Ueda, K. Suzuki, and F. Sato, MEDI-333, *230th ACS National Meeting*, Washington DC, August, 2005.

[62] M. L. Zani, S. M. Nobar, S. A. Lacour, S. Lemoine, C. Boudier, J. G. Bieth and T. Moreau, *Eur. J. Biochem.*, 2004, **271**, 2370.

[63] Proteo fact sheet, http://www.proteo.de/download_en/2007/Proteo-fact_sheet-11-2007.pdf.

[64] European Commission grants Orphan Drug Status to Elafin for the treatment of PAH, Proteo, Inc./Proteo Biotech AG press release, March 29, 2007, http://www.proteo.de/pages/presse. php?DATUM = 2007-03-29&BACK = /pages/home.php&lang = en.

[65] F. Buhling, N. Waldburg, A. Reisenauer, A. Heimburg, H. Golpon and T. Welte, *Eur. Resp. J.*, 2004, **23**, 620.

[66] F. Lecaille, J. Kaleta and D. Bromme, *Chem. Rev.*, 2002, **102**, 4459.

[67] V. Turk, B. Turk and D. Turk, *EMBO J.*, 2001, **20**, 4629.

[68] H. Takahashi, K. Ishidoh, D. Muno, A. Ohwada, T. Nukiwa, E. Kominami and S. Kira, *Am. Rev. Resp. Dis.*, 1993, **147**, 1562.

[69] C. C. Taggart, C. M. Greene, T. P. Carroll, S. J. O'Neill and N. G. McElvaney, *Am. J. Resp. Crit. Care Med.*, 2005, **171**, 1070.

[70] F. Veillard, F. Lecaille and G. Lalmanach, *Int. J. Biochem. Cell Biol.*, 2008, **40**, 1079.

[71] C. Desmazes, L. Galineau, F. Gauthier, D. Bromme and G. Lalmanach, *Eur. J. Biochem.*, 2003, **270**, 171.

[72] W. C. Black and M. D. Percival, in *"Annual Reports in Medicinal Chemistry"* (ed. J. E. Macor), Vol. 42, Academic Press, New York, 2007, p. 111.

[73] R. W. Marquis, in *"Annual Reports in Medicinal Chemistry"* (ed. A. Doherty), Vol. 39, Academic Press, New York, 2004, p. 79.

[74] R. W. Marquis, in *"Annual Reports in Medicinal Chemistry"* (eds W. Hagmann and A. Doherty), Vol. 35, Academic Press, New York, 2000, p. 309.

[75] J. O. Link and S. Zipfel, *Curr. Opin. Drug Discov. Devel.*, 2006, **9**, 471.

[76] S. Gupta, R. K. Singh, S. Dastidar and A. Ray, *Expert Opin. Ther. Targets*, 2008, **12**, 291.

[77] E. Sokolova and G. Reiser, *Pharmacol. Ther.*, 2007, **115**, 70.

[78] R. W. Marquis, I. James, J. Zeng, R. E. Trout, S. Thompson, A. Rahman, D. S. Yamashita, R. Xie, Y. Ru, C. J. Gress, S. Blake, M. A. Lark, S. M. Hwang, T. Tomaszek, P. Offen, M. S. Head, M. D. Cummings and D. F. Veber, *J. Med. Chem.*, 2005, **48**, 6870.

[79] D. C. Tully, H. Liu, A. K. Chatterjee, P. B. Alper, R. Epple, J. A. Williams, M. J. Roberts, D. H. Woodmansee, B. T. Masick, C. Tumanut, J. Li, G. Spraggon, M. Hornsby, J. Chang, T. Tuntland, T. Hollenbeck, P. Gordon, J. L. Harris and D. S. Karanewsky, *Bioorg. Med. Chem. Lett.*, 2006, **16**, 5112.

[80] D. J. Gustin, C. A. Sehon, J. Wei, H. Cai, S. P. Meduna, H. Khatuya, S. Sun, Y. Gu, W. Jiang, R. L. Thurmond, L. Karlsson and J. P. Edwards, *Bioorg. Med. Chem. Lett.*, 2005, **15**, 1687.

[81] J. Wei, B. A. Pio, H. Cai, S. P. Meduna, S. Sun, Y. Gu, W. Jiang, R. L. Thurmond, L. Karlsson and J. P. Edwards, *Bioorg. Med. Chem. Lett.*, 2007, **17**, 5525.

[82] Safety and Effectiveness Study of RWJ-445380 Cathepsin-S Inhibitor in Patients With Active Rheumatoid Arthritis Despite Methotrexate Therapy, http://clinicaltrials.gov/ct2/show/ NCT00425321?term = RWJ-445380&rank = 1.

[83] Study to Investigate the Safety, Tolerability, Absorption, Distribution, Metabolism, and Elimination of RWJ-445380 Administered to Patients With Plaque Psoriasis, http://clinicaltrials. gov/ct2/show/NCT00396422?term = RWJ-445380&rank = 2.

[84] Celera Genomics press release, September 20, 2005, http://www.celera.com/celera/pr_1127168705.

[85] S. V. Culpitt, D. F. Rogers, S. L. Traves, P. J. Barnes and L. E. Donnelly, *Resp. Med.*, 2005, **99**, 703.

[86] J. Hu, P. E. Van den Steen, Q. X. Sang and G. Opdenakker, *Nat. Rev. Drug Discov.*, 2007, **6**, 480.

[87] P. Geraghty, C. M. Greene, M. O'Mahony, S. J. O'Neill, C. C. Taggart and N. G. McElvaney, *J. Biol. Chem.*, 2007, **282**, 33389.

[88] A. Churg, R. Wang, X. Wang, P. O. Onnervik, K. Thim and J. L. Wright, *Thorax*, 2007, **62**, 706.

[89] K. M. Bottomley and M. G. Belvisi, in *"Annual Reports in Medicinal Chemistry"* (ed. A. Doherty), Vol. 37, Academic Press, New York, 2002, p. 209.

[90] A. J. M. Richards, R. M. Bannister and S. A. Chaplin, *Patent Application US* 2003/0099600, 2003.

[91] R. Hanemaaijer, H. Visser, P. Koolwijk, T. Sorsa, T. Salo, L. M. Golub and V. W. van Hinsbergh, *Adv. Dent. Res.*, 1998, **12**, 114.

[92] Effect of Macrolide Antibiotics on Airway Inflammation in People With Chronic Obstructive Pulmonary Disease (COPD), http://clinicaltrials.gov/ct2/show/NCT00549445?term = NCT00549445&rank = 1.

[93] N. A. Rizvi, J. S. Humphrey, E. A. Ness, M. D. Johnson, E. Gupta, K. Williams, D. J. Daly, D. Sonnichsen, D. Conway, J. Marshall and H. Hurwitz, *Clin. Cancer Res.*, 2004, **10**, 1963.

[94] L. Devel, V. Rogakos, A. David, A. Makaritis, F. Beau, P. Cuniasse, A. Yiotakis and V. Dive, *J. Biol. Chem.*, 2006, **281**, 11152.

[95] H. Wallberg, M. H. Xu, G. Q. Lin, X. S. Lei, P. Sun, K. Parkes, T. Johnson and B. Samuelsson, *Patent Application WO 2007/068474-A1*, 2007.

[96] R. Morales, S. Perrier, J. M. Florent, J. Beltra, S. Dufour, M. De, I. P. Manceau, A. Tertre, F. Moreau, D. Compere, A. C. Dublanchet and M. O'Gara, *J. Mol. Biol.*, 2004, **341**, 1063.

[97] C. A. Kontogiorgis, P. Papaioannou and D. J. Hadjipavlou-Litina, *Curr. Med. Chem.*, 2005, **12**, 339.

[98] Q. X. Sang, Y. Jin, R. G. Newcomer, S. C. Monroe, X. Fang, D. R. Hurst, S. Lee, Q. Cao and M. A. Schwartz, *Curr. Top. Med. Chem.*, 2006, **6**, 289.

[99] R. Hayashi, X. Jin and G. R. Cook, *Bioorg. Med. Chem. Lett.*, 2007, **17**, 6864.

[100] E. Breuer, J. Frant and R. Reich, *Exp. Opin. Ther. Patents*, 2005, **15**, 253.

[101] B. O. J.Nordén, I. Shamovsky, B. Gabos, M. M. Rosenschöld, M. Lepistö, G. Carlström, J. Evenäs, D. Musil, and K. Stenvall, COMP-387, *234th ACS National Meeting*, Boston, MA, August, 2007.

PART IV:
Oncology

Editor: David C. Myles
Combithera, Inc.
San Francisco
California

CHAPTER **12**

mTOR Inhibitors in Oncology

Jeroen Verheijen, Ker Yu and Arie Zask

1. INTRODUCTION

The mammalian target of rapamycin (mTOR) is the founding member of a family of unconventional high molecular weight serine/threonine protein kinases termed phosphoinositide-3-kinase (PI3K)-related kinases (PIKKs) (reviewed in [1]). PIKKs play diverse roles in cell growth and surveillance of both the genome and transcriptome. The catalytic sites of the PIKK family resemble those of PI3K but differ from those of the broad-spectrum conventional protein kinases. These distinctive structural features coupled with the essential biological function and scarcity of PIKKs in the entire human kinome of approximately 500 kinases

Wyeth Research, Pearl River, New York 10965

Annual Reports in Medicinal Chemistry, Volume 43
ISSN 0065-7743, DOI 10.1016/S0065-7743(08)00012-2

highlight mTOR and the PIKK family as exciting drug targets for the development of potent and selective inhibitor therapy.

Molecular and biochemical characterization of mTOR uncovered an important signaling network that regulates fundamental aspects of cell growth, metabolism, and proliferation in response to growth factors, nutrients, and energy supply (reviewed in [2,3]). In human cells, mTOR primarily resides in two functional complexes, mTOR complex 1 (mTORC1) and mTOR complex 2 (mTORC2), which are differentially formed through complex-specific binding partners, and are believed to dictate subcellular mTOR functions and/or substrate specificity. mTORC1 is composed of mTOR, Raptor, mLST8/GβL, and PRAS40, while mTORC2 contains mTOR, Rictor, mLST8/GβL, and mSIN1. A dominant role in promoting cellular translation is well established for mTORC1 through its direct phosphorylation of the ribosomal protein S6 kinase 1 (S6K1) and eukaryotic translation initiation factor 4E-binding protein 1 (4E-BP1). Both S6K1 and 4E-BP1 are regulatory proteins in translation machinery and cell growth. The recent discovery of mTORC2 has elucidated new aspects of mTOR in cancer biology. mTORC2 phosphorylates the serine/threonine kinase AKT, leading to an increased cell survival and resistance to chemotherapy. mTORC2 is also predicted to modulate the cytoskeletal network in human cells through biochemical mechanisms yet to be identified. These mTORC2-related functions are vital to the maintenance and progression of malignant and metastatic cancer cells [2,3].

Although the mTOR gene locus is not known to be mutated or amplified in cancer, mTOR signaling contributes to tumorigenic effects by numerous oncogenic proteins such as PI3K, AKT, EGFR, HER2/neu, and BCR-Abl as well as the effects due to loss of tumor-suppressor genes such as the phosphatase and tensin homolog (PTEN), tuberous sclerosis complex (TSC), von Hippel-Lindau (VHL), and neurofibromatosis type I (NF1) (reviewed in [4,5]). In preclinical models of these diseases, inhibition of mTOR signaling often correlates with anti-tumor activity. Heightened mTOR activity, as indicated by an elevated phosphorylation of its downstream substrates phospho-S6K1, phospho-S6, and phospho-AKT, has frequently been observed in clinical samples of various solid tumors as well as hematopoietic malignancies. There is strong preclinical and some clinical evidence that certain tumors with deregulated PI3K/AKT/mTOR signaling are particularly susceptible to mTOR inhibition (reviewed in [6]).

2. MECHANISM OF mTOR INHIBITION

2.1 Inhibition of mTORC1

Rapamycin (**1**), at single digit nanomolar concentrations, forms a tight complex with the 12 kDa FK506-binding protein (FKBP12) that in turn binds with high affinity to the FKBP12-rapamycin-binding domain (FRB domain) adjacent to the catalytic domain of mTOR (Figure 1) [7–9]. The resulting ternary complex may alter the composition and/or conformation of mTORC1, thereby interfering with

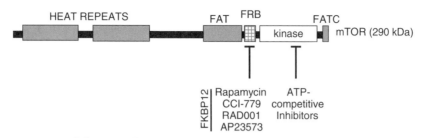

Figure 1 Structural domains of mTOR and molecular sites targeted by mTOR inhibitors.

its phosphotransferase activity. Intriguingly, both *in vitro* and *in vivo* studies indicate that the FRB domain in mTORC2 is not accessible to rapamycins as illustrated by the lack of suppression of phosphorylation of the mTORC2 substrate AKT. Rapamycins have single agent anti-tumor activity in various tumor models, particularly those with a heightened PI3K/AKT/mTOR status or deregulated angiogenesis signaling [6]. In some cell types, activation of mTORC1 leads to repression of PI3K/AKT signaling. This negative-feedback loop can be inhibited by the binding of rapamycins to mTORC1 resulting in increased PI3K-AKT activity, a phenomenon that may not be desirable for cancer therapy [3,6].

2.2 Inhibition of mTORC1 and mTORC2

In vitro anti-proliferative effects of the rapamycins are generally modest and variable in cancer cells, in part due to their inaccessibility to mTORC2 and the feedback-activation of PI3K/AKT signaling. In contrast, ATP-competitive inhibitors of mTOR kinase targeting both mTORC1 and mTORC2 (Figure 1), suppress mTOR signaling globally in cancer cells and in elements of the tumor microenvironment, and minimize the feedback activation of PI3K signaling. These properties of ATP-competitive inhibitors may provide new opportunities for more robust anti-tumor efficacy in a broader range tumor types.

3. RAPAMYCIN ANALOGS IN THE CLINIC

The discovery of the immunosuppressive activity of rapamycin in the 1980s, coupled with its unique mechanism of action, led to extensive structure–activity relationship (SAR) investigations (reviewed in [10, 11]). Most of these early investigations were done by semi-synthesis. Modifications of the 42-hydroxy group gave rise to three new analogs currently in the clinic, temsirolimus (**2**, CCI-779, Torisel®), everolimus (**3**, RAD001), and deforolimus (**4**, AP23573, MK-8669) (Figure 2).

3.1 Temsirolimus (CCI-779, Torisel®)

Temsirolimus (**2**) is a soluble 42-[2,2-bis(hydroxymethyl)]-propionic ester of rapamycin. The Food and Drug Administration (FDA) approved the use

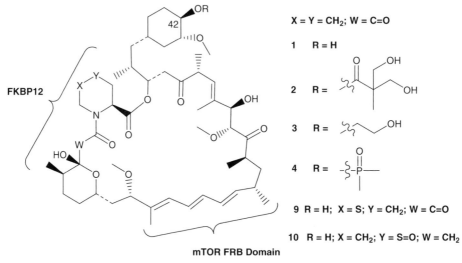

X = Y = CH$_2$; W = C=O

1   R = H

2   R = [structure with OH groups]

3   R = [structure with OH]

4   R = [phosphate structure]

9   R = H; X = S; Y = CH$_2$; W = C=O

10   R = H; X = CH$_2$; Y = S=O; W = CH$_2$

mTOR FRB Domain

Figure 2   Rapamycin (**1**), temsirolimus (**2**), everolimus (**3**), and deforolimus (**4**). Precursor directed biosynthesis derived analogs **9** and **10**.

of temsirolimus for the treatment of advanced renal cell carcinoma (RCC) in May 2007. In a phase III trial with 626 RCC patients, single-agent temsirolimus was associated with a statistically significant improvement in overall survival [12]. Phase II studies evaluating temsirolimus in a broad range of tumors have also been reported. The most promising activity has been seen in mantle cell lymphoma (MCL) [13] and endometrial carcinoma [14] with objective tumor response rates of 30%–40%. Moderate activity was reported for metastatic breast cancer [15] and recurrent glioblastoma multiforme [16]. Minimal activity was reported for metastatic melanoma [17]. In patients with advanced RCC, temsirolimus response was associated with the phosphorylation of mTOR pathway markers phospho-AKT and phospho-S6 [18]. In a breast carcinoma study, the loss of PTEN and/or HER2 overexpression was also linked to temsirolimus response [19]. However, while frequent loss of PTEN occurs in melanoma and endometrial cancers, the lack of temsirolimus activity in melanoma versus a strong response in endometrial cancer patients indicates that molecular mechanisms other than PTEN status may determine the degree of response.

3.2 Everolimus (RAD001)

Everolimus (**3**), 42-*O*-(2-hydroxyethyl)rapamycin, was developed to improve the oral bioavailability of rapamycin [20]. Oral formulations of everolimus are being evaluated in several late stage phase III trials in patients with pancreatic islet cell tumors, and in phase II studies in patients with breast, lung, gastrointestinal, and hematologic cancers [21]. In a phase II study of metastatic RCC, a partial response rate of 33% was observed in the everolimus-treated patients [22]. Results of phase I studies in hematologic, breast, nonsmall cell lung, and pediatric solid cancers have been reported [23–26]. In preclinical studies, everolimus demonstrated anti-tumor activity against MCL [27], pancreatic neuroendocrine tumors [28], and ovarian tumors [29].

3.3 Deforolimus (AP23573, MK-8669)

Deforolimus (**4**), a dimethylphosphinate-modified rapamycin analog, is being evaluated in a broad range of cancer trials, with promising results reported for several tumor types. In a dose escalation phase I study, 22/29 patients (76%) experienced stable disease or partial responses [30]. In a phase II study of patients with advanced soft tissue or bone sarcomas, 54/193 (28%) achieved a clinical benefit response (CBR) [31]. In two ongoing phase II studies in patients with refractory hematologic malignancies, 41% and 55% had at least stable disease [32,33]. In an ongoing phase II trial of advanced endometrial cancer, 7 out of the first 19 patients (37%) achieved CBR [34].

4. PRE-CLINICAL RAPAMYCIN ANALOGS

Rapamycin analogs have been prepared primarily by semi-synthesis (reviewed in [10,11]). Rapamycin has also been prepared by long and complex total synthesis (e.g. [35]). Modifications of rapamycin by enzymatic methods, by exploitation of the biosynthetic pathway, and by genetic manipulation have also been utilized ([36] and references therein).

Recently, precursor directed biosynthesis has been applied to the genera-tion of new rapamycin analogs [37]. This approach, also known as mutasynthesis, couples chemical synthesis with molecular biology and is especially useful for modification of complex natural products, such as rapamycin, whose lengthy and complex total synthesis precludes ready lead optimization. A mutant strain of *Streptomyces hygroscopicus* (MG2-10) that does not generate 4,5-dihydroxycyclohex-1-ene carboxylic acid (DHCHC), the source of the dihydroxycyclohexane moiety of rapamycin, allowed for the incorporation of novel starter units in the biosynthesis of pre-rapamycin (**5**) and pre-rapamycin analogs **6–8** (Figure 3) [37–39]. Another approach to precursor-directed biosynthesis utilized the observation that nipecotic acid inhibits the biosynthesis of rapamycin, while concurrent feeding with l-pipecolate restores production. Thus, feeding of sulfur-containing pipecolate analogs to cultures of *S. hygroscopicus* along with nipecotic acid led to production of two new sulfur-containing pipecolate analogs (**9, 10**) (Figure 2) of rapamycin [36]. Both analogs were found to bind several orders of magnitude less tightly to FKBP12 than did rapamycin.

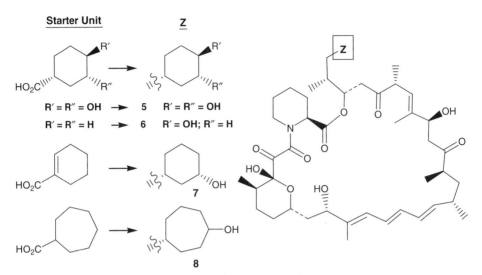

Figure 3 Precursor directed biosynthesis of rapamycin analogs.

5. ATP COMPETITIVE mTOR INHIBITORS

5.1 Mixed mTOR/PI3K inhibitors

As described in the preceding section, the majority of reports on mTOR inhibitors have dealt with rapamycin and its analogs. Small molecules that interact with the ATP-binding site of mTOR have also recently been described. As outlined in the introduction, these types of molecules would be expected to inhibit both mTORC1 and mTORC2 complexes, whereas rapamycins inhibit predominantly mTORC1. Unlike rapamycin, which due to its unique ternary complex formation is a very selective inhibitor of mTORC1, most mTOR active site inhibitors reported to date also inhibit one or more related kinases. For example, SF1126 (**11**, Figure 4) is a vascular-targeted conjugate of the well-characterized PI3K/mTOR/DNA-PK inhibitor LY294002 [40]. The structure of SF1126 shown in Figure 4 is based on an X-ray crystal structure [41]. The previously published structure of SF1126 showed the tripeptide linked to LY294002 through the morpholine nitrogen [40]. In 2007, Phase I clinical trials studying SF1126 in patients with solid tumors and multiple myeloma were initiated [41].

LY303511 (**12**, Figure 4), historically considered an inactive analog of LY294002 due to its lack of PI3K inhibition, displayed a biomarker profile in A549 cells suggestive of mTOR inhibition without PI3K inhibition [42]. In addition to mTOR, **12** inhibited casein kinase 2. Despite its relatively low potency (micromolar concentrations were required for inhibition of mTOR biomarkers in tumor cells), LY303511 inhibited tumor growth in a xenograft model of human adenocarcinoma (PC-3) following *i.p.* administration at 10 mg/kg, *q.d.*

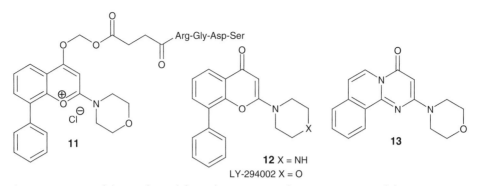

Figure 4 mTOR inhibitors derived from the prototypical pan-PI3K/mTOR inhibitor LY294002.

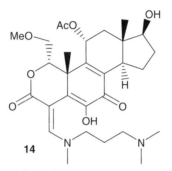

Figure 5 mTOR inhibitor derived from the pan-PI3K/mTOR inhibitor wortmannin.

Another analog of LY294002, **13** (Figure 4), inhibited DNA-PK and mTOR, without inhibiting PI3K [43]. In Rat-1 fibroblasts, **13** inhibited the activity of both mTORC1 and mTORC2 as demonstrated by inhibition of the phosphorylation of S6K1 Thr-389 and AKT Ser-473.

Analogs of another well-known pan-PI3K/mTOR inhibitor, wortmannin, were recently reported [44]. Ring-opening of the furan ring of 17-hydroxywort-mannin with secondary amines led to analogs with improved stability, toxicity, and aqueous solubility versus wortmannin. As with wortmannin, these analogs (e.g. **14**) (Figure 5) inhibited mTOR, albeit several orders of magnitude less potently than PI3K-alpha.

Mixed inhibitors of mTOR and PI3K have also been developed from new scaffolds. For example, it was recently shown that the PI3K inhibitor PI-103 (**15**, Figure 6) inhibits both mTORC1 and mTORC2 at low nanomolar concentrations (IC$_{50}$ values of 20–80 nM) [45,46]. PI-103 was more effective *in vivo* in glioma xenograft models as compared to selective PI3K-alpha inhibitors, which was ascribed to its additional effects on mTOR [46].

NVP-BEZ-235 (**16**, Figure 6), a potent mixed inhibitor of PI3K and mTOR with low nanomolar IC$_{50}$ values against both enzymes [47,48] was active in A549 (lung) and BT474 (breast) xenograft models following oral dosing. Compound **16** is reportedly in phase I clinical trials.

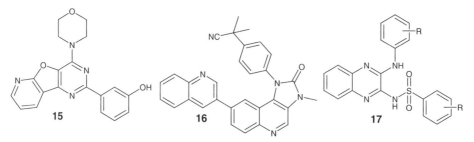

Figure 6 Mixed PI3K and mTOR inhibitors in advanced stages of development.

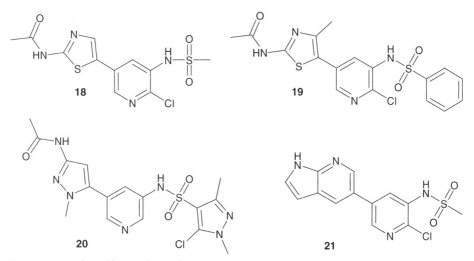

Figure 7 3-Aryl-5-sulfonamidopyridine mixed mTOR/PI3K inhibitors.

XL-765 (structure undisclosed) is also a mixed mTOR/PI3K inhibitor reported to be in clinical development. XL-765 inhibits the various isoforms of PI3K with IC_{50} values of 9–113 nM and inhibits mTOR with an IC_{50} of 157 nM [49]. Exelixis has recently filed a patent application on substituted *N*-[3-aryl-quinoxalin-2-yl]-benzenesulfonamides (cf. **17**, Figure 6) as PI3K-alpha inhibitors [50].

The recent patent literature contains several reports on 3-aryl-5-sulfonami-dopyridines as mixed PI3K/mTOR inhibitors (Figure 7). Both the methanesulfo-nyl derivative **18** and the benzenesulfonyl analog **19** inhibited PI3K-alpha and mTOR with IC_{50} values of 10 nM [51,52]. The orientation of the sulfonamide bond was important for mTOR inhibitory activity as reversal of the sulfonamide bond significantly decreased the mTOR potency. Compound **19** gave 20% tumor regression following 0.5 mg/kg *p.o.* dosing in nude mice in a PC3 tumor xenograft model [53].

Two additional patent applications disclosed 3-aryl-5-sulfonamidopyridines as well, although very limited biological data were presented. Thus, compound

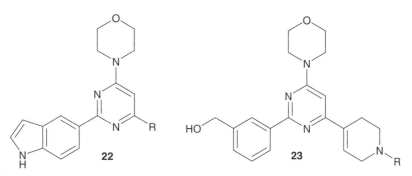

Figure 8 Substituted pyrimidine mixed PI3K/mTOR inhibitors.

20 had IC_{50} values against PI3K-alpha and mTOR of 0.5–1 µM [54]. Compound **21** inhibited mTOR with an IC_{50} of 2 µM and was slightly more potent against PI3K-alpha ($IC_{50} = 0.5$ µM) [55].

Patent applications disclosing substituted pyrimidines have appeared. Although no specific inhibitory activities are provided, compounds such as **22** (Figure 8) are claimed to inhibit PI3K and/or mTOR with IC_{50} values between 1 and 500 nM [56]. Similarly, compounds such as **23** possess potent inhibitory activity against mTOR and PI3K-alpha [57].

5.2 Selective mTOR inhibitors

Recently, several patent applications claimed mTOR inhibitors without claims of PI3K activity. 4-Morpholin-4-yl-pyrido[2,3-*d*]pyrimidines inhibited mTOR at nanomolar concentrations [58,59]. For example, compound **24** (Figure 9) had an mTOR IC_{50} of 43 nM. Substituted morpholino-triazines, such as **25**, with IC_{50} values against mTOR below 1.5 µM, have also been reported [60]. KU-0063794 (structure undisclosed) was reported to be a highly potent inhibitor of mTOR ($IC_{50} = 16$ nM) with >100-fold selectivity versus other PIKK members (e.g. PI3Kalpha, DNA-PK, ATM, ATR) [61].

3-Alkyl-1-alkynyl-imidazo[1,5-*a*]pyrazin-8-ylamines, such as compound **26**, have been disclosed as mTOR inhibitors with IC_{50} values below 10 µM [62]. Several 3-alkyl-1-aryl-imidazo[1,5-*a*]pyrazin-8-ylamines, such as **27**, possessed IC_{50} values below 10 nM [63]. The effect on related kinases was not reported in these patent applications. Replacement of the imidazopyrazine core of the above compounds with a 1*H*-pyrazolo[3,4-d]pyrimidine group led to analogs (e.g. **28** in Figure 9) with mTOR IC_{50} values below 1 µM that also inhibited several other kinases (cKIT, Tie2, FLT3, PDGFR, RET, and IR) with nanomolar IC_{50} values [64]. OXA-01, an imidazopyrazine (structure undisclosed), inhibited mTORC1 and mTORC2 with IC_{50} values of 29 and 7 nM, respectively [65]. In an MDA-MB-231 xenograft model, 100% tumor growth inhibition was seen with 75 mg/kg *p.o.* bid dosing of OXA-01 for 14 days.

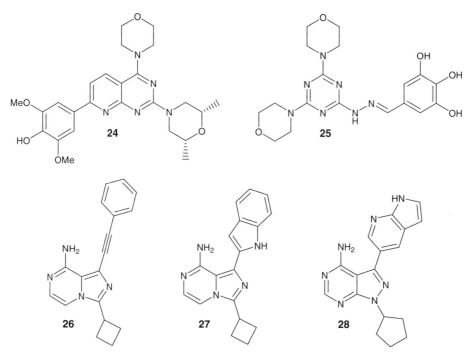

Figure 9   mTOR inhibitors from recent patent literature.

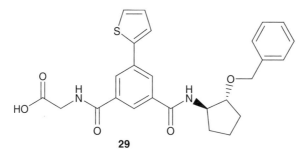

Figure 10   HTS-1, a non-rapamycin derived FRB domain binder.

6. OTHER mTOR INHIBITORS

Compound HTS-1 (**29**) (Figure 10) was obtained from a high-throughput screen for binders to the FRB domain [66]. NMR solution structural studies revealed that the sites on the mTOR FRB domain that interact with HTS-1 closely match those that are responsible for rapamycin binding. The dissociation constant for compounds of this type is in the low micromolar range [66].

Other publications claim inhibition of mTOR signaling pathways, but do not show evidence of direct inhibition of mTOR. Hence, inhibition of upstream effectors, rather than mTOR itself, cannot be excluded. For example,

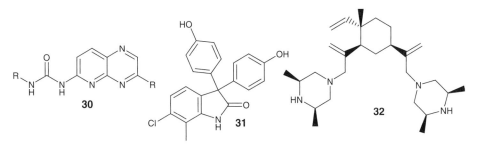

Figure 11   mTOR signaling pathway inhibitors.

pyrido[2,3-*b*]pyrazin-6-yl-ureas (cf. **30**, Figure 11) that inhibit signaling pathways and enzymes, including mTOR, were disclosed [67,68]. 3,3-Diaryl-1,3-dihydro-indol-2-one inhibitors of mTOR pathway activation (e.g. **31**) showed efficacy in xenograft models of human tumors [69]. *Beta*-elemene derivatives such as **32** were reported to have an anti-proliferative effect on tumor cells at low micromolar concentrations, due to their inhibition of mTOR activity [70].

7. CONCLUSION

Through the success of rapamycin analogs (i.e. **2**, **3**, **4**) in the clinic, mTOR has been firmly established as a therapeutic target for the treatment of cancer. The unique mechanism of mTOR inhibition by rapamycin and its analogs through binding to the FRB domain and formation of a ternary complex with FKBP12 make these compounds extremely selective for the complex mTORC1 with relatively little inhibition of mTORC2. Recently, ATP competitive inhibitors of mTOR have been shown to inhibit both complexes of mTOR and may offer clinical advantages in treating tumors that are not sensitive to rapamycin analogs.

REFERENCES

[1] R. T. Abraham, *DNA Repair (Amst.)*, 2004, **8**, 883.

[2] S. Wullschleger, R. Loewith and M. N. Hall, *Cell*, 2006, **124**, 471.

[3] D. A. Guertin and D. M. Sabatini, *Cancer Cell*, 2007, **12**, 9.

[4] K. Inoki, M. N. Corradetti and K. L. Guan, *Nat. Genet.*, 2005, **37**, 19.

[5] G. G. Chiang and R. T. Abraham, *Trends Mol. Med.*, 2007, **13**, 433.

[6] R. T. Abraham and J. J. Gibbons, *Clin. Cancer Res.*, 2007, **13**, 3109.

[7] A. K. C. Vézina and S. N. Sehgal, *J. Antibiot. (Tokyo)*, 1975, **28**, 721.

[8] J. Choi, J. Chen, S. L. Schreiber and J. Clardy, *Science*, 1996, **273**, 239.

[9] J. C. J. Liang and J. Clardy, *Acta Crystallogr. D Biol. Crystallogr.*, 1999, **55**(Pt 4), 736.

[10] C. E. Caufield, *Curr. Pharm. Des.*, 1995, **1**, 145.

[11] A. Farooq, S. Anjum and A. Ur-Rahman, *Curr. Org. Chem.*, 1998, **2**, 281.

[12] G. Hudes, M. Carducci, P. Tomczak, J. Dutcher, R. Figlin, A. Kapoor, E. Staroslawska, J. Sosman, D. McDermott, I. Bodrogi, Z. Kovacevic, V. Lesovoy, I. G. Schmidt-Wolf, O. Barbarash, E. Gokmen, T. O'Toole, S. Lustgarten, L. Moore and R. J. Motzer, *N. Engl. J. Med.*, 2007, **356**, 2271.

[13] T. E. Witzig, S. M. Geyer, I. Ghobrial, D. J. Inwards, R. Fonseca, P. Kurtin, S. M. Ansell, R. Luyun, P. J. Flynn, R. F. Morton, S. R. Dakhil, H. Gross and S. H. Kaufmann, *J. Clin. Oncol.*, 2005, **23**, 5347.

[14] A. M. Oza, L. Elit, J. Biagi, W. Chapman, M. Tsao, D. Hedley, C. Hansen, J. Dancey and E. Eisenhauer, *J. Clin. Oncol.*, 2006, **24**(18S), 3003.

[15] S. Chan, M. E. Scheulen, S. Johnston, K. Mross, F. Cardoso, C. Dittrich, W. Eiermann, D. Hess, R. Morant, V. Semiglazov, M. Borner, M. Salzberg, V. Ostapenko, H. J. Illiger, D. Behringer, N. Bardy-Bouxin, J. Boni, S. Kong, M. Cincotta and L. Moore, *J. Clin. Oncol.*, 2005, **23**, 5314.

[16] E. Galanis, J. C. Buckner, M. J. Maurer, J. I. Kreisberg, K. Ballman, J. Boni, J. M. Peralba, R. B. Jenkins, S. R. Dakhil, R. F. Morton, K. A. Jaeckle, B. W. Scheithauer, J. Dancey, M. Hidalgo and D. J. Walsh, *J. Clin. Oncol.*, 2005, **23**, 5294.

[17] K. Margolin, J. Longmate, T. Baratta, T. Synold, S. Christensen, J. Weber, T. Gajewski, I. Quirt and J. H. Doroshow, *Cancer*, 2005, **104**, 1045.

[18] D. Cho, S. Signoretti, S. Dabora, M. Regan, A. Seeley, M. Mariotti, A. Youmans, A. Polivy, L. Mandato, D. McDermott, E. Stanbridge and M. Atkins, *Clin. Genitourin. Cancer*, 2007, **5**, 379.

[19] J. E. Dancey, *Cancer Biol. Ther.*, 2006, **5**, 1065.

[20] G. I. Kirchner, I. Meier-Wiedenbach and M. P. Manns, *Clin. Pharmacokinet.*, 2004, **43**, 83.

[21] Investigational Drugs Database 13340 Update Date: 2007-12-20 http://www.thomson-pharma.com

[22] R. J. Amato, A. Misellati, M. Khan and S. Chiang, *J. Clin. Oncol.*, 2006, **24**(18S), 4530.

[23] K. W. L. Yee, Z. Zeng, M. Konopleva, S. Verstovsek, F. Ravandi, A. Ferrajoli, D. Thomas, W. Wierda, E. Apostolidou, M. Albitar, S. O'Brien, M. Andreeff and F. J. Giles, *Clin. Cancer Res.*, 2006, **12**, 5165.

[24] A. Awada, F. Cardoso, C. Fontaine, L. Dirix, J. De Greve, C. Sotiriou, J. Steinseifer, C. Wouters, C. Tanaka, U. Zoellner, P. Tang and M. Piccart, *Eur. J. Cancer*, 2008, **44**, 84.

[25] D. T. Milton, G. J. Riely, C. G. Azzoli, J. E. Gomez, R. T. Heelan, M. G. Kris, L. M. Krug, W. Pao, B. Pizzo, N. A. Rizvi and V. A. Miller, *Cancer*, 2007, **110**, 599.

[26] M. Fouladi, F. Laningham, J. Wu, M. A. O'Shaughnessy, K. Molina, A. Broniscer, S. L. Spunt, I. Luckett, C. F. Stewart, P. J. Houghton, R. J. Gilbertson and W. L. Furman, *J. Clin. Oncol.*, 2007, **25**, 4806.

[27] T. Haritunians, A. Mori, J. O'Kelly, Q. T. Luong, F. J. Giles and H. P. Koeffler, *Leukemia*, 2007, **21**, 333.

[28] K. Zitzmann, E. N. De Toni, S. Brand, B. Goke, J. Meinecke, G. Spottl, H. H. Meyer and C. J. Auernhammer, *Neuroendocrinology*, 2007, **85**, 54.

[29] S. Mabuchi, D. A. Altomare, M. Cheung, L. Zhang, P. I. Poulikakos, H. H. Hensley, R. J. Schilder, R. F. Ozols and J. R. Testa, *Clin. Cancer Res.*, 2007, **13**, 4261.

[30] M. M. Mita, A. C. Mita, Q. S. Chu, E. K. Rowinsky, G. J. Fetterly, M. Goldston, A. Patnaik, L. Mathews, A. D. Ricart, T. Mays, H. Knowles, V. M. Rivera, J. Kreisberg, C. L. Bedrosian and A. W. Tolcher, *J. Clin. Oncol.*, 2008, **26**, 361.

[31] S. P. Chawla, A. W. Tolcher, A. P. Staddon, S. M. Schuetze, G. Z. D'Amato, J. Y. Blay, K. K. Sankhala, S. T. Daly, V. M. Rivera and G. D. Demetri, *J. Clin. Oncol.*, 2006, **24**(18S), 9505.

[32] D. A. Rizzieri, E. Feldman, J. O. Moore, G. J. Roboz, J. F. DiPersio, N. Gabrail, W. Stock, V. M. Rivera, M. Albitar, C. L. Bedrosian and F. Giles, *Blood*, 2005, **106** (*ASH Annual Meeting Abstracts*) 2980.

[33] E. Feldman, F. Giles, G. Roboz, K. Yee, T. Curcio, V. M. Rivera, M. Albitar, R. Laliberte and C. L. Bedrosian, *J. Clin. Oncol.*, 2005, **23**(18S), 6631.

[34] N. Colombo, S. McMeekin, P. Schwartz, J. Kostka, C. Sessa, P. Gehrig, R. Holloway, P. Braly, D. Matei and M. Einstein, *J. Clin. Oncol.*, 2007, **25**(18S), 5516.

[35] M. L. Maddess, M. N. Tackett, H. Watanabe, P. E. Brennan, C. D. Spilling, J. S. Scott, D. P. Osborn and S. V. Ley, *Angew. Chem. Int. Ed.*, 2007, **46**, 591.

[36] E. I. Graziani, F. V. Ritacco, M. Y. Summers, T. M. Zabriskie, K. Yu, V. Bernan, M. Greenstein and G. T. Carter, *Org. Lett.*, 2003, **5**, 2385.

[37] K. J. Weissman, *Trends Biotechnol.*, 2007, **25**, 139.

[38] R. J. M. Goss, S. E. Lanceron, N. J. Wise and S. J. Moss, *Org. Biomol. Chem.*, 2006, **4**, 4071.

[39] M. A. Gregory, H. Petkovic, R. E. Lill, S. J. Moss, B. Wilkinson, S. Gaisser, P. F. Leadlay and R. M. Sheridan, *Angew. Chem. Int. Ed.*, 2005, **44**, 4757.

[40] J. R. Garlich, P. De, N. Dey, J. D. Su, X. Peng, A. Miller, R. Murali, Y. Lu, G. B. Mills, V. Kundra, H. K. Shu, Q. Peng and D. L. Durden, *Cancer Res.*, 2008, **68**, 206.

[41] J. R. Garlich, J. Su and X. Peng, LB105, *AACR Annual Meeting*, San Diego, CA April 2008.

[42] A. S. Kristof, G. Pacheco-Rodriguez, B. Schremmer and J. Moss, *J. Pharmacol. Exp. Ther.*, 2005, **314**, 1134.

[43] L. M. Ballou, E. S. Selinger, J. Y. Choi, D. G. Drueckhammer and R. Z. Lin, *J. Biol. Chem.*, 2007, **282**, 24463.

[44] A. Zask, J. Kaplan, L. Toral-Barza, I. Hollander, M. Young, M. Tischler, C. Gaydos, M. Cinque, J. Lucas and K. Yu, *J. Med. Chem.*, 2008, **51**, 1319.

[45] M. Hayakawa, H. Kaizawa, H. Moritomo, T. Koizumi, T. Ohishi, M. Yamano, M. Okada, M. Ohta, S. Tsukamoto, F. I. Raynaud, P. Workman, M. D. Waterfield and P. Parker, *Bioorg. Med. Chem. Lett.*, 2007, **17**, 2438.

[46] Q. W. Fan, Z. A. Knight, D. D. Goldenberg, W. Yu, K. E. Mostov, D. Stokoe, K. M. Shokat and W. A. Weiss, *Cancer Cell*, 2006, **9**, 341.

[47] F. Stauffer, C. Garcia-Echeverria, P. Furet, H.-G. Capraro, P. Holzer and M. Maira, *Proceedings of the American Association for Cancer Research (AACR)*, Los Angeles, CA, April 2007, p. 3953.

[48] F. Stauffer, C. Garcia-Echeverria, P. Furet, C. Schnell, S. Ruetz, M. Maira and C. M. Fritsch, *Proceedings of the American Association for Cancer Research (AACR)*, Los Angeles, CA, April 2007, p. 269.

[49] D. Laird, B250, AACR-NCI-EORTC Internationl Conference Molecular Targets Cancer Therapy, San Francisco, CA, October 2007.

[50] W. Bajjalieh, L. C. Bannen, S. D. Brown, P. Kearney, M. Mac, C. K. Marlowe, J. M. Nuss, Z. Tesfai, Y. Wang and W. Xu, *WO Patent* WO07044729, 2007.

[51] J. C. Arnould, K. M. Foote and E. J. Griffen, *WO Patent* WO07129044-A1, 2007.

[52] M. Bengtsson, J. Larsson, G. Nikitidis, P. Storm, J. P. Bailey, E. J. Griffen, J.-C. Arnould and T. G. C. Bird, *WO Patent* WO06051270, 2006.

[53] S. C. Cosulich and E. J. Griffen, *Keystone Symposia: PI3-Kinase Signaling Pathways in Disease*, Santa Fe, NM, February, 2007.

[54] L. David, K. M. Foote and A. Lisius, *WO Patent* WO07129052, 2007.

[55] K. M. Foote and E. J. Griffen, *WO Patent* WO07135398, 2007.

[56] K. G. Pike, M. R. V. Finlay, S. M. Fillery and A. P. Dishington, *WO Patent* WO07080382, 2007.

[57] S. P. Mutton and M. Pass, *WO Patent* WO07066103, 2007.

[58] M. G. Hummersone, S. Gomez, K. A. Menear, X. F. Cockcroft, P. Edwards, V. J. M. L. Loh and G. C. M. Smith, *WO Patent* WO06090169, 2006.

[59] M. G. Hummersone, S. Gomez, K. A. Menear, G. C. M. Smith, K. Malagu, H. M. E. Duggan, X. F. Cockcroft and G. J. Hermann, *WO Patent* WO07060404, 2007.

[60] M. G. Hummersone, S. Gomez, K. A. Menear, X. F. Cockcroft and G. C. M. Smith, *WO Patent* WO06090167, 2006.

[61] I. Hickson, L. M. Smith, S. J. Maguire, H. M. E. Duggan, K. Malagu, M. G. Hummersone, K. A. Menear, C. M. Chresta, B. Davies, M. Pass, N. M. Martin and G. C. M. Smith, *AACR Annual Meeting*, San Diego, CA, April 2008.

[62] A. P. Crew, D. S. Werner and P. A. R. Tavares, *WO Patent* WO07087395, 2007.

[63] X. Chen, H. Coate, A. P. Crew, H.-Q. Dong, A. Honda, M. J. Mulvihill, P. A. R. Tavares, J. Wang, D. S. Werner, K. M. Mulvihill, K. W. Siu, B. Panicker, A. Bharadwaj, L. D. Arnold, M. Jin, B. Volk, Q. Weng and J. D. Beard, *WO Patent* WO07061737, 2007.

[64] K. M. Shokat, Z. A. Knight, and B. Aspel, *WO Patent* WO07114926, 2007.

[65] S. V. Bhagwat, A. P. Crew, P. C. Gokhale, A. Cooke, J. Kahler, Y. Yao, A. Chan, P. A. Tavares, B. Panicker, C. Mantis, J. Workman, D. Landfair, M. Bittner, R. Wild, L. D. Arnold, D. M. Epstein and J. A. Pachter, *AACR Annual Meeting*, San Diego, CA, April 2008.

[66] V. Ververka, T. Crabbe, I. Bird, G. Lennie, F. W. Muskett, R. J. Taylor and M. D. Carr, *Oncogene*, 2008, **27**, 585.

[67] E. Claus, I. Seipelt, E. Guenther, E. Polymeropoulos, M. Czech and T. Schuster, *WO Patent* WO07054556, 2007.

[68] E. Claus, I. Seipelt, E. Guenther, E. Polymeropoulos, M. Czech and T. Schuster, *Eur. Patent* *EP*1785423, 2007.

[69] J. Felding, H. C. Pedersen, C. Krog-Jensen, M. Praestegaard, S. P. Butcher, V. Linde, T. S. Coulter, C. Montalbetti, M. Uddin and S. Reignier, *WO Patent* WO05097107, 2005.

[70] L. Xu, S. Tao, X. Wang, Z. Yu, M. Wang, D. Chen, Y. Jing and J. Dong, *Bioorg. Med. Chem.*, 2006, **14**, 5351.

CHAPTER 13

Oncology Drug Targets in the Sphingomyelin-Signaling Pathway

William Garland, Amy Cavalli and **Geneviève Hansen**

1. INTRODUCTION

1.1 Emerging role of sphingosyl-LPLs as signaling agents

Lysophospholipids (LPLs) are phospholipids featuring a free hydroxyl group not conjugated to a fatty acid. In the late 19th century, LPLs were characterized as amphipathic, organic solvent-extractable components of cells and tissues that are found in the physiologic fluids of mammals at μmolar concentrations. Two classes of LPLs are best known. The first class, the glyceryl-LPLs, exemplified by lysophosphatidic acid (LPA, **1**), predominate quantitatively among lipids serving

Lpath, Inc., San Diego, CA 92121, USA

Annual Reports in Medicinal Chemistry, Volume 43
ISSN 0065-7743, DOI 10.1016/S0065-7743(08)00013-4

as structural constituents of cell membranes. The other LPL class, sphingosyl-LPLs represented by sphingosine 1-phosphate (S1P, **2**), share a common sphingoid backbone and are quantitatively less prominent than glyceryl-LPLs. These two classes plus sterols represent the three main classes of lipids present in cell membranes.

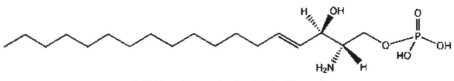

1 (Stearoyl LPA; myristoyl, palmitoyl, oleoyl & arachidonoyl acyl side chains also common)

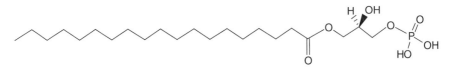

2 (S1P; only one structural variant known)

LPLs in the membranes of organelles provide a framework for embedded proteins (e.g., receptors and ion channels), and the function of sphingosyl-LPL was, until recently, thought to be solely to provide structural order to plasma lipids and proteins within the plasma membrane bilayer [1,2]. Sphingosyl-LPLs are now recognized as a functionally complex class of biological regulators that act through specific receptors or enzymes and thereby play a critical role as intercellular and intracellular molecules participating in physiological and pathological signaling associated with cellular survival, proliferation, differentiation and adhesion [3–6].

This chapter summarizes recent advances in targeting sphingosyl-LPLs to provide treatments for cancer. Sphingosyl-LPLs include sphingomyelins, ceramides, sphingosines and sphingosine-1-phosphate. They are ubiquitous membrane components of essentially all eukaryotic cells and are abundantly located in all plasma membranes as well as in some intracellular organelles (endoplasmic reticulum, Golgi complex and mitochondria). Sphingomyelin-associated sphingosyl-LPLs are of particular interest in anti-cancer therapy because of three metabolites of sphingomyelin: ceramide and sphingosine are potent inducers of apoptosis and produce cell cycle arrest, whereas sphingosine-1-phosphate is anti-apoptotic and promotes cell growth and migration [7–11]. Thus, the metabolism of sphingomyelin generates signal molecules important to the modulation of cell growth and proliferation, differentiation and apoptosis (cell survival) — processes critical to the progression of cancer and resistance to cancer therapy.

Understanding the biochemistry of LPLs is complicated by the bewildering combination of specific head groups including head groups containing various carbohydrates and hydrocarbon (fatty acid) tails [12,13] often assigned the same general name, e.g., ceramide. The reason(s) for this structural diversity is not understood but is likely significant with respect to biologic function.

1.2 Elements of the sphingomyelin-signaling pathway

The key elements of the sphingomyelin-associated signaling pathway including the bioactive lipid mediators ceramide, sphingosine and sphingosine-1-phosphate (S1P) are presented in Figure 1. These mediators are derived from sphingomyelin which is present in the plasma membranes of all mammalian cells.

Table 1 provides a list of the enzymes and other processes responsible for forming and clearing the three sphingomyelin metabolites. Each of these enzymes or processes provides a potential target for cancer therapy. For instance, ceramide is produced by cancer cells in response to exposure to radiation and chemotherapeutic agents, and is an intracellular second messenger that activates enzymes leading to apoptosis. Because of its central role in apoptosis, pharmacologic manipulation of intracellular ceramide levels results in attenuation or enhancement of drug resistance. This may be achieved through direct application of sphingolipids or by the inhibition/activation of the enzymes or other processes that either produce or use ceramide.

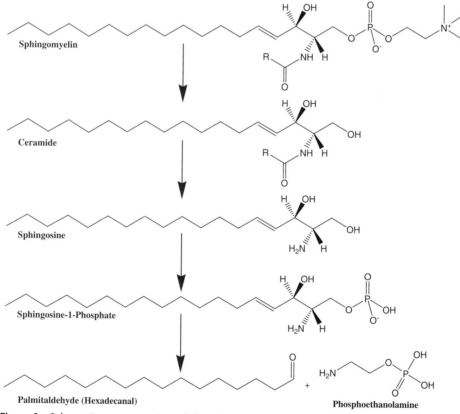

Figure 1 Schematic representation of the sphingomyelin-associated signaling cascade.

Table 1 Sphingolipid cascade: sphingomyelin → ceramide → sphingosine → sphingosine-1-phosphate

Element	Structure	Formation	Clearance
Sphingomyelin		Sphingomyelin (SM) synthase [14,15]	Sphingomyelinase [16–18]
Ceramide		Dihydroceramide desaturase [19–21] & sphingomyelinase [16–18], ceramide 1-phosphate phosphatase, glucosylceramide β-glucosidase [22]	SM synthase [23], ceramidase [24–26], ceramide kinase [27], 1-O-acylceramide synthase [28], ceramide glucosyltransferase [29], ceramide galactosyltransferase [30]
Sphingosine		Ceramidase [31], endocytic recycling of plasma membrane [32,33], S1P phosphatase [34]	Sphingosine kinase [35,36]
Sphingosine-1-phosphate (S1P)		Sphingosine kinase [35,36]	S1P phosphatase [34], S1P lyase [37,41]
Palmitaldehyde phospho-ethanolamine		S1P lyase [37]	Metabolized to glycerolphospholipid & alk(en)yl-phospholipid [38]

1.3 Tumor biology of the sphingomyelin-signaling pathway

1.3.1 Sphingomyelin

Sphingomyelin is found in all animal cell membranes particularly in the myelin sheath that surrounds nerve cell axons. The structure of all sphingomyelins, like all sphingolipids, consists of sphingosine bonded to one fatty acid and one polar head group that is typically a phosphocholine. Sphingomyelin serves as a reservoir for ceramide formation via hydrolysis by sphingomyelinases (SMases). SMases — specialized enzymes with phospholipase C activity — hydrolyze the phosphodiester bond of sphingomyelin [39]. Sphingomyelin's metabolites, but not sphingomyelin itself, appear involved in signaling.

Several isoforms of SMase can be distinguished by their different pH optima and some, but not all, of these molecules have been structurally identified and characterized. Neutral SMase (nSMase) and acid SMase (aSMase) are rapidly activated by diverse stress stimuli and promote an increase in cellular ceramide levels over a period of minutes to hours. Alkaline SMase, found in the intestinal mucosa and bile, does not appear to participate in signal transduction.

Sphingomyelin can be produced from ceramide. Sphingomyelin synthase (SMS) transfers the phosphocholine group of phosphatidylcholine to generate sphingomyelin and diacylglycerol [40]. Two variants of this membrane integral protein, SMS1 and SMS2, exist in man.

1.3.2 Ceramide

Ceramides, products of sphingomyelin hydrolysis by SMase, are present in the subcellular fractions of many human tissues. *Ceramide is implicated as a key mediator in signaling pathways leading to suppression of growth, cellular senescence, differentiation, and apoptosis* [41–43]. Another biochemical route for ceramide synthesis is de novo from the serine palmitoyltransferase (SPT) catalyzed condensation of serine and palmitoyl CoA to produce 3-ketodihydrosphingosine. This intermediate is subsequently reduced to sphingosine followed by fatty acid acylation by dihydroceramide synthase to form dihydroceramide. Subsequent double bond formation to produce ceramide is catalyzed by dihydroceramide desaturase [44,45]. Dihydroceramides may have a direct anti-cancer effect separate from their conversion to ceramides [46].

Ceramide is also the key biochemical in the biosynthesis of glycosphingolipids and gangliosides. Like sphingomyelin, glycosphingolipids are also ubiquitous membrane constituents. While sphingomyelins contain a phosphorylcholine group, glycosphingolipids contain a carbohydrate. Two classes of glycosphingo-lipids are known: neutral (gangliosides) or acidic (sulfatides). Gangliosides influence cell death, invasion and metastasis and are also potential diagnostic markers and therapeutic targets for cancer [47–50]. One glycosphingolipid, glucosylceramide, formed from ceramide by the action of glucosylceramide synthase, plays a significant role in multi-drug resistance including increased resistance to the apoptotic effects of cytotoxics such as doxorubicin through regulation of the cellular P-glycoprotein efflux pump [51–53]. In addition, sphingo-lipid galactosyltransferases can convert glucosylceramide to lactosylceramide, a precursor mono-sialyllactosylceramide that itself can be converted to di-sialyllactosylceramides. Several of these complex gangliosides products such as GM3 and GD3 are strongly pro-apoptotic [54].

The formation and clearance of ceramide is complex. Ceramide-1-phosphate (C1P), a metabolite of ceramide produced by ceramide kinase, mediates cell survival and activation and translocation of cPLA2 [55,56], a process that is associated with certain cancers [57,58] and can even lead to increased production of LPLs [59]. Ceramidase converts ceramide to sphingosine, and sphingomyelin synthase converts ceramide in a coupled reaction to sphingomyelin with phosphatidylcholine as the source of the phosphocholine. In the conversion, phosphatidylcholine is converted to diacylglycerol, a known activator of the strongly pro-growth α form of protein kinase C.

1.3.3 Sphingosine

Sphingosine and its reduced derivative sphinganine are the major sphingolipid bases in mammals. Both molecules feature an amino alcohol with a long unsaturated hydrocarbon chain. The biological effects of sphingosine may vary among cell types but are associated with negative (pro-apoptotic) effects on cell growth and survival, sphingosine has been implicated as an inhibitor of protein kinase A as well as a protein kinase that disrupts the activity of the profoundly anti-apoptotic 14-3-3 protein [60,61]. Inhibitors of both protein kinase A and Protein kinase C have been clinically evaluated as anti-tumor agents [62].

1.3.4 Sphingosine 1-phosphate

S1P is a product of sphingosine phosphorylation and has potent bioactivity. Generated by sphingosine kinase, S1P controls numerous aspects of cell physiology including cell survival and mammalian inflammatory responses [63,64]. While ceramide and sphingosine are associated with apoptosis, S1P is typically viewed as a mediator of cell proliferation and activation of survival pathways [65,66]. Importantly, S1P acts counter to ceramide to mediate cell growth and survival, as well as influencing directed cell movement. The apparent role of S1P in angiogenesis [67–71] and tumor biology [72–75], including migration and invasion of cancer and angiogenesis-associated endothelial cells [76–79], is now well recognized. S1P also affects other endogenous pro-angiogenic growth factors like VEGF [80–82] and PDGF [83]. In addition, S1P transactivates receptors important to tumor growth like EGF [84]. Finally, S1P also regulates the production of eicosanoids, an important class of inflammatory mediators. For example, S1P, but not dihydrosphingosine-1-phosphate, induces cyclooxygenase-2 [85].

In adults, S1P is released from platelets and mast cells [86–89] to create a local pulse of S1P sufficient enough to exceed the Kd of the S1P receptors to promote wound healing and participate in the inflammatory response. Under normal conditions, the total S1P in the plasma is quite high (300–500 nM). However, S1P is 'buffered' by serum proteins, particularly albumin and HDLs, so that the bio-available or free fraction of S1P is not sufficient to appreciably activate S1PRs. If this were not the case, inappropriate angiogenesis and inflammation would result.

These effects — primarily vasculogenesis, cell growth/survival/migration and inflammation [90–92] — are mediated via a family of G protein-coupled receptors (GPCRs) known as endothelial differentiation genes (EDG). Five GPCRs have been identified as high-affinity S1P receptors: S1PR1/EDG-1, S1PR2/EDG-5, S1PR3/EDG-3, S1PR4/EDG-6 and S1PR5/EDG-8 [93–97]. The S1PRs, particularly S1PR1, are required for normal embryonic development, particularly for maturation of the vasculature [98]. Many responses evoked by S1P are sensitive to the pertussis toxin indicating that the S1P receptors are coupled to different heterotrimeric Gi/o (Gq-, Gi-, G12-13) proteins and the small GTPases of the Rho family. S1P also has intracellular targets that are less well characterized [99].

S1P appears to play a unique role in regulating lymphocyte trafficking through interaction at the S1PR1 receptor. There are two models for the control of lymphocyte trafficking by the S1P–S1PR1 axis that can be envisaged on the basis of a tentative integration of the existing evidence. In the first model, which is known as the stromal-gate control model, lymphocytes are passive bystanders, and their egress from the lymph nodes is altered by S1P-induced changes in the permeability of the endothelium in blood vessels and lymphoid vascular sinuses without considerable changes in the motility of the lymphocytes themselves. In the second model, known as the S1P–S1PR1 control-of-lymphocytes model, site-specific variable suppression of S1P-mediated lymphocyte chemotaxis is the principal mechanism for the reduced egress of lymphocytes from the lymph nodes, with little or no contribution from changes in the endothelium. This immunomodulatory action of S1P has been utilized using the above mentioned FTY720 as a potential treatment of multiple sclerosis, an autoimmune disease [100].

Sphingosine kinase (SPHK), a purported oncogene, phosphorylates sphingosine to generate S1P and thereby plays a pivotal role in promoting tumor growth [101]. The sole reaction product of SPHK is S1P and SPK is the rate-limiting step in its synthesis. SHK is highly conserved from protozoa to mammals and is ubiquitous in living tissues [102]. The enzyme is induced by transforming growth factor-β and mediates TIMP-1 up-regulation [103,104]. SPHK degradation, accompanied by cell apoptosis because of decreased S1P, is an important response to relative oxygen species (ROS) produced during ischemia/reperfusion injury [105,106].

There are two isoforms of SPHK: SPHK1 and SPHK2. Both isoforms and their respective splice variants have been characterized [107]. SPHK1 is found in the cytosol of eukaryotic cells, and migrates to the plasma membrane upon activation; SPHK2 is localized to the nucleus. SPHK1 is slightly more efficient than SPHK2 in phosphorylating sphingosine, whereas SPHK2 is significantly more efficient toward unnatural substrates such as the immunomodulatory drug FTY720. SPHK1 has been shown to regulate a wide variety of cellular processes, including promotion of cell proliferation, survival and motility. In contrast, SPHK2 has been shown to inhibit DNA synthesis and also to induce apoptosis in a variety of cell types. Significantly, nuclear localization is essential to affect the anti-proliferative effects of SPHK2.

SPHK appears to play an important role in maintaining health, playing a double role regulating cellular function. At the basal catalytic level, SPHK is involved in sphingomyelin and glycosphingolipid turnover, and is responsible for clearing the cell of sphingosine and ceramide. Upon activation with agonists, SPHK1 functions as a component of the sphingomyelin signal transduction pathway. For example, TNF-α induced activation of SPHK1 is required for the TRAF2-dependent activation of NF-κB. Furthermore, activation of SPHK1 and its translocation to the membrane is mediated by the ERK1/2-induced phosphorylation at Ser-225. Phosphorylation of SPHK1 by ERKs is dependent on its association with TRAF2. SPHK1 is not only an effector of ERK1/2, but is also involved in the sphingomyelin-signaling pathway leading to ERK1/2 activation. It has been shown that activation of ERK1/2 by tumor necrosis factor-α, vascular endothelial growth factor and 17β-estradiol requires SPHK1. The pleiotropic effects of SPK appear mediated by the product of the enzymes action, S1P and perhaps dihydro-S1P.

S1P appears to be cleared primarily by S1P lyase to palmitaldehyde (hexadecanal) and O-phosphoethanolamine, two glycerophospholipids precursors or by one or more sphingophospholipid phosphatases back to sphingosine.

2. THERAPEUTIC TARGETS

An analysis of the tumor biology associated with the sphingomyelin pathway suggests the targets listed in Table 2 for possible effect on cancer (modulation of growth, angiogenesis, metastasis or maintenance of efficacy).

The direct targets for the elements in the sphingolipid cascade are provided in Table 3.

Selected, known inhibitors of sphingolipid production, degradation and function are provided in Table 4.

One final target deserves special mention. SPHK and S1P have been implicated in the development of resistance to standard cancer chemotherapy [188,189] and agents that inhibit SPHK and/or S1P might be used to overcome this treatment-limiting feature of most current chemotherapeutic treatments. This application offers simplicity in clinical trial design not available in most other oncology applications.

3. TREATMENT MODALITIES

No treatment modalities based on modulating sphingomyelin pathway biochemicals are approved for marketing for oncology or any other therapeutic indication. The agent closest to NDA approval is FTY720 for multiple sclerosis which is in Phase III clinical trials. However, many agents are in development and examples are surveyed below. Only selected examples are provided.

Table 2 Predicted effect on cancer with modulation of activity enzymes associated with the sphingolipid cascade

Target	Modulation	Expected biochemical effect	Effect
Sphingomyelin (SM) synthase	Inhibition	Less SM, less diacylglycerol & more ceramide	+
	Potentiation	More SM, more diacylglycerol & less ceramide	−
Sphingomyelinase (SMase)	Inhibition	Less ceramide	+
	Potentiation	More ceramide	−
Serine palmitoyltransferase (SPT)	Inhibition	Less ceramide	+
	Potentiation	More ceramide	−
Dihydroceramide synthase	Inhibition	Less ceramide	+
	Potentiation	More ceramide	−
Dihydroceramide desaturase	Inhibition	Less ceramide & more dihydroceramide	?
	Potentiation	More ceramide & less dihydroceramides	?
Cerebrosidase	Inhibition	Increases ceramide	+
	Potentiation	Decreases ceramide	−
Glucosylceramide synthase	Inhibition	Increases ceramide	+
	Potentiation	Decreases ceramide	−
Galactosyltransferases	Inhibition	Decrease GM3 and GD3	−
	Potentiation	Increase GM3 and GD3	+
Ceramide kinase	Inhibition	Increases ceramide & decreases C1P	+
	Potentiation	Decreases ceramide & increases C1P	−
C1P phosphatase	Inhibition	Increases ceramide & decreases C1P	+
	Potentiation	Decreases ceramide & increases C1P	−
Ceramidase	Inhibition	Increases ceramide	+
	Potentiation	Decreases ceramide	−
Ceramide synthase	Inhibition	Increases sphingosine & decreases ceramide	?
	Potentiation	Decreases sphingosine & increases ceramide	?
Sphingosine kinase	Inhibition	Increases ceramide & decreases S1P	+
	Potentiation	Decreases ceramide and increases S1P	−

Table 2 *(Continued)*

Target	Modulation	Expected biochemical effect	Effect
S1P phosphatases	Inhibition	Increase sphingosine and decrease S1P	?
	Potentiation	Decrease sphingosine and increase S1P	?
S1P lyase	Inhibition	Increases S1P	−
	Potentiation	Decreases S1P	+

Table 3 Direct sphingolipid targets

Element	Target
Sphingomyelin	No direct targets identified
Ceramide	Ceramide-activated protein phosphatase 1 & 2 (CAPP) [108,109], ceramide-activated protein kinase (CAPK), kinase suppressor of Ras (KRS) [110], cathepsin C [111,112], PKC-α,δ & ξ [113,114], membrane rafts [115], c-Raf [116], MEKK-1[117] & PLA2 [118]
Ceramide 1-phosphate	PLA$_2$ [119] & acid sphingomyelinase [120]
1-O-acylceramide	No direct targets identified
Glucosylceramide	No direct targets identified building block for synthesis of complex GSLs [121]
Galactosylceramide	No direct targets identified building block for synthesis of complex GSLs [121]
Sphingosine	PKC [122,123], voltage-operated calcium channels (VOCC) [124] & calcium release-activated calcium current (ICRAC) [125]
S1P	S1PR$_{1-5}$ [126–129], SCaMPER [130] & unknown intracellular targets [131,132]
Palmitaldehyde	No direct targets known-metabolized to phospholipid
Phosphoethanolamine	No direct targets known

3.1 Therapeutic antibodies

The presence of multiple receptors appears to be a common characteristic for signaling lipids. For this reason targeting the enzymatic product rather than the receptor may offer a promising approach. Two examples to date provide experimental findings to support this concept.

- A murine anti-S1P mAb demonstrated pronounced anti-growth and anti-angiogenic activity [186,190]. The use of an antibody in this application

Table 4 Selected, known inhibitors of sphingolipid production, degradation and function

Target	Action	Drug name	References
SM synthase	Inhibition	MS-209, D609 (& analogs) & fenretinide	[133–136]
Acidic SMase	Inhibition	PtdIns3,5P2, PtdIns3,4,5P3, desipramine, SR33557, NB6, mangostin	[137–143]
Neutral SMase	Inhibition	C11AG, GW4869, scyphostatin, macquarimicin A, alutenusin, chlorogentisylquinone, manumycin A, CerP difluoromethylene analogues, siRNA, gluthathione	[144–151]
	Activation	Danorubicin, D-erythro-MAPP, B13	[14, 152–156]
Ceramide synthase	Inhibition	Fumonisin B, AAL-toxin, australifungins, erythro- & threo-2-amino-3-hydroxy-2-amio-3,5-dihydroxyctadecanes	[157–159]
	Activation	Doxorubicin, vincristine/ vinblastine, paclitaxel, tamoxifen, 4-HPR	[160–165]
Ceramide Glusosyltransferase	Inhibition	D-t-PDMP, P4, antisense oligonucleotide, tamoxifen, verapamil, cyclosporine A	[166,167]
Ceramidase	Inhibition	B13	[168]
Serine Palitoyltransferase	Inhibition	Sphingofungins, lipoxaycine, myriocin, virdiofungins, K-cycloserine, B-cCholror-l-alanine	[123, 169, 170]
Dihydroceramide Desaturase	Inhibition	GT-11 (& analogs), dihydroceamide analogues	[171,172]
Ceramidase	Inhibition	D-MAPP, Cer/SPH Stereoisomers, AD2646, B13	[173–175]
Sphingosine kinase	Inhibition	Safingol, DMS, SG-14, S-15183, B5354, phenoxodiol (PXD), docetaxel, camptothecin	[176–184]
S1P receptors	Inhibition	FTY720	[185]
S1P	Neutralization	Sphingomab: sonepcizumab (humanized)/LT1002 (murine)	[186]
S1P lyase	Inhibition	1-Desoxysphinganine-1-phosphate	[187]

provides selectivity with regard to location of the S1P (extracellular S1P only not intracellular S1P) but also potentially lowers S1P activity at all extracellular receptors.

- One of the characteristics of tumor cells is an enhanced expression of various gangliosides, and recently, a chimeric anti-GD3 antibody (KM871) was tested in phase I clinical trials in patients with metastatic melanoma [191], and another was evaluated in a phase III clinical trial against small cell lung cancers [192].

Besides antibodies, the approach of using macromolecules to bind and neutralize lipid signaling molecules could also potentially be used with aptamers [193–195] and receptor decoys [196] targeting the lipid signaling molecules.

3.2 Selected small molecule enzyme inhibitors

Sphingomyelinase [197–198]

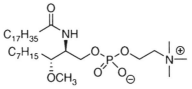

Example of sphingomyelin analog inhibitor[198]

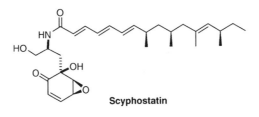

Scyphostatin

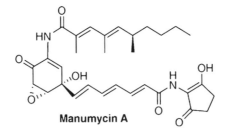

Manumycin A

Serine palmitoyltransferase [199]

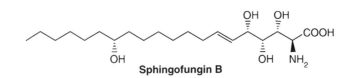

Sphingofungin B

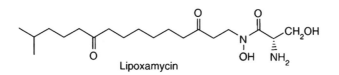

Lipoxamycin

Dihydroceramide desaturase [123,200]

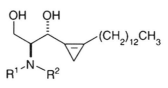

GT11 Analogs (For GT11, R1=H and R2=CO(CH₂)₈CH₃

Glucosylceramide synthase [201]

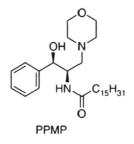

PPMP

Ceramide kinase [202]

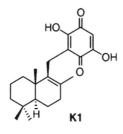

K1

Ceramidase [173]

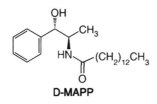

D-MAPP

Ceramide synthase [157,203]

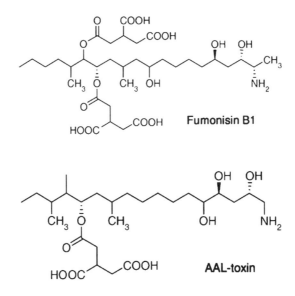

Fumonisin B1

AAL-toxin

Sphingosine kinase [204,205]

Safingol

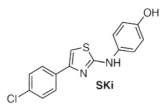

SKi

S1P lyase [187,206,207]

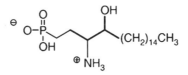

1 desoxysphinganine-1-phosphonate

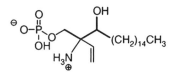

2-vinylsphinganine-1-phosphate

Also FTY720 (Structure in following section)

The scientific literature does not provide compelling examples of inhibitors of sphingomyelin synthase, dihydroceramide synthase. cerebrosidase, C1P phosphatase(s), galactosyltransferases and S1P phosphatase(s).

3.3 Small molecule receptor antagonists and agonists

FTY720 is a sphingosine-like drug that is phosphorylated intracellularly by SPHK to become an S1P-receptor agonist that has significant immunosuppressive activity. FTY720 was the result of a systematic SAR exploration of myriocin, a fermentation product of *Isaria sinclairi* that demonstrated significant immunosuppressive activity.

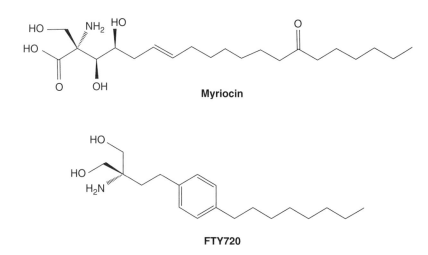

Myriocin

FTY720

Both FTY720 and phospho-FTY720 function to down-regulate the S1P-receptor S1PR1 and other S1P receptors. In animal models of human cancer, FTY720 has demonstrated a significant inhibitory effect on tumor vascularization [208–211].

Structures of purported specific inhibitors of S1PR1 (AUY-954 and W146), S1P2 (JTE013), S1PR3, S1PR4, S1PR4/5 and S1P R1/5 are available on-line [212].

3.4 Antisense/siRNA

Although no examples are available in the scientific literature, the use of antisense and/or siRNA to inhibit the production the enzymes responsible for the production or clearance of the sphingosyl-LPLs must be considered a realistic approach [213]. In this regard, recent advances in modifying the structures of antisense molecules to provide more chemical stability [214] and specificity [215] suggest progress overcoming the problems associated with earlier antisense-based anti-cancer drug candidates.

3.5 Other

Although no examples are available in the scientific literature, enzyme supplement/replacement therapy appears a potentially useful approach to treat cancer by modulating concentrations of enzymes either producing or clearing signaling sphingosyl-LPLs important to the progression of the disease. For example, cancer patients could be treated with human S1P lyase to lower systemic S1P concentrations and thereby mitigate the effect of this pro-growth and pro-angiogenesis/vasculogenesis lipid. There are many successful examples of this approach in treating non-oncologic conditions [216,217]. Gene therapy is a similar approach that could provide the same result locally [218,219].

4. OPPORTUNITY AND THREAT

The sphingosyl-LPLs as a class offer a great opportunity for the treatment of cancer. Cancer is defined by six abnormal characteristics: self-sufficiency in growth factors (amplification of external growth cues or tumor-produced growth factors), insensitivity to anti-growth signals from surrounding tissue, evasion of apoptosis, limitless replicative potential, sustained neovascularization, invasiveness and motility [220]. The tumor biology of sphingosyl-LPL presented previously in this chapter suggests that this class of molecules is involved in all aspects of the six defining characteristics of cancer. However, the critical involvement of these same agents in normal biochemical function suggests that assuring adequate safety of any sphingosyl-LPL based therapeutic must be carefully evaluated before progressing the agent into clinical development.

5. DISCUSSION AND CONCLUSIONS

The now rapidly emerging knowledge on the role of sphingosyl-LPLs in tumor biology illuminates the importance of these molecules in the regulation of tumor development and progress as well as cancer cell death. For this reason, a substantial increase in interest in discovering chemicals to modulate this class of molecules is anticipated. The effort will most likely result in many new important anti-cancer drugs. However, the effort will require overcoming a few difficulties intrinsic to sphingosyl-LPLs.

- Because the sphingosyl-LPLs often contain side chains with varying lengths, the optimum agents may need to target a specific isoform rather than the generic sphingosyl-LPL. Conversely, an individual agent may only effectively target one or a multiple isoforms when targeting all isoforms is in fact needed to obtain optimal efficacy.
- Similarly, multiple isoforms of the enzymes responsible for producing and clearing a specific sphingosyl-LPL typically exist and sometimes the same isoform is present in very different physiologic compartments such as extracellular, cytoplasmic and nuclear. The issues of what isoform and in what location to target greatly increases the complexity of the program to discover new agents related to sphingosyl-LPL dependent biology.
- The concentration of sphingosyl-LPLs is often in the micromole range and this relatively high concentration can be perceived as an obstacle if the drug targets the lipid directly as with a neutralizing agent like an antibody.
- Sphingosyl-LPLs are typically highly protein bound and this characteristic can be perceived as an obstacle if the drug needs to remove all, or most of, the sphingosyl-LPL to achieve adequate efficacy.
- Most of the model compounds discovered to date which inhibit either the enzymes responsible for the production or clearance of sphingosyl-LPLs or the sphingosyl-LPL receptors are natural products that possess many of the amphipathic structural characteristics of the natural sphingosyl-LPL. These molecules may not be optimal candidates for oral activity and may not have reasonably long half-lives following administration.
- Successful small molecule inhibitors of the receptors for sphingosyl-LPLs must typically overcome the complexity of multiple receptors with different selectivity that also features considerable cost crosstalk among the receptors.
- A final difficulty with the search for small molecule inhibitors of the receptors is the lack of X-ray crystallography data for any of the receptors and the limited data available comparing the structure of the receptors across species.

However, relative to many drug discovery programs, the search for sphingosyl-LPL related treatment modalities does benefit from a few technical characteristics of the sphingosyl-LPLs. The most important is that sphingosyl-LPLs are structurally identical across species. In addition, a limited amount of X-ray crystallography data are now emerging for a few of the enzymes responsible for producing and clearing sphingosyl-LPLs [221,222].

In summary, sphingosyl-LPLs provide many exciting new targets for oncology drug discovery. A number of significant class-specific challenges will, however, need to be overcome to produce therapeutics that will realize the potential of the rapidly emerging knowledge concerning the tumor biology associated with the sphingosyl-LPLs.

REFERENCES

[1] F. M. Goni and A. Alonso, *Biochim. Biophys. Acta*, 2006, **1758**, 1902.

[2] W. Zheng, J. Kollmeyer, H. Symolon, A. Momin, E. Munter, E. Wang, S. Kelly, J. C. Allegood, Y. Liu, Q. Peng, H. Ramaraju, M. C. Sullards, M. Cabot and A. H. Merrill Jr., *Biochim. Biophys. Acta*, 2006, **1758**, 1864.

[3] S. E. Gardell, A. E. Dubin and J. Chun, *Trends Mol. Med.*, 2006, **12**, 65.

[4] W. H. Moolenaar, *Exp. Cell Res.*, 1999, **253**, 230.

[5] E. J. Goetzl, M. Graeler, M. C. Huang and G. Shankar, *Sci. World J.*, 2002, **2**, 324.

[6] E. Birgbauer and J. Chun, *Cell Mol. Life Sci.*, 2006, **63**, 2695.

[7] J. M. Padron, *Curr. Med. Chem.*, 2006, **13**, 755.

[8] D. E. Modrak, D. V. Gold and D. M. Goldenberg, *Mol. Cancer Ther.*, 2006, **5**, 200.

[9] B. Segui, N. Andrieu-Abadie, J. P. Jaffrezou, H. Benoist and T. Levade, *Biochim. Biophys. Acta*, 2006, **1758**, 2104.

[10] B. Ogretmen, *FEBS Lett.*, 2006, **580**, 5467.

[11] B. Ogretmen and Y. A. Hannun, *Nat. Rev. Cancer*, 2004, **4**, 604.

[12] A. H. Futerman and Y. A. Hannun, *EMBO J. Rep.*, 2004, **5**, 777.

[13] T. S. Worgall, *Curr. Opin. Clin. Nutr. Metab. Care*, 2007, **10**, 149.

[14] R. Bose, M. Verheij, A. Haimovitz-Friedman, K. Scotto, Z. Fuks and R. Kolesnick, *Cell*, 1995, **82**, 405.

[15] D. K. Perry, J. Carton, A. K. Shah, F. Meredith, D. J. Uhlinger and Y. A. Hannun, *J. Biol. Chem.*, 2000, **275**, 9078.

[16] E. H. Schuchman, M. Suchi, T. Takahashi, K. Sandhoff and R. J. Desnick, *J. Biol. Chem.*, 1991, **266**, 8531.

[17] S. Tomiuk, K. Hofmann, M. Nix, M. Zumbansen and W. Stoffel, *Proc. Natl. Acad. Sci., USA*, 1998, **95**, 3638.

[18] K. Hofmann, S. Tomiuk, G. Wolff and W. Stoffel, *Proc. Natl. Acad. Sci., USA*, 2000, **97**, 5895.

[19] G. S. Dbaibo, W. El-Assaad, A. Krikorian, B. Lui, K. Diab, N. Z. Idriss, M. El-Sabban, T. A. Driscoll, D. K. Perry and Y. A. Hannun, *FEBS Lett.*, 2001, **503**, 7.

[20] P. Ternes, S. Franks, U. Zahringer, P. Sperling and E. Heinz., *J. Biol. Chem.*, 2002, **277**, 25512.

[21] C. Michel, G. van Echten-Deckert, J. Rother, K. Sandhoff, E. Wang and A. H. Merrill, *J. Biol. Chem.*, 1997, **272**, 22432.

[22] M. W. Ho and J. S. O'Brien, *Proc. Natl. Acad. Sci., USA*, 1971, **68**, 2810.

[23] K. Huitema, J. Van den Dikenberg, J. F. Brouwers and J. C. Holthuis, *EMBO J.*, 2004, **23**, 33.

[24] J. Koch S. Gartner, C. M. Li, L. E. Quintern, K. Bernardo, Q. Levran, D. Schnabel, R. J. Desnick, E. H. Schuchman and K. Sandhoff, *J. Biol. Chem.*, 1996, **271**, 33110.

[25] S. El Bawab, P. Roddy, T. Qian, A. Bielawska, J. J. Lemasters and Y. A. Hannun, *J. Biol. Chem.*, 2000, **275**, 21508.

[26] M. Tani, N. Okino, S. Mitsutake, T. Tanigawa, H. Izu and M. Izu, *J. Biol. Chem.*, 2000, **275**, 11229.

[27] M. Sugiura, K. Kono, H. Liu, T. Shimizugawa, H. Minekura, S. Spiegel and T. Kohama, *J. Biol. Chem.*, 2002, **277**, 23294.

[28] M. Hiraoka, A. Abe and J. A. Shayman, *J. Biol. Chem.*, 2002, **277**, 10090.

[29] S. Ichikawa, H. Sakiyama, G. Suzuki, H. I. Hidari and Y. Hirabayashi, *Proc. Natl. Acad. Sci., USA*, 1996, **93**, 12654.

[30] S. Schulte and W. Stoffel, *Proc. Natl. Acad. Sci., USA*, 1993, **90**, 10265.

[31] A. Huwiler, T. Kolter, J. Pfeilschifter and K. Sandhoff, *Biochim. Biophys. Acta*, 2000, **1485**, 63.

[32] H. Grassme, A. Jekle, A. Riehle, H. Schwarz, J. Berger, K. Sandhoff, R. Kolesnick and E. Gulbins, *J. Biol. Chem.*, 2001, **276**, 20589.

[33] E. Gulbins and R. Kolesnick, *Oncogene*, 2003, **22**, 7070.

[34] H. Le Stunff, C. Peterson, H. Liu, S. Milstien and S. Spiegel, *J. Biol. Chem.*, 2002, **277**, 8920.

[35] M. Maceyka, H. Sankala, N. C. Hait, H. Le Stunff, H. Liu, R. Toman, C. Collier, M. Zhang, L. S. Satin, A. H. Merrill, S. Milstien and S. Spiegel, *J. Biol. Chem.*, 2005, **280**, 37118.

[36] S. Spiegel and S. Milstien, *Biochem. Soc. Trans.*, 2003, **31**(Pt 6), 1216.

[37] A. H. Futerman and H. Riezman, *Trends Cell Biol.*, 2005, **15**, 312.

[38] P. P. Van Veldhoven and G. P. Mannaerts, *J. Biol. Chem.*, 1991, **266**, 12502.
[39] R. Kolesnick, *J. Clin. Invest.*, 2002, **110**, 3.
[40] F. G. Tafesse, P. Ternes and J. C. Holthuis, *J. Biol. Chem.*, 2006, **281**, 29421.
[41] T. A. Taha, T. D. Mullen and L. M. Obeid, *Biochim. Biophys. Acta*, 2006, **1758**, 2027.
[42] C. F. Lin, C. L. Chen and Y. S. Lin, *Curr. Med. Chem.*, 2006, **13**, 1609.
[43] G. S. Dbaibo and Y. A. Hannun, *Apoptosis*, 1998, **3**, 317.
[44] A. H. Merrill Jr., E. M. Schmelz, D. L. Dillehay, S. Spiegel, J. A. Shayman, J. J. Schroeder, R. T. Riley, K. A. Voss and E. Wang, *Toxicol. Appl. Pharmacol.*, 1997, **142**, 208.
[45] A. Delgado, J. Casas, A. Llebaria, J. L. Abad and G. Fabrias, *Biochim. Biophys. Acta*, 2006, **1758**, 1957.
[46] J. M. Kraveka, L. Li, Z. M. Szulc, J. Bielawski, B. Ogretmen, Y. A. Hannun, L. M. Obied and A. Bielawska, *J. Biol. Chem.*, 2007, **282**, 16718.
[47] K. Furukawa, K. Hamamura, W. Aixinjueluo and K. Furukawa, *Ann. N.Y. Acad. Sci.*, 2006, **1086**, 185.
[48] X. Zhang and F. L. Kiechle, *Ann. Clin. Lab. Sci.*, 2004, **34**(Winter), 3.
[49] M. Bektas and S. Spiegel, *Glycoconj. J.*, 2004, **20**, 39.
[50] S. Birkle, G. Zeng, L. Gao, R. K. Yu and J. Aubry, *Biochimie.*, 2003, **85**, 455.
[51] A. Breier, M. Barancik, Z. Sulova and B. Uhrik, *Curr. Cancer Drug Targets*, 2005, **5**, 457.
[52] B. Ogretmen and Y. A. Hannun, *Drug Resist. Updat.*, 2001, **4**, 368.
[53] V. Gouaze-Andersson and M. C. Cabot, *Biochim. Biophys. Acta*, 2006, **1758**, 2096.
[54] P. E. Lovat, F. Di Sano, M. Corazzari, B. Fazi, R. P. Donnorso, A. D. Pearson, A. G. Hall, C. P. Redfern and M. J. Piacentini, *Natl. Cancer Inst.*, 200, **96**, 1288.
[55] C. E. Chalfant and S. Spiegel, *J. Cell Sci.*, 2005, **118**, 4605.
[56] B. J. Pettus, K. Kitatani, C. E. Chalfant, T. A. Taha, T. Kawamori, J. Bielawski, L. M. Obeid and Y. A. Hannun, *Mol. Pharmacol.*, 2005, **68**, 330.
[57] V. Panel, P. Y. Boelle, J. Ayala-Sanmartin, A. M. Jouniaux, R. Hamelin, J. Masliah, G. Trugnan, J. F. Flejou and D. Wendum, *Cancer Lett.*, 2006, **243**, 255.
[58] M. Gorovetz, M. Baekelandt, A. Berner, C. G. Trope, B. Davidson and R. Reich, *Gynecol. Oncol.*, 2006, **103**, 831.
[59] I. Kudo and M. Murakami, *Prostag. Other Lipid Mediat.*, 2002, **3**, 68–69.
[60] J. Woodcock, *IUBMB Life*, 2006, **58**, 462.
[61] O. Cuvillier, *Biochim. Biophys. Acta*, 2002, **1585**, 153.
[62] A. R. Yuen and B. I. Sikic, *Front. Biosci.*, 2000, **5**, D588.
[63] O. Cuvillier, *Anticancer Drugs*, 2007, **18**, 105.
[64] S. Spiegel and S. Milstien, *Nat. Rev. Mol. Cell. Biol.*, 2003, **4**, 397.
[65] S. An and N. Y. Ann, *Acad. Sci.*, 2000, **905**, 25.
[66] H. Zhang, N. N. Desai, A. Olivera, T. Seki, G. Brooker and S. Spiegel, *J. Cell. Biol.*, 1991, **14**, 155.
[67] S. S. Chae, J. H. Paik, H. Furneaux and T. Hla, *J. Clin. Invest.*, 2004, **114**, 1082.
[68] B. Annabi, S. Thibeault, Y. T. Lee, N. Bousquet-Gagnon, N. Eliopoulos, S. Barrette, J. Galipeau and R. Beliveau, *Exp. Hematol.*, 2003, **31**, 640.
[69] H. Ozaki, T. Hla and M. J. Lee, *J. Atheroscler. Thromb.*, 2003, **10**, 125.
[70] M. E. Skaznik-Wikiel, T. Kaneko-Tarui, A. Kashiwagi and J. K. Pru, *Biol. Reprod.*, 2006, **74**, 569.
[71] S. Von Otte, J. R. Paletta, S. Becker, S. Konig, M. Fobker, R. R. Greb, L. Kiesel, G. Assmann, K. Diedrich and J. R. Nofer, *J. Biol. Chem.*, 2006, **281**, 5398.
[72] R. A. Sabbadini, *Br. J. Cancer*, 2006, **95**, 1131.
[73] M. Murph, T. Tanaka, S. Liu and G. B. Mills, *Clin. Cancer Res.*, 2006, **12**, 6598.
[74] T. Hla, *Semin. Cell Dev. Biol.*, 2004, **5**, 513.
[75] D. N. Brindley, *J. Cell. Biochem.*, 2004, **92**, 900.
[76] S. Langlois, D. Gingras and R. Beliveau, *Blood*, 2004, **103**, 3020.
[77] Y. Smicun, S. Reierstad, F. Q. Wang, C. Lee and D. A. Fishman, *Gynecol. Oncol.*, 2006, **103**, 952.
[78] M. Meriane, S. Duhamel, L. Lejeune, J. Galipeau and B. Annabi, *Stem Cells*, 2006, **24**, 2557.
[79] D. H. Walter, U. Rochwalsky, J. Reinhold, F. Seeger, A. Aicher, C. Urbich, I. Spyridopoulos, J. Chun, V. Brinkmann, P. Keul, B. Levkau, A. M. Zeiher, S. Dimmeler and J. Haendeler, *Arterioscler. Thromb. Vasc. Biol.*, 2007, **27**, 275.

[80] J. Igarashi, P. A. Erwin, A. P. Dantas, H. Chen and T. Michel, *Proc. Natl. Acad. Sci., USA*, 2003, **100**, 10664.

[81] C. B. Fieber, J. Eldridge, T. A. Taha, L. M. Obeid and R. C. Muise-Helmericks, *Exp. Cell Res.*, 2006, **312**, 1164.

[82] C. Barthomeuf, S. Lamy, M. Blanchette, D. Boivin, D. Gingras and R. Beliveau, *Free Radic. Biol. Med.*, 2006, **40**, 581.

[83] S. Usui, N. Sugimoto, N. Takuwa, S. Sakagami, S. Takata, S. Kaneko and Y. Takuwa, *J. Biol. Chem.*, 2004, **279**, 12300.

[84] D. Shida, J. Kitayama, H. Yamaguchi, H. Yamashita, K. Mori, T. Watanabe, Y. Yatomi and H. Nagawa, *FEBS Lett.*, 2004, **577**, 333.

[85] S. Bu, M. Yamanaka, H. Pei, A. Bielawska, J. Bielawski, Y. A. Hannun, L. Obeid and M. Trojanowska, *FASEB J.*, 2006, **20**, 184.

[86] P. S. Jolly, M. Bektas, K. R. Watterson, H. Sankala, S. G. Payne, S. Milstien and S. Spiegel, *Blood*, 2005, **105**, 4736.

[87] N. Murata, K. Sato, J. Kon, H. Tomura, M. Yanagita, A. Kuwabara, M. Ui and F. Okajima, *Biochem. J.*, 2000, **352**(Pt 3), 809.

[88] Y. Yatomi, Y. Igarashi, L. Yang, N. Hisano, R. Qi, N. Asazuma, K. Satoh, Y. Ozaki and S. Kume, *J. Biochem. (Tokyo)*, 1997, **121**, 969.

[89] M. Tani, T. Sano, M. Ito and Y. Igarashi, *J. Lipid Res.*, 2005, **46**, 2458.

[90] B. J. Pettus, C. E. Chalfant and Y. A. Hannun, *Curr. Mol. Med.*, 2004, **4**, 405.

[91] M. El Alwani, B. X. Wu, L. M. Obeid and Y. A. Hannun, *Pharmacol. Ther.*, 2006, **112**, 171.

[92] S. G. Payne, S. Milstien, S. E. Barbour and S. Spiegel, *Semin. Cell Dev. Biol.*, 2004, **15**, 521.

[93] D. Meyer zu Heringdorf and K. H. Jakobs, *Biochim. Biophys. Acta*, 2007, **1768**, 923.

[94] B. Anliker and J. Chun, *Semin. Cell. Dev. Biol.*, 2004, **15**, 457.

[95] B. Anliker and J. Chun, *J. Biol. Chem.*, 2004, **279**, 20555.

[96] M. J. Kluk and T. Hla, *Biochim. Biophys. Acta*, 2002, **1582**, 72.

[97] T. Hla, M. J. Lee, N. Ancellin, J. H. Paik and M. J. Kluk, *Science*, 2001, **294**, 1875.

[98] M. Kono, Y. Mi, Y. Liu, T. Sasaki, M. L. Allende, Y. P. Wu, T. Yamashita and R. L. Proia, *Biol. Chem.*, 2004, **279**, 29367.

[99] J. R. Van Brocklyn, M. J. Lee, R. Menzeleev, A. Olivera, L. Edsall, O. Cuvillier, D. M. Thomas, P. J. Coopman, S. Thangada, C. H. Liu, T. Hla and S. Spiegel, *J. Cell Biol.*, 1998, **142**, 229.

[100] N. Cooke and F. Zecri, *Ann. Rep. Med. Chem.*, 2007, **42**, 245.

[101] S. Spiegel and S. Milstien, *J. Biol. Chem.*, 2007, **282**, 2125.

[102] T. H. Kee, P. Vit and A. J. Melendez, *Clin. Exp. Pharmacol. Physiol.*, 2005, **32**, 153.

[103] M. Yamanaka, D. Shegogue, H. Pei, S. Bu, A. Bielawska, J. Bielawski, B. Pettus, Y. A. Hannun, L. Obeid and M. Trojanowska, *J. Biol. Chem.*, 2004, **279**, 53994.

[104] O. Cuvillier, *Anticancer Drugs*, 2007, **18**, 105.

[105] M. Maceyka, S. Milstien and S. Spiegel, *Circ. Res.*, 2007, **100**, 7.

[106] D. Pchejetski, O. Kunduzova, A. Dayon, D. Calise, M. H. Seguelas, N. Leducq, I. Seif, A. Parini and O. Cuvillier, *Circ. Res.*, 2007, **100**, 41.

[107] N. C. Hait, C. A. Oskeritzian, S. W. Paugh, S. Milstien and S. Spiegel, *Biochim. Biophys. Acta*, 2006, **1758**, 2016.

[108] R. T. Dobrowsky, C. Kamibayashi, M. C. Mumby and Y. A. Hannun, *J. Biol. Chem.*, 1993, **268**(21), 15523.

[109] R. T. Dobrowsky and Y. A. Hannun, *J. Biol. Chem.*, 1992, **267**(8), 5048.

[110] Y. Zhang, B. Yao, S. Delikat, S. Bayoumy, X. H. Lin, S. Basu, M. McGinley, P. Y. Chan-Hui, H. Lichenstein and R. Kolesnick, *Cell*, 1997, **89**(1), 63.

[111] M. Heinrich, M. Wickel, W. Schneider-Brachert, C. Sandberg, J. Gahr, R. Schwandner, T. Weber, P. Saftig, C. Peters, J. Brunner, M. Kronke and S. Schutze, *EMBO J.*, 1999, **18**(19), 5252.

[112] M. Heinrich, J. Neumeyer, M. Jakob, C. Hallas, V. Tchikov, S. Winoto-Morbach, M. Wickel, W. Schneider-Brachert, A. Trauzold, A. Hethke and S. Schutze, *Cell Death Differ.*, 2004, **11**(5), 550.

[113] N. A. Bourbon, J. Yun and M. Kester, *J. Biol. Chem.*, 2000, **275**(45), 35617.

[114] A. Huwiler, D. Fabbro and J. Pfeilschifter, *Biochemistry*, 1998, **37**(41), 14556.

[115] H. Grassme, A. Jekle, A. Riehle, H. Schwarz, J. Berger, K. Sandhoff, R. Kolesnick and E. Gulbins, *J. Biol. Chem.*, 2001, **276**(23), 20589.

[116] A. Huwiler, J. Brunner, R. Hummel, M. Vervoordeldonk, S. Stabel, H. Van Den Bosch and J. Pfeilschifter, *Proc. Natl. Acad. Sci., USA*, 1996, **93**(14), 6959.

[117] A. Huwiler, C. Xin, A. K. Brust, V. A. Briner and J. Pfeilschifter, *Biochim. Biophys. Acta*, 2004, **1636**(2–3), 159.

[118] A. Huwiler, B. Johansen, A. Skarstad and J. Pfeilschifter, *FASEB J.*, 2001, **15**(1), 7.

[119] B. J. Pettus, A. Bielawska, P. Subramanian, D. S. Wijesinghe, M. Maceyka, C. C. Leslie, J. H. Evans, J. Freiberg, P. Roddy, Y. A. Hannun and C. E. Chalfant, *J. Biol. Chem.*, 2004, **279**(12), 11320.

[120] A. Gomez-Munoz, *FEBS Lett.*, 2004, **562**(1-3), 5.

[121] T. Kolter, T. Doering, G. Wilkening, N. Werth and K. Sandhoff, *Biochem. Soc. Trans.*, 1999, **27**(4), 409.

[122] Y. A. Hannun, C. R. Loomis, A. H. Merrill and R. M. Bell, *J. Biol. Chem.*, 1986, **261**(27), 12604.

[123] K. Hanada, *Biochim. Biophys. Acta*, 2003, **1632**(1–3), 16.

[124] A. Titievsky, I. Titievskaya, M. Pasternack, K. Kaila and K. Tornquist, *J. Biol. Chem.*, 1998, **273**(1), 242.

[125] C. Mathes, A. Fleig and R. Penner, *J. Biol. Chem.*, 1998, **273**(39), 25020.

[126] M. J. Lee, *Science*, 1998, **279**(5356), 1552.

[127] P. S. Jolly, M. Bektas, A. Olivera, C. Gonzalez-Espinosa, R. L. Proia, J. Rivera, S. Milstien and S. Spiegel, *J. Exp. Med.*, 2004, **199**(7), 959.

[128] S. K. Goparaju, P. S. Jolly, K. R. Watterson, M. Bektas, S. Alvarez, S. Sarkar, L. Mel, I. Ishii, J. Chun, S. Milstien and S. Spiegel, *Mol. Cell Biol.*, 2005, **25**(10), 4237.

[129] E. J. Goetzl and S. An, *FASEB J.*, 1998, **12**(15), 1589.

[130] A. L. Cavalli, N. W. O'Brien, S. B. Barlow, R. Betto, C. C. Glembotski, P. T. Palade and R. A. Sabbadini, *Am. J. Cell Physiol.*, 2003, **284**(3), C780.

[131] D. M. Zu Heringdorf, M. E. Vincent, M. Lipinski, K. Danneberg, U. Stropp, D. A. Wang, G. Tigyi and K. H. Jakobs, *Cell. Signal.*, 2003, **15**(7), 677.

[132] T. Blom, J. P. Slotte, S. M. Pitson and K. Tornquist, *Cell. Signal.*, 2005, **17**(7), 827.

[133] J. Robert, *Curr. Opin. Invest. Drugs*, 2004, **5**(12), 1340.

[134] A. Meng, C. Luberto, P. Meier, A. Bai, X. Yang, Y. A. Hannun and D. Zhou, *Exp. Cell Res.*, 2004, **292**(2), 385.

[135] A. Gonzalez-Roura, J. Casas and A. Llebaria, *Lipids*, 2002, **37**(4), 401.

[136] E. Amtmann and M. Zoller, *Biochem. Pharmacol.*, 2005, **69**(8), 1141.

[137] R. Hurwitz, K. Ferlinz, G. Vielhaber, H. Moczall and K. Sandhoff, *J. Biol. Chem.*, 1994, **269**(7), 5440.

[138] M. Kolzer, C. Arenz, K. Ferlinz, N. Werth, H. Schulze, R. Klingenstein and K. Sandhoff, *Biol. Chem.*, 2003, **384**(9), 1293.

[139] D. X. Zhang, F. X. Yi, A. P. Zou and P. L. Li, *Am. J. Physiol. Heart Circ.*, 2002, **283**(5), 785.

[140] J. P. Jaffrezou, T. Levade, A. Bettaieb, N. Andrieu, C. Bezombes, N. Maestre, S. Vermeersch, A. Rousse and G. Laurent, *EMBO J.*, 1996, **15**(10), 2417.

[141] H. P. Deigner, R. Claus, G. A. Bonaterra, C. Gehrke, N. Bibak, M. Blaess, M. Cantz, J. Metz and R. Kinscherf, *FASEB J.*, 2001, **15**(3), 807.

[142] F. D. Testai, M. A. Landek and G. Dawson, *J. Neurosci. Res.*, 2004, **75**(1), 66.

[143] C. Okudaira, Y. Ikeda, S. Kondo, S. Furuya, Y. Hirabayashi, T. Koyano, Y. Saito and K. Umezawa, *J. Enzyme Inhib.*, 2000, **15**(2), 129.

[144] E. Amtmann and M. Zoller, *Biochem. Pharmacol.*, 2005, **69**(8), 1141.

[145] C. Luberto, D. F. Hassler, P. Signorelli, Y. Okamoto, H. Sawai, E. Boros, D. J. Hazen-Martin, L. M. Obeid, Y. A. Hannun and G. K. Smith, *J. Biol. Chem.*, 2002, **277**(43), 41128.

[146] M. Tanaka, F. Nara, Y. Yamasato, S. Masuda-Inoue, H. Doi-Yoshioka, S. Kumakura, R. Enokita and T. Ogita, *J. Antibiot. (Tokyo)*, 1999, **52**(7), 670.

[147] R. Uchida, H. Tomoda, Y. Dong and S. Omura, *J. Antibiot. (Tokyo)*, 1999, **52**(6), 572.

[148] C. Arenz, M. Thutewohl, O. Block, H. Waldmann, H. J. Altenbach and A. Giannis, *Chembiochem.*, 2001, **2**(2), 141.

[149] C. De Palma, E. Meacci, C. Perrotta, P. Bruni and E. Clementi, *Arterioscler. Thromb. Vasc. Biol.*, 2006, **26**(1), 99.

[150] N. Marchesini, W. Osta, J. Bielawski, C. Luberto, L. M. Obeid and Y. A. Hannun, *J. Biol. Chem.*, 2004, **279**(24), 25101.

[151] B. Liu, N. Andrieu-Abadie, T. Levade, P. Zhang, L. M. Obeid and Y. A. Hannun, *J. Biol. Chem.*, 1998, **273**(18), 11313.

[152] J. P. Jaffrezou, T. Levade, A. Bettaieb, N. Andrieu, C. Bezombes, N. Maestre, S. Vermeersch, A. Rousse and G. Laurent, *EMBO J.*, 1996, **15**(10), 2417.

[153] K. J. Turnbull, B. L. Brown and P. R. Dobson, *Leukemia*, 1999, **13**(7), 1056.

[154] D. Chauvier, H. Morjani and M. Manfait, *Int. J. Oncol.*, 2002, **20**(4), 855.

[155] M. Selzner, A. Bielawska, M. A. Morse, H. A. Rudiger, D. Sindram, Y. A. Hannun and P. A. Clavien, *Cancer Res.*, 2001, **61**(3), 1233.

[156] P. Giussani, M. Maceyka, H. Le Stunff, A. Mikami, S. Lepine, E. Wang, S. Kelly, A. H. Merill, Jr., S. Milstien and S. Spiegel, *Mol. Cell. Biol.*, 2006, **26**(13), 5055.

[157] C. K. Winter, D. G. Gilchrist, M. B. Dickman and C. Jones, *Adv. Exp. Med. Biol.*, 1996, **392**, 307.

[158] S. M. Mandala, R. A. Thornton, B. R. Frommer, J. E. Curotto, W. Rozdilsky, M. B. Kurtz, R. A. Giacobbe, G. F. Bills, M. A. Cabello and I. Martin, *J. Antibiot. (Tokyo)*, 1995, **48**(5), 349.

[159] F. Yi, A. Y. Zhang, J. L. Janscha, P. L. Li and A. P. Zou, *Kidney Int.*, 2004, **66**(5), 1977.

[160] J. Turnbull, B. L. Brown and P. R. Dobson, *Leukemia*, 1999, **13**(7), 1056.

[161] A. Lucci, T. Y. Han, Y. Y. Liu, A. E. Giuliano and M. C. Cabot, *Int. J. Oncol.*, 1999, **15**(3), 541.

[162] J. Zhang, N. Alter, J. C. Reed, C. Borner, L. M. Obeid and Y. A. Hannun, *Proc. Natl. Acad. Sci., USA*, 1996, **93**(11), 5325.

[163] Y. Lavie, H. Cao, A. Volner, A. Lucci, T. Y. Han, V. Geffen, A. E. Giuliano and M. C. Cabot, *J. Biol. Chem.*, 1997, **272**(3), 1682.

[164] S. Mehta, D. Blackinton, I. Omar, N. Kouttab, D. Myrick, J. Klostergaard and H. Wanebo, *Cancer Chemother. Pharmacol.*, 2000, **46**(2), 85.

[165] B. J. Maurer, L. S. Metelitsa, R. C. Seeger, M. C. Cabot and C. P. Reynolds, *J. Natl. Cancer Inst.*, 1999, **91**(13), 1138.

[166] M. Selzner, A. Bielawska, M. A. Morse, H. A. Rudiger, D. Sindram, Y. A. Hannun and P. A. Clavien, *Cancer Res.*, 2001, **61**(3), 1233.

[167] K. M. Nicholson, D. M. Quinn, G. L. Kellett and J. R. Warr, et al., *Br. J. Cancer*, 1999, **81**(3), 423.

[168] L. Samsel, G. Zaidel, H. M. Drumgoole, D. Jelovac, C. Drachenberg, J. G. Rhee, A. M. Brodie, A. Bielawska and M. J. Smyth, *Prostate*, 2004, **58**(4), 382.

[169] K. S. Sundaram and M. Lev, *J. Neurochem.*, 1984, **42**(2), 577.

[170] K. A. Medlock and A. H. Merrill, Jr., *Biochemistry*, 1988, **27**(18), 7079.

[171] G. Triola, G. Fabrias and A. Llebaria, *Angew. Chem. Int. Ed. Engl.*, 2001, **40**(10), 1960.

[172] S. De Jonghe, I. Van Overmeire, J. Gunst, A. De Bruyn, C. Hendrix, S. Van Calenbergh, R. Busson, D. De Keukeleire, J. Philippe and P. Herdewijn, *Bioorg. Med. Chem. Lett.*, 1999, **9**(21), 3159.

[173] A. Bielawska, M. S. Greenberg, D. Perry, S. Jayadev, J. A. Shayman, C. McKay and Y. A. Hannun, *J. Biol. Chem.*, 1996, **271**(21), 12646.

[174] J. Usta, S. El Bawab, P. Roddy, Z. M. Szulc, A. Yusuf Hannun and A. Bielawska, *Biochemistry*, 2001, **40**(32), 9657.

[175] A. Dagan, C. Wang, E. Fibach and S. Gatt, *Biochim. Biophys. Acta*, 2003, **1633**(3), 161.

[176] Y. Yatomi, F. Ruan, T. Megidish, T. Tovokuni, S. Hakomori and Y. Igarashi, *Biochemistry*, 1996, **35**(2), 626.

[177] K. Endo, Y. Igarashi, M. Nisar, Q. H. Zhou and S. Hakomori, *Cancer Res.*, 1991, **51**(6), 1613.

[178] J. W. Kim, Y. W. Kim, Y. Inagaki, Y. A. Hwang, S. Mitsutake, Y. W. Ryu, W. K. Lee, H. J. Ha, C. S. Park and Y. Igarashi, *Bioorg. Med. Chem.*, 2005, **13**(10), 3475.

[179] K. Kono, M. Tanaka, T. Mizuno, K. Kodama, T. Ogita and T. Kohama, *J. Antibiot. (Tokyo)*, 2000, **53**(8), 753.

[180] J. R. Gamble, P. Xia, C. N. Hahn, J. J. Drew, C. J. Drogemuller, D. Brown and M. A. Vadas, *Int. J. Cancer*, 2006, **118**(10), 2412.

[181] D. Pchejetski, M. Golzio, E. Bonhoure, C. Calvet, N. Doumerc, V. Garcia, C. Mazerolles, P. Rischmann, J. Teissie, B. Malavaud and O. Cuvillier, *Cancer Res.*, 2005, **65**(24), 11667.

[182] G. K. Schwartz, D. Ward, L. Saltz, E. S. Casper, T. Spiess, E. Mullen, J. Woodworth, R. Venuti, P. Zervos, A. M. Storniolo and D. P. Kelsen, *Clin. Cancer Res.*, 1997, **3**(4), 537.

[183] H. Liu, M. Sugiura, V. E. Nava, L. C. Edsall, K. Kono, S. Poulton, S. Milstien, T. Kohama and S. Spiegel, *J. Biol. Chem.*, 2000, **275**(26), 19513.

[184] A. Olivera, T. Kohama, Z. Tu, S. Milstien and S. Spiegel, et al., *J. Biol. Chem.*, 1998, **273**(20), 12576.

[185] M. Matloubian, C. G. Lo, G. Cinamon, M. J. Lesneski, Y. Xu, V. Brinkmann, M. L. Allende, R. L. Proia and J. G. Cyster, *Nature*, 2004, **427**(6972), 355.

[186] B. Visentin, J. A. Vekich, B. J. Sibbald, A. L. Cavalli, K. M. Moreno, R. G. Matteo, W. A. Garland, Y. Lu, H. S. Hall, V. Kundra, G. B. Mills and R. A. Sabbadini, *Cancer Cell*, 2006, **9**(3), 225.

[187] W. Stoffel and M. Grol, *Chem. Phys. Lipids*, 1974, **13**(4), 372.

[188] E. Bonhoure, D. Pchejetski, N. Aouali, H. Morjani, T. Levade, T. Kohama and O. Cuvillier, *Leukemia*, 2006, **20**, 95.

[189] Y. Baran, A. Salas, C. E. Senkal, U. Gunduz, J. Bielawski, L. M. Obeid and B. Ogretmen, *J. Biol. Chem.*, 2007, **13**(282), 10922–10934.

[190] S. Milstien and S. Spiegel, *Cancer Cell*, 2006, **9**, 148.

[191] A. M. Scott, Z. Liu, C. Murone, T. G. Johns, D. MacGregor, F. E. Smyth, F. T. Lee, J. Cebon, I. D. Davis, W. Hopkins, A. J. Mountain, A. Rigopoulos, N. Hanai and L. J. Old, *Cancer Immun.*, 2005, **5**, 3.

[192] P. B. Chapman, *Curr. Opin. Invest. Drugs*, 2003, **4**, 710.

[193] G. Kaur and I. Roy, *Expert Opin. Invest. Drugs*, 2008, **17**, 43.

[194] S. M. Nimjee, C. P. Rusconi and B. A. Sullenger, *Annu. Rev. Med.*, 2005, **56**, 555.

[195] C. R. Ireson and L. R. Kelland, *Mol. Cancer Ther.*, 2006, **5**, 2957.

[196] K. Rajarathnam, *Curr. Pharm. Des.*, 2002, **8**, 2159.

[197] V. Wascholowski and A. Giannis, *Angew. Chem. Int. Ed.*, 2006, **45**, 827.

[198] M. D. Lister, Z. S. Ruan and R. Bittman, *Biochim. Biophys. Acta*, 1995, **1256**, 25.

[199] K. Hanada, *Biochim. Biophys. Acta*, 2003, **1632**, 16.

[200] G. Triola, G. Fabrias, J. Casas and A. Llebaria, *J. Org. Chem.*, 2003, **68**, 9924.

[201] R. R. Vunnam and N. S. Radin, *Chem. Phys. Lipids*, 1980, **26**, 265.

[202] J. W. Kim, Y. Inagaki, S. Mitsutake, N. Maezawa, S. Katsumura, Y. W. Ryu, C. S. Park, M. Taniguchi and Y. Igarashi, *Biochim. Biophys. Acta*, 2005, **1738**(1–3), 82.

[203] K. Desai, M. C. Sullards, J. Allegood, E. Wang, E. M. Schmelz, M. Hartl, H. U. Humpf, D. C. Liotta, Q. Peng and A. H. Merrill, *Biochim. Biophys. Acta*, 2002, **1585**, 188.

[204] B. M. Buehrer and R. M. Bell, *J. Biol. Chem.*, 1992, **267**, 3154.

[205] K. J. French, R. S. Schrecengost, B. D. Lee, Y. Zhuang, S. N. Smith, J. L. Eberly, J. K. Yun and C. D. Smith, *Cancer Res.*, 2003, **63**, 5962.

[206] P. Bandhuvula, Y. Y. Tam, B. Oskouian and J. D. Saba, *J. Biol. Chem.*, 2005, **280**, 33697.

[207] A. Boumendjel and S. P. F. Miller, *Tetrahedron Lett.*, 1994, **35**, 819.

[208] K. LaMontagne, A. Littlewood-Evans, C. Schnell, T. O'Reilly, L. Wyder, T. Sanchez, B. Probst, J. Butler, A. Wood, G. Liau, E. Billy, A. Theuer, T. Hla and J. Wood, *Cancer Res.*, 2006, **66**, 221.

[209] G. Schmid, M. Guba, I. Ischenko, A. Papyan, M. Joka, S. Schrepfer, C. J. Bruns, K. W. Jauch, C. Heeschen and C. Graeb, *J. Cell. Biochem.*, 2007, **101**(1), 259.

[210] T. K. Lee, K. Man, J. W. Ho, X. H. Wang, R. T. Poon, Y. Xu, K. T. Ng, A. C. Chu, C. K. Sun, I. O. Ng, H. C. Sun, Z. Y. Tang, R. Xu and S. T. Fan, *Clin. Cancer Res.*, 2005, **11**, 8458.

[211] J. W. Man, C. K. Sun, T. K. Lee, R. T. Poon and S. T. Fan, *Mol. Cancer Ther.*, 2005, **4**, 1430.

[212] http://www.nature.com/nrm/posters/lipidsignalling-disease

[213] F. Eckstein, *Expert Opin. Biol. Ther.*, 2007, **7**, 1021.

[214] V. Wacheck and U. Zangemeister-Wittke, *Crit. Rev. Oncol. Hematol.*, 2006, **59**, 65.

[215] E. Henke, J. Perk, J. Vider, P. De Candia, Y. Chin, D. B. Solit, V. Ponomarev, L. Cartegni, K. Manova, N. Rosen and R. Benezra, *Nat. Biotechnol.*, 2008, **26**, 91.

[216] T. A. Burrow, R. J. Hopkin, N. D. Leslie, B. T. Tinkle and G. A. Grabowski, *Curr. Opin. Pediatr.*, 2007, **19**, 628.

[217] E. H. Davies, A. Erikson, T. Collin-Histed, E. Mengel, A. Tylki-Szymanska and A. Vellodi, *J. Inherit Metab. Dis.*, 2007, **30**, 935.

[218] K. Park, W. J. Kim, Y. H. Cho, Y. I. Lee, H. Lee, S. Jeong, E. S. Cho, S. I. Chang, S. K. Moon, B. S. Kang, Y. J. Kim and S. H. Cho, *Front Biosci.*, 2008, **13**, 2653.

[219] H. Kinoh and M. Inoue, *Front Biosci.*, 2008, **13**, 2327.

[220] D. Hanahan and R. A. Weinberg, *Cell*, 2000, **100**, 57.

[221] A. E. Openshaw, P. R. Race, H. J. Monzó, J. A. Vázquez-Boland and M. J. Banfield, *J. Biol. Chem.*, 2005, **280**, 35011.

[222] S. P. Zela, M. F. Fernandes-Pedrosa, M. T. Murakami, S. A. De Andrade, R. K. Arni and D. V. Tambourgi, *Acta Crystallogr. D Biol. Crystallogr.*, 2004, **60**, 1112.

PART V:
Infectious Diseases

Editor: John L. Primeau
Astrazeneca Pharmaceuticals LP
35 Gatehouse Drive
Waltham
Massachusetts

Inhibitors of Respiratory Syncytial Virus

Malcolm Carter and **G. Stuart Cockerill**

Contents

1. INTRODUCTION

Human respiratory syncytial virus (RSV) is the most important respiratory pathogen that causes lower respiratory tract infections such as bronchiolitis and pneumonia in infants and young children, resulting in up to 125,000 hospitalizations annually in the United States [1]. The infants most at risk of severe disease are those under 6 weeks of age, those with bronchopulmonary dysplasia, congenital heart disease, or immunodeficiency and those born prematurely. Hospital admission rates in these groups range between 5% and 30% [2]. The mortality rate among children admitted to hospital is approximately 3% for those with heart and lung problems and up to 1% for those without these risk factors [2,3]. In adults and the elderly, RSV pneumonia is increasingly

Arrow Therapeutics, Britannia House, London, SE1 1DB, UK.

Annual Reports in Medicinal Chemistry, Volume 43
ISSN 0065-7743, DOI 10.1016/S0065-7743(08)00014-6

recognized as a significant cause of morbidity and mortality, being associated with more than 17,000 deaths annually between 1991 and 1998 [4,5]. In the hospitalized elderly, mortality can be as high as 10–20% and in the severely immunocompromised with RSV pneumonia it can be in the order of 50–70% [6]. There is, therefore, an urgent and unmet medical need for novel therapies to deal with infections caused by this virus.

Although research into the prevention and treatment of RSV infection has been ongoing for almost 40 years, vaccine development is difficult [7,8] and to date there is no clinically approved vaccine. The development of RSV vaccines for use in young infants has been complicated by reduced immune responses in this age group due to immunologic immaturity and the immunosuppressive effects of maternal antibodies. Passive immunization with the monoclonal antibody palivizumab (Synagis®) has provided about 50% protection to high-risk children [9]. These include infants born prematurely and those with congenital conditions. The antibody has to be given prophylactically and to date, its use has been limited to developed countries. The effectiveness of ribavirin 1 [10], the only licensed small molecule for treatment of RSV, has been challenged [11]. Ribavirin has to be given by a prolonged aerosol and there are certain doubts as to its safety versus its efficacy in treatment of RSV infection. The unmet need for additional effective and safe treatments for RSV is paramount.

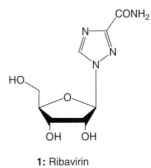

1: Ribavirin

RSV, a pneumovirus of the paramyxovirus family, is an enveloped non-segmented, negative-stranded RNA virus [12]. The 15.2 Kb genome has been fully sequenced and contains 10 mRNAs encoding 11 distinct proteins (Figure 1). The genome is encapsidated by the nucleocapsid (N) protein, which forms a helical nucleocapsid and protects the RNA from ribonucleases. The N-protein is also associated with the viral polymerase, phosphoprotein and the M2-1 protein which together form the transcriptase complex. The ribonucleoprotein (RNP) is essential for transcription, and naked RNA does not provide a template for the viral polymerase. As with all single stranded RNA viruses, the virus does not have a proofreading mechanism during replication, resulting in a relatively high error rate and frequent mutations. Inhibitors targeting the fusion event, for example, have been slow to progress partly due to concerns associated with the rapid emergence of resistant mutants mapping to the F gene [13].

PARAMYXOVIRIDAE: Sub-family PNEUMOVIRINAE

Respiratory Syncytial Virus

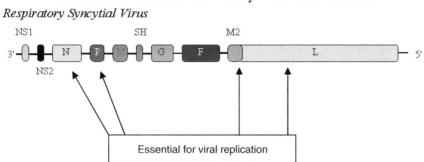

Figure 1 RSV essential genes. N, nucleocapsid, P, phosphoprotein, and L, long protein including polymerase. (Acknowledgement, Andrew Easton, University of Warwick.) (See Color Plate 14.1 in Color Plate Section.)

A more effective antiviral compound might ideally target the proteins derived from essential replication genes as these are often more highly conserved due to their functional role [14].

The development of new therapeutics for RSV, with an emphasis on the biological and clinical aspects of their progress has recently been reviewed [14,15].

2. FUSION INHIBITORS

The disruption of viral attachment and entry to cells is a common strategy in the design of antiviral therapies, as evidenced by successful approaches to influenza and HIV inhibition [16]. In the case of the Pneumovirinae subfamily, which includes RSV, a distinguishing feature is the absence of hemagglutinin/ neuraminidase fusion proteins and the dependence on the F protein alone for virus binding and cell entry [17]. The associated G glycoprotein appears to facilitate viral attachment. This pivotal role for the fusion protein has inspired a number of approaches over the years and led to compounds suitable for clinical evaluation.

Several compound types have been described and two are reported as having progressed into early clinical development. Leading the field currently are a series of benzimidazole pyridines [18]. The initial lead described, **2** (JNJ-2408068), showed nanomolar potency in RSV *in vitro* assays and, despite rapid elimination from the plasma, displayed overly long tissue retention times in the lung (153 h). A substructure-based assessment of the distribution and elimination properties of JNJ-2408086 (Figure 2) concluded that the aminoethyl piperidine fragment was responsible for the slow elimination of this compound from lung tissue [19].

Modification of the aminoethyl piperidine to a morpholinopropylamino functionality as in **3** did indeed provide the desired drop in tissue retention time but with an unwelcome loss of antiviral activity. Potency was recovered by incorporation of a hydroxypropyl phenyl group, as shown in the clinical

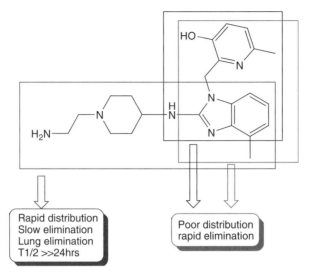

Figure 2 JNJ-2408086 distribution and elimination assessment by substructure. (See Color Plate 14.2 in Color Plate Section.)

candidate TMC-353121 (**4**), which was reported to access an additional binding pocket in the binding space [18].

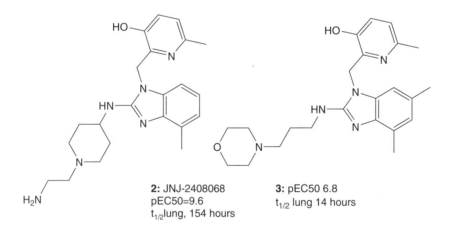

2: JNJ-2408068
pEC50=9.6
$t_{1/2}$lung, 154 hours

3: pEC50 6.8
$t_{1/2}$ lung 14 hours

TMC-353121 (**4**) [18,20] displays sub-nanomolar RSV *in vitro* activity with a lung tissue residence half-life of 25 h. This compound encapsulates what is now a traditional profile of fusion inhibitors both *in vitro* and *in vivo*. All fusion inhibitors described thus far exhibit acute time of addition dependence. Dosing prior to infection *in vitro* is preferred and is an absolute requirement *in vivo*, although this is in part associated with the limited replication capability of the animal models. Compounds are only active within the first 2–3 h of the replication cycle *in vitro*. Cotton rat data is described for inhaled, intravenous and oral routes of administration for this compound. The greatest efficacy is observed via the inhalation route when an aerosol was administered to animals over a 1-hour period, 1 h prior to infection. A 1.6 log reduction in virus titre is observed at 20 mg/mL. A range of *i.v.* studies investigated the effect on efficacy of dosing up to 96 h before infection, and up to 24 h post infection. These studies concluded that a halving of efficacy in this model is observed when dosing took place 24 h either side of infection. A significant drop in viral log titre is also observed upon oral dosing. The use of animal models in the evaluation of fusion inhibitors would seem reasonable, despite the clear doubts regarding replication in these models [14,20], as limited inhibition of replication would be expected with fusion inhibitors.

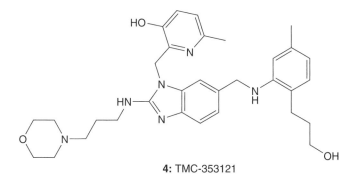

4: TMC-353121

TMC-353121 (**4**) displays a similar resistance profile to its predecessor compound **2** (JNJ-2408068), although with lower factors of resistance claimed. Mutant viruses raised against these compounds showed sequence changes in the fusion protein in close proximity (Figure 3) or identical to those observed with a series described by Meanwell and Krystal [21]. This structural supposition was supported when a photoaffinity probe compound **5** was prepared and shown to bind to peptides derived from the heptad repeat HR-1 structure within the fusion protein. Peptide sequences and point changes within the putative binding site were selected by reference to modelled structures of the F protein [18].

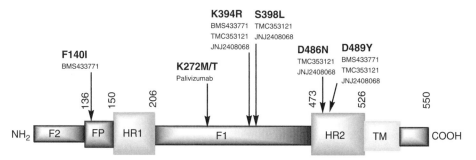

Figure 3 Reported fusion protein point mutations to inhibitors. FP, fusion peptide, HR, heptad repeat and TM, transmembrane. (See Color Plate 14.3 in Color Plate Section.)

TMC-353121 is described as undergoing evaluation in clinical trials.

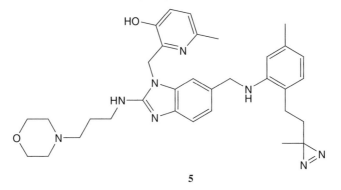

5

Meanwell and Krystal have documented their work [13,21] which has taken the low affinity benzimidazole **6** ($EC_{50} = 470\,nM$), derived from high-throughput screening, and demonstrated efficacy in animal models with the unstable benzoate **7**. Subsequent lead optimization produced BMS-433771 (**8**) as a clinical candidate.

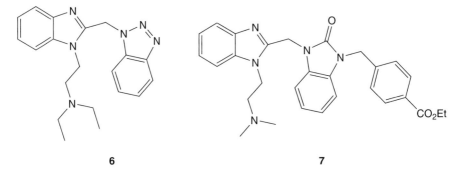

6   7

BMS-433771 (**8**) is highly potent, non-toxic, orally bioavailable, and shows an EC_{50} of ca. 24 nM over a range of both laboratory and clinically relevant strains of the virus. As described previously, time-dependent dosing was critical. The resistance profile of this compound has been reviewed [21] and is summarized in Figure 3. BMS-433771 was progressed into pre-clinical evaluation.

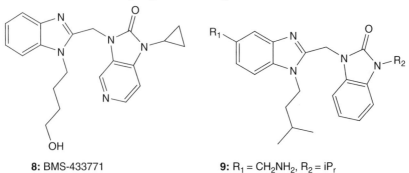

8: BMS-433771   **9:** $R_1 = CH_2NH_2$, $R_2 = iP_r$
                              10: $R_1 = C(=NH)NH_2$, R_2 = isopropenyl

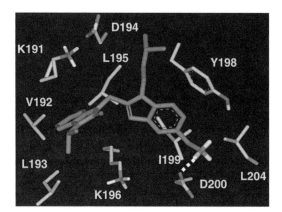

Figure 4 Binding pocket showing 5-aminomethylene interaction with an aspartic acid (D200). (See Color Plate 14.4 in Color Plate Section.)

Recently, the results of a substituent scan around the benzimidazole nucleus have been reported [22]. A clear dependence for substitution at the 5th position was observed. Specifically, basic groups like aminomethylene and amidine in compounds **9** and **10** exhibited nanomolar potencies alongside the parent BMS-433771 (**8**). Most interestingly, modification of the resistance profile was observed with the aminomethylene analogue **9**, where an EC_{50} of 20 nM was demonstrated against the K394R, BMS-433771-resistant, mutant virus. This effect has been rationalized by the proposed association of basic groups in this position with an aspartic acid (D200, Figure 4) which forms part of the pocket identified from previously reported photoaffinity labelling studies [23].

More recent work from this group [24] has focussed on modifications to the benzimidazolone fragment, specifically ring expanding to a 6,6-fused system. Pharmacokinetic properties were improved relative to BMS-433771 with microsomal half-lives in excess of 100 minutes observed (cf. 36 min for BMS-433771). Virus titres were reduced in the BALB/c mouse model (by 1.6 $\log_{10}$ units with **11**) when dosed prior to infection. However only **12** demonstrated *in vitro* potency (20 nM) versus the K394R mutant as described above, perhaps a reflection of the reduced intrinsic potency of this 6,6-bicylic series over the benzimidazolone system of BMS-433771 and its analogues.

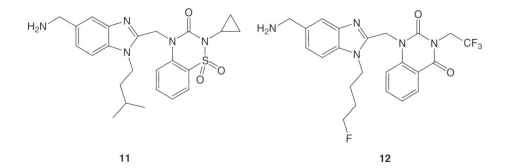

11   12

An example of a more recent class of tricyclic imidazolines, **13**, is shown below. This compound was reported to inhibit RSV in cell culture at 100–250 ng/mL. No further information has been released on these compounds since 2005 [25].

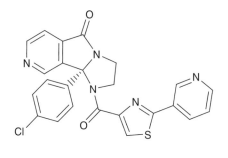

13

Other inhibitors of fusion have been reported and reviewed previously [26]. Despite their advantageous *in vitro* profiles, synthetic complexity and drugability issues have precluded progress.

3. N-PROTEIN INTERACTING COMPOUNDS

The RSV genome is encapsidated by the nucleocapsid (N) protein, which forms a helical nucleocapsid and protects the RNA from ribonucleases. The N-protein is highly conserved within RSV strains and is associated with the viral polymerase, M2-1 and phosphoprotein, which together form the transcriptase complex, essential for viral replication. Compounds that bind to, or otherwise adversely effect the formation of this complex are therefore likely to inhibit viral growth.

A series of 1,4-benzodiazepine derivatives have recently been described [27] which inhibit viral replication and have been shown to give rise to mutations in the N-protein. Initial library screening in a full virus assay led to the identification of the acetamide **14** with an IC_{50} of 8 μM. This activity was confirmed in both ELISA and plaque-reduction assays. Structural modifications to **14** showed that the unsubstituted benzodiazepine template was optimal for antiviral activity but that potency could be markedly improved by modification of the amide moiety. Ortho substituted aromatic groups were particularly good, (e.g. **15**; $IC_{50} = 2$ μM). Urea derivatives were also potent. Resolution of the pendant amide bond into the (S)-configuration demonstrated enhanced activity with this enantiomer [28]. The ortho-fluorophenyl urea **16** was identified as a development candidate (RSV-604, $IC_{50} = 600$ nM). This compound was equipotent versus both A- and B-strains of RSV and active against a large range of clinical isolates. Time of addition studies showed that this compound acted late in the replication cycle. This was well demonstrated by the fact that the IC_{50} was virtually unchanged when given prior or post infection in either the ELISA or plaque assays.

Mutant virus was generated which was 40-fold less sensitive to RSV-604 compared to wild-type virus. The entire viral genome was then sequenced and compared to wild-type [29]. Three mutations were seen: K107N, I129L and L139I, all in the N-gene. The relevance of these mutations was confirmed by reverse genetic experiments. When incorporated into recombinant viruses, I129L; L139I; I129L and L139I point changes were found to show resistance to RSV-604 (5-fold for the single mutants and over 40-fold for the double mutant).

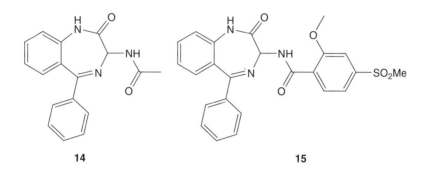

14   15

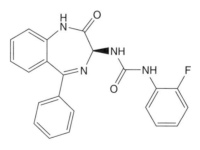

16: RSV-604

Further evidence of the therapeutic potential of this molecule was demonstrated in a model of the human airway epithelium [30]. In this model, the apical surface of the cells was treated with RSV. RSV-604 was then dosed to the medium surrounding the basolateral surface. A concentration of 10 μM of RSV-604 was able to eliminate viral infection when dosed either concomitantly or 24 h post infection. Even when given 24 h post infection at this dose only 1–2 infected cells could be observed (utilizing a green fluorescent tagged virus), thus demonstrating the compounds ability to translate to the site of infection and inhibit RSV replication in human epithelium.

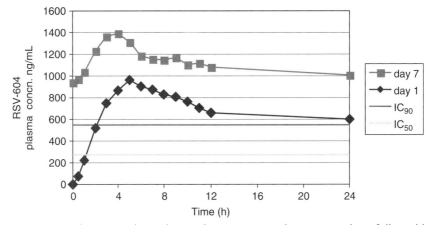

Figure 5   RSV-604 human PK data, where volunteers received 600 mg on day 1 followed by 450 mg on days 2–7. (See Color Plate 14.5 in Color Plate Section.)

RSV-604 (**16**) had an oral bioavailability of > 90% in a lipid-based formulation in the rat, low clearance (7.2 mL/min/kg) and reasonable half-lives (2.6 h *p.o.* and 3.8 h *i.v.*). This ensured that RSV-604 was taken into pre-clinical safety studies. The outcome of these was favourable and the compound was subsequently dosed to healthy human volunteers [31].

A food effect was observed in volunteers given a high-fat meal prior to dosing. RSV-604 was well-tolerated in all dosing regimens with no serious adverse events reported. Plasma concentrations of RSV-604 above the *in vitro* IC_{90} values were achieved. Subjects given a loading dose of 600 mg RSV-604 followed by 6 days at 450 mg exhibited trough concentrations above the *in vitro* IC_{90} at the end of day 1 and were $\geq 4 \times IC_{50}$ and $\geq 2 \times IC_{90}$ for the following 6 days (Figure 5). RSV-604 is now in Phase II Clinical Trials.

4. OTHER MECHANISMS

4.1 Ribavirin analogues

Ribavirin **1** was the first and remains the only small molecule inhibitor of RSV approved for clinical use. A nucleoside analogue, it was characterized as a broad spectrum antiviral more than 30 years ago [32] and its mechanism of action still remains a source of debate. Ribavirin **1** has been described as an inhibitor of the polymerase of a number of viruses [33–37], its 5' triphosphate **17** has been shown to block GTP-dependent viral mRNA capping via the inhibition of the guanyl transferase [38] and the monophosphate has been shown to inhibit inosine monophosphate dehydrogenase (IMPDH) [39].

Despite reports of possible Ribavirin analogues and alternate approaches, none have successfully progressed through clinical evaluation versus RSV. [40–44]

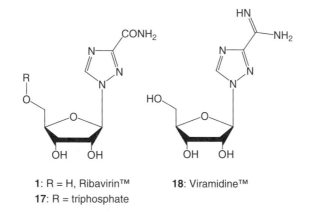

1: R = H, Ribavirin™
17: R = triphosphate

18: Viramidine™

Viramidine **18**, an ammoniated prodrug of ribavirin is still progressing, albeit slowly. Viramidine was designed to address the toxicity profile of its parent. This compound is under development primarily for the oral treatment of hepatitis C virus (HCV) but also other viral infections including RSV. Initial Phase I trials had demonstrated lower plasma levels than Ribavirin and an associated lack of adverse events [41]. However, two Phase III HCV trials failed to meet their efficacy endpoints. Enrolment in a Phase IIb trial to evaluate higher doses of the drug [42] began in March 2007. Progress of this compound as an RSV treatment lags behind its development for HCV.2

4.2 PPARγ agonists

More recently Arnold and König [45] have investigated the effect of PPARγ agonists in cell culture, specifically Rosiglitazone™ **19** and Ciglitazone™ **20**. These compounds effectively protected A549, Hep-2 cells and normal human bronchial epithelial cells against RSV induced cytotoxicity and syncytia formation. In addition a reduction in viral mRNA and release of progeny virus was observed. These effects were observable both in pre- and post-infection dosing schedules. These effects were not observed with PPARα-selective Bezafibrate™. No modification of toll-like receptor 2 (TLR-2), or cellular dehydrogenase levels was observed with these compounds in the presence of RSV infections indicative of a lack of a general cellular cytopathic effect. These observations, coupled with previously published observations on the anti-inflammatory capacity of PPARγ agonists suggest the possibility of a beneficial effect on the course of RSV infection.

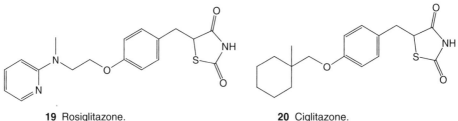

19 Rosiglitazone.   **20** Ciglitazone.

4.3 Benzazepine inhibitors

Sudo and co-workers have described a series of benzazepine thiophenes with submicromolar activity in plaque-reduction assays against RSV [46]. YM53403 (**21**) displays an EC_{50} of 200 nM with a selectivity index of 412. The compounds were identified from high-throughput screening and data is presented for a limited number of analogues. The compounds appeared to be selective for A- and B-strains of RSV with no activity observed against influenza, measles and HSV (herpes simplex virus). Variations in substitution at the amide have a profound effect on activity with, for example, the ethyl ester **22** being completely inactive.

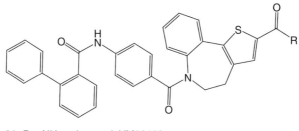

21: R = NH-cyclopropyl, YM53403
22: R = OCH$_2$CH$_3$

Further mechanism of action data was described for YM53403. Time of addition studies showed that activity was observed up to 8 h post infection. Subsequent mutant generation and sequencing identified one mutation in the L (polymerase) gene, Y361H. No mutations were observed in the remainder of the genome.

No further progress with this series has been reported since, particularly in terms of further information on the potential binding interactions of the lead compound.

4.4 Quinazoline-dione inhibitors

Mason, Liuzzi and co-workers established a novel poly A capture assay utilizing a crude RNP complex isolated from RSV infected Hep-2 cells [47]. They were subsequently able to run a high-throughput screen targeting inhibitors of this essential virus specific replication complex. As a result of this screening, quinazoline-diones exemplified by **23**, were identified as inhibitors. Indeed **23** not only inhibited the replication complex with an $IC_{50} = 4.5\,\mu M$, but antiviral activity was demonstrated in RSV infected cells ($EC_{50} = 1.3\,\mu M$) utilizing an MTT cell viability read out. Selectivity was relatively poor in this assay with a CC_{50} observed at $7.7\,\mu M$.

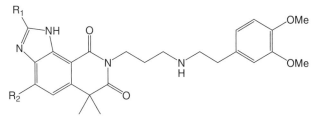

23: R_1 = 2-MeOPh, R_2 = H
24: R_1 = 4-Me, 3-NO₂Ph, R_2 = NH₂

A lead optimization programme was able to significantly improve potency in both replicase and antiviral assays. Compound **24** exhibited an IC_{50} of 89 nM in the RSV MTT assay with a 400-fold selectivity. Confirmation of this antiviral effect was observed in an F protein ELISA ($IC_{50} = 21$ nM) as well as a plaque-reduction assay ($IC_{50} \sim 42$ nM). This compound was progressed into the BALB/c mouse model of infection, dosing intranasally at 0.4 mg/kg/day post infection, initially at +3 and +6 h then *t.i.d* for four days. A rather modest, but significant, 0.6 log reduction in virus titer was observed for compound **24** in this model.

Mechanism of action studies showed that efficacy could be maintained against the virus up to 9 h post infection. The compounds also showed activity ($IC_{50} = 33$ nM) in an RSV minigenome Hep-2 cell based assay which contains only N, P, M2-1 and L genes. This eliminated the possibility of a fusion inhibition mechanism. No cross-resistance was observed with nucleotides and resistant mutants generated against compound **24** showed mutations in the L-protein (I1381S, E1269D and L1421F). A mechanism associated with the inhibition of guanylation of mRNA transcripts was proposed [48], a process performed by the multifunctional polymerase protein.

A novel mechanism and demonstrating efficacy *in vivo*, this novel class of compound would appear to be a good starting point for the development of complementary anti-RSV therapeutics.

4.5 Transcription and regulation inhibitors

Olivo and co-workers have described two series of compounds in the patent literature. They were identified from a minigenome assay of RSV, which would preclude the possibility of fusion inhibition [49,50]. Thienopyrimidine **25** is described as having an $EC_{50} < 25\,\mu M$ and quinoline **26** an EC_{50} of $1.46\,\mu M$. No further progress on these structures has been reported.

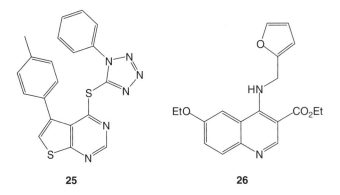

25   26

5. ANTISENSE COMPOUNDS AND siRNA

In the last few years the theoretical selectivity of using nucleic acid sequences to inhibit viral functions has provoked a lot of interest. Off-target activity [55] coupled with the technical difficulty of delivery to target organs has contributed to unsuccessful antisense approaches thus far. However, a positive result in 2002 [51] was claimed in the African green monkey model of RSV infection.

The recent advent of siRNA has rekindled interest. The first demonstration of the antiviral effects of siRNA were claimed for RSV by Bitko and Baril in a reverse genetics study [52]. Extension of this work to *in vivo* studies showed an apparent effect of the siRNA in the mouse model of RSV infection [53]. In this approach the authors used siRNAs that had *in vitro* IC_{50}s of 15–20 nM simply delivered to the mouse nasal passages. Their experiments seem to demonstrate excellent control of infection in both prophylatic and therapeutic modes. Importantly no controls were done to look at carry-over from the treated mouse to the virology assays *in vitro*.

In another approach Zhang [54] used a DNA vector derived siRNAs targeting the NS1 protein. Nanoparticles were used to deliver the DNA vectors *in vivo*. The potency of the different siRNAs was not determined but a significant reduction in virus yield was seen in interferon producing A549 cells but not in Vero cells. The authors think these RNAs work by abrogating the anti-interferon response of NS1. The *in vivo* effects were again assayed in the mouse model. The results were less impressive and less specific than those seen previously [53]. Given reservations concerning this mouse model, based around inoculum size and evidence of replication, a more rational host infection system data study, e.g. pneumonia virus of mice infection, would have been more convincing.

Most recently, the outcome of a Phase I study of the siRNA ALN-RSV01 was reported [55]. A randomized, double-blind, placebo-controlled trial in healthy adult volunteers was designed to assess the safety, tolerability, and pharmaco-kinetics of inhaled agent administered via nebulizer. All major objectives of the trial were met, including definition of a safe and well-tolerated dose and regimen for advancement of ALN-RSV01 into further Phase II development. This initial data provides for some encouragement but work is required to answer or confirm the major concerns over this approach.

6. FUTURE PROSPECTS

The prospect for a treatment for RSV infection which is both more efficacious and more widely available to a number of patient populations appears at this time to be more likely than ever. The search for mechanistically sound inhibitors of the virus has progressed through a fairly common "potent but unprogressable" inhibitor stage to the point where a number of drug-like inhibitors aimed at viral specific targets are being identified.

The portfolio of fusion inhibitors balanced with inhibitors targeting the viral replication mechanism looks promising. The more recent approaches to siRNA may bear fruit, despite the documented reservations associated with this approach.

REFERENCES

[1] T. G. Boyce, B. G. Mellen, E. F. Mitchel, P. F. Wright and M. R. Griffin, *J. Pediatr.*, 2000, **137**, 865.
[2] E. E. L. Wang, B. J. Law, J. L. Robinson, S. Dobson, S. Al Jumaah, D. Stephens, F. D. Boucher, J. McDonald, I. Mitchell and N. E. MacDonald, *Paediatrics*, 1997, **99**, 3.E9.
[3] D. E. Fixler, *Pediatr. Cardiol.*, 1996, **17**, 163.
[4] A. R. Falsey, P. A. Hennessey, M. A. Formica, C. Cox and E. E. Walsh, *New Engl. J. Med.*, 2005, **352**, 1749; P. L. Collins and B. S. Graham, *J. Virol.*, 2008, **82**, 2040.
[5] W. W. Thompson, D. K. Shay, E. Weintraub, L. Brammer, N. Cox, L. J. Anderson and K. Fukuda, *JAMA*, 2003, **289**, 179.
[6] A. R. Falsey and E. E. Walsh, *Clin. Microbiol. Rev.*, 2000, **13**, 371.
[7] P. L. Collins and B. R. Murphy, in *"Respiratory Syncytial Virus: Perspectives in Medical Virology"* (ed. P. Cane), Vol. 13, Elsevier BV Publishers, Amsterdam, Holland, 2006. p. 233.
[8] C. B. Hall, *Rev. Inf. Dis.*, 1980, **2**, 384.
[9] The Impact-RSV Study Group., *Pediatrics*, 1998, **102**, 531.
[10] J. F. Hruska, J. M. Bernstein, R. G. Douglas, Jr. and C. B. Hall, *Antimicrob. Agents Chemother.*, 1980, **17**, 770.
[11] K. Ventre and A. G. Randolph, *Syst. Rev.*, 2004, October 18(4), CD000181.
[12] P. L. Collins, R. M. Chanock and B. R. Murphy, in *Field's Virology* (eds D. M Knipe and P. M. Howley), 4th Edition, Lippincott Williams and Wilkins Publishers, Philadelphia, PA, 2002, p. 1443.
[13] C. Cianci, K. L. Yu, K. Combrink, N. Sin, B. Pearce, A. Wang, R. Civiello, S. Voss, G. Luo, K. Kadow, E. V. Genovesi, B. Venables, H. Gulgeze, A. Trehan, J. James, L. Lamb, I. Medina, J. Roach, Z. Yang, L. Zadjura, R. Colonno, J. Clark, N. Meanwell and M. Krystal, *Antimicrob. Agents Chemother.*, 2004, **48**, 413.
[14] K. L. Powell and D. G. Alber, in *"Respiratory Syncytial Virus: Perspectives in Medical Virology"* (ed. P. Cane), Vol. 13, Elsevier BV Publishers, Amsterdam, Holland, 2006. p. 279.
[15] K. Maggon and S. Barik, *Rev. Med. Virol.*, 2004, **14**, 149.

[16] Y. S. Babu, P. Chand and P. L. Kotian, *Annu. Rep. Med. Chem.*, 2006, **42**, 287; S. J. Little, G. Drusano, R. Schooley, D. W. Haas, P. Kumar, S. Hammer, D. McMahon, K. Squires, R. Asfour, D. Richman, J. Chen, A. Saah, R. Leavitt, D. Hazuda and B.-Y. Nguyen, Protocol 004 Study Team, 12th Conference on Retroviruses and Opportunistic Infections, February 22–25, 2005, Boston, MA, Abstract 161.

[17] R. A. Karron, D. A. Buonagurio and A. F. Georgiu, *Proc. Natl. Acad. Sci. USA*, 1997, **94**, 13961.

[18] T. Gevers, R. Willebrords, C. Sommen, J. Lacrampe, F. Janssens and P. R. Wyde, *Antiviral Res.*, 2003, **60**, 209.

[19] J.-F. Bonfanti, F. Doublet, J. Fortin, J. Lacrampe, J. Guillemont, P. Muller, L. Queguiner, E. Arnoult, T. Gevers, P. Janssens, H. Szel, R. Willebrords, P. Timmerman, K. Wuyts, F. Janssens, C. Sommen, P. Wigerinck and K. Andries, *J. Med. Chem.*, 2007, **50**, 4572.

[20] J.-F. Bonfanti, C. Meyer, F. Doublet, J. Fortin, P. Muller, L. Queguiner, T. Gevers, P. Janssens, H. Szel, R. Willebrords, P. Timmerman, K. Wuyts, P. van Remoortere, F. Janssens, P. Wigerinck and K. Andries, *J. Med. Chem.*, 2008, **51**, 875.

[21] N. A. Meanwell and M. Krystal, *Drugs Future*, 2007, **32**, 441.

[22] X. A. Wang, C. W. Cianci, K.-L. Yu, K. D. Combrink, J. W. Thuring, Y. Zhang, R. L. Civiello, K. F. Kadow, J. Roach, Z. Li, D. R. Langley, M. Krystal and N. A. Meanwell, *Bioorg. Med. Chem. Lett.*, 2007, **17**, 4592.

[23] C. Cianci, D. R. Langley, D. D. Dischino, Y. Sun, K.-L. Yu, A. Stanley, J. Roach, Z. Li, R. Dalterio, R. Colonno, N. A. Meanwell and M. Krystal, *Proc. Natl. Acad. Sci. U.S.A.*, 2004, **101**, 15046; X. Zhao, M. Singh, V. N. Malashkevich and P. S. Kim, *Proc. Natl. Acad. Sci. U.S.A.*, 2000, **97**, 14172.

[24] K. D. Combrink, H. B. Gulgeze, J. W. Thuring, K.-L. Yu, R. L. Civiello, Y. Zhang, B. C. Pearce, Z. Yin, D. R. Langley, K. F. Kadow, C. W. Cianci, Z. Li, J. Clarke, E. V. Genovesi, I. Medina, L. Lamb, Z. Yang, L. Zadjura, M. Krystal and N. A. Meanwell, *Bioorg. Med. Chem. Lett.*, 2007, **17**, 4784.

[25] S. Bond, V. A. Sanford, J. N. Lambert, C. Y. Lim, J. P. Mitchell, A. G. Draffan, and R. H. Nearn, *WO Patent* 2005061513, 2005.

[26] J. L. Douglas, *Expert Rev. Anti. Infect. Ther.*, 2004, **2**, 625.

[27] M. C. Carter, D. G. Alber, R. C. Baxter, S. K. Bithell, J. Budworth, A. Chubb, G. S. Cockerill, V. C. L. Dowdell, E. A. Henderson, S. J. Keegan, R. D. Kelsey, M. J. Lockyer, J. N. Stables, L. J. Wilson and K. L. Powell, *J. Med. Chem.*, 2006, **49**(7), 2311.

[28] E. A. Henderson, D. G. Alber, R. C. Baxter, S. K. Bithell, J. Budworth, M. C. Carter, A. Chubb, G. S. Cockerill, V. C. L. Dowdell, I. J. Fraser, R. A. Harris, S. J. Keegan, R. D. Kelsey, J. A. Lumley, J. N. Stables, N. Weerasekera, L. J. Wilson and K. L. Powell, *J. Med. Chem.*, 2007, **50**(7), 1685.

[29] J. Chapman, E. Abbott, D. G. Alber, R. C. Baxter, S. K. Bithell, E. A. Henderson, M. C. Carter, P. Chambers, A. Chubb, G. S. Cockerill, P. L. Collins, V. C. L. Dowdell, S. J. Keegan, R. D. Kelsey, M. J. Lockyer, C. Luongo, P. Najarro, R. J. Pickles, M. Simmonds, D. Taylor, S. Tyms, L. J. Wilson and K. L. Powell, *Antimicrob. Agents Chemother.*, 2007, **51**(9), 3346.

[30] L. Zhang, M. E. Peeples, R. C. Boucher, P. L. Collins and R. J. Pickles, *J. Virol.*, 2002, **76**, 5654.

[31] M. A. Francisco, Double-Blind, Randomized, Placebo-Controlled Study to Evaluate the Safety and Efficacy of RSV604 in Adults with Respiratory Syncytial Virus Infection Following Stem Cell Transplantation, IX International Symposium on Respiratory Viral Infections, Hong Kong, March 6, 2007.

[32] R. W. Sidwell, J. H. Huffman, G. P. Khare, L. B. Allen, J. T. Witkowski and R. K. Robins, *Science*, 1972, **177**, 705.

[33] B. Eriksson, E. Helgstrand and N. G. Johansson, *Antimicrob. Agents Chemother.*, 1977, **11**, 946.

[34] S. K. Wray, B. E. Gilbert and V. Knight, *Antiviral Res.*, 1985, **5**, 39.

[35] P. Toltzis, K. O'Connell and J. L. Patterson, *Antimicrob. Agents Chemother.*, 1988, **32**, 492.

[36] L. F. Cassidy and J. L. Patterson, *Antimicrob. Agents Chemother.*, 1989, **33**, 2009.

[37] J. T. Rankin, S. B. Eppes, J. B. Antczak and W. K. Joklik, *Virology*, 1989, **168**, 147.

[38] B. B. Goswami, E. Borek, O. K. Sharma, J. Fujitaki and R. A. Smith, *Biochem. Biophys. Res. Commun.*, 1979, **89**, 830.

[39] B. E. Gilbert and V. Knight, *Antimicrob. Agents Chemother.*, 1986, **30**, 201.

[40] S. Crotty, D. Maag and J. J. Arnold, *Nature Med.*, 2000, **6**, 1375.

[41] C. C. Lin, L. Philips, C. Xu and L. T. Yeh, *J. Clin. Pharmacol.*, 2004, **44**, 265.

[42] Valeant Pharmaceuticals, Press Release, September 12, 2006.

[43] E. De Clercq, M. Cools, J. Balzarini, R. Snoeck, G. Andrei, M. Hosoya, S. Shigeta, T. Ueda, N. Minakawa and A. Matsuda, *Antimicrob. Agents Chemother.*, 1991, **35**, 679; Y. Kosugi, Y. Saito, S. Mori, J. Watanabe, M. Baba and S. Shigeta, *Antiviral Chem. Chemother.*, 1994, **5**, 366.

[44] D. A. Harki, J. D. Graci and V. S. Korneeva, *Biochemistry*, 2002, **41**, 9026.

[45] R. Arnold and W. König, *Virol.*, 2006, **350**, 335.

[46] K. Sudo, Y. Miyazaki, N. Kojima, M. Kobayashi, H. Suzuki, M. Shintani and Y. Shimizu, *Antiviral Res.*, 2005, **65**, 125.

[47] S. W. Mason, C. Lawetz, Y. Gaudette, F. Do, E. Scouten, L. Lagace, B. Simoneau and M. Liuzzi, *Nucleic Acids Res.*, 2004, **32**, 4758.

[48] M. Liuzzi, S. W. Mason, M. Cartier, C. Lawetz, R. S. McCollum, N. Dansereau, G. Bolger, N. Lapeyre, Y. Gaudette, L. Lagace, M.-J. Massariol, F. Do, P. Whitehead, L. Lamarre, E. Scouten, J. Bordeleau, S. Landry, J. Rancourt, G. Fazal and B. Simoneau, *J. Virol.*, 2005, **79**, 13105.

[49] P. D. Olivo, B. A. Buscher, J. Dyall, J. I. Jocket-Balsarotti, A. K. O'Guin, R. M. Roth, Y. Zhou, G. W. Franklin and G. W. Starkey, *WO Patent* 2006121767, 2006.

[50] P. D. Olivo, B. A. Buscher, J. Dyall, J. I. Jocket-Balsarotti, A. K. O'Guin, R. M. Roth, Y. Zhou, G. W. Franklin and G. W. Starkey, *WO Patent* 2006093518, 2006.

[51] D. W. Leaman, F. J. Longano, J. R. Okicki, K. F. Soike, P. F. Torrence, R. H. Silverman and H. Cramer, *Virol.*, 2002, **292**, 70.

[52] V. Bitko and S. Baril, *BMC Microbiol.*, 2001, **1**, 34.

[53] V. Bitko, A. Musiyenko, O. Shulyayeva and S. Barik, *Nat. Med.*, 2005, **11**, 50.

[54] W. Zhang, H. Yang, X. Kong, S. Mohapatra, H. S. Juan-Vergara, G. Hellermann, S. Behera, R. Singam, R. F. Lockey and S. S. Mohapatra, *Nat. Med.*, 2005, **11**, 56.

[55] F. Eckstein, *Trends Biochem. Sci.*, 2005, **30**, 445.

CHAPTER 15

Recent Developments in β-Lactamases and Inhibitors

Tarek S. Mansour, Patricia A. Bradford and
Aranapakam M. Venkatesan

Contents

1. INTRODUCTION

At the beginning of the 21st century, we now find ourselves experiencing a taste of what life was like prior to the advent of the antibiotic age in the 20th century. We are again faced with life-threatening infections for which there are very few

Wyeth Research, 401 North Middletown Road, Pearl River, New York 10965

Annual Reports in Medicinal Chemistry, Volume 43
ISSN 0065-7743, DOI 10.1016/S0065-7743(08)00015-8

antibiotic options for treatment. However, unlike the previous century, these infections are often caused by Gram-negative pathogens [1].

One of the important factors in the increase in antibiotic resistance is intensive care units (ICUs), which have been deemed "factories for creating, disseminating and amplifying resistance to antibiotics [2]." A majority of patients in the ICU receive antibiotics during their stay, often in combinations in an attempt to circumvent the development of resistance. For example, in the latest report from the National Nosocomial Infections Surveillance (NNIS) System, there was a 47% increase in *Klebsiella pneumoniae* resistant to third generation cephalosporins isolated in the ICUs in the USA in 2003 compared to the previous four years [3]. Furthermore, this study revealed a 20% increase in quinolone resistance in *Pseudomonas aeruginosa*.

Several multidrug-resistant pathogens are of specific concern. Whereas prior to the 1990s *Acinetobacter baumannii* were almost universally susceptible to broad-spectrum antibiotics, during this decade they became increasingly resistant to penicillins, cephalosporins, fluoroquinolones and aminoglycosides [4]. Thus in recent years, many of these antimicrobials are no longer reliable treatments of infections caused by this organism. Most notable is the increase in resistance to the carbapenems which are caused by a variety of β-lactamases and changes in penicillin-binding proteins (PBPs) [5]. There are now reports of multidrug-resistant *A. baumannii* strains that are susceptible only to polymixin B and colistin [6].

Multidrug-resistant Gram-negative pathogens, including *Escherichia coli* and *K. pneumoniae* expressing extended-spectrum β-lactamases (ESBLs) are on the recent list of six microbial pathogens of concern according to the Infectious Diseases Society of America [7,8]. In addition, inappropriate antibiotic therapy for patients with isolates harboring an ESBL has been correlated with increased mortality in patients [9,10]. Therefore, a strategy to restore the susceptibility of these pathogens through the search for new β-lactamase inhibitors is warranted.

Production of β-lactamase is the most common resistance mechanism against β-lactam antibiotics in Gram-negative bacteria. These enzymes hydrolyze the β-lactam ring of all classes of β-lactam antibiotics, thus inactivating the drug. As most β-lactamases share an active site motif (Ser-XX-Lys) with penicillin-binding proteins (PBPs), it has been proposed that serine β-lactamases evolved from PBPs as protection against β-lactam antibiotics produced by molds in the environment. TEM-1 has spread worldwide and is now found in many different species of Enterobacteriaceae, *P. aeruginosa*, *Haemophilus influenzae* and *Neisseria gonorrhoeae*.

Over the last 20 years many new β-lactam antibiotics have been developed that were specifically designed to be resistant to hydrolysis by β-lactamases. However, with each new class of β-lactam antibiotics that has been used to treat patients, new β-lactamases have emerged that caused resistance to that drug.

This review will describe advances made during the last five years [11] in the inhibition of β-lactamases, mechanisms of inactivation and the use of biophysical techniques and X-ray crystallography to elucidate active sites structures of β-lactamases for the design of new inhibitors. Crystallography efforts have made excellent progress in generating over 50 structures for both apo enzymes and complexes with inhibitors/substrates.

2. COMBINATION OF INHIBITORS WITH ANTIBIOTICS

An important strategy that has been successfully utilized for overcoming β-lactamase-mediated resistance to β-lactams has been the co-administration of the β-lactam antibiotic together with an enzyme inhibitor [12–14]. The inhibitors bind to β-lactamase enzymes in an irreversible manner (in most cases) that inactivates the β-lactamase from destroying the accompanying β-lactam antibiotic [14]. There are three β-lactamase inhibitors that are currently commercially available: clavulanate, which is used in combination with amoxicillin or ticarcillin; sulbactam, which is used in combination with ampicillin; and tazobactam, which is used in combination with piperacillin. These three inhibitors are effective only against class A serine β-lactamases [13]. The β-lactam/β-lactamase inhibitor combinations have been used successfully for over 20 years for treatment of serious infections caused by β-lactamase producing Gram-positive and Gram-negative pathogens. However, with the changing landscape of the type of β-lactamases produced by pathogens found routinely in hospitals, their continued usefulness is in jeopardy. Importantly, interest in this area has grown considerably and most notably from pharmaceutical companies as they have responded to what is a clear medical need for broad-spectrum inhibitors that include activity against class C and class D enzymes.

3. β-LACTAMASES

This section will discuss four types of β-lactamases that make a significant contribution to resistance to β-lactams in Gram-negative pathogens.

3.1 Extended-spectrum β-Lactamases

Resistance to expanded-spectrum β-lactam antibiotics in Enterobacteriaceae is often due to the presence of an ESBL, which is most often a derivative of TEM or SHV enzymes, belonging to molecular class A or functional group 2be (Table 1) [15–17]. With both of TEM and SHV enzymes a few point mutations at selected loci within the structural gene encoding the enzyme give rise to the extended-spectrum phenotype. TEM- and SHV-type ESBLs are most often found in *E. coli* and *K. pneumoniae*, however they are also found in *Proteus* spp., *Enterobacter* spp. and other members of the Enterobacteriaceae [16].

As noted, ESBLs continue to be a problem in hospitalized patients worldwide. The incidence of ESBLs in strains of *E. coli* and *K. pneumoniae* varies from 0% to 40%, depending on the country [18]. In the United States, the overall incidence is around 3%; however, it can be as high as 40% in some institutions [19]. Many of the patients infected with ESBL-producing pathogens are in ICUs, although they can also occur in surgical wards also. Organisms with ESBLs are also being isolated with increasing frequency from patients in extended care facilities [20,21]. The newest family of plasmid-mediated ESBLs that has arisen (CTX-M) preferentially hydrolyzes cefotaxime [15,17,22–25]. These enzymes are not closely

Table 1 Classification of β-lactamases[a]

Functional group	Molecular class	Descriptor	Preferred substrates	Inhibited by		Representative enzymes
				CA	EDTA	
1	C	AmpC-type β-lactamases	Cephalosporins	−	−	Inducible enzymes from *Enterobacter, Citrobacter, Serratia* and *Pseudomonas*; plasmid-mediated enzymes ACT-1, CMY-type and FOX-type
2a	A	Gram-positive penicillinases	Penicillins	+	−	*Staphylococcus aureus* PC1 penicillinase
2b	A	Broad-spectrum enzymes	Penicillin and narrow-spectrum cephalosporins	+	−	TEM-1, TEM-2, SHV-1, OHIO-1
2be	A	Extended-spectrum β-lactamases	Penicillins, all cephalosporins and monobactams	+	−	TEM-3, TEM-10, TEM-26, etc.; SHV-2, SHV-5, SHV-7 etc.; CTX-M-1, etc.; *K. oxytoca* K1
2br	A	Inhibitor-resistant β-lactamases	Penicillin and narrow-spectrum cephalosporins	−	−	TEM-30 to TEM-36, etc.; SHV-10
2c	A	Carbenicillin hydrolyzing β-lactamases	Penicillins including carbenicillin	+	−	PSE -1, PSE-3 and PSE-4; *Aeromonas* AER-1; CARB-3 and CARB-4

Group	Molecular class	Enzyme family	Preferred substrates			Representative enzymes
2d	D	Oxacillin-hydrolyzing β-lactamases	Penicillins including cloxacillin	±	−	OXA-1 to OXA-30
2e	A	Cephalosporinases inhibited by clavulanic acid	Cephalosporins	+	−	Bacteroides fragilis CepA, Proteus vulgaris FPM-1, S. maltophilia L2
2f	A	Non-metallo carbapenemases	Penicillins, cephalosporins and carbapenems	+	−	IMI-1, NMC-A, Sme-1
3a	B	Metallo-β-lactamases	Penicillins, cephalosporins and carbapenems	−	+	B. cereus II, CcrA, IMP-1, L1, VIM-1, VIM-2. SPM-1
3b	B	Metallo-β-lactamases	Carbapenems	−	+	CphA, AsbM1, ASA-1, ImiS
3c	B	Metallo-β-lactamases	Cephalosporins, carbapenems	−	+	β-Lactamase from Legionella gormanii
4	Not determined	Penicillinase not inhibited by clavulanate	Penicillins	−	−	β-Lactamase from Burkholderia cepacia, SAR-2

[a] Adapted from [13,41,63].

related to TEM or SHV β-lactamases, but instead share a high homology to the class A chromosomally encoded β-lactamase of *Klebsiella oxytoca*. Another unique feature of these enzymes is that they are inhibited more effectively by tazobactam than the other β-lactamase inhibitors sulbactam and clavulanate [26,27]. Another growing family of ESBLs is the OXA-type enzymes, which may provide weak resistance to oxyimino-cephalosporins. These enzymes differ from the TEM and SHV enzymes in that they belong to molecular class D and functional group 2d (Table 1) [15]. To date, at least 45 OXA variants have been identified, although not all of these possess the ESBL phenotype [17,28].

The overall structure of wild-type Toho-1, an ESBL encoded by a plasmid and produced in *E. coli* (classified into the CTX-M group) was determined at 1.65 Å resolution [29]. Several important features of this structure are noteworthy. The Ω loop, which contains Glu166, moves outward to widen the active sites and two alternative conformations of Lys73 were reported. In conformation 1, the side chain of Lys73 points toward Glu166, Ser70, Asn132 and a water molecule, Wat165. However, in conformation 2, the ammonium group of Lys73 points toward the hydroxyl group of Ser130, Ser70 and Asn132. This conformation appears to be similar to that in most class A β-lactamases. It was proposed based on the structural analysis that several proton-relay pathways occur in the acylation step involving Lys73. The ammonium group of Lys73 transfers a proton to Glu166 carboxylate, which, in turn, activates Ser70 via a catalytic water molecule for acylation. Alternatively, Ser130 upon transfer of a proton to the substrate would accept a proton from Lys73 ammonium group and, in the final step, the now neutralized Lys73 would activate Ser70 [30]. The Toho-1 active site resembles that of TEM-1 and importantly, did not feature the enlarged active sites of ESBL mutants (TEM-52 and TEM-64). In the crystal structures of CTX-M-9 (0.98 Å) and CTX-M-14 (1.10 Å), Lys73 adopts two distinct conformations [31] in accordance with the Toho-1 structure. In one conformation, Lys73 activates the hydrolytic attack of Ser70 on the substrate. Subsequently, it is proposed to activate Ser130, which is thought to act as a general acid-promoting departure of the β-lactam nitrogen. Another key finding in this study is that CTX-M-16 (1.74 Å), which carries Val231Ala and Asp240Gly substitutions in the B3 strand, enhances substrate specificity [31].

Strong evidence for the acylation mechanism in CTX-M β-lactamase was deduced in a separate study from the high-resolution (0.88 Å) crystal structure of CTX-M-9. It was proposed that in the apo enzyme at physiological pH, Ser70 donates a proton to the catalytic water, which in turn transfers its proton to Glu166 anion [32]. This event makes Ser70 anionic for attack on the β-lactam substrate. Lys73 in this structure adopts two different conformations based on the protonation state of Glu166.

3.2 Class A carbapenemases

Reports of carbapenem-hydrolyzing β-lactamases have been increasing over the last few years. This phenotypic grouping of enzymes is comprised of a heterogeneous group of β-lactamases belonging to either molecular class A, which has a serine in the active site of the enzyme, or molecular class B, the metallo-β-lactamases (MBLs) which have zinc in the active site (Table 1) [33]. The class A, functional group 2f carbapenem-hydrolyzing β-lactamases have been found in *Enterobacter cloacae* (IMI-1 and NMC-A) [34,35], *Serratia marcescens* (Sme-1-3) [36–38] *K. pneumoniae* (KPC-1-3) [39,40], *P. aeruginosa* (GES-2) [41] and *A. baumannii* [42]. They are inhibited by clavulanate, but not by EDTA. While these enzymes hydrolyze imipenem (IPM) well enough to provide resistance, they show much higher hydrolysis rates for ampicillin and cephaloridine and have been found in strains that also possess an AmpC β-lactamase. This combination of enzymes provides resistance to a broad range of β-lactams [38]. In recent years, there have been many reports of *Klebsiella* spp., *Enterobacter* spp. and *Salmonella enterica* ser Cubana expressing KPC-type enzymes in cities in the Eastern USA [40,43–48].

The structure of KPC-2 at 1.85 Å resolution was determined to shed light on the widespread emergence of IPM resistance in *K. pneumoniae*. The active site in this carbapenemase is distinctively smaller than that of SHV-1 or TEM-1 and Ser70 is repositioned into the active site. Another key finding in this study indicates that Ser70 is positioned in a shallow position suggesting that the β-lactam ring of carbapenems does not have to enter the catalytic site as deeply as in other enzymes [49].

3.3 Metallo-β-lactamases

The class B metallo-enzymes have zinc in the active site, and have been described as naturally occurring, chromosomally mediated enzymes in diverse genera of Gram-positive and Gram-negative bacteria [38]. MBLs can hydrolyze IPM at a measurable rate; however, they differ in their abilities to hydrolyze other β-lactam substrates [38]. All MBLs have two distinct zinc-binding sites (Zn1, Zn2), which closely resemble those found in zinc proteins, however, MBLs require just one metal ion for catalytic activity.

A particularly disturbing event that occurred in Japan in the early 1990s was the discovery of an MBL, IMP-1, that was encoded on plasmids. These β-lactamase-producing plasmids were readily transferable to other strains. This enzyme was originally found in *S. marcescens*, but has also been found in *K. pneumoniae* and *P. aeruginosa* [50,51]. For a number of years, this enzyme was contained within that country. However, in recent years, the IMP-1 β-lactamase and at least 18 related enzymes have been identified, first in other Asian countries, then recently in Europe and finally in North America [17,33]. IMP-1 from *P. aeruginosa* has two Zn^{2+} ions that are exchangeable by exogenous Zn^{2+} ions as demonstrated by using inductively coupled plasma mass spectrometry (ITV-ICP-MS) in conjunction with electrospray-ionization mass spectrometry (ESI-MS) [52]. Other plasmid-mediated MBLs include the VIM-type enzymes, that have been reported throughout Europe as well as South East Asia and North America in strains of *P. aeruginosa*, *K. pneumoniae*, *E. cloacae* [16,33,53,54]. VIM-type enzymes are often harbored in gene cassettes and are also associated with integrons [54,55].

The overall structure of another MBL originating in Brazil, SPM-1, from *P. aeruginosa* (1.90 Å) conforms closely to the MBL B1 subgroup containing a single zinc ion coordinated by His116, His118, His196 and a water molecule, Wat333, in a tetrahedral arrangement in Zn1 site. The Zn2, anchored by Cys221-His263-Asp120, contains no metal ion [56]. The structure highlights the importance of the Asp120 positioning for optimal binding in Zn2. The three crystal structures of a B2 subgroup, MBL CphA, from *Aeromonas hydrophilia*; wild-type, Asn220Gly mutant and in complex with biapenem (BIPM) reveal that this enzyme under physiological conditions and in the presence of a substrate, has the zinc ion located in the Zn2 site. Moreover, the Asn233 residue rotation orients the amide into the active site, resulting in a hydrophilic wall from the N-domain with a mobile C loop which traps off the carbapenem antibiotic in the active site pocket [57].

The role of Asp120 in MBLs was also elucidated by studying the B3 subclass enzyme L1 from *Stenotrophomonas maltophilia*, carrying the Asp120Asn and Asp120Cys mutations that retain dinuclear zinc clusters [58]. The L1 Asp120Asn and Asp120Cys structures determined to resolutions of 1.76 and 2.25 Å retain the Zn1 tetrahedral structure [58]. However, the Zn2 site differs from that of the mutant enzymes in comparison to the wild type. In the Asp120Asn mutant, the Zn2 site is tetrahedral, while that of the Asp120Asn mutant is octahedral and the two metal ions are bridged by a water molecule. In a separate study aimed at elucidating the role of Asp120 [59], comparison of the structures of L1 complexed with oxacephem moxalactam led to the conclusion that the primary role of Asp120 is to position Zn2 for catalytically important interactions with the charged amide nitrogen of the substrate, in addition to its role in proton transfer processes [59]. Evidence that a water molecule situated between the two metal ions as the nucleophile in the hydrolytic reaction was proposed with a dual role for the Zn2 in binding to substrate β-lactam carboxylate and facilitating catalysis. A related study clarifying the role of Asp120 on four mutants of the *Bacillus cereus* enzymes BcII carrying Asp120Asn, Asp120Ser, Asp120Glu and Asp120Gln mutations concluded that the conserved Asp120 acts as a strong Zn2 ligand for optimal substrate stabilization [60].

3.4 AmpC

AmpC β-lactamases are chromosomally encoded enzymes found naturally in *Enterobacter* spp., *Citrobacter* spp., *Serratia* spp., *P. aeruginosa* and *Hafnei alvei*. In these genera, the β-lactamase is normally expressed at low levels and is induced after exposure to some β-lactam antibiotics. AmpC-type β-lactamases, belonging to molecular class C and functional group 1, generally confer resistance to oxyimino-cephalosporins such as ceftazidime, cefotaxime and ceftriaxone, and cephamycins such as cefoxitin and cefotetan (Table 1) [15]. As a group, class C enzymes are not inhibited by the currently available β-lactamase inhibitors clavulanate, sulbactam and tazobactam.

In the 1990s AmpC β-lactamases found their way onto mobile elements, especially plasmids, and are being expressed constitutively at high levels in *K. pneumoniae* and *E. coli*. These β-lactamases appear to have originated from the

AmpC enzyme of *E. cloacae*, *Citrobacter freundii*, *P. aeruginosa*, *Morganella morganii* and *H. alvei* [15,61–68]. Most of the plasmid-mediated AmpC β-lactamases have lost the *ampR* regulatory gene and are therefore produced constitutively at high levels, resulting in resistance to oxyimino-cephalosporins.

In addition to these plasmid-mediated AmpC enzymes conferring general resistance to cephalosporins and cephamycins, they can also contribute to carbapenem resistance when combined with the loss of outer-membrane porin proteins in *K. pneumoniae* [61,69]. These strains are particularly disturbing because IPM has been considered to be the drug of choice for treatment of infections caused by Enterobacteriaceae that are resistant to multiple antibiotics.

4. INHIBITORS BASED ON CRYSTAL STRUCTURES OF β-LACTAMASES

Important advances in tracking reaction intermediates inside protein crystals containing bound inhibitors were elegantly achieved using Raman spectroscopy. The Case Western investigators demonstrated the application of this technology on all three inhibitors in clinical use by using the Glu166Ala variant of SHV-1, which is regarded as a deacylation-deficient enzyme [70–72].

It was found that tazobactam forms a stoichiometric *trans*-enamine intermediate 1 in 1.63 Å resolution, which is different from the imine structure observed in the wild-type complex. This intermediate is also referred to as a "waiting-room" intermediate to reinforce its transient nature and stability [70]. The *trans*-enamine intermediates were also observed for sulbactam and clavulanic acid in complexes with the Glu166Ala mutant enzyme at 1.34 and 1.43 Å resolution [71]. From a structural perspective, the C5–C6 dihedral angle for all inhibitors was in the range of 181–188°, the O8 atom for all inhibitors was positioned in the oxyanion hole but some differences in interacting with the side chain of Asn170 were noted. The Raman crystallographic measurements of the intermediates relied on line widths which point out to the presence of close-lying conformations for both sulbactam and clavulanic acid. The latter exists in two conformations, the *cis*-enamine 2 and a decarboxylated *trans*-enamine intermediate 3. Based on these findings, a penam sulfone inhibitor **SA2-13** (**4**, $K_D = 1.7\,\mu M$) was designed by replacing the triazolyl moiety of tazobactam with a negatively charged carboxylate side chain to stabilize the *trans*-enamine structure [72]. The atomic structure of SA2-13 in complex with wild-type SHV-1 (1.28 Å resolution) was reported to be similar to the Glu166Ala tazobactam structure, thus validating this strategy for designing inhibitors but with a notable difference regarding Asn170 conformations.

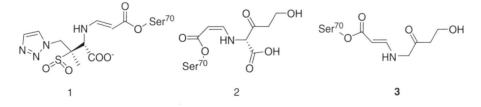

1   2   3

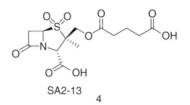

SA2-13

4

Based on the findings that Toho-1, CTX-M-9, CTX-M-14, CTX-M-16 and CTX-M-27 (1.20 Å) have characteristic small active sites, a set of glycylboronic acids bearing side chains reminiscent of cephalosporins and penicillins **5–8** were synthesized [73]. These compounds proved to be reversible and competitive inhibitors of CTX-M enzymes. The 2-aminothiazole inhibitor **7** had K_i values of 4 and 15 nM for CTX-M-16 and CTX-M-9, respectively. The crystal structures of the three inhibitors **5, 6** and **7** and that of cefoxitin in complexes with CTX-M-9 were determined to have resolutions of 1.16 and 1.70 Å. With compound **7**, the co-crystal structures consisted of molecules in both the apo and complex states. It was found that as progression along the reaction coordinate occurs, Lys73 and Glu166 move. Specifically, the side chain of Lys73 switches from two conformations in the apo state to either one or two conformations in the transition state analogue. The Glu166 side chain changes conformation as well [73], *albeit* less dramatically. The authors attributed these conformational changes to the general notion of Lys73 and Glu166 as possible catalytic bases for the reaction. In the cefotoxin structure, the 7α substituent prevents the formation of a transition state and thus blocks the progress of the reaction beyond the acyl intermediate.

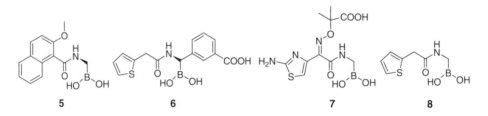

| 5 | 6 | 7 | 8 |

The design of inhibitors of MBLs is an active area of investigation [74]. Competitive inhibitors were reported from the *N*-arylsulfonyl hydrazone class with the tricyclic compound **9** having a K_i value of 1.7 μM [75]. The structure-activity relationship (SAR) in this series indicated a requirement for a bulky substituent on each side of the hydrazone moiety. Irreversible inhibitors of IMP-1 based on 3-(3-mercaptopropionylsulfanyl)-propionic acid pentafluorophenyl esters **10, 11** were designed based on targeting Lys224 for hydrolysis and binding [76]. The co-crystal structure of **11** with IMP-1 at a resolution of 2.63 Å indicated that the thiolate group bridges the two Zn^{II} ions in the active site. MBL inhibitors were identified from a structure-based pharmacophore design based on reported structures of B1 subclass enzymes with succinic acid and mercapto-carboxylic acids. The CATALYST program searched the ACD database and selection was based on GOLD scores. The study led to the identification of new scaffolds with IC_{50} values below 12 μM [77].

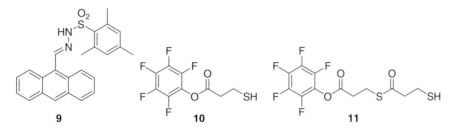

9   **10**   **11**

A series of DansylC$_n$SH inhibitors of IMP-1, derived from N-(aminoalkyl)-5-dimethylaminonaphthaline-1-sulfonamide and 3-thiopropionic acid, **12–15** were reported to be fluorescent probes and inhibitors of IMP-1 and VIM-2. In the co-crystal structure of DansylC$_4$SH **13** and IMP-1, the naphthyl ring makes an edge-to-face interaction with Trp64, and the thiol group is coordinated with two Zn^{II} ions [78].

Chelating agents such as 2-picolinic acid **16** and pyridine 2,4-dicarboxylic acid (PDCA) **17** were reported to be competitive inhibitors of CphA with K_i values of 46 and 17 μM, respectively at pH 7.0 [79]. The zinc ion in the co-crystal structure of CphA and PDCA (1.86 Å) is coordinated to five ligands, Asp120-Cys-221-His263, pyridyl nitrogen and one oxygen of the carboxylate group of PDCA.

Recently, Meiji chemists reported a number of substituted maleic acid derivatives as class B MBL inhibitors [80,81]. Compound **18** designated as **ME 1071**-(CP3242),{IC_{50} (IMP-1) 2.5 μM and (VIM-2) 12.6 μM} when combined with BIPM or IPM reduced the minimum inhibitory concentration (MIC) values of these antibiotics from resistant level (MIC 256 μg/mL for IPM and 62–128 μg/mL BIPM) to 1–4 μg/mL against IMP-1 producing *P. aeruginosa*. Some of these derivatives were also co-crystallized with IMP-1 and VIM-2 enzymes. The co-crystal structures showed that the two zinc atoms at the active site were chelated to the carboxy portion of the inhibitors.

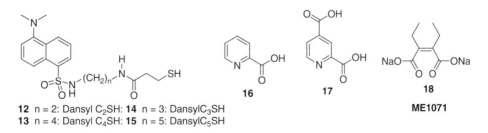

16   **17**   **18**
ME1071

12 n = 2: Dansyl C$_2$SH: **14** n = 3: DansylC$_3$SH
13 n = 4: Dansyl C$_4$SH: **15** n = 5: DansylC$_5$SH

The crystal structure of OXA-24, a class D β-lactamase from *A. baumannii* was resolved at resolution of 2.50 Å [82]. The active site has an overall positive charge and contains an active Ser81 residue in addition to Thr82, Phe83 and Lys84. Based on mutation studies, it was found that Tyr112 and Met223 are key residues for antibiotic specificity. In comparison to other OXA enzymes, the OXA-24 N-terminus closely resembles that of TEM-1 and will be used as a basis for designing new inhibitors.

5. BROAD-SPECTRUM INHIBITORS

Several-substituted spirocyclopropylpenam sulfones (**19, 20**) were found to be potent, mechanism-based inhibitors of TEM-1 and AmpC enzymes [83]. Mechanistically, the cyclopropyloxy group can promote subsequent chemical events after initial acylation, as for example to unravel the aldehyde or oxycarbenium species for further cross-linking with the enzyme. Among the various diastereomers prepared, compounds **19** (IC_{50} TEM-1 = 0.65 µM; AmpC = 0.02 µM) and **20** (IC_{50} TEM-1 = 0.51 µM; AmpC = 0.03 µM) were found to be the most potent and had good MIC_{50} values in combination with cefotaxime and piperacillin. Modification of the 6-position of the penicillanate core with mercapto functionality as shown in **21** yielded potent class B inhibitors with marginal activity against TEM-1 and AmpC enzymes (IC_{50} TEM-1 = 6.8 µM; AmpC = 10.5 µM; L1, class B = 0.10 µM) [84].

A series of tricyclic carbapenems were reported to be potent inhibitors of class A and C enzymes with **22** (LK-157) being the lead compound (IC_{50} TEM-1 = 55 nM, AmpC = 62 nM) [85]. The structure of LK-157 complexed with AmpC from *E. cloacae* indicates the formation of a covalent bond with Ser64; the β-lactam carbonyl oxygen is oriented in the oxyanion hole and hydrophobic interactions with Leu109 and Leu293 were reported [85]. In this complex, the carboxylic group is directed out of the active site and has no direct interactions with the enzyme. LK-157 has similar activity to tazobactam and clavulanate against TEM-1 and SHV-1, but is least 300-fold more potent against AmpC.

Depsipeptides *O*-aryloxycarbonyl hydroxamates such as **23** and **24** irreversibly inhibited the class C β-lactamase of *E. cloacae* P99 in a time-dependent manner. The co-crystal structures and ESI-MS studies provided evidence for cross-linking of the active site to form a cyclic carbamate **25** involving a Lys315 side chain [86]. This work was based on earlier kinetic and structural findings of α-substituted phenaceturates **26** and **27** as substrates of P99 enzymes [87].

The non-β-lactam inhibitor **28**, **NXL104** (formerly AVE1330A) represents a novel class of bridged bicyclic [3.2.1] diazabicyclo-octanones. The compound was designed to inhibit class C β-lactamases better than the currently available β-lactamase inhibitors. **NXL104** is active against both TEM-1, a class A enzyme and P99, a class C β-lactamase with reported IC_{50} values of 8 and 80 nM, respectively [88]. When used in combination with ceftazidime, **NXL104** restored the susceptibility of *E. coli* and *K. pneumoniae* expressing ESBLs, with MIC_{90} values of ceftazidime reduced from >64 to 2 µg/mL in the presence of the inhibitor [88]. The activity of ceftazidime/**NXL-104** against class A also extends to the CTX-M-type ESBLs and KPC-type carbapenemases [89,90]. **NXL104** is currently undergoing phase 1 clinical trials and is being developed for use with ceftazidime by Novexel and for use with ceftaroline by Forrest Laboratories.

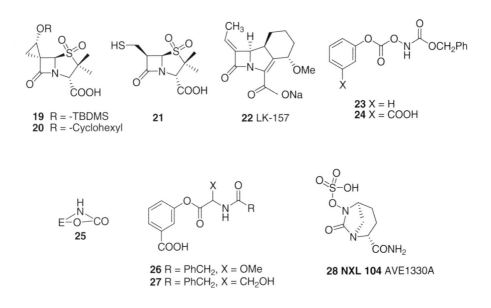

19 R = -TBDMS
20 R = -Cyclohexyl

21

22 LK-157

23 X = H
24 X = COOH

25

26 R = PhCH$_2$, X = OMe
27 R = PhCH$_2$, X = CH$_2$OH

28 NXL 104 AVE1330A

There are very few examples of class D inhibitors reported to date. In a series comprised of diphosphates and phosphonates, dibenzoyl phosphate stood out as a potent inhibitor with a K_i value of 94 nM against the OXA-1 enzymes [91]. In this class of inhibitors the acylation rate of the Ser67 residue is influenced by the hydrophobic substituents whereas the deacylation reaction is rather slow and is influenced by the electronic effects of substituents [92]. Of interest here is that all serine β-lactamases are inhibited by this class. The 6-methylidene penems including **BRL-42715** are also potent inhibitors of class D enzymes and have demonstrated this activity in susceptibility tests with class D-resistant enzymes. An important consideration for future design of such inhibitor is the relative similarity in the active sites between TEM-1 and OXA-1.

6. 6-METHYLIDENE PENEMS

6.1 Monocyclic inhibitors

The 6-methylidene penem class was recently proven to represent a set of broad-spectrum inhibitors. Over a decade ago **BRL-42715** and **SB-206999z** were reported to be effective inhibitors in protecting β-lactam antibiotics against wide range of clinically important β-lactamases [93,94].

The design of new inhibitors was initially based on docking studies in the crystal structure of apo TEM-1 and apo AmpC enzymes. The key assumptions were Lys234 (TEM-1) or Lys315 (AmpC) Nz-carboxylate O: 2.4 Å, Ser OG-lactam

O: 2.2–2.5 Å with maximum interaction in TEM-1 residues Glu104 and Tyr105 and AmpC residues Leu119 and Gln120 [95]. To demonstrate the validity of this approach a set of 6-methylidene penem inhibitors bearing monocyclic appendages were synthesized and tested [96]. It was found that the 2-benzyl-substituted imidazole derivative **29** inhibited TEM-1 and AmpC with excellent potency ($IC_{50} = 0.4$ nM; 2 nM) and MIC_{50} values in antibacterial susceptibility studies, when combined with piperacillin. In a murine acute lethal infection model, compound **29** in combination with piperacillin enhanced piperacillin activity ($ED_{50} = 43$ mg/kg). Molecular modeling suggested that the benzyl substituent forms a $\pi-\pi$-stacking interaction with Tyr 105 in TEM-1 active site [95].

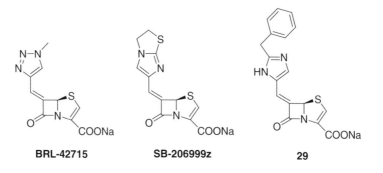

BRL-42715   SB-206999z   **29**

6.2 Bicyclic inhibitors

This class of inhibitors contains either a [5,5] or a [5,6] bicyclic heterocycles in place of the monocyclic 6-methylidene substituent. In the [5,5] fused systems, the heterocycles were chosen to contain pyrazolo or imidazo rings [97]. Compound **30** was found to be a potent inhibitor of IC_{50} TEM-1 = 0.75 nM; IC_{50} AmpC = 1 nM and **31** (IC_{50} TEM-1 = 0.4 nM; AmpC = 0.5 nM), displayed excellent MIC_{50} values in combination with piperacillin against a number of resistant strains. Incidentally, the corresponding E isomer of **30** was found to be 640 times less active against TEM-1 and AmpC confirming the SAR trend of these penems. The *in vivo* activity of **30** in an acute lethal infection model in combination with piperacillin (4:1 Pip: **30**) administered subcutaneously was found to be effective against E. coli-LSU 80 (a piperacillin-resistant TEM-1 strain, $ED_{50} = 18.9$ mg/kg), E. coli (ESBL, $ED_{50} = 49$ mg/kg) and Enterobacter aerogenes (Amp C, $ED_{50} = 98.1$ mg/kg) [98].

The related [5,6] bicyclic heterocycles also emerged as a potent class of broad-spectrum inhibitors [99]. Based on the attachment of the heterocycles, the [5,6] fused compounds can be classified into 6-methylidene penems directly attached to thiophene, imidazole and pyrazole rings [99] with the latter two subclasses exhibiting greater *in vitro* activities as illustrated for **32** (**BLI-489**) (IC_{50} TEM-1 = 0.4 nM; IC_{50} AmpC = 2 nM), **33** (IC_{50} TEM-1 = 3 nM; IC_{50} AmpC = 3 nM) and **34** (IC_{50} TEM-1 = 0.5 nM; IC_{50} AmpC = 1 nM). Penems **31** and **34** ($K_i = 73$ and 45 nM, respectively) are also potent inhibitors of OXA-1 (class D) β-lactamase [100], which is a highly active penicillinase whose structure is known [101].

The MIC$_{50}$ values of piperacillin in the presence of **BLI-489** (tested at a constant concentration of 4 µg/mL) and determined against panels of piper-acillin-resistant Gram-negative bacterial pathogens expressing known, well-characterized β-lactamases of all four molecular classes (A, B, C and D) demonstrated synergistic activity, in that the MICs of piperacillin were reduced from resistant (>64 µg/mL) to susceptible levels for many pathogenic bacteria [102]. For some recent isolates of *E. coli* and *K. pneumoniae* from urinary tract infection (UTI), the MIC$_{90}$ values were 8 and 16 µg/mL for piperacillin plus **BLI-489**. In contrast, the MIC$_{90}$ values were >64 µg/mL for both species when tested with piperacillin plus tazobactam.

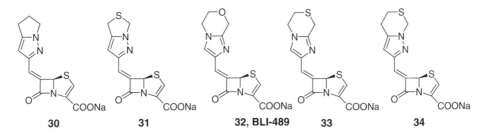

30	**31**	**32, BLI-489**	**33**	**34**

For isolates producing ESBLs, piperacillin in combination with **BLI-489** was active against 88% of the *E. coli*, 78% of the *K. pneumoniae* and 90% of the *Enterobacter* spp. at 16 µg/mL piperacillin or less, compared to inhibition of 71% and 30%, respectively with piperacillin plus tazobactam at these concentrations. Against AmpC-producing strains, piperacillin plus **BLI-489** inhibited 88% of strains at 16 µg/mL or less. *In vivo*, compound **BLI-489** when combined with piperacillin at 8:1 ratio, had an ED$_{50}$ values of 12–13 and 32–38 mg/kg against *E. coli* (TEM-1-producing strain, class A) and *E. cloacae* (AmpC-producing strain, class C), respectively [103].

The co-crystal structures of **BLI-489** in complex with SHV-1 and GC1 determined to be 1.1 and 1.38 Å, respectively reveal the formation of a 7-membered ring thiazepine ring and clearly establish the formation of covalent bonds with Ser70 and Ser 64, respectively [104]. In the SHV-1 complex, the binding site was more open than in apoenzyme Tyr105 is rotated 20° and Glu166/W501 are closer to inhibitor. Specific interactions involved Ser130, Asn132, Ala237 and π-stacking with Tyr 105. Glu166 does not make any specific interactions with the inhibitor. In the GC1 structure, major changes involving, Tyr224 displacement ca. ~6 Å from apoenzyme and the thiazepine binding mode is rotated ca. 180° with interactions involving Asn152 and Tyr150. It was also noted that in class A complex, the acyl ester bond was stabilized to hydrolysis because, the buried water molecule on the α-face of the ester bond appears to be loosely bound or absent. In the class C complex, the hydrolytic water molecule on the β-face was disordered and poorly activated for hydrolysis.

6.3 Tricyclic inhibitors

The extension of the terminal heterocyclic ring of the 6-methylidene substituent led to the discovery of potent, broad-spectrum inhibitors. These compounds in general offer greater solution stability and desirable lipophilicity for *in vivo* evaluation. The SAR studies indicated that a saturated terminal ring is preferred over an aromatic one, although in the latter case electron-withdrawing substituents can restore activity [105]. Among the various derivatives prepared in this class, compound **35** was reported to have good enzyme potency (IC_{50} TEM-1 = 1.4 nM; AmpC = 2.1 nM). Saturated 6-membered rings such as **36** (IC_{50} TEM-1 = 1.9 nM; AmpC = 0.62 nM) including pyranosyl compound **37** (IC_{50} TEM-1 = 2.8 nM; AmpC = 1.5 nM) exhibited comparable potency relative to **35** [106]. Based on the effectiveness of the *in vitro* activity, compound **35** was tested in a murine acute lethal infection model against TEM-1 producing *E. coli* and AmpC producing *E. cloacae*. When compound **35** was mixed with piperacillin in 4:1 ratio to piperacillin and administered subcutaneously, it exhibited marked reduction in the ED_{50} value of piperacillin (ED_{50} value 19.2 mg/kg against *E. coli* and 42 mg/kg against *E. cloacae*) [98].

The co-crystal structures of **35** in complex with SHV-1 and GC1, determined to be 2.0 and 1.38 Å, respectively, reveal the formation of a 7-membered thiazepine ring and clearly establish the formation of covalent bonds with Ser70 and Ser64, respectively [105]. In the SHV-1 complex, the binding site was more open than in apoenzyme and overall H-bonds were essentially identical to those seen with **BLI-489**. However, clear differences were seen in the GC1 structure particularly in the formation of two thiazepines in the co-crystal structure differing at the C7 configuration and the binding mode of the major C7-(*S*)-isomer [105].

Relative to imipenem, the 6-methylidene penems are stable to hydrolysis by mouse and human renal dihydropeptidases particularly **35** and **BLI-489** [105].

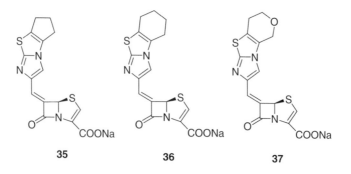

6.4 Mechanism of inhibition

The kinetics of the reaction of several 6-methylidene penems with TEM-1, SHV-1 and AmpC were characterized by ESI-MS [107]. Gas-phase ESI-MS/MS and T-6 trypsin digestion studies established the covalent binding involving Ser70 and Ser64. This analysis also confirmed the formation of thiazepines **38** and **39** by comparison to synthetic samples. The crystal structure of **BRL-42715** in complex with *E. cloacae* 908R class C β-lactamase reveals the formation of the C7-(S)-triazolyl-1,4-dihydrothiazepine **40** [108]. The binding of the thiazepine indicates π-stacking with Tyr221. The possibility of influencing the stereochemistry of the C7 heterocycle by π-stacking with the tyrosine residue led to further evaluation of R and S thiazepines in the bicyclic and tricyclic series by interaction energy calculations. Corroborative data from the X-ray co-crystal structures shed additional light on key residues in contact with the inhibitors and the stereochemical findings regarding the C7 configuration. Such studies on [5,6] penems by computational methods revealed a consistent stereochemical preference for C7 in SHV-1 and GC1 [99] relative to **BLI-489**. However, in the case of [5,5] penems [97], the predicted C7 configuration was opposite to that of [5,6] penems in SHV-1 but similar in GC1. The formation of 1,4-dihydrothiazepine was explained by the initial attack by the catalytic serine residue followed by β-lactam ring opening and subsequent conformational changes culminating in a *7-endo trig* cyclization to yield the 1,4-dihydrothiazepine. 6-Methylidene penems appear to be a unique class of inhibitors that share the same mechanism of inactivation by all classes of β-lactamases and hence understanding this mechanism is an important strategy for designing new inhibitors in this class.

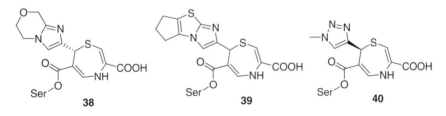

7. CONCLUSION

The search for new β-lactamase inhibitors continues to be a highly desirable endeavor owing to the clear medical need; however, it is steadily faced with an increase in the number of β-lactamases from all classes that cause resistance. The strategy for targeting several types of β-lactamases by a single chemical class offers future promise and is likely the path forward for the serine β-lactamases. Insights from X-ray crystallography and other spectroscopic methods such as Raman, ESI-MS and ITV-ICP-MS will prove useful in the future for elucidating mechanisms of inactivation and the design of new inhibitors.

REFERENCES

[1] A. M. Hujer, C. R. Bethel, K. M. Hujer and R. A. Bonomo, *Clin. Lab. Med.*, 2004, **24**, 343–361.
[2] J. Carlet, A. Ben Ali and A. Chalfine, *Curr. Opin. Infect. Dis.*, 2004, **17**, 309–316.

[3] CDC, *Am. J. Infect. Contr.*, 2004, **32**, 470–485.

[4] F. Perez, A. M. Hujer, K. M. Hujer, B. K. Deker, P. N. Rather and R. A. Bonomo, *Antimicrob. Agents Chemother.*, 2007, **51**, 3471–3484.

[5] P. Nordmann and L. Poirel, *Clin. Microbiol. Infect.*, 2002, **8**, 321–331.

[6] R. Zarrilli, R. Casillo, A. Di Popolo, M. F. Tripodi, M. Bagattini, S. Cuccurullo, V. Crivaro, E. Ragone, A. Mattei, N. Galdieri, M. Triassi and R. Utili, *Clin. Microbiol. Infect.*, 2007, **13**, 481–489.

[7] IDSA, Infectious Diseases Society of America, Alexandria, VA, 2004.

[8] G. H. Talbot, J. Bradley, J. E. Edwards, Jr., D. Gilbert, M. Scheld and J. G. Bartlett, *Clin. Infect. Dis.*, 2006, **42**, 657–668.

[9] D. L. Paterson, W. C. Ko, A. Von Gottberg, S. Mohapatra, J. M. Casellas, H. Goossens, L. Mulazimoglu, G. Trenholme, K. P. Klugman, R. A. Bonomo, L. B. Rice, M. M. Wagener, J. G. McCormack and V. L. Yu, *Clin. Infect. Dis.*, 2004, **39**, 31–37.

[10] M. J. Schwaber and Y. Carmeli, *J. Antimicrob. Chemother.*, 2007, **60**, 913–920.

[11] V. P. Sandanayaka and A. S. Prashad, *Curr. Med. Chem.*, 2002, **9**, 1145–1165.

[12] J. D. Buynak, *Biochem. Pharm.*, 2006, **7**, 930–940.

[13] L. A. Miller, K. Ratnam and D. J. Payne, *Curr. Opin. Pharm.*, 2001, **1**, 451–458.

[14] K. Bush, *Clin. Microbiol. Rev.*, 1988, **1**, 109–123.

[15] K. Bush, G. A. Jacoby and A. A. Medeiros, *Antimicrob. Agents Chemother.*, 1995, **39**, 1211–1233.

[16] P. A. Bradford, *Clin. Microbiol. Rev.*, 2001, **14**, 933–951.

[17] G. A. Jacoby and K. Bush, *J. Clin. Microbiol.*, 2005, **43**, 6220.

[18] A. A. Shah, F. Hasan, S. Ahmed and A. Hameed, *Res. Microbiol.*, 2004, **155**, 409–421.

[19] G. A. Jacoby, *N. Engl. J. Med.*, 2005, **352**, 380–389.

[20] M. H. Nicolas-Chanoine and V. P. Jarlier, *Clin. Microbiol. Infect.*, 2008, **14**(Suppl. 1), 111–116.

[21] J. Wiener, J. P. Quinn, P. A. Bradford, R. V. Goering, C. Nathan, K. Bush and R. A. Weinstein, *J. Am. Med. Assoc.*, 1999, **281**, 517–523.

[22] R. Canton, A. Novais, A. Valverde, E. Machado, L. Peixe, F. Baquero and T. M. Coque, *Clin. Microbiol. Infect.*, 2008, **14**(Suppl. 1), 144–153.

[23] J. Delmas, Y. Chen, F. Prati, F. Robin, B. K. Shoichet and R. Bonnet, *J. Mol. Biol.*, 2008, **375**, 192–201.

[24] L. Brasme, P. Nordmann, F. Fidel, M. F. Lartigue, O. Bajolet, L. Poirel, D. Forte, V. Vernet-Garnier, J. Madoux, J. C. Reveil, C. Alba-Sauviat, I. Baudinat, P. Bineau, C. Bouquigny-Saison, C. Eloy, C. Lafaurie, D. Simeon, J. P. Verquin, F. Noel, C. Strady and C. DeChamps, *J. Antimicrob. Chemother.*, 2007, **60**, 956–960.

[25] L. Pallecchi, A. Bartoloni, C. Fiorelli, A. Mantella, T. Di Maggio, H. Gamboa, E. Gotuzzo, G. Kronvall, F. Paradisi and G. M. Rossolini, *Antimicrob. Agents Chemother.*, 2007, **51**, 2720–2725.

[26] P. A. Bradford, Y. Yang, D. Sahm, I. Grope, D. Gardovska and G. Storch, *Antimicrob. Agents Chemother.*, 1998, **42**, , 1980–1984.

[27] L. Ma, Y. Ishii, M. Ishiguro, H. Matsuzawa and K. Yamaguchi, *Antimicrob. Agents Chemother.*, 1998, **42**, 1181–1186.

[28] T. Naas, L. Poirel and P. Nordmann, *Clin. Microbiol. Infect.*, 2008, **14**(Suppl. 1), 42–52.

[29] A. S. Ibuka, Y. Ishii, M. Galleni, M. Ishiguro, K. Yamaguchi, J.-M. Frére, H. Matsuzawa and H. Sakai, *Biochemistry*, 2003, **42**, 10634–10643.

[30] M. Ishiguro and S. Imajo, *J. Med. Chem.*, 1996, **39**, 2207–2218.

[31] Y. Chen, J. Delmas, J. Sirot, B. Shoichet and R. Bonnet, *J. Mol. Biol.*, 2005, **348**, 349–362.

[32] Y. Chen, R. Bonnet and B. K. Shoichet, *J. Am. Chem. Soc.*, 2007, **129**, 5378–5380.

[33] T. R. Walsh, M. A. Toleman, L. Poirel and P. Nordmann, *Clin. Microbiol. Rev.*, 2005, **18**, 306–325.

[34] B. A. Rasmussen, K. Bush, D. Keeney, Y. Yang, R. hare, C. O'Gara and A. A. Medeiros, *Antimicrob. Agents Chemother.*, 1996, **40**, 2080–2086.

[35] P. Nordmann, S. Mariotte, T. Nass, R. labia and M.-H. Nicolas, *Antimicrob. Agents Chemother.*, 1993, **37**, 939–946.

[36] Y. Yang, P.-C. Wu and D. M. Livermore, *Antimicrob. Agents Chemother.*, 1990, **34**, 755–758.

[37] J. Walther-Rasmussen and N. Hoiby, *J. Antimicrob. Chemother.*, 2007, **60**, 470–482.

[38] A. M. Queenan and K. Bush, *Clin. Microbiol. Rev.*, 2007, **20**, 440–458.

[39] H. Yigit, A. M. Queenan, G. J. Anderson, A. Domenech-Sanchez, J. W. Biddle, C. D. Steward, S. Alberti, K. Bush and F. C. Tenover, *Antimicrob. Agents Chemother.*, 2001, **45**, 1151–1161.

[40] P. A. Bradford, S. Bratu, C. Urban, M. Visalli, N. Mariano, D. L. Landman, J. J. Rahal, S. Brooks, S. Cebular and J. Quale, *Clin. Infect. Dis.*, 2004, **39**, 55–60.

[41] L. Poirel, G. F. Weldhagen, T. Naas, C. D. Champs, M. G. Dove and P. Nordmann, *Antimicrob. Agents Chemother.*, 2001, **45**, 2598–2603.

[42] M. Afzal-Shah, H. E. Villar and D. M. Livermore, *J. Antimicrob. Chemother.*, 1999, **43**, 127–131.

[43] H. Yigit, A. M. Queenan, J. K. Rasheed, J. W. Biddle, A. Domenech-Sanchez, S. Alberti, K. Bush and F. C. Tenover, *Antimicrob. Agents Chemother.*, 2003, **47**, 3881–3889.

[44] A. Hossain, M. J. Ferraro, R. M. Pino, R. B. Dew III, E. S. Moland, T. J. Lockhart, K. S. Thomson, R. V. Goering and N. D. Hanson, *Antimicrob. Agents Chemother.*, 2004, **48**, 4438–4440.

[45] N. Woodford, P. M. Tierno, Jr., K. Young, L. Tysall, M.-F. I. Palepou, E. Ward, R. E. Painer, D. F. Suber, D. Shungu, L. L. Silver, K. Inglima, J. Kornblum and D. M. Livermore, *Antimicrob. Agents Chemother.*, 2004, **48**, 4793–4799.

[46] S. Bratu, D. L. Landman, M. Alam, E. Tolentino and J. Quale, *Antimicrob. Agents Chemother.*, 2005, **49**, 776–778.

[47] V. Miriagou, L. S. Tzouvelekis, S. Rossiter, E. Tzelepi, F. J. Angulo and J. M. Whichard, *Antimicrob. Agents Chemother.*, 2003, **47**, 1297–1300.

[48] S. Bratu, D. L. Landman, R. Haag, R. Recco, A. Eramo, M. Alam and J. Quale, *Antimicrob. Agents Chemother.*, 2005, **165**, 1430–1435.

[49] W. Ke, C. R. Bethel, J. M. Thomson, R. A. Bonomo and F. van den Akker, *Biochemistry*, 2007, **46**, 5732–5740.

[50] E. Osano, Y. Arakawa, R. Wacharotayankun, M. Ohta, T. Horii, H. Ito, F. Yosimura and N. Kato, *Antimicrob. Agents Chemother.*, 1994, **38**, 71–78.

[51] S. Fukigai, J. Alba, S. Kimura, T. Iida, N. Nishikura, Y. Ishii and K. Yamaguchi, *Int. J. Antimicrob. Chemother.*, 2007, **29**, 306–310.

[52] S. Siemann, H. R. Badiei, V. Karanassios, T. Visamanatha and G. I. Dmitrienko, *Chem. Commun.*, 2006. 532–534.

[53] K. Lee, J. B. Lim, J. H. Yum, D. Yong, Y. Chong, J. M. Kim and D. M. Livermore, *Antimicrob. Agents Chemother.*, 2002, **46**, 1053–1058.

[54] M. Psichogiou, P. T. Tassios, A. Avlamis, I. Stefanou, C. Kosmidis, E. Platsouka, O. Paniara, A. Xanthaki, M. Toutouza, G. L. Daikos and L. S. Tzouvelekis, *J. Antimicrob. Chemother.*, 2008, **61**, 59–63.

[55] P. Giakkoupi, G. Petrikkos, L. S. Tzouvelekis, S. Tsonas, N. J. Legakis and A. C. Vatopoulos, *J. Clin. Microbiol.*, 2003, **41**, 822–825.

[56] T. A. Murphy, L. E. Catto, S. E. Halford, A. T. Hadfield, W. Minor, T. R. Walsh and J. Spencer, *J. Mol. Biol.*, 2006, **357**, 890–903.

[57] G. Garau, C. Bebrone, C. Anne, M. Galleni, J.-M. Frére and O. Dideberg, *J. Mol. Biol.*, 2005, **345**, 785–795.

[58] J. Crisp, R. Conners, J. D. Garrity, A. L. Carenbauer, M. W. Crowder and J. Spencer, *Biochemistry*, 2007, **46**, 10664–10674.

[59] J. Spencer, J. Read, R. B. Sessions, S. Howell, G. M. Blackburn and S. J. Gamblin, *J. Am. Chem. Soc.*, 2005, **127**, 14439–14444.

[60] L. I. Llarrull, S. M. Fabiane, J. M. Kowalski, B. Bennett, B. J. Sutton and A. J. Vila, *J. Biol. Chem.*, 2007, **282**, 18276–18285.

[61] P. A. Bradford, C. Urban, N. Mariano, S. J. Projan, J. J. Rahal and K. Bush, *Antimicrob. Agents Chemother.*, 1997, **41**, 563–569.

[62] R. C. da Silva Dias, A. A. Borges-Neto, G. I. D'Almeida Ferraiuoli, M. P. de-Oliveira, L. W. Riley and B. M. Moreira, *Diagn. Microbiol. Infect. Dis.*, 2008, **60**, 79–87.

[63] Y. J. Park, J. K. Yu, S. Lee, E. J. Oh and G. J. Woo, *J. Antimicrob. Chemother.*, 2007, **60**, 368–371.

[64] M. B. Zaidi, V. Leon, C. Canche, C. Perez, S. Zhao, S. K. Hubert, J. Abbott, K. Blickenstaff and P. F. McDermott., *J. Antimicrob. Chemother.*, 2007, **60**, 398–401.

[65] G. Bou, A. Oliver, M. Ljeda, C. Monzon and J. Martinez-Beltran, *Antimicrob. Agents Chemother.*, 2000, **44**, 2549–2553.

[66] J. D. Pitout, D. B. Gregson, D. L. Church, K. B. Laupland and G. Barnaud, *Emerg. Infect. Dis.*, 2007, **13**, 443–448.

[67] L. M. Deshpande, R. N. Jones, T. R. Fritsche, H. S. Sader, A. Bauernfeind, I. Schneider, R. Jungwirth, H. Sahly and U. Ullmann, *Int. J. Antimicrob. Chemother.*, 2006, **28**, 578–581.

[68] D. Girlich, T. Naas, S. Bellais, L. Poirel, A. Karim and P. Nordman, *Antimicrob. Agents Chemother.*, 2000, **44**, 1470–1478.

[69] K. Lee, D. Yong, Y. S. Choi, J. H. Yum, J. M. Kim, N. Woodford and D. M. Livermore, *Int. J. Antimicrob. Chemother.*, 2007, **29**, 201–206.

[70] P. S. Padayatti, M. S. Helfand, M. A. Totir, M. P. Carey, A. M. Hujer, P. R. Carey, R. A. Bonomo and F. van den Akker, *Biochemistry*, 2004, **43**, 843–848.

[71] P. S. Padayatti, M. S. Helfand, M. A. Totir, M. P. Carey, P. R. Carey, R. A. Bonomo and F. van den Akker, *J. Biol. Chem.*, 2005, **280**, 34900–34907.

[72] P. S. Padayatti, A. Sheri, M. A. Totir, M. S. Helfand, M. P. Carey, V. E. Anderson, P. R. Carey, C. R. Bethel, R. A. Bonomo, J. D. Buynak and F. van den Akker, *J. Am. Chem. Soc.*, 2006, **128**, 13235–13242.

[73] Y. Chen, B. Shoichet and R. Bonnet, *J. Am. Chem. Soc.*, 2005, **127**, 5423–5434.

[74] J. H. Toney and J. G. Moloughney, *Curr. Opin. Investig. Drugs*, 2004, **5**, 823–826.

[75] S. Siemann, D. P. Evanoff, L. Marrone, A. J. Clarke, T. Visamanatha and G. I. Dmitrienko, *Antimicrob. Agents Chemother.*, 2002, **46**, 2450–2457.

[76] H. Kurosaki, Y. Yamaguchi, T. Higashi, K. Soga, S. Matsueda, H. Yumoto, S. Misumi, Y. Yamata, Y. Arakawa and M. Goto, *Angew. Chem. Int. Ed. Engl.*, 2005, **44**, 3861–3864.

[77] L. Olsen, S. Jost, H.-W. Adolph, I. Pettersson, L. Menningsen and F. S. Jorgensen, *Bioorg. Med. Chem.*, 2006, **14**, 2627–2635.

[78] H. Kurosaki, Y. Yamaguchi, H. Yasuzawa, W. Jin, Y. Yamagata and Y. Arakawa, *ChemMedChem*, 2006, **1**, 969–972.

[79] L. E. Horsfall, G. Garau, B. M. R. Lienard, O. Dideberg, C. J. Schofield, J. M. Frére and M. Galleni, *Antimicrob. Agents Chemother.*, 2007, **51**, 2136–2142.

[80] T. Abe, K. Chikauchi, Y. Hiraiwa, J. Saito, M. Yamada, T. Watanabe, A. Morinaka, M. Kurazono, T. Kudo and M. Hikida, Poster # F1-329, in 47th ICAAC, 2007.

[81] Y. Osaki, A. Morinaka, T. Mikuniya, Y. Sanbongi, T. Ida, T. Yoshida and M. Yonezawa, Poster # F1-332, in 47th ICAAC, 2007.

[82] E. Santillana, A. Beceiro, G. Bou and A. Romero, *Proc. Natl. Acad. Sci. USA*, 2007, **104**, 5354–5359.

[83] V. P. Sandanayaka, A. S. Prashad, Y. Yang, R. T. Williamson, Y. I. Lin and T. S. Mansour, *J. Med. Chem.*, 2003, **46**, 2569–2571.

[84] J. D. Buynak, H. Chen, L. Vogeti, V. R. Ghadachanda, C. A. Buchanan, T. Palzkill, R. W. Shaw, J. Spencer and T. R. Walsh, *Bioorg. Med. Chem. Lett.*, 2004, **14**, 1299–1304.

[85] I. Plantan, L. Selic, T. Mesar, P. S. Anderluh, M. Oblak, A. Prezelj, L. Hesse, M. Andrejasic, M. Vilar, D. Turk, A. Kocijan, T. Prevec, G. Vilfan, D. Kocjan, A. Copar, U. Releb and T. Solmajer, *J. Med. Chem.*, 2007, **50**, 4113–4121.

[86] P. N. Wyrembak, K. Babaoglu, R. B. Pelto, B. K. Shoichet and R. F. Pratt, *J. Am. Chem. Soc.*, 2007, **129**, 9548–9549.

[87] S. A. Adediran, D. Cabaret, R. R. Flavell, J. A. Sammons, M. Wakselman and R. F. Pratt, *Bioorg. Med. Chem.*, 2006, **14**, 7023–7033.

[88] A. Bonnefoy, C. Dupuis-Hamelin, V. Steier, C. Delachaume, C. Seys, T. Stachyra, M. Fairley, M. Guitton and M. Lampilas, *J. Antimicrob. Chemother.*, 2004, **54**, 410–417.

[89] S. Mustaq, M. Warner, C. Miossec, N. Woodford and D. M. Livermore, Paper presented at the 47th ICAAC, Chicago, IL, 2007.

[90] T. Stachrya, M.-C. Pechereu, S. Petrella, N. Ziental-Gelus, W. Sougakoff, M. Claudon, M. Black, P. Levasseur, A.-M. Girard and C. Miossec, Paper presented at the 47th ICAAC, Chicago, IL, 2007.

[91] S. Majumdar, S. A. Adediran, M. Nukaga and R. F. Pratt, *Biochemistry*, 2005, **44**, 16121–16129.

[92] S. A. Adediran, M. Nukaga, S. Baurin, J.-M. Frére and R. F. Pratt, *Antimicrob. Agents Chemother.*, 2005, **49**, 4410–4412.

[93] I. Bennett, N. J. P. Broom, G. Bruton, S. Calvert, B. P. Clarke, K. Coleman, R. Edmondson, P. Edwards, D. Jones, N. F. Osborne and G. Walker, *J. Antibiotics*, 1991, **44**, 331–337.

[94] N. J. P. Broom, T. H. Farmer, N. F. Osborne and J. W. Tyler, *J. Chem. Soc. Chem. Commun.*, 1992, 1663–1664.

[95] A. M. Venkatesan, A. Agarwal, T. Abe, H. Ushirogochi, I. Yamamura, T. Kumagai, P. J. Petersen, W. J. Weiss, E. Lenoy, Y. Yang, D. M. Shlaes and T. S. Mansour, *Bioorg. Med. Chem.*, 2004, **12**, 5807–5817.

[96] T. Abe, C. Sato, H. Ushirogochi, K. Sato, T. Takasaki, T. Isoda, A. Mihira, I. Yamamura, K. Hayashi, T. Kumagai, S. Tamai, M. Shiro, A. M. Venkatesan and T. S. Mansour, *J. Org. Chem.*, 2004, **69**, 5850–5860.

[97] T. S. Mansour, A. Agarwal, A. M. Venkatesan, T. Abe, A. Mihira, T. Takasaki, K. Sato, H. Ushirogochi, I. Yamamura, T. Isoda, Z. Li, Y. Yang and T. Kumagai, *ChemMedChem*, 2007, **2**, 1713–1716.

[98] W. J. Weiss, P. J. Petersen, T. M. Murphy, L. Tardio, Y. Yang, P. A. Bradford, A. M. Venkatesan, T. Abe, T. Isoda, A. Mihira, H. Ushirogochi, T. Takasake, S. Projan, J. O'Connell and T. S. Mansour, *Antimicrob. Agents Chemother.*, 2004, **48**, 4589–4596.

[99] A. M. Venkatesan, A. Agarwal, T. Abe, H. Ushirogochi, I. Yamamura, A. Mihira, T. Takasaki, O. Santos, Y. Gu, F. W. Sum, Z. Li, G. Francisco, Y. I. Lin, P. J. Petersen, Y. Yang, T. Kumagai, W. J. Weiss, D. M. Shlaes, J. R. Knox and T. S. Mansour, *J. Med. Chem.*, 2006, **49**, 4623–4637.

[100] C. R. Bethel, A. M. Distler, M. W. Ruszczycky, M. P. Carey, A. M. Hujer, M. Taracila, M. S. Helfand, J. M. Thomson, M. Kalp, V. E. Anderson, D. A. Leonard, K. M. Hujer, T. Abe, A. M. Venkatesan, T. S. Mansour and R. A. Bonomo, *Antimicrob. Agents Chemother.*, 2008, **52**, 267. in-print.

[101] T. Sun, M. Nukaga, K. Mayama, E. H. Braswell and J. R. Knox, *Protein Sci.*, 2003, **12**, 82–91.

[102] P. J. Petersen, C. H. Jones and P. A. Bradford, Poster FI-322 presented at the 47th ICAAC, Chicago, IL, 2007.

[103] P. J. Petersen, C. H. Jones and P. A. Bradford, Poster FI-323 presented at the 47th ICAAC, Chicago, IL, 2007.

[104] M. Nukaga, T. Abe, A. M. Venkatesan, T. S. Mansour, R. A. Bonomo and J. R. Knox, *Biochemistry*, 2003, **42**, 13152–13159.

[105] A. M. Venkatesan, Y. Gu, O. Santos, T. Abe, A. Agarwal, Y. Yang, P. J. Petersen, W. J. Weiss, T. S. Mansour, M. Nukaga, A. M. Hujer, R. A. Bonomo and J. R. Knox, *J. Med. Chem.*, 2004, **47**, 6556–6568.

[106] A. M. Venkatesan, A. Agarwal, T. Abe, H. Ushirogochi, A. Mihira, T. Takasaki, O. Santos, Z. Li, G. Francisco, Y. I. Lin, P. J. Petersen, Y. yang, W. J. Weiss, D. M. Shlaes and T. S. Mansour, *Bioorg. Med. Chem.*, 2008, **16**, 1890–1902.

[107] K. Tabei, X. Feng, A. M. Venkatesan, T. Abe, U. Hideki, T. S. Mansour and M. M. Siegel, *J. Med. Chem.*, 2004, **47**, 3674–3688.

[108] C. Michaux, P. Charlier, J.-M. Frére and J. Wouters, *J. Am. Chem. Soc.*, 2005, **127**, 227–231.

CHAPTER **16**

Recent Development in the Treatment of *Clostridium difficile* Associated Disease (CDAD)

Xicheng Sun, Joseph Guiles and **Nebojsa Janjic**

1. INTRODUCTION

Clostridium difficile has emerged in recent years as a significant health concern among elderly patients in hospitals and other healthcare facilities. This chapter will focus on reviewing the existing treatments for the disorders caused by this bacterial pathogen and highlight the current efforts to develop new treatments. *C. difficile* is a Gram-positive spore-forming bacillus first described more than 70 years ago [1] and was later found to be the leading cause of

Replidyne, Inc., 1450 Infinite Dr., Louisville, CO 80027, USA

Annual Reports in Medicinal Chemistry, Volume 43
ISSN 0065-7743, DOI 10.1016/S0065-7743(08)00016-X

antibiotic-associated pseudomembrane colitis and its milder form, *C. difficile* associated disease (CDAD) [2,3]. This disease is induced by the disruption of the colonic flora through the administration of antibiotics such as clindamycin, ampicillin, or cephalosporins [4]. When antibiotics are administrated to destroy pathogens that are causing disease they can also disrupt the protective microflora present in the intestinal tract. One function of the normal intestinal microflora is to inhibit colonization and subsequent infection by opportunistic pathogens. When this protective barrier is disrupted, the host may become susceptible to pathogens like *C. difficile* [5,6]. After exposure to antibiotics, especially broad-spectrum agents that disrupt a wide variety of anaerobic microflora, there is a long period of susceptibility to colonization; it may take as long as three months for the normal microflora to fully recover after antibiotic exposure. Another risk factor is exposure to *C. difficile*, which typically occurs in hospitals [7–9] via asymptomatic carriers and symptomatic hospital patients who shed *C. difficile* cells and spores into the hospital environment. Outbreaks and new sporadic cases occur through patient-to-patient contact, usually by hand transmission, or indirectly through exposure to a contaminated environment. *C. difficile* has the ability to form spores that are resistant to antibiotics, antiseptic agents, and extreme environmental conditions. As a result, the spores of *C. difficile* can remain in the hospital environment for months, providing a reservoir for new infections. The capability of the host's immune system to produce protective antibodies against the toxins of *C. difficile* is also a risk factor for the manifestation of the disease.

Recently, there has been renewed interest in *C. difficile*, reflecting an increased incidence of a disease form that is more frequent, more severe, and more refractory to existing therapies [10]. It has been estimated that CDAD affects more than 300,000 patients per year in the United States [11–13]. Some studies have suggested that CDAD increases the length of hospital stay by as long as two weeks for the average patient [14] and it has been estimated that the hospital costs attributed to CDAD in the United States exceed $4,000 per case [15].

C. difficile has become a substantial financial burden, in healthcare settings and this problem has been compounded with the alarming emergence of a more virulent strain of *C. difficile* (BI/NAP1). BI/NAP1 is capable of producing 16–23 times the amount of damaging toxins (TcdA and TcdB) compared to other toxin-producing strains as well as a binary toxin (CDT). In addition, BI/NAP1 is more resistant to a fluoroquinolone and exhibits increased sporulation capacity [10,16–18].

2. ROLE OF TOXINS IN CDAD

C. difficile colonizes the large bowel of patients undergoing antibiotic treatment and produces two toxins, toxin A and toxin B [19]. Clostridial toxin A and B are glucosyltransferases that target the Ras superfamily of GTPases. Toxin A is an enterotoxin responsible for diarrhea and causes the release of inflammatory cytokines, recruitment of polymorphonuclear cells, and stimulation of prostaglandin synthesis. Toxin B, a cytotoxin, is responsible for cytopathic changes to enterocytes. Another toxin, binary toxin, has also been described, but its role in disease is not yet fully understood [20,21].

3. REVIEW ON CURRENT TREATMENTS

For patients with CDAD, treatment recommendations include the discontinua-tion of antibiotic regimens. Studies show that 15–23% of the patients with CDAD had spontaneous resolution of symptoms within 2–3 days of stopping the antibiotics that were associated with the disease [8,21,22]. However, the severity of the patient's condition from the original infection and the risk to the patient of terminating antibiotic therapy is difficult to assess and therefore it is often difficult to apply this approach in practice.

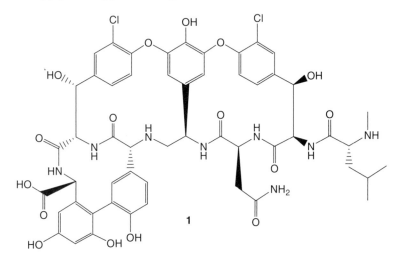

1

Vancomycin **1** is a broad-spectrum antibiotic with MIC_{90} of 0.75–2.0 μg/ml [25–31] against *C. difficile*. Vancomycin is a highly water soluble agent with very low oral bioavailability and is administered orally to treat CDAD. Over the past four decades this approach has resulted in a resolution rate of approximately 90% [22,32]. Vancomycin is currently the only FDA approved therapy to treat CDAD. The recommended dose of vancomycin is 125 mg/kg (PO, QID).

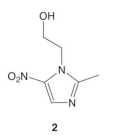

2

Metronidazole **2** is not approved for the treatment of CDAD, but it is the first-line therapy, ahead of vancomycin, for reasons of cost as well as concern about the development of vancomycin resistance due to overuse [23,24]. The cost differential between similar courses of metronidazole and oral vancomycin can be more than tenfold.

Both metronidazole and oral vancomycin have very potent *in vivo* activity against *C. difficile*. For patients with severe *C. difficile* colitis, oral vancomycin or intravenous metronidazole is used. However, recent reports suggest increasing treatment failure rates with metronidazole [33] and vancomycin is generally preferred for patients who are seriously ill or who fail to respond to metronidazole [34–36].

4. AGENTS IN DEVELOPMENT

4.1 Rifaximin

Rifaximin **3** is a semisynthetic analog of the rifamycin group of antibiotics. The mechanism of action of rifaximin is inhibition of the initiation of RNAs synthesis by binding to the β-subunit of the RNA polymerase [37]. The Food and Drug administration (FDA) approved this agent for the treatment of travelers' diarrhea caused by non-invasive strains of *Escherichia coli* in patients who are 12 years of age or older.

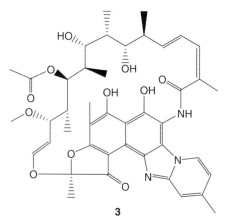

3

Rifaximin is one of the most potent ($MIC_{90} = 0.015\,\mu g/ml$, 110 clinical isolates) agents against *C. difficile* [30,38]. The addition of a benzimidazole ring makes rifaximin essentially non-absorbed with a bioavailability of $<0.4\%$ [39].

An open-label study indicated that rifaximin was found to be effective in 9 out of 10 patients suffering from CDAD, while vancomycin was successful in all 10 patients who received it [40]. In other cases researchers found that rifaximin was successfully used for CDAD when other agents failed. Overall, rifaximin appears to be a valid alternative for the treatment and management of CDAD, although further, larger studies are needed to clearly define its role.

4.2 Ramoplanin

Ramoplanin **4** is a lipoglycopeptide antibiotic produced from an *Actinoplanes* strain and is bactericidal for a wide variety of Gram-positive bacteria including *streptococci, enterococci,* and methicillin-resistant *staphylococci,* and there is no cross resistance with other antibiotics in clinical use [41,42]. Like vancomycin, ramoplanin inhibits bacterial cell wall biosynthesis. However, unlike vancomycin, which binds to the terminal dipeptide (D-Ala-D-Ala) component of Lipid II (a substrate for bacterial cell wall biosynthesis), ramoplanin directly inhibits the enzyme, transglycosylase, which is responsible for generating glycosidic linkages within the cell wall.

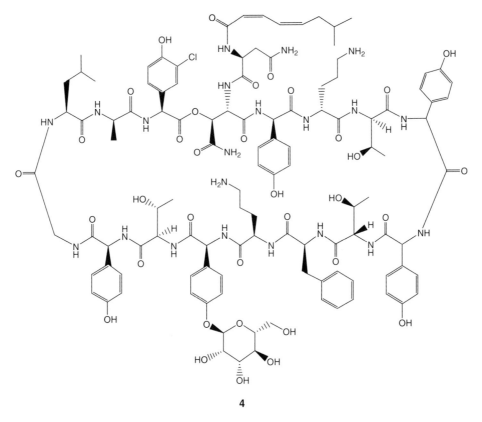

4

Currently, ramoplanin is an investigational new drug being studied for the treatment of CDAD. The primary endpoint was a response rate of 71% with ramoplanin at 400 mg versus 78% with vancomycin at the test-of-cure visit. An end-of-therapy response rate was 85.2% with ramoplanin 400 mg and 85.7% with vancomycin. Treatment failures were seen in one patient receiving ramoplanin and three patients receiving vancomycin. A dose–response relationship with ramoplanin was documented, with a higher response rate seen with 400 mg [43].

4.3 Difimicin (Opt-80)

Difimicin **5**, also known as tiacumicin B, is an 18-membered macrolide antibiotic that was discovered in the fermentation broth of strain AB718C-41. Study on the biological properties of tiacumicin B indicated that this compound has moderate activity against pathogenic strains of *Staphylococcus aureus*, *Streptococcus pyogenes*, and *Enterococcus faecium* and demonstrated that it has limited, but potent, activity against several anaerobic bacteria [44]. The mechanism of action of this class of natural products is inhibition of RNA synthesis [45]. Further studies indicated tiacumicin B has MICs against 15 strains of *C. difficile* of 0.12–0.25 µg/ml [46]. The Gram-positive only spectrum, potent activity against *C. difficile in vitro* and *in vivo* [47,48], and low or no systemic absorption has led to the current clinical development of this compound for the treatment of CDAD. No clinical data have been published yet.

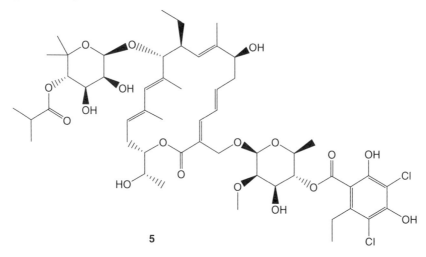

5

4.4 Nitazoxanide

Nitazoxanide **6** is an FDA approved drug that is marketed in the United States and has been widely used throughout the world to treat parasitic diseases of the gastrointestinal tract; several million children have been treated with this drug during the past decade. The drug acts by interfering with anaerobic metabolic pathways [49,50].

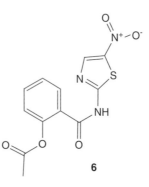

6

Nitazoxanide has been shown to have excellent *in vitro* and *in vivo* activity against *C. difficile* [51,52]. After oral administration to humans, two-thirds of the drug is excreted in the feces [53]. The properties of nitazoxanide, together with increasing documentation of treatment failure with metronidazole and ongoing reluctance to use oral vancomycin, motivated a controlled, double-blind clinical trial undertaken comparing nitazoxanide with metronidazole in treating *C. difficile* colitis. The study demonstrated the non-inferiority of nitazoxanide. The findings are encouraging enough to warrant further examination in larger studies, especially in light of the possibly of a higher success rate with longer-duration nitazoxanide therapy (i.e., with 10 days versus 7 days of therapy). A larger study is now being designed to compare vancomycin and nitazoxanide therapy, each given for 10–14 days, for treatment of this disease [54].

4.5 Rifalazil

Rifalazil **7**, also known as KRM-1648 or benzoxazinorifamycin, is a semisynthetic rifamycin [55]. Rifalazil possesses significant antimicrobial activity against a wide range of Gram-positive and Gram-negative bacteria, including mycobacteria [56], *chlamydiae* [57], and *Helicobacter pylori* [58]. *In vivo* animal studies indicate that rifalazil possesses significant activity against tuberculosis-causing bacteria [59].

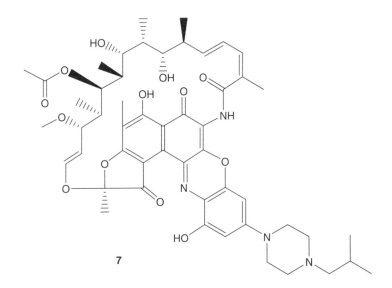

7

A recently published study indicated that rifalazil is highly active against *C. difficile* with MIC_{90} of 0.04 µg/ml (37 clinical isolates) and was successful when administered prophylactically, as well as therapeutically, in *C. difficile* infected hamsters. In the hamster model, rifalazil, displayed no reappearance of *C. difficile* infection following its discontinuation. This is in contrast to vancomycin, which, upon discontinuation, displayed high levels of recurrence. This indicates that rifalazil may be superior to vancomycin as a first line of treatment for CDAD [60].

4.6 Human monoclonal antibody

CDA-1 (MDX-066) and MDX-1388 are novel, fully human antibodies that are in development to target and neutralize the effects of Toxin A and Toxin B, respectively. CDA1 and MDX-1388 were tested in the well-established hamster model of *C. difficile* associated disease. CDA1 alone resulted in a statistically significant reduction of mortality in hamsters; however, the combination treatment offered enhanced protection. Compared to controls, combination therapy reduced mortality from 100% to 45% ($P<0.0001$) in the primary disease hamster model and from 78% to 32% ($P<0.0001$) in the less stringent relapse model [61]. CDA1 and MDX1388 have been chosen for testing in clinic for treatment of CDAD or prevention of relapses.

4.7 *C. difficile* toxoid vaccine

Vaccines have also been studied as a method of prophylaxis for CDAD. A parenteral *C. difficile* formalin-inactivated toxoid A and B vaccine has been shown to be highly immunogenic in clinical studies in 30 healthy adults when given as 4 intramuscular inoculations on days 1, 8, 30, and 60 [62]. The vaccine elicited fourfold increase in neutralizing antibodies to toxins A and B in all but two of the subjects, and serum antitoxin A IgA and IgG antibody responses were seen in all but one subject.

A pilot study on the *C. difficile* toxoid vaccine in recurrent CDAD was conducted [63]. The study concluded that a *C. difficile* toxoid vaccine induced immune responses to toxins A and B in patients with CDAD and was associated with resolution of recurrent diarrhea. The results of this study support the feasibility of active vaccination against *C. difficile* and its toxins in high-risk individuals. This result was later validated in a larger, randomized, double-blind, and placebo-controlled trial.

4.8 Probiotics

Probiotics are living microorganisms that affect the host in a beneficial manner by modulating mucosal and systemic immunity and improving nutritional and microbial balance in the intestinal tract. There has also been growing interest in probiotics that use non-pathogenic organisms to repopulate the colonic microflora, and thus, presumably, restrict the growth of toxigenic *C. difficile*. Agents that have been studied include a non-toxigenic strain of *C. difficile* [64], *Saccharomyces boulardii*, and *Lactobacillus* [65,66]. A double-blinded, randomized placebo-controlled efficacy clinical trial of *L. acidophilus* and *L. casei* is underway.

4.9 Tolevamer

Tolevamer is a soluble, high-molecular weight, anionic polymer (1,400 kDa) that non-covalently binds *C. difficile* toxins A and B [67,68]. Syrian hamsters infected with *C. difficile* were durably cured of infection and protected from recurrence by toxin-binding therapy with tolevamer [67]. It is hypothesized that tolevamer's unique mechanism of action, toxin neutralization rather than antibacterial activity, should resolve CDAD, and by allowing restoration of the normal microflora in the absence of antibiotic suppression, should result in a lower recurrence rate. A multicenter, randomized, double-blind, active-controlled study compared doses of 3 g per day and 6 g per day regimens of tolevamer with vancomycin for the treatment of CDAD. In this study, it was concluded that the 6 g per day regimen of tolevamer was no less effective than vancomycin administered at 500 mg per day with respect to time to resolve diarrhea and was also associated with a strong trend towards a lower recurrence rate. The potential for reducing antibiotic resistance, improving primary outcomes, and reducing recurrence merits further clinical development of tolevamer as a non-antibiotic treatment for CDAD [69].

5. CONCLUSIONS

Amid increased incidence of CDAD in many healthcare facilities, including those cases caused by hypervirulent *strains*, there are currently few clinical options for the treatment of this serious intestinal infection. In many clinical settings, the first-line therapy is still oral metronidazole. Vancomycin is generally reserved for patients refractory to metronidazole. A major deficiency of current treatments is the relatively high rate of recurrence, which occurs in up to 25% of metronidazole, or vancomycin treated patients. These high levels of recurrence are reflective of an important unmet medical need for new antibiotics to be developed. The new agents described in this review are characterized into classical antibiotic or more novel non-antibiotic approaches. The new antibiotics in development are highly potent against *C. difficile*. These agents are generally of low oral bioavailability, thus limiting systemic exposure and concentrating drug in the gut. However, they are in general broad-spectrum agents that also attack beneficial intestinal flora. Non-antibiotic therapies are showing significant promise for the treatment of CDAD. These include agents that inactivate the clostridial toxins, agents that stimulate a targeted immune response against *C. difficile*, or probiotic treatments that restore the protective effect of gut microflora. Such non-antibiotic therapies represent independent options for the treatment of CDAD that could be used either alone or in combination with antibiotics.

REFERENCES

[1] I. C. Hall and E. O'Toole, *Am. J. Dis. Child*, 1935, **49**, 390.
[2] J. G. Bartlett, T. Chang, N. S. Taylor and A. B. Onderdonk, *Rev. Infect. Dis.*, 1979, **1**, 370.
[3] L. Kyne, R. J. Farrell and C. P. Kelly, *Gastroenterol. Clin. N. Am.*, 2001, **30**, 753.
[4] J. G. Bartlett, *Johns Hopkins Med. J.*, 1981, **149**, 6.
[5] L. V. McFarland, *Microb. Ecol. Health Dis.*, 2000, **12**, 193.

[6] C. J. Donskey, *Clin. Infect. Dis.*, 2004, **39**, 219.
[7] L. V. McFarland, M. E. Mulligan, R. Y. Y. Kwok and W. E. Stamm, *N. Engl. J. Med.*, 1989, **320**, 204.
[8] M. M. Olson, C. J. Shanholtzer, J. T. Lee, Jr. and D. N. Gerding, *Infect. Control Hosp. Epidemiol.*, 1994, **15**, 371.
[9] F. W. Shek, B. S. F. Stacey, J. Rendell, M. D. Hellier and P. J. V. Hanson, *J. Hosp. Infect.*, 2000, **45**, 235.
[10] L. C. McDonald, G. E. Killgore and A. Thompson, et al., *N. Engl. J. Med.*, 2005, **353**, 2433.
[11] S. Johnson, C. R. Clabots, F. V. Linn, M. M. Olson, L. R. Peterson and D. N. Gerding, *Lancet*, 1990, **336**, 97.
[12] L. V. McFarland, M. E. Mulligan, R. Y. Kwok and W. Stamm, *N. Engl. J. Med.*, 1989, **320**, 204.
[13] T. V. Riley, G. L. O'Neill, R. A. Bowman and C. L. Golledge, *Epidemiol. Infect.*, 1994, **113**, 13.
[14] T. V. Riley, J. P. Codde and I. L. Rouse, *Lancet*, 1995, **345**, 455.
[15] L. Kyne, M. B. Hamel, R. Polavaram and C. P. Kelly, *Clin. Infect. Dis.*, 2002, **34**, 346.
[16] R. C. Owens, *Pharmacotherapy*, 2006, **26**, 299.
[17] N. B. O'Mara, *Pharmacist Lett.*, 2006, **22**, 2201.
[18] V. G. Loo, L. Poirier and M. A. Miller, et al., *N. Engl. J. Med.*, 2005, **353**, 2442.
[19] K. J. Ryan and C. G. Ray (eds), *Sherris Medical Microbiology*, 4th edition, McGraw Hill, 2004, p. 322.
[20] H. Barth, K. Aktories, M. Popoff and B. Stiles, *Microbiol. Mol. Biol. Rev.*, 2004, **68**, 373.
[21] D. G. Teasley, D. N. Gerding and M. M. Olson, et al., *Lancet*, 1983, **2**, 1043.
[22] J. G. Bartlett, *Rev. Infect. Dis.*, 1984, **6**(Suppl. 1), S235.
[23] L. C. McDonald, G. E. Killgore and A. Thompson, et al., *Am. J. Health Syst. Pharm.*, 1998, **55**, 1407.
[24] Recommendations for preventing the spread of vancomycin resistance: Recommendations of the Hospital Infection Control Practices Advisory Committee (HICPAC)., *MMWR Morb. Mortal. Wkly. Rep.*, 1995, **44**(RR-12), 1.
[25] J. Dzink and J. G. Bartlett, *Antimicrob. Agents Chemother.*, 1980, **17**, 695.
[26] F. Biavasco, E. Manso and P. E. Varaldo, *Antimicrob. Agents Chemother.*, 1991, **35**, 195.
[27] S. S. Wong, P. C. Woo, W. K. Luk and K. Y. Yuen, *Microbiol. Infect. Dis.*, 1999, **34**, 1.
[28] S. H. Cheng, F. Y. Chu, S. H. Lo and J. J. Lu, *J. Microbiol. Immunol. Infect.*, 1999, **32**, 116.
[29] F. Barbut, D. Decre and B. Burghoffer, et al., *Antimicrob. Agents Chemother.*, 1999, **43**, 2607.
[30] A. Marchese, A. Salerno, A. Pesce, E. A. Debbia and G. C. Schito, *Chemotherapy*, 2000, **46**, 253.
[31] W. Y. Jamal, E. M. Mokaddas, T. L. Verghese and V. O. Rotimi, *Int. J. Antimicrob. Agents*, 2002, **20**, 270.
[32] J. F. Wallace, R. H. Smith and R. G. Petersdorf, *N. Engl. J. Med.*, 1965, **272**, 1014.
[33] D. M. Musher, S. Aslam and N. Logan, et al., *Clin. Infect. Dis.*, 2005, **40**, 1586.
[34] D. N. Gerding, *Clin. Infect. Dis.*, 2005, **40**, 1598.
[35] J. Pepin, S. Routhier, S. Gagnon and I. Brazeau, *Clin. Infect. Dis.*, 2006, **42**, 758.
[36] R. Fekety and A. B. Shah, *JAMA*, 1993, **269**, 71.
[37] L. Gerard, K. W. Garey and H. L. Dupont, *Expert Rev. Anti. Infect. Ther.*, 2005, **3**, 201.
[38] D. N. Gerding, S. Johnson, J. R. Osmolski, S. P. Sambol and D. W. Hecht, *Abstracts of the 45th Annual Interscience Conference on Antimicrobial Agents and Chemotherapy*, Washington, DC, American Society for Microbiology, 2005, p. 79, abstract E-1439.
[39] C. Scarpignato and I. Pelosini, *Chemotherapy*, 2005, **51**(Suppl. 1), 36.
[40] M. Boero, E. Berti and A. Morgando, et al., *Microbiol. Med.*, 1990, **5**, 74.
[41] B. Cavalleri, H. Pagnn, G. Volpe, E. Selva and F. Parenti, *J. Antibiot.*, 1984, **37**, 309.
[42] R. Pallanza, M. Berti, R. Scotti, E. Randisi and V. Arioli, *J. Antibiot.*, 1984, **37**, 318.
[43] D. K. Farver, D. D. Hedge and S. C. Lee, *Ann. Pharmacother.*, 2005, **39**, 863.
[44] R. J. Theriault, J. P. Karwowski, M. Jackson, R. L. Girolami, G. N. Sunga, C. M. Vojtko and L. J. Coen, *J. Antibiot.*, 1987, **40**, 567.
[45] S. Sergio, G. Pirali, R. White and F. Parenti, *J. Antibiot.*, 1975, **28**, 543.
[46] R. N. Swanson, D. J. Hardy, N. L. Shipkowitz, C. W. Hanson, N. C. Ramer, P. B. Fernandes and J. J. Clement, *Antimicrob. Agents Chemother.*, 1991, **35**, 1108.
[47] S. M. Finegold, D. Molitoris, M. L. Vaisanen, Y. Song, C. Liu and M. Bolanos, *Antimicrob. Agents Chemother.*, 2004, **48**, 4898.
[48] G. Ackermann, B. Loffler, D. Adler and A. C. Rodloff, *Antimicrob. Agents Chemother.*, 2004, **48**, 2280.

[49] C. A. White, Jr., *Expert Rev. Anti. Infect. Ther.*, 2004, **2**, 43.

[50] L. M. Fox and L. D. Saravolatz, *Clin. Infect. Dis.*, 2005, **40**, 1173.

[51] L. Dubreuil, I. Houcke, Y. Mouton and J. F. Rossignol, *Antimicrob. Agents Chemother.*, 1996, **40**, 2266.

[52] C. S. Mcvay and R. D. Rolfe, *Antimicrob. Agents Chemother.*, 2000, **44**, 2254.

[53] J. Broekhuysen, A. Stockis, R. L. Lins, J. De Graeve and J. F. Rossignol, *Int. J. Clin. Pharmacol. Ther.*, 2000, **38**, 387.

[54] D. M. Musher, N. Logan, R. J. Hamill, H. L. DuPont, A. Lentnek, A. Gupta and J. Rossignol, *Clin. Infect. Dis.*, 2006, **43**, 421.

[55] H. Saito, H. Tomioka, K. Sato, M. Emori, T. Yamane, K. Yamashita, K. Hosoe and T. R. Hidaka, *Antimicrob. Agents Chemother.*, 1991, **35**, 542.

[56] J. Luna-Herrera, M. V. Reddy and P. R. Gangadharam, *Antimicrob. Agents Chemother.*, 1995, **39**, 440.

[57] D. M. Rothstein, A. D. Hartman, M. H. Cynamon and B. I. Eisenstein, *Exp. Opin. Investig. Drugs*, 2003, **12**, 255.

[58] K. Akada, M. Shirai, K. Fujii, K. Okita and T. Nakazawa, *Antimicrob. Agents Chemother.*, 1999, **43**, 1072.

[59] T. Hirata, H. Saito, H. Tomioka, K. Sato, J. Jidoi, K. Hosoe and T. Hidaka, *Antimicrob. Agents Chemother.*, 1995, **39**, 2295.

[60] P. M. Anton, M. O'Brien, E. Kokkotou, B. Eisenstein, A. Michaelis, D. Rothstein, S. Paraschos, C. P. Kelly and C. Pothoulakis, *Antimicrob. Agents Chemother.*, 2004, **48**, 3975.

[61] G. J. Babcock, T. J. Broering, H. J. Hernandez, R. B. Mandell, K. Donahue, N. Boatright, A. M. Stack, I. Lowy, R. Graziano, D. Molrine, D. M. Ambrosino and W. D. Thomas, Jr., *Infect. Immun.*, 2006, **74**, 6339.

[62] K. L. Kotloff, S. S. Wasserman and G. A. Losonsky, et al., *Infect. Immun.*, 2001, **69**, 988.

[63] S. Sougioultzis, L. Kyne, D. Drudy, S. Keates, S. Maroo, C. Pothoulakis, P. J. Giannasca, C. K. Lee, M. Warny, T. P. Monath and C. P. Kelly, *Gastroenterology*, 2005, **128**, 764.

[64] D. Seal, S. P. Borriello, F. Barclay, A. Welch, M. Piper and M. Bonnycastle, *Eur. J. Clin. Microbiol.*, 1987, **6**, 51.

[65] M. Pochapin, *Am. J. Gastroenterol.*, 2000, **95**(Suppl. 1), S11.

[66] M. Wullt and M. L. Hagslatt, *Scand. J. Infect. Dis.*, 2003, **35**, 65.

[67] C. B. Kurtz, E. P. Cannon and A. Brezzani, et al., *Antimicrob. Agents Chemother.*, 2001, **45**, 2340.

[68] W. Braunlin, Q. Xu and P. Hook, et al., *Biophys. J.*, 2004, **87**, 534.

[69] T. J. Louie, J. Peppe and C. K. Watt, et al., *Clin. Infect. Dis.*, 2006, **43**, 411.

CHAPTER 17

Recent Advances in the Discovery of Hybrid Antibacterial Agents

Michael R. Barbachyn

1. INTRODUCTION

1.1 Resistance situation and medical need

Bacterial resistance to currently available therapeutic antibacterial agents is increasing in prevalence [1]. Multidrug-resistant (MDR) Gram-positive bacteria of particular concern include methicillin-resistant *Staphylococcus aureus* (MRSA), glycopeptide-intermediate *S. aureus* (GISA), vancomycin-resistant *S. aureus* (VRSA), vancomycin-resistant enterococci (VRE) and penicillin- and macrolide-resistant *Streptococcus pneumoniae* [2]. In addition, Gram-negative bacilli resistant to extended-spectrum β-lactam antibiotics (including carbapenems), fluoroquinolones, aminoglycosides and other antibacterial agents have become

AstraZeneca R & D Boston, Infection Discovery, 35 Gatehouse Drive, Waltham, MA 02451, USA

Annual Reports in Medicinal Chemistry, Volume 43
ISSN 0065-7743, DOI 10.1016/S0065-7743(08)00017-1

increasingly prevalent in the healthcare setting. Particularly problematic Gram-negative organisms include extended-spectrum β-lactamase-expressing strains of *Klebsiella pneumoniae*, as well as the non-fermenters *Pseudomonas aeruginosa*, *Acinetobacter baumannii* and *Stenotrophomonas maltophilia* [3]. Management of nosocomial infections associated with all of the above organisms presents a difficult therapeutic challenge for critical care physicians. Indeed, infections caused by these bacteria are often associated with considerable morbidity and mortality. Because of this, there is an urgent need to identify new therapeutic agents with the ability to target these MDR bacterial strains.

1.2 Dual-action/targeting concept as a solution to resistance development

There are a variety of potential solutions available to address the growing bacterial resistance problem [4]. One approach that has been used with some clinical success has been the combination of two antibacterial agents employing complementary mechanisms of action. In this combination approach, one agent can target any bacterial strains that possess intrinsic resistance or develop resistance to the partner agent and vice versa. In yet another variation of this theme, the two partner compounds can be covalently linked to form a single hybrid agent with the ability to modulate two essential targets [5–8]. In this scenario, the linkage can, in turn, be either a cleavable, labile moiety, which really can be considered a prodrug approach, or a robust, stable entity. The latter category of dual-action, hybrid antibacterial agents is the principal focus of this report.

Some possible advantages of a dual-action, hybrid antibacterial agent might include: (1) an expanded spectrum of activity, including coverage of resistant organisms, (2) greatly reduced potential for spontaneous mutations and resistance development, (3) synergistic activity superior to simple 1+1 combinations of the individual partner agents and (4) reduction of potential toxicity for a constituent agent.

A key, under-appreciated aspect of the design of any hypothetical hybrid antibacterial agent must be the careful consideration of the mechanism of action of each partner agent. Minimally, a partner agent should not antagonize the antibacterial effect of its intramolecular collaborator. For example, a protein synthesis inhibitor can sometimes attenuate the bactericidal activity of a DNA synthesis inhibitor such as a fluoroquinolone. Overall, this effect manifests itself by significantly altering the rapid killing kinetics usually observed for the fluoroquinolone. Ideally, the partner agents should work together in a synergistic sense to promote a strong antibacterial effect and stop resistance development.

1.3 Pioneering efforts in the area of hybrid antibacterial agents

Before highlighting recent developments in the dual-action, hybrid anti-bacterial agent field it is appropriate to acknowledge the genesis of the concept and seminal efforts in this area. The first example (**1**) of a dual-action antibacterial agent was reported by workers at Glaxo in 1976 [9]. This compound was comprised of cefamandole linked at its 3-position to 2-mercaptopyridine-*N*-oxide (omadine, an antiseptic). Compound **1** has appreciable antibacterial activity in its own right but when it encounters bacterial strains expressing a β-lactamase, the resulting hydrolysis of the β-lactam ring releases the omadine fragment, which can then exert its antibacterial effect.

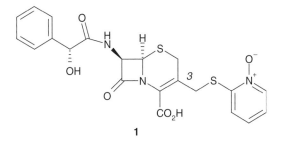

1

RO 23-9424 (**2**) was another important hybrid antibacterial agent that triggered many additional efforts in the dual-action agent area [10]. This compound, combining the cephalosporin desacetylcefotaxime with the fluoro-quinolone fleroxacin, exhibits a broad spectrum of activity, reflecting both mechanisms of action [11]. RO 23-9424 acts primarily as a cephalosporin but once the β-lactam either acylates the target penicillin-binding protein or is hydrolyzed by a β-lactamase then fluoroquinolone activity (DNA gyrase and topoisomerase IV inhibition) dominates. Ultimately, solubility and chemical and/or metabolic stability problems limited the practical utility of this hybrid antibacterial agent. To date, no fluoroquinolone/β-lactam hybrid agent has progressed beyond Phase 1 clinical trials.

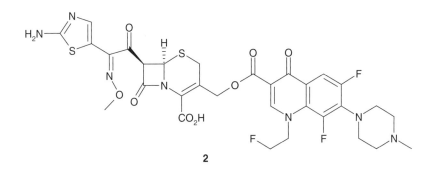

2

The remainder of this annual report will detail recent developments in the hybrid antibacterial field over the last year.

2. NOVEL HYBRID/DUAL-ACTION ANTIBACTERIAL AGENTS

2.1 Rifamycin-quinolone hybrids

The rifamycins, bacterial RNA polymerase inhibitors, have demonstrated good clinical utility against a variety of Gram-positive bacteria and *Mycobacterium tuberculosis* [12]. However, the spontaneous mutation frequency observed in bacteria treated with the rifamycins is such that bacterial resistance arises rapidly. Therefore, use of these agents is primarily limited to combination approaches [13]. Workers at Cumbre Pharmaceuticals have recently disclosed some novel hybrid rifamycin derivatives wherein the macrocyclic core is linked to an analog of the fluoroquinolone class [14]. Fluoroquinolones, of course, are well-established, broad-spectrum antibacterial agents which target bacterial topo-isomerases. The most promising compound prepared thus far is CBR-2092 (**3**) [15]. CBR-2092 incorporates a fluorinated *4H*-4-oxoquinolizine (an advanced fluoroquinolone analog) at the 3-position of its rifamycin core, tethered via a hydrazone linkage.

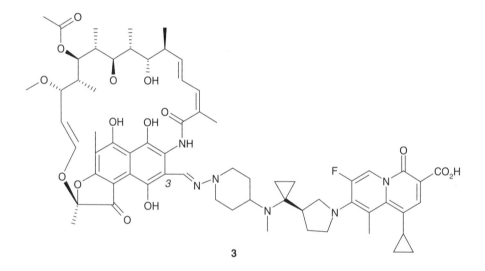

3

CBR-2092 exhibits broad-spectrum *in vitro* activity that is significantly improved over that observed with various rifampin/fluoroquinolone monomer combinations, suggesting synergy [16]. The compound also displays rapid, enhanced bactericidal killing against relevant Gram-positive organisms. Most importantly, CBR-2092 displays a very low-level frequency of spontaneous mutation ($<10^{-12}$), suggesting that resistance to this agent will be very slow to develop [17]. Further, CBR-2092 was found to provide superior efficacy in multiple *in vitro* models of staphylococcal biofilm states [18]. Finally, the exemplary *in vitro* activity profile of CBR-2092 also extends to the *in vivo* setting. The compound was found to provide outstanding efficacy in standard rodent infection models and in a rabbit endocarditis model where the causative organism was MRSA [19]. It should be noted that CBR-2092 is administered either intravenously or intraperitoneally, presumably because of poor oral bioavailability. CBR-2092 is currently undergoing further assessment in Phase 1 clinical trials.

2.2 Oxazolidinone-quinolone heterodimers

Fluoroquinolone-derived dual-action agents are not limited to the combinations described above. The independent discovery of hybrid antibacterial agents comprised of a fluoroquinolone subunit and a member of the oxazolidinone class has also been described [20]. The oxazolidinone partner compounds are relatively new antibacterial agents that employ protein synthesis inhibition as their mechanism of action [21]. These hybrid analogs also incorporate chemically and metabolically stable linkages between the two partner antibiotics. One of these heterodimeric agents, MCB 3681 (4), has progressed into human clinical trials as a phosphate ester prodrug, MCB 3837 (5) [22].

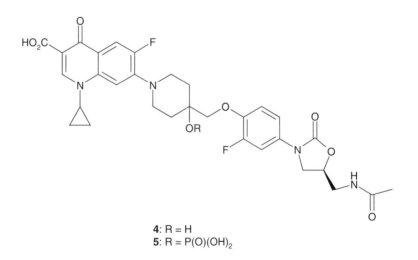

4: R = H
5: R = P(O)(OH)$_2$

MCB 3681 (**4**) exhibits broad-spectrum activity, including coverage of Gram-positive, most Gram-negative (but not *P. aeruginosa*) and anaerobic organisms. This reflects the parent compound's dual targeting of DNA gyrase and ribosomal-mediated protein synthesis inhibition. The compound exhibits activity against both linezolid- and fluoroquinolone-resistant strains of pathogenic Gram-positive bacteria, including MRSA, GISA, VRE and MDR *S. pneumoniae* (MDRSP). MCB 3837 was recently administered (IV) to healthy human volunteers in a single ascending dose study without serious adverse events at doses up to 3 mg/kg [23].

2.3 Cephalosporin-glycopeptide conjugates

Workers at Theravance have described novel hybrid vancomycin derivatives bearing a covalently linked cephalosporin antibacterial agent [24]. Conceptually, this represents an attractive combination of mechanisms of action, since the physical proximity of the targeted proteins (D-alanyl-D-alanine terminus of the intermediate pentapeptide for the glycopeptide and penicillin-binding proteins for the cephalosporin partner), as well as their sequential role in the overall cell wall biosynthetic pathway, suggests that a single molecule that inhibits both targets might confer some advantages. Chief among these dual-action hybrid analogs is TD-1792 (6).

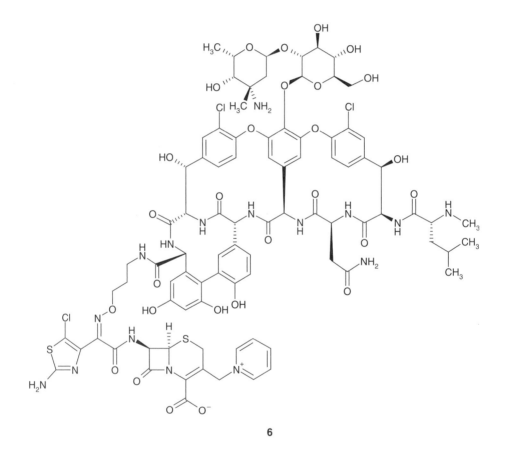

6

The heterodimer TD-1792 exhibits very potent *in vitro* activity against both sensitive and problematic MDR Gram-positive bacteria, with the notable exception of vancomycin-resistant *Enterococcus faecium* (VanA resistance phenotype), where TD-1792 was found to be ineffective (minimum inhibitory concentration or MIC > 100μg/mL) [25]. The MICs of TD-1792 are 14- to >100-fold more potent than equimolar combinations of the relevant monomeric partner agents, suggesting that considerable synergy is expressed in this hybrid antibacterial agent [26]. TD-1792 also displayed this synergistic effect *in vivo* in a mouse thigh model, being considerably more efficacious against MRSA than the monomers, dosed either separately or in combination. TD-1792 is currently in Phase 2 clinical trials examining its utility in treating complicated skin and skin structure infections (cSSIs).

2.4 DNA polymerase IIIC-quinolone hybrids

Inhibition of bacterial DNA replication is a bactericidal event. Therefore, the combination of two mechanistically distinct DNA replication inhibitors in a unified hybrid agent has intrinsic appeal. To this end, workers at Microbiotix have covalently linked a typical fluoroquinolone antibacterial agent, targeting bacterial DNA gyrase and topoisomerase IV, with a DNA polymerase IIIC inhibitory fragment, namely a substituted anilinouracil, to provide hybrid heterodimeric agents such as racemic MBX-500 (**7**) [27].

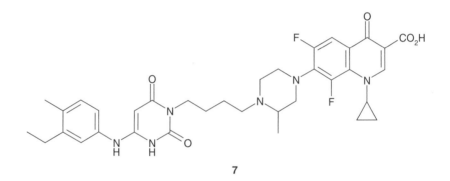

7

Hybrids such as **7** are potent nanomolar-level inhibitors of bacterial pol IIIC and have outstanding selectivity profiles with, for example, indices of >25,000 against mammalian pol-alpha and >36,000 vs. pol-gamma [28]. MBX-500 is also active against *B. subtilis* topoisomerase IV and gyrase, but at a lower level than either the monomeric fluoroquinolone partner or a control fluoroquinolone, ciprofloxacin. MBX-500 primarily targets Gram-positive bacteria and MDR strains, including fluoroquinolone-resistant isolates, are within its spectrum. Like other successful hybrid agents, MBX-500 is significantly more potent than the corresponding combination of constituent agents. Spontaneous mutation frequency studies with MBX-500 also revealed that the hybrid agent had a lower frequency of resistance development than did the individual monomers. Similar to many other hybrid agents, one challenge with the MBX-500 series is obtaining adequate aqueous solubility from these high-molecular-weight compounds to permit convenient intravenous dosing. A recent disclosure described modifications of MBX-500 via an ester prodrug approach to address this shortcoming [29]. Unfortunately, this preliminary study failed to resolve the IV formulation issue due to incomplete cleavage of the ester moiety in human plasma.

3. CONCLUSIONS

The area of dual-action, hybrid antibacterial agents is one of marked contrasts. On the one hand, there is a considerable body of scientific evidence that *in vitro* antibacterial activity and *in vivo* efficacy can be greatly potentiated by employing this concept. Coverage of problematic MDR organisms can be achieved and resistance development minimized. When the appropriate mechanisms of action are combined into a hybrid, the above effects are superior to those observed when simple 1+1 combinations of the individual constituents making up the hybrid are examined. However, the reality is that to date no hybrid antibacterial agent has progressed beyond Phase 2 clinical trials (TD-1792). What are the liabilities limiting the development of hybrid agents? Physicochemical properties can represent a significant challenge. High-molecular-weight hybrid agents tend to have poor oral bioavailability and may be limited to IV and/or topical administration routes. Some of these hybrids also have poor aqueous solubility, making it difficult to develop a satisfactory IV formulation, a requisite for the serious infections these agents are targeting in the hospital setting. The dual mechanisms of action employed by these agents may also create some challenges from a scientific and regulatory perspective. Many measurable parameters just tend to become more complicated. Chemical and metabolic stability may be more problematic with these large, often structurally complex molecules. Drug–drug interactions might also be inherently more challenging to assess and understand. Manufacturing of a hybrid antibacterial agent may be an issue. For many of these agents, the complexity of the chemistry and an increase in the number of synthetic steps may lead to an unacceptable cost of goods situation. Regulators may ask whether the hybrid motif is needed at all. Why not just go with two separate agents in either an optimized fixed-dose combination or just as separate agents? Scientific justification for a hybrid agent, most likely involving a demonstration of synergy superior to that provided by the simple co-administration of individual agents, may be required.

Obviously, a decision to enter the hybrid, dual-action antibacterial agent area should not be taken lightly. Nevertheless, in an era of increasing bacterial resistance, with few truly new antibacterial agents on the horizon, and fewer companies focused on antibacterial research, the hybrid, dual-action strategy should definitely be considered. It seems only a matter of time before a hybrid antibacterial agent ultimately succeeds in the clinical setting and justifies the considerable promise of this concept.

REFERENCES

[1] D. Cardo, T. Horan, M. Andrus, M. Dembinski, J. Edwards, G. Peavy, J. Tolson and D. Wagner, *Am. J. Infect. Control*, 2004, **32**, 470.
[2] (a) R. M. Klevens, M. A. Morrison, J. Nadle, S. Petit, K. Gershman, S. Ray, L. H. Harrison, R. Lynfield, G. Dumyati, J. M. Townes, A. S. Craig, E. R. Zell, G. E. Fosheim, L. K. McDougal, R. B. Carey and S. K. Fridkin, *JAMA*, 2007, **298**, 1763. (b) P. Nordmann, T. Naas, N. Fortineau and L. Poirel, *Curr. Opin. Microbiol.*, 2007, **10**, 436.

[3] (a) A. M. Ferrara, *Int. J. Antimicrob. Agents*, 2006, **27**, 183. (b) D. A. Enoch, C. L. Birkett and H. A. Ludlam, *Int. J. Antimicrob. Agents*, 2007, **29**(Suppl. 3), S33. (c) L. B. Rice, *Clin. Infect. Dis.*, 2006, **43**(Suppl. 2), S100.

[4] L. L. Silver and K. A. Bostian, *Antimicrob. Agents Chemother.*, 1993, **37**, 377.

[5] R. Morphy and Z. Rankovic, *J. Med. Chem.*, 2005, **48**, 6523.

[6] J. B. Bremner, J. I. Ambrus and S. Samosorn, *Curr. Med. Chem.*, 2007, **14**, 1459.

[7] L. L. Silver, *Nat. Rev. Drug Discov.*, 2007, **6**, 41.

[8] A. Bryskier, *Exp. Opin. Invest. Drugs*, 1997, **6**, 1479.

[9] C. H. O'Callaghan, R. B. Sykes and S. E. Staniforth, *Antimicrob. Agents Chemother.*, 1976, **10**, 245; D. Greenwood and F. O'Grady, *Antimicrob. Agents Chemother.*, 1976, **10**, 249.

[10] H. A. Albrecht, G. Beskid, K.-K. Chan, J. G. Christenson, R. Cleeland, K. H. Deitcher, N. H. Georgopapadakou, D. D. Keith, D. L. Pruess, J. Sepinwall, A. C. Specian, Jr., R. L. Then, M. Weigele, K. F. West and R. Yang, *J. Med. Chem.*, 1990, **33**, 77.

[11] N. H. Georgopapadakou, A. Bertasso, K. K. Chan, J. S. Chapman, R. Cleeland, L. M. Cummings, B. A. Dix and D. D. Keith, *Antimicrob. Agents Chemother.*, 1989, **33**, 1067.

[12] H. G. Floss and T. W. Yu, *Chem. Rev.*, 2005, **105**, 621.

[13] R. E. Chaisson, *Antimicrob. Agents Chemother.*, 2003, **47**, 3037.

[14] C. Z. Ding, Z. Ma, J. Li, S. Harran, Y. He, K. P. Minor, I. H. Kim, J. C. Longgood, Y. Jin and K. D. Combrink, *US Patent Application Publication*, 0019986, 2006.

[15] A. S. Lynch, E. J. Bonventre, T. B. Doyle, Q. Du, L. Duncan, G. T. Robertson and E. D. Roche, F1-2101, in 47th Interscience Conference on Antimicrobial Agents and Chemotherapy, Chicago, September, 2007.

[16] (a) G. T. Robertson, E. J. Bonventre, T. B. Doyle, Q. Du, L. Duncan, T. W. Morris, E. D. Roche, D. Yan and A. S. Lynch, *Antimicrob. Agents Chemother.*, 2008, **52**, 2324. (b) G. T. Robertson, E. J. Bonventre, T. B. Doyle, Q. Du, L. Duncan, E. D. Roche and A. S. Lynch, F1-2102, in 47th Interscience Conference on Antimicrobial Agents and Chemotherapy, Chicago, September, 2007.

[17] (a) G. T. Robertson, E. J. Bonventre, T. B. Doyle, Q. Du, L. Duncan, T. W. Morris, E. D. Roche, D. Yan and A. S. Lynch, *Antimicrob. Agents Chemother.*, 2008, **52**, 2313. (b) Q. Du, T. B. Doyle, L. Duncan, G. T. Robertson and A. S. Lynch, F1-2103, in 47th Interscience Conference on Antimicrobial Agents and Chemotherapy, Chicago, September, 2007.

[18] T. B. Doyle, E. J. Bonventre, Q. Du, G. T. Robertson and E. D. Roche and A. S. Lynch, F1-2104, in 47th Interscience Conference on Antimicrobial Agents and Chemotherapy, Chicago, September, 2007.

[19] (a) P. J. Renick, T. W. Morris, P. M. Nguyen, M. E. Pulse and W. J. Weiss, F1-2105, in 47th Interscience Conference on Antimicrobial Agents and Chemotherapy, Chicago, September, 2007. (b) W. J. Weiss, D. Bao, G. A. Miller, T. W. Morris, P. M. Nguyen, M. E. Pulse, P. J. Renick and T. Truong, F1-2106, in 47th Interscience Conference on Antimicrobial Agents and Chemotherapy, Chicago, September, 2007. (c) Y. Q. Xiong, W. J. Weiss, T. B. Doyle, G. A. Abdelsayed and A. S. Bayer, F1-2107, in 47th Interscience Conference on Antimicrobial Agents and Chemotherapy, Chicago, September, 2007.

[20] (a) M. F. Gordeev, C. Hackbarth, M. R. Barbachyn, L. S. Banitt, J. R. Gage, G. W. Luehr, M. Gomez, J. Trias, S. E. Morin, G. E. Zurenko, C. N. Parker, J. M. Evans, R. J. White and D. V. Patel, *Bioorg. Med. Chem. Lett.*, 2003, **13**, 4213. (b) C. Hubschwerlen, J.-L. Specklin, D. K. Baeschlin, Y. Borer, S. Haefeli, C. Sigwalt, S. Schroeder and H. H. Locher, *Bioorg. Med. Chem. Lett.*, 2003, **13**, 4229. (c) C. Hubschwerlen, J.-L. Specklin, C. Sigwalt, S. Schroeder and H. H. Locher, *Bioorg. Med. Chem. Lett.*, 2003, **11**, 2313.

[21] (a) S. J. Brickner, *Curr. Pharm. Des.*, 1996, **2**, 175. (b) M. R. Barbachyn and C. W. Ford, *Angew. Chem. Int. Ed.*, 2003, **42**, 2010. (c) D. K. Hutchinson, *Curr. Topics Med. Chem.*, 2003, **3**, 1021.

[22] C. P. Gray, M. W. Cappi and N. Frimodt-Moller, F-513, in 45th Interscience Conference on Antimicrobial Agents and Chemotherapy, Washington, DC, December, 2005.

[23] A. Dalhoff, F-638, in 47th Interscience Conference on Antimicrobial Agents and Chemotherapy, Chicago, September, 2007.

[24] J. L. Pace and G. Yang, *Biochem. Pharmacol.*, 2006, **71**, 968.

[25] D. D. Long, J. Aggen, J. Chinn, S.-K. Choi, B. G. Christensen, P. Fatheree, D. Green, J. K. Judice, S. S. Hegde, K. M. Krause, M. Ledbetter, M. Linsell, D. Marquess, E. Moran, M. Nodwell, J. L. Pace, S. Trapp and S. D. Turner, F1-2109, in 47th Interscience Conference on Antimicrobial Agents and Chemotherapy, Chicago, September, 2007.

[26] S. Difuntorum, J. Blais, S. S. Hegde, R. Skinner, N. Reyes, J. Trumbull, S. D. Turner and D. Marquess, F1-2110, in 47th Interscience Conference on Antimicrobial Agents and Chemotherapy, Chicago, September, 2007.

[27] C. Zhi, Z. Long, A. Manikowski, J. Comstock, W.-C. Xu, N. C. Brown, P. M. Tarantino, Jr., K. A. Holm, E. J. Dix, G. E. Wright, M. H. Barnes, M. M. Butler, K. A. Foster, W. A. LaMarr, B. Bachand, R. Bethell, C. Cadilhac, S. Charron, S. Lamothe, I. Motorina and R. Storer, *J. Med. Chem.*, 2006, **49**, 1455.

[28] M. M. Butler, W. A. LaMarr, K. A. Foster, M. H. Barnes, D. J. Skow, P. T. Lyden, L. M. Kustigian, C. Zhi, N. C. Brown, G. E. Wright and T. L. Bowlin, *Antimicrob. Agents Chemother.*, 2007, **51**, 119.

[29] G. E. Wright, S. Dvoskin, M. Yanachkova, W. Xu, I. B. Yanachkov, E. Dix, N. C. Brown, M. M. Butler, L. Kustigian and T. Bowlin, F1-2111, in 47th Interscience Conference on Antimicrobial Agents and Chemotherapy, Chicago, September, 2007.

PART VI:
Topics in Biology

Editor: John Lowe
Pfizer Inc.
Groton
Connecticut

CHAPTER **18**

Recent Renaissance of Prostaglandin Research

Robert M. Burk

1. INTRODUCTION

Prostaglandins (PGs) are a class of naturally occurring hormone-like substances that have been known for over 70 years. PGs are derived from 20-carbon essential fatty acid lipids and as a result possess a highly oxygenated C_{20} backbone. They are produced in response to various stimuli in cells throughout the body and are

Allergan, Inc., Department of Medicinal Chemistry, 2525 Dupont Drive, Irvine, CA 92612, USA

Annual Reports in Medicinal Chemistry, Volume 43
ISSN 0065-7743, DOI 10.1016/S0065-7743(08)00018-3

synthesized, released and then act primarily in the vicinity of their formation. Their *in vivo* activities were first discovered in 1930 from lipid fractions isolated from semen-induced contraction and relaxation of the human uterus [1]. After further work supporting this discovery these substances were mistakenly assumed to be produced by the prostate gland and thus given the name prostaglandins [2]. In the early 1960s the first pure PGs, PGE_1 and $PGF_{1\alpha}$, were isolated and their structures elucidated [3]. Further research identified PG production as a result of the arachidonic acid (AA) cascade which generated PGD_2, PGE_2, $PGF_{2\alpha}$, PGI_2 and TxA_2, via conversion of the intermediate endoperoxide PGH_2 by the appropriate PG synthase [4]. PGA_2, PGB_2 and PGC_2 were later identified but were believed to be artifacts resulting from extraction procedures [5,6]. Eventually it was determined that all PGs were biosynthesized from three fatty acid precursors, dihomo-γ-linolenic acid, AA and timodonic acid, yielding the 1-, 2- and 3-series PGs, respectively [7]. AA is recognized as the most important precursor therefore the 2-series PGs are the most abundant and produce an extremely wide range of biological activities *in vivo*.

The prostanoid family of receptors are G protein-coupled receptors with seven transmembrane domains which are ~20–30% homologous within the family. The structures, properties and function of PG receptors have been extensively reviewed [8,9]. Early on it was evident that PGs were involved in a diverse range of biological functions. However, research was complicated by chemical and metabolic instability of the PG analogs and difficult structure–activity-relationship (SAR) analysis due to functional assays based on mixed receptor systems. Remarkably, in the early 1980s a non-definitive but "working hypothesis" classification was proposed for DP, EP, FP, IP and TP receptors according to the binding affinities, potencies and selectivities of endogenous PGD_2, PGE_2, $PGF_{2\alpha}$, PGI_2 and TxA_2 [10,11]. Additional pharmacological studies and molecular biology techniques later corroborated this classification [12,13] and also assisted in subdivision of the EP-receptor into EP_1, EP_2, EP_3 and EP_4 [14,15]. Further use of molecular biology led to development of single receptor cell-based assays and advanced the understanding of PG pharmacology, thus promoting a revival in PG biological utilities. This renewed interest is evident from an almost doubling of published articles since 1990 compared with the previous 20 years and the recent introduction of PG-based drugs to the market.

2. SELECTIVE PROSTANOID AGONISTS AND ANTAGONISTS

Over the past 30 years significant advances in PG synthesis [16–18] have facilitated the ability of medicinal chemists to design PG analogs. Much of the initial receptor pharmacology was performed with the DP-agonist BW245C [19], FP-agonists fluprostenol and cloprostenol [20], IP-agonist iloprost [21] and the TP-agonist U-46619 [22]. Earlier EP-receptor pharmacological studies were conducted with the EP_1-agonist 17-phenyl PGE_2 [23], a weak EP_2-agonist 11-deoxy PGE_1 [24] and sulprostone, a mixed EP_1/EP_3 agonist one order of magnitude more potent for the EP_3 receptor [20]. As a result of recent investigations in defining PG biological functions, numerous new potent, selective DP-, FP-, IP- and TP-receptor agonists have been discovered which exhibit subnanomolar potency at the target receptor with at least 1,000-fold selectivity over off-target receptors [8]. While lacking in the past, selective agonists now exist for all four EP-receptor subtypes, such as ONO-DI-004 (EP_1), ONO-AE1-259 (EP_2), ONO-AE-248 (EP_3) and ONO-AE1-329 (EP_4) [9]. The advancement in the discovery of potent, selective agonists of PG receptors will now allow pharmacologists to expand the definition of specific functional responses coupled to activation of each receptor by the use of both isolated tissues and cell lines expressing the respective endogenous or cloned PG receptor.

Of the now recognized eight PG receptor subtypes, there exist the following experimentally useful antagonists, possessing the required potency and selectivity: BWA868C (DP) [25], ONO-8713 (EP_1) [8], ONO-AE3-208 (EP_4) [8], RO1138452 (IP) [26] and SQ-29,548 (TP) [27]. ONO-AE3-240 has been reported as a potent, selective EP_3-antagonist; unfortunately no structure was revealed [9]. AL-8810 has been reported as an FP-antagonist which also acts as a partial agonist [28]. AH-6809 has been defined as a weak, non-selective EP_2-antagonist [29]. Finally, N-alkyl-N'-[2-(aryloxy)-5-nitro benzenesulfonyl]-ureas have been reported as selective antagonists of the TP_β- relative to TP_α-isoforms [30].

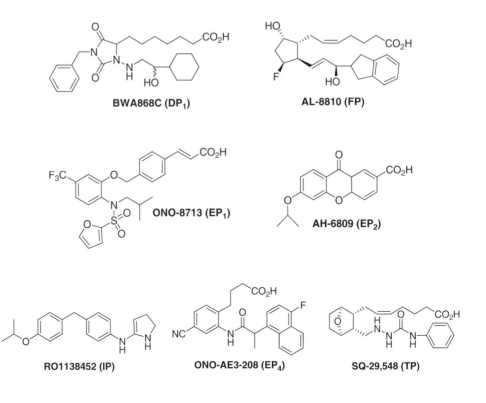

BWA868C (DP$_1$)

AL-8810 (FP)

ONO-8713 (EP$_1$)

AH-6809 (EP$_2$)

RO1138452 (IP)

ONO-AE3-208 (EP$_4$)

SQ-29,548 (TP)

3. RECENT NOVEL PROSTANOID STRUCTURE DESIGN AND ACTIVITY

Recently DP$_1$-antagonist research has been summarized [31] and PG patent literature has been reported [32]. While extremely active in the past, TP-antagonist and IP-mimetic research has diminished as a result of the PGI$_2$/TxA$_2$ imbalance theory [33]. The majority of more recent pioneering medicinal chemistry research of PGs has focused primarily on FP-, EP$_2$- and EP$_4$-agonists and EP$_1$-antagonists and DP-agonists and EP$_3$- and EP$_4$-antagonists to a lesser extent. For the most part EP$_2$-agonists were pursued using two past reported EP$_2$-agonists, butaprost **1** and AH-13205 **2**, both being ω-chain variants of PGE$_1$. Success of EP$_4$-agonist design has been realized mainly with use of heterocyclic replacement of the cyclopentyl ring of PGE$_1$ and appropriate α-chain variations. More recent FP-agonist design has been based on two potent, non-selective FP-agonists, 17-phenyl-PGF$_{2\alpha}$ **3** and 16-phenoxy-PGF$_{2\alpha}$ **4**. Antagonist leads have been derived from non-prostanoid structures and the majority is devoid of stereochemical complexity.

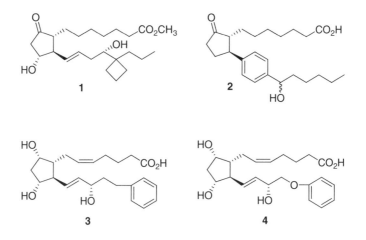

3.1 EP$_2$-agonists

The obvious difficulty confronting medicinal chemists when designing EP-agonists based on PGE$_2$ as a lead structure is the inherent chemical and metabolic instability of analogs possessing the 9-keto-11-hydroxy functionality. Successful modification was initially realized using the analog butaprost **1** [34]. Although butaprost **1** was selective for the EP$_2$-receptor among all EP receptors, it did demonstrate some off-target IP-receptor activity. More recent chemical modifications of butaprost focusing on the α- and ω-chains provided the 5,6-*cis* unsaturated butaprost analog **5** ($K_i = 92$ nM, EP$_2$) devoid of IP-receptor activity [35]. However, analog **5** required the labile 11-hydroxy-9-keto functionality for potency and selectivity. Replacement of the 9-hydroxy group of PGF$_{2\alpha}$ with a 9β-chloro group provided compound **6** with EP$_2$-receptor activity ($K_i = 1.4$ nM).

Combination of the 9β-chloro group and the 5,6-*cis* unsaturation in butaprost resulted in identification of the highly, selective and potent EP$_2$-agonist **7** (K_i = 4.2 nM) [36]. Additional SAR analysis around the 9-position, unsaturations and the ω-chain tail identified analog **8** (K_i = 3.3 nM, EC$_{50}$ = 3.8 nM) as the most potent, selective EP$_2$-agonist of this compound class [36,37].

Recently a similar strategy has been applied in designing analogs of AH-13205 [38] in which 9β-chloro substitution along with addition of the 5,6-*cis* unsaturation has provided potent and selective EP$_2$-agonists [39]. 7-(*N*-alkylmethanesulfonamido)heptanoic acid **9**, a seco-prostanoid, had been previously reported as having PG-like activity [40]. Recent modification and SAR of seco-prostanoid compounds have been exploited in an approach to potent, selective EP$_2$-agonists targeting the treatment of osteoporosis [41]. Compound **10** (K_i = 8 nM, cAMP K_b = 3 nM) [42] and other cinnamic acid derivatives have recently been reported as EP$_2$-agonists [43].

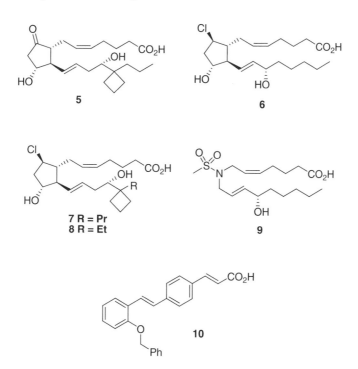

3.2 EP$_4$-agonists

While ω-chain variants of PGE$_2$ have provided EP$_2$-receptor agonistic selectivity, successful EP$_4$-receptor selectivity has been obtained by variation of the α-chain of PGE$_2$. Initially, the 3,7-dithia-PGE$_1$ derivative **11** was reported as a lead compound eliciting subnanomolar EP$_4$-receptor potency (K_i = 0.7 nM), however, possessing submicromolar potency at other EP-receptor subtypes [39]. Addition of a 16-phenyl group [44] increased selectivity and eventual substitution of the 16-phenyl groups provided compound **12** showing marked improvement in potency and selectivity [45]. Unfortunately, use of this class of compounds as clinical candidates was hampered by the inherent self-degradation of the β-hydroxyketone functionality and the easy epimerization at the sulfur substituted position-8. Discovery of a 5-thia α-chain analog **13** ($K_i = 0.7$ nM, $EC_{50} = 1.6$ nM) avoided C-8 epimerization [46]. However, combination of the 9β-chloro, to avoid elimination, with the 5-thia substituent unexpectedly resulted in an unstable compound [47].

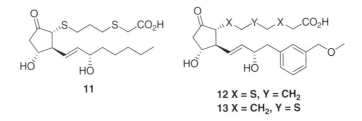

11

12 X = S, Y = CH$_2$
13 X = CH$_2$, Y = S

Replacement of the cyclopentyl moiety of the PGE class has been successful in identifying novel selective EP$_4$-receptor agonists. Originally the 1-heptanoic acid-2-pyrrolidinone **14** was identified as having prostanoid-like activity [47] and was later found to exhibit nanomolar binding and functional agonism at the EP$_4$ receptor with micromolar activity at the other PG receptors [48]. SAR around **14** seemed to indicate that the 11-hydroxy is not necessary for potency and that its removal results in increased EP$_4$ receptor selectivity. Eventual optimization of **14** by 17-arylation and bioisosteric tetrazolyl replacement of the C1-CO$_2$H led to compound **15** which exhibited a much higher degree of potency ($K_i = 1.2$ nM) and selectivity, as well as a superior *in vivo* half-life than PGE$_2$ [48]. Additional SAR around 17-aryl analogs of **14** has identified compound **16** (IC$_{50} = 6$ nM, $EC_{50} = 0.5$ nM) [49].

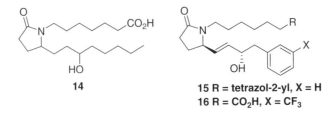

14

15 R = tetrazol-2-yl, X = H
16 R = CO$_2$H, X = CF$_3$

Modifications of both the α- and ω-chain of **14** have also been investigated. Initially, ω-chain variations using substituted 15-biphenyl groups and removal of the metabolically labile 13,14-unsaturation produced highly selective and potent EP$_4$-agonists such as **17** ($K_i = 3.1$ nM, $EC_{50} = 0.8$ nM) [50]. α- and ω-Chain modifications of **14** with ethylbenzoic acid and the substituted moiety known to impart EP$_4$ subtype selectivity [40] provided **18** ($K_i = 0.7$ nM, EC$_{50} = 5.9$ nM) [51]. A more recent report discloses that α-chain benzoic acid derivatives in combination with dienyl ω-chains also produced both mixed EP$_2$/EP$_4$-agonists and potent, selective EP$_4$-agonists [52].

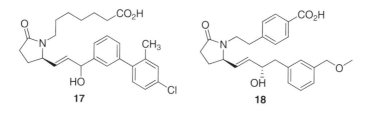

Use of a pyrazolidinone head group has been reported to generate both mixed EP$_2$/EP$_4$ agonists and selective EP$_4$-agonists [53]. Expansion of the pyrrolidinone (γ-lactam) to a piperidone (δ-lactam) has also provided analogs with solely EP$_4$-receptor selective nanomolar activity [54]. Further extension of the six-member ring class has been extended to synthesis 1,3-oxazinanes as potent, selective EP$_4$-agonists [55].

3.3 FP-agonists

Past SAR of FP-agonist design focused on synthesis of analogs of PGF$_{2\alpha}$. More recent analog development centered on modifications of 17-phenyl PGF$_{2\alpha}$ **3** and 16-phenoxy PGF$_{2\alpha}$ **4**. These two analogs are an order of magnitude more potent than PGF$_{2\alpha}$, however, they exhibit significant off-target PG receptor activity. A thorough investigation of substituted phenyl analogs of **3** has revealed that a slight increase in FP potency [56] can be obtained. Removal of the 13,14-unsaturation provided a better separation between ocular side effects and IOP lowering effect leading to identification of latanoprost **19**, the first prostanoid introduced to market for treatment of glaucoma. Bioisosteric replacement of **3** with various 17-thienyl substituents has provided highly potent (EC$_{50} < 1$ nM) and selective FP-agonists [57].

SAR analysis of 16-phenoxy analogs initially indicated that removal of the 5,6- and 13,14-unsaturations resulted in a 10-fold loss of potency, yet provided a better selectivity ratio over other PG receptors [58]. However, one analog **20** was shown to retain potency and improve selectivity [58]. 15-(S)-fluoro substitution of the stereochemically sensitive (S)-15-hydroxy of **4** has provided compounds with moderate to weak binding affinities for the FP-receptor, however, analog **21** (EC$_{50} = 4$ nM) did exhibit acceptable functional activity and comparable ocular *in vivo* effects as other prostanoids [59]. Compound **22**, the difluoro analog of **4**, has been reported to be a potent FP-agonist and, surprisingly, did not cause the ocular iridial pigmentation side effect in monkeys as witnessed with typical FP-agonists [60]. Finally, replacement of the 11-hydroxy substituent with an 11-oxa group in combination with 16-(substituted)phenoxy ω-chains slightly lessens FP potency as demonstrated in analog **23** ($K_i = 147\,\text{nM}$, $EC_{50} = 23\,\text{nM}$) [61].

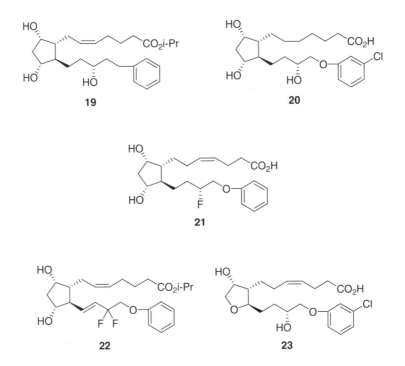

3.4 DP-agonists

Although in the keto-hydroxy functionality of PGD$_2$ is the reverse of PGE$_2$, the 9β-chloro cyclopentyl moiety has been shown to afford the DP-receptor selective agonist ZK-110,841 **24** and its metabolically more stable analog **25** [62]. SAR reported on variants of **25** has identified a less potent 13,14-dihydro analog AL-6556 **26**, whose i-Pr ester prodrug **27** is a highly efficacious ocular hypotensive agent [63].

A noteworthy example of design of DP-receptor selective agonists was demonstrated by "flip-flopping" the α- and ω-chains of the EP$_4$-agonist pyrrolidinone template (γ-lactam). Synthesis of an analog with both the pentanor-15-cyclohexyl appendage and a propylbenzoic acid appendage provided **28** ($K_i = 2.4\,\text{nM}$, $EC_{50} = 0.00072\,\text{nM}$) [64].

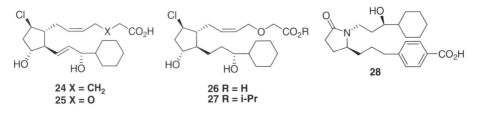

24 X = CH$_2$
25 X = O

26 R = H
27 R = i-Pr

3.5 EP$_1$-antagonists

PGE$_2$ has been shown to be a major pro-inflammatory mediator and studies have implicated the EP$_1$-receptor in PGE$_2$-generated allodynia and inflammatory pain [8]. Thus, a considerable effort has followed in identification of potent, selective EP$_1$-antagonists. The phenylsulfonamide ZD-6419 **29** has been reported as a potent, selective EP$_1$-antagonist, but no further details are available. However, a new class of aryl sulfonamides was found through in-house compound screening [65]. N-Alkylation of the sulfonamide and replacement of the amido group with an oxamethylene group proved essential in providing selective EP$_1$-antagonist leads with subnanomolar receptor affinity [65]. Replacement of the phenylsulfonyl moiety with a hydrophilic heteroarylsulfonyl moiety results in increased potency and efficacy in animal models [66]. Isosteric replacement of the carboxylic acid increased potency, but without inhibiting P450 [67]. Optimization of this class of compounds was realized by incorporation of a thiazolyl carboxylic acid affording a pre-clinical candidate **30** ($K_i = 0.39$ nM, IC$_{50} = 4.3$ nM) [68]. Modifications of 2,3-diaryl thiophene **31**, a reported EP$_1$-antagonist ($K_i = 15$ nM) and 50–100-fold selective over other PG receptors, have been investigated [69]. Other EP$_1$-antagonists have been reported in which the thienyl moiety of **31** has been replaced with pyrrolyl [70,71], cyclopentenyl [72] or various heterocyclic groups [73]. Use of azole acid as motifs in conjunction with partial pharmacophoric sections of **29** in chemical arrays has led to identification of pyrazole acid EP$_1$-anatagonists **32** [74]. A novel glycine sulfonamide EP$_1$-antagonist lead has been disclosed, however, this structural type suffers from poor metabolic stability [75]. Bioisosteric replacement of the amide of the glycine sulfonamides with a 1,3,4-oxadiazole provided more potent EP$_1$-antagonists but did provide sufficiently improved metabolic stability [76].

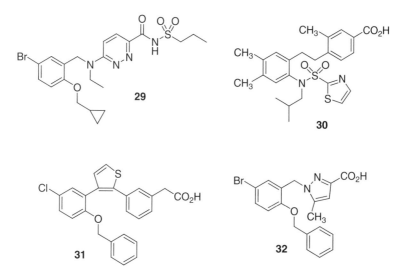

3.6 EP₃-antagonists

Reported EP$_3$-antagonists are modified cinnamic acid derivatives. Compound **33** ($K_i = 3$ nM, cAMP $K_b = 5$ nM) and analogs were discovered by varying and altering the substitution patterns of known EP$_1$-antagonists [42,77]. Minor modifications of substituents and/or variation can dramatically shift the EP-receptor selectivity of this class of compounds [42]. A second class of compounds include **34** ($K_i = 0.7$ nM, cAMP $K_b = 0.7$ nM) [77] discovered by SAR studies of a lead identified by screening a library designed from an in-house sample hit [78,79].

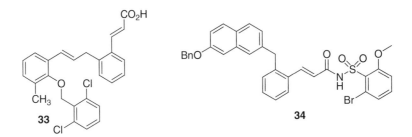

3.7 EP₄-antagonists

Only very recently have claims of EP$_4$-antagonists appeared in the literature. Optimized diphenyloxazole **35** ($K_i = 0.3$ nM) and Nδ-Z-ornithine **36** ($K_i = 0.91$ nM) were reported and are also highly selective [80]. Benzoisoindole **37** (p$K_b = 7.9$) was reported as a valuable pharmacological tool [81]. CJ-42794 **38** was recently disclosed in a report investigating the possible role of EP$_4$-antagonists in GI ulcers and healing responses [82]. The structure of a purported EP$_4$-antagonist **39** was disclosed in a bone resorption study [83].

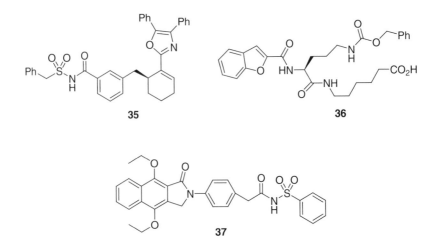

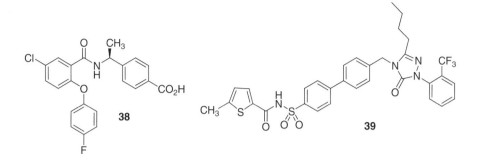

4. CURRENT RESEARCH AIMED AT PROSTANOID-DERIVED THERAPEUTICS

A recent review has comprehensively covered PG receptor signaling in disease and the resulting therapeutic implications [84]. From the studies discussed it is postulated that selective manipulation of individual receptors may be beneficial in the treatment of acute inflammation, pain, fever, arthritis, inflammatory bowel disease, pulmonary fibrosis as well as the effects on dendritic cells, thymocyte development, development and function of mature lymphocytes and B lymphocyte class switching. Recent drug discovery studies and on-going clinical development of PGs are presented later.

4.1 Glaucoma-intraocular pressure lowering (IOP) agents

4.1.1 FP-agonists

A major effort has focused on the development of FP-agonists as therapeutics aimed at lowering high intraocular pressure (IOP) in the treatment of glaucoma. To date three new drugs have entered the market, all of which are C1-ester prodrugs of potent and selective FP-agonists. The first was latanoprost **19**, an i-propyl ester prodrug marketed by Pfizer in 1996. At this time latanoprost **19** became a first-line treatment and the most efficacious IOP lowering drug. However, it was not completely devoid of hyperemic side effects in human patients. It also caused an unexpected adverse advent of iridial pigmentation which was observed after one to three months of treatment. Concerns about the origin of this event were studied and its occurrence was later dismissed as a safety issue.

In 2002, Alcon introduced travoprost [85] to the market. This drug is simply the i-propyl ester prodrug of the highly potent FP-agonist fluprostenol [86] which had been discovered over 20 years ago. Based on the effects witnessed with $PGF_{2\alpha}$-i-propyl ester and 17-phenyl isopropyl ester, it would have been predicted that a severe hyperemic effect would be produced with the i-propyl ester of fluprostenol. However, clinical use of a dose about 10-fold less than latanoprost provides slightly better efficacy and a similar hyperemic response in comparison to latanoprost [87].

Finally tafluprost, the i-Pr ester prodrug of **22**, recently demonstrated efficacy and tolerability in healthy volunteers [88] and is currently awaiting marketing approval. Substitution of the stereochemically sensitive 15(S)-hydroxy group of cloprostenol [89] with a difluoro group remarkably retained potency and did not promote iridial pigmentation in *in vivo* studies [90]. Similar to travoprost, tafluprost is also dosed 10-fold lower in patients than latanoprost, however, as witnessed with the other FP-agonists, this compound is not completely devoid of hyperemic side effects. In summary, several highly efficacious PG derived drugs have been delivered to the market and are currently the most efficacious therapeutics for lowering IOP in the treatment of glaucoma and their high efficacy appears to offset the minimal hyperemia witnessed in some patients.

4.1.2 DP-agonists

ZK-118,182 **25**, a highly potent DP-agonist, has been reported as a highly efficacious ocular hypotensive in animals. However, it caused intense conjunctival hyperemia [62,91]. Synthesis and SAR evaluation around compound **25** discussed earlier led to the identification of AL-6556 **26** which exhibited interesting pharmacological results [63]. While compound **26** was about 10-fold less potent in binding affinity for the DP-receptor than **25**, its i-Pr ester prodrug **27** demonstrated a pronounced effect in lowering IOP $>50\%$ in animals. In 2004, AL-6598 **27** was reported to be in phase II clinical trials, however, as of 2005 no disclosure of outcome or further progress has been reported [92].

4.2 Bone disease

4.2.1 EP-agonists/antagonists

The role of PGE_2-based analogs in bone resorption have been studied for over four decades. The initial lack of success was attributed to poor models of skeletal metabolism. As research progressed it became apparent that systemic administration of EP-receptor selective subtype agonists in a clinical study would most likely be hampered by side effects. The strategy for drug delivery evolved to that of controlled, targeted delivery. As of mid-2007 CP-533,536, a non-prostanoid derived highly potent, selective EP_2-agonist [93] is in phase II clinical trials as a topical gel [95].

CJ-023423 has been reported as a highly selective EP_4-antagonist which inhibits PGE_2-evoked elevated intracellular cAMP at human EP_4-receptors with a pA_2 of 8.3 [94]. In late 2006 it was reported that CJ-023423 is in phase II clinical studies for osteoarthritis [95]. ONO-4819, an EP_4-agonist, the methyl ester of compound **13**, was under development for vertebral fracture and discontinued when the phase II study indicated no evidence of efficacy [96].

4.2.2 FP-agonists

FP-agonists have been investigated as agents for producing a bone anabolic response. A series of 16-oxaaryl substituted FP-agonists, lacking the 5,6-*cis* and

13,14-*trans* unsaturations, have been reported as being both potent and highly selective [58]. However, development of this class of compounds was reported to have ceased at the pre-development stage due to fatal toxicity in certain animal models [97].

4.3 Pain and inflammation

4.3.1 EP$_1$-antagonists

A few EP-1 antagonists, ZD-4953, ZD-6416 and ZD-6809 have been reported as under clinical development for hyperalgesia [98]. ZD-4953 and ZD-6416 studies ceased at phase II and the current status of ZD-6809 is unreported. The fate of the earlier two candidates has not been discussed; however, it has been reported that in women ZD-6416 has a previously unknown potential for teratogenicity [99].

4.3.2 EP$_4$-antagonists

CJ-023423, described earlier, is also reported to be in late stage phase II clinical studies for osteoarthritis [94].

4.4 Pre-term labor

PGs have been studied for years for use in pre-term labor [100]. ONO-8815Ly, a highly selective and potent 9-chloro analog of butaprost, was discontinued in 2005 as an EP$_2$-agonist for treatment of imminent premature labor [101].

4.5 Gastrointestinal disease

An EP$_4$-agonist, ONO-4819, was reported to be in phase II clinical study for the treatment of ulcerative colitis as of February 2006 [102].

5. CONCLUSION

From the number of recent publications regarding PGs, an apparent resurgence in research interest has occurred. Molecular biology has assisted pharmacologists in characterizing and developing stably expressed cell-based assays. The availability of single cell assays and much improved functional assays has greatly assisted medicinal chemists in identifying both selective PG agonists and antagonists. Three novel drugs have recently reached the market and several others are currently in clinical trials. However, as in the past, newer drugs appear to suffer minor side effects that may be prostanoid-receptor based. The continued unveiling of new PG receptor isoforms will allow pharmacologists to explore new targets but may indicate that designing selective prostanoid ligands devoid of off-target activity will still continue to be a challenge.

REFERENCES

[1] R. Kurzrok and C. C. Lieb, *Proc. Soc. Exp. Biol. Med.*, 1930, **28**, 268.

[2] U. S. von Euler, *Arch. Exp. Pathol. Pharmacol.*, 1934, **175**, 78.

[3] S. Bergstrom, R. Ryhage, B. Samuelsson and J. Sjovall, *Acta Chem. Scand.*, 1962, **16**, 501.

[4] M. Hamberg and B. Samuelsson, *Proc. Nat. Acad. Sci. USA*, 1973, **70**, 899.

[5] W. P. Schneider, J. E. Pike and F. P. Kupiecki, *Biochim. Biophys. Acta*, 1966, **125**, 611.

[6] E. W. Horton, in *Chemistry, Biochemistry and Pharmacological Activity of Prostaglandins* (eds S. M. Roberts and F. Scheinmann), Pergamon Press, Oxford, UK, 1979.

[7] D. A. von Dorp, R. K. Beethuis, D. H. Nugteren and H. Vonkeman, *Nature*, 1964, **203**, 839.

[8] S. Narumiya, Y. Sugimoto and F. Ushikubi, *Physiol. Rev.*, 1999, **79**, 1193.

[9] Y. Sugimoto and S. Narumiya, *J. Biol. Chem.*, 2007, **282**, 11613.

[10] I. Kennedy, R. A. Coleman, P. P. A. Humphrey, G. P. Levy and P. Lumley, *Prostaglandins*, 1982, **24**, 667.

[11] R. A. Coleman, P. P. A. Humphrey, I. Kennedy and P. Lumley, *Trends Pharmacol. Sci.*, 1984, **5**, 303.

[12] R. A. Coleman, W. L. Smith and S. Narumiya, *Pharmacol. Rev.*, 1994, **46**, 205.

[13] S. Narumiya, in *Prostanoid Receptors: Structure, Function and Distribution, in Cellular Generation, Transport, and Effects of Eicosanoids. Biological Roles and Pharmacological Intervention* (eds R. A. Lewis and M. Rola-Pleszczynski), The New York Academy of Sciences, New York, 1994.

[14] R. A. Coleman, I. Kennedy, P. P. A. Humphrey, K. Bunce and P. Lumley, in *Comprehensive Medicinal Chemistry. Membranes and Receptors* (ed. J. C. Emmett), Pergamon Press, Oxford, UK, 1990.

[15] R. A. Coleman, S. P. Grix, S. A. Head, J. B. Louttit, A. Mallett and R. L. G. Sheldrick, *Prostaglandins*, 1994, **47**, 151.

[16] P. H. Bentley, *Chem. Soc. Rev.*, 1973, **2**, 29.

[17] E. J. Corey and X.-M. Cheng, in *The Logic Of Chemical Synthesis*, Wiley, New York, 1989, p. 250.

[18] F. C. Biaggio, A. R. Rufino, M. H. Zaim, C. Y. H. Ziam, M. A. Bueno and A. Rodrigues, *Curr. Org. Chem.*, 2005, **9**, 419.

[19] M.-H. Town, J. Casal-Stenzel and E. Schillinger, *Prostaglandins*, 1983, **25**, 13.

[20] R. A. Coleman, P. P. A. Humphrey and I. Kennedy, in *Trends Autonomic Pharmacology* (ed. S. Kalsner), Vol. 3, Taylor and Francis, London, 1985.

[21] K. Schror, H. Darius, R. Matzky and R. Ohlendorf, *Naunyn-Schmeideberg's Arch. Pharmacol.*, 1981, **316**, 252.

[22] R. A. Coleman, P. P. A. Humphrey, I. Kennedy, G. P. Levy and P. Lumley, *Br. J. Pharmacol.*, 1981, **73**, 773.

[23] W. L. Miller, J. R. Weeks, J. W. Lauderdale and K. T. Kirton, *Prostaglandins*, 1975, **9**, 9.

[24] Y. J. Dong, R. L. Jones and N. H. Wilson, *Br. J. Pharmacol.*, 1986, **87**, 97.

[25] H. Giles, P. Leff, M. L. Bolofo, M. G. Kelly and A. D. Robertson, *Br. J. Pharmacol.*, 1989, **96**, 291.

[26] K. R. Bley, A. Bhattacharya, D. V. Daniels, J. Gever, A. Jahangir, C. O'Yang, S. Smith, D. Srinivasan, A. P. D. W. Ford and M.-F. Jett, *Br. J. Pharmacol.*, 2006, **147**, 335.

[27] M. L. Ogletree, D. N. Harris, R. Greenberg, M. F. Haslanger and M. Nakane, *J. Pharmacol. Exp. Ther.*, 1985, **234**, 435.

[28] B. W. Griffin, P. Klimko, J. Y. Crider and N. A. Sharif, *J. Pharmacol. Exp. Ther.*, 1999, **290**, 1278.

[29] M. Abramovitz, M. Adam, Y. Boie, M. Carriere, D. Denis, C. Godbout, S. Lamontagne, C. Rochette, N. Sawyer, N. M. Tremblay, M. Belley, M. Gallant, C. Dufresne, Y. Gareau, R. Ruel, H. Juteau, M. Labelle, N. Ouimet and K. M. Metters, *Biochim. Biophys. Acta.*, 2000, **1483**, 285.

[30] J. Hanson, J.-M. Dogne, J. Ghiotto, A.-L. Moray, B. T. Kinsella and B. Piroette, *J. Med. Chem.*, 2007, **50**, 3928.

[31] J. C. Medina and J. Liu, *Ann. Rep. Med. Chem.*, 2006, **41**, 221.

[32] K. K. Ebenezar and F. G. Smith, *Expert Opin.*, 2007, **17**, 1131.

[33] J.-M. Dogne, J. Hanson and D. Pratico, *Trends Pharmacol. Sci.*, 2005, **26**, 639.

[34] P. J. Gardiner, *Br. J. Pharmacol.*, 1986, **87**, 45.

[35] K. Tani, A. Naganawa, A. Ishida, K. Sagawa, H. Harada, M. Ogawa, T. Maruyama, S. Ochuchida, H. Nakai, K. Kondo and M. Toda, *Bioorg. Med. Chem.*, 2002, **10**, 1093.

[36] K. Tani, A. Naganawa, A. Ishida, H. Egashira, K. Sagawa, H. Harada, M. Ogawa, T. Maruyama, S. Ochuchida, H. Nakai, K. Kondo and M. Toda, *Bioorg. Med. Chem. Lett.*, 2001, **11**, 2025.

[37] K. Tani, A. Naganawa, A. Ishida, H. Egashira, K. Sagawa, H. Harada, M. Ogawa, T. Maruyama, S. Ochuchida, H. Nakai, K. Kondo and M. Toda, *Bioorg. Med. Chem.*, 2002, **10**, 1107.

[38] A. T. Nials, C. J. Vardey, L. H. Denyer, M. Thomas, S. J. Sparrow, G. D. Shepard and R. A. Coleman, *Cardiovasc. Drug Rev.*, 1993, **11**, 165.

[39] Y. Donde and J. Nguyen, *US Patent* 7091231, 2006.

[40] J. H. Jones, W. J. Holtz, J. B. Bicking and E. J. Cragoe, Jr., *J. Med. Chem.*, 1977, **20**, 1299.

[41] K. O. Cameron, H. Z. Ke, B. A. Lefker, R. L. Rosati and D. D. Thompson, *US Patent* 6998423, 2006.

[42] M. Belley, M. Gallant, B. Roy, K. Houde, N. Lachance, M. LaBelle, L. A. Trimble, N. Chauret, C. Li, N. Sawyer, N. Tremblay, S. Lamontagne, M.-C. Carriere, D. Denis, G. M. Grieg, D. Slipetz, K. M. Metters, R. Gordon, C. C. Chan and R. J. Zamboni, *Bioorg. Med. Chem. Lett.*, 2005, **15**, 527.

[43] A. W. Oxford, R. J. Davis, R. A. Coleman, K. L. Clark, N. V. Harris, G. Fenton, G. Hynd, K. A. J. Stuutle, J. M. Sutton, M. R. Ashton, E. A. Boyd and S. A. Brunton, *US Patent* 7326732 B2, 2008.

[44] T. Maruyama, M. Asada, T. Shiraishi, H. Egashira, H. Yoshida, T. Maruyama, S. Ochuchida, H. Nakai, K. Kondo and M. Toda, *Bioorg. Med. Chem.*, 2002, **10**, 975.

[45] T. Maruyama, M. Asada, T. Shiraishi, A. Ishida, H. Yoshida, T. Maruyama, S. Ochuchida, H. Nakai, K. Kondo and M. Toda, *Bioorg. Med. Chem.*, 2002, **10**, 989.

[46] T. Maruyama, M. Asada, T. Shiraishi, H. Yoshida, H. Yoshida, T. Maruyama, S. Ochuchida, H. Nakai, K. Kondo and M. Toda, *Bioorg. Med. Chem.*, 2002, **10**, 1743.

[47] R. L. Smith, T. Lee, N. P. Gould and E. J. Cragoe, Jr., *J. Med. Chem.*, 1977, **20**, 1292.

[48] X. Billot, A. Chateauneuf, N. Chauret, D. Denis, G. Grieg, M.-C. Mathieu, K. M. Metters, D. M. Slipetz and R. N. Young, *Bioorg. Med. Chem. Lett.*, 2003, **13**, 1129.

[49] K. O. Cameron, B. A. Lefker, M. Y. Chu-Moyer, D. T. Crawford, P. DaSilva Jardine, S. L. DeNinno, S. Gilbert, W. A. Grasser, H. Ke, B. Lu, T. A. Owen, V. M. Paralkar, H. Qi, D. O. Scott, D. D. Thompson, C. M. Tjoa and M. P. Zawistoski, *Bioorg. Med. Chem. Lett.*, 2006, **16**, 1799.

[50] T. R. Elworthy, D. J. Kertesz, W. Kim, M. G. Roepel, L. Quattrocchio-Setti, D. B. Smith, J. L. Tracy, A. Chow, F. Li, E. R. Brill, L. K. Lach, D. McGee, D. S. Yang and S.-S. Chiou, *Bioorg. Med. Chem. Lett.*, 2004, **14**, 1655.

[51] T. R. Elworthy, E. R. Brill, S.-S. Chiou, F. Chu, J. R. Harris, R. T. Hendricks, J. Huang, W. Kim, L. K. Lach, T. Mirzadegan, C. Yee and K. A. M. Walker, *J. Med. Chem.*, 2004, **47**, 6124.

[52] Y. Xiao, G. L. Araldi, Z. Zhao, N. Brugger, S. Karra, D. Fischer and E. Palmer, *Bioorg. Med. Chem. Lett.*, 2007, **17**, 4323.

[53] Z. Zhao, G. L. Araldi, Y. Xiao, A. P. Reddy, Y. Liao, S. Karra, N. Brugger, D. Fischer and E. Palmer, *Bioorg. Med. Chem. Lett.*, 2007, **17**, 6572.

[54] T. R. Elworthy, E. R. Brill, C. C. Caires, W. Kim, L. K. Lach, J. L. Tracy and S.-S. Chiou, *Bioorg. Med. Chem. Lett.*, 2005, **15**, 2523.

[55] J. Colucci, Y. Han and J. A. Farand, *WO* 2007/014462 A1, 2007.

[56] C. Liljebris, G. Selen, B. Resul, J. Stjernschantz and U. Hacksell, *J. Med. Chem.*, 1995, **38**, 289.

[57] R. Burk, *US Patent* 6037364, 2000.

[58] Y. Wang, J. A. Wos, M. J. Dirr, D. L. Soper, M. A. deLong, G. E. Mieling, B. De, J. S. Amburgey, E. G. Suchanek and C. J. Taylor, *J. Med. Chem.*, 2000, **43**, 945.

[59] P. Klimko, M. Hellberg, M. Mclaughlin, N. Sharif, B. Severns, G. Williams, K. Haggard and J. Liao, *Bioorg. Med. Chem.*, 2004, **12**, 3451.

[60] T. Nakajima, T. Matsugi, W. Goto, M. Kageyama, N. Mori, Y. Matsumura and H. Hara, *Biol. Pharm. Bull.*, 2003, **26**, 1691.

[61] R. D. Selliah, M. R. Hellberg, N. A. Sharif, M. A. McLaughlin, G. W. Williams, D. A. Scott, D. Earnest, K. S. Haggard, W. D. Dean, P. Delgado, M. S. Gaines, R. E. Conrow and P. G. Klimko, *Bioorg. Med. Chem. Lett.*, 2004, **14**, 4525.

[62] K.-H. Thierauch, C. St. Sturzebecher, E. Schillinger, H. Rehwinkel, B. Raduchel, W. Skuballa and H. Vorbruggen, *Prostaglandins*, 1988, **35**, 855.

[63] M. R. Hellberg, R. E. Conrow, N. A. Sharif, M. A. McLaughlin, J. E. Bishop, J. Y. Crider, W. D. Dean, K. A. Dewolf, D. R. Pierce, V. L. Sallee, R. D. Selliah, B. S. Severns, S. J. Sproull, G. W. Williams, P. W. Zinke and P. G. Klimko, *Bioorg. Med. Chem.*, 2002, **10**, 2031.

[64] G.-L. Araldi, *WO* 20061061366 A1, 2006.

[65] A. Naganawa, T. Saito, Y. Nagao, H. Egashira, M. Iwahashi, T. Kambe, M. Koketsu, H. Yamamoto, M. Kobayashi, T. Maruyama, S. Ochuchida, H. Nakai, K. Kondo and M. Toda, *Bioorg. Med. Chem.*, 2006, **14**, 5562.

[66] A. Naganawa, T. Matsui, T. Saito, M. Ima, T. Tatsumi, S. Yamamoto, M. Murota, H. Yamamoto, T. Maruyama, S. Ochuchida, H. Nakai, K. Kondo and M. Toda, *Bioorg. Med. Chem.*, 2006, **14**, 6639.

[67] A. Naganawa, T. Matsui, M. Ima, T. Saito, M. Murota, Y. Aratani, H. Kijima, H. Yamamoto, T. Maruyama, S. Ochuchida, H. Nakai and M. Toda, *Bioorg. Med. Chem.*, 2006, **14**, 7121.

[68] A. Naganawa, T. Matsui, M. Ima, K. Yoshida, H. Tsurata, S. Yamamoto, H. Yamamoto, H. Okada, T. Maruyama, H. Nakai, K. Kondo and M. Toda, *Bioorg. Med. Chem.*, 2006, **14**, 7774.

[69] Y. Ducharme, M. Blouin, M.-C. Carriere, A. Chateauneuf, B. Cote, D. Denis, R. Frenette, G. Greig, S. Kargman, S. LaMontagne, E. Martins, F. Nantel, G. O'Neill, N. Sawyer, K. M. Metters and R. W. Friesen, *Bioorg. Med. Chem. Lett.*, 2005, **15**, 1155.

[70] A. Hall, S. Atkinson, S. H. Brown, I. P. Chessell, A. Chowdhury, N. M. Clayton, T. Coleman, G. M. P. Giblin, R. J. Gleave, B. Hammond, M. P. Healy, M. R. Johnson, A. D. Michel, A. Naylor, R. Novelli, D. J. Spalding and S. P. Tang, *Bioorg. Med. Chem. Lett.*, 2006, **16**, 3657.

[71] A. Hall, S. H. Brown, I. P. Chessell, A. Chowdhury, N. M. Clayton, T. Coleman, G. M. P. Giblin, B. Hammond, M. P. Healy, M. R. Johnson, A. Metcalf, A. D. Michel, A. Naylor, R. Novelli, D. J. Spalding and J. Sweeting, *Bioorg. Med. Chem. Lett.*, 2007, **17**, 732.

[72] G. M. P. Giblin, R. A. Bit, S. H. Brown, H. M. Chaignot, A. Chowdhury, I. P. Chessell, N. M. Clayton, T. Coleman, A. Hall, B. Hammond, D. N. Hurst, A. D. Michel, A. Naylor, R. Novelli, T. Scoccitti, D. Spalding, S. P. Tang, A. W. Wilson and R. Wilson, *Bioorg. Med. Chem. Lett.*, 2007, **17**, 385.

[73] A. Hall, R. A. Bit, S. H. Brown, H. M. Chaignot, I. P. Chessell, T. Coleman, G. M. P. Giblin, D. N. Hurst, I. R. Kilford, X. Q. Lewell, A. D. Michel, S. Mohamed, A. Naylor, R. Novelli, L. Skinner, D. Spalding, S. P. Tang and R. Wilson, *Bioorg. Med. Chem. Lett.*, 2006, **16**, 2666.

[74] S. C. McKeown, A. Hall, G. M. P. Giblin, O. Lorthioir, R. Blunt, X. Q. Lewell, R. J. Wilson, S. H. Brown, A. Chowdhury, S. P. Watson, I. P. Chessell, A. Pipe, N. Clayton and P. Goldsmith, *Bioorg. Med. Chem. Lett.*, 2006, **16**, 4767.

[75] S. C. McKeown, A. Hall, R. Blunt, S. H. Brown, I. P. Chessell, A. Chowdhury, G. M. P. Giblin, M. P. Healy, M. R. Johnson, O. Lorthioir, A. D. Michel, A. Naylor, X. Lewell, S. Roman, S. P. Watson, W. J. Winchester and R. J. Wilson, *Bioorg. Med. Chem. Lett.*, 2007, **17**, 1750.

[76] A. Hall, S. H. Brown, A. Chowdhury, G. M. P. Giblin, M. Gibson, M. P. Healy, D. G. Livermore, R. J. MacArthurWilson, D. A. Rawlings, S. Roman, E. Ward and C. Willay, *Bioorg. Med. Chem. Lett.*, 2007, **17**, 4450.

[77] M. Belley, C. C. Chan, Y. Gareau, M. Gallant, H. Juteau, K. Houde, N. Lachance, M. LaBelle, N. Sawyer, N. Tremblay, S. Lamontagne, M.-C. Carriere, D. Denis, G. M. Grieg, D. Slipetz, R. Gordon, N. Chauret, C. Li, R. J. Zamboni and K. M. Metters, *Bioorg. Med. Chem. Lett.*, 2006, **16**, 5639.

[78] H. Juteau, Y. Gareau, M. LaBelle, C. F. Sturino, N. Sawyer, N. Tremblay, S. LaMontagne, M.-C. Carriere, D. Denis and K. M. Metters, *Bioorg. Med. Chem.*, 2001, **9**, 1977.

[79] M. Gallant, M. C. Carriere, A. Chateauneuf, D. Denis, Y. Gareau, C. Godbout, G. Greig, H. Juteau, N. LaChance, P. Lacombe, S. Lamontagne, K. M. Metters, C. Rochette, R. Ruel, D. Slipetz, N. Sawyer, N. Tremblay and M. LaBelle, *Bioorg. Med. Chem. Lett.*, 2002, **12**, 2583.

[80] K. Hattori, A. Tanaka, N. Fujii, H. Takasugi, Y. Tenda, M. Tomita, S. Nakazato, K. Nakano, Y. Kato, Y. Kono, H. Murai and K. Sakane, *J. Med. Chem.*, 2005, **48**, 3103.

[81] R. J. Wilson, G. M. P. Giblin, S. Roomans, S. A. Rhodes, K.-A. Cartwright, V. J. Shield, J. Brown, A. Wise, J. Chowdhury, S. Pritchard, J. Coote, L. S. Noel, T. Kenakin, C. L. Burns-Kurtis, V. Morrison, D. W. Gray and H. Giles, *Br. J. Pharmacol.*, 2006, **148**, 326.

[82] K. Takeuchi, A. Tanaka, S. Kato, E. Aihara and K. Amagese, *J. Pharmacol. Exp. Ther.*, 2007, **322**, 903.

[83] M. Tomita, X. Li, Y. Okada, F. N. Woodiel, R. N. Young, C. C. Pilbeam and L. G. Raisz, *Bone*, 2002, **30**, 159.

[84] T. Matsuoka and S. Naumiya, *Scientific World Journal*, 2007, **7**, 1329.

[85] J. Waugh and B. Jarvis, *Drugs Aging*, 2002, **19**, 465.

[86] M. Abramovitz, Y. Boie, T. Nguyen, T. H. Rushmore, M. A. Bayne, K. M. Metters, D. M. Slipetz and R. Grygorczyk, *J. Biol. Chem.*, 1994, **269**, 2632.

[87] P. Denis, A. Lafuma, B. Khoshnood, V. Mimaud and G. Berdeaux, *Curr. Med. Res. Opin.*, 2007, **23**, 601.

[88] A. Sutton, A. Gilvarry and A. Ropo, *J. Ocul. Pharmacol. Ther.*, 2007, **23**, 359.

[89] M. Dukes, W. Russell and A. L. Walpole, *Nature*, 1974, **250**, 330.

[90] Y. Takagi, T. Nakajima, A. Shimazaki, M. Kageyama, T. Matsugi, Y. Matsumura, B. T. Gabelt, P. L. Kaufmann and H. Hara, *Exp. Eye Res.*, 2004, **78**, 767.

[91] B. G. Schluz, B. Beckman, B. Muller, G. Schroder, B. Maass, K.-H. Thierauch, B. Buchmann, P. F. J. Verhallen and W. Frohlich, *Adv. Prostaglandin Thromboxane Leukot. Res.*, 1991, **21**(21), B591.

[92] M. R. Hellberg, M. A. McLaughlin, T. R. Dean, L. Desantis, N. Sharif, A. Kothe and V. L. Sallee, *Invest. Ophthalmol. Vis. Sci.*, 2001, **42**, S837.

[93] V. M. Paralkar, F. Borovecki, H. Z. Ke, K. O. Cameron, B. Lefker, W. A. Grasser, T. A. Owen, M. Li, P. DaSilva-Jardine, M. Zhou, R. L. Dunn, F. Dumont, R. Korsmeyer, P. Krasney, T. A. Brown, D. Plowchalk, S. Vukicevic and D. D. Thompson, *Proc. Natl. Acad. Sci. USA*, 2003, **100**, 6736.

[94] K. Nakao, A. Murase, H. Ohshiro, T. Okumura, K. Tanguchi, Y. Murata, M. Masuda, T. Kato, Y. Okumura and J. Takada, *J. Pharmacol. Exp. Ther.*, 2007, **322**, 686.

[95] Pfizer, company pipeline, 31 July 2007.

[96] ONO, company pipeline, March 2006.

[97] J. Wos, *6th International Conference on Bioactive Lipids in Cancer, Inflammation and Related Diseases*, Boston, MA, September 12, 1999.

[98] *Expert Opin. Ther. Pat.*, 2004, **14**, 435.

[99] S. Sarkar, A. R. Hobson, A. Hughes, J. Growcott, C. J. Woolf, D. G. Thompson and Q. Aziz, *Gastroenterology*, 2003, **124**, 18.

[100] A. A. Calder, *Reprod. Fert. Develop.*, 1990, **2**, 553.

[101] ONO, Pharmaceutical Press Release 2005, February 04.

[102] Kyoto University, clinicaltrial.gov, NCT00296556, February 2006.

CHAPTER 19

Human Exposure and Dose Projections

Punit H. Marathe, Christine Huang and A. David Rodrigues

1. INTRODUCTION

Increasing drug development costs, and competitive pressure to bring new drugs to market, has forced pharmaceutical companies to ascertain new opportunities to improve productivity and increase profitability. Toward this end, scientists are

Bristol-Myers Squibb, Department of Metabolism and Pharmacokinetics, PO Box 4000, Princeton, NJ 08543, USA

Annual Reports in Medicinal Chemistry, Volume 43
ISSN 0065-7743, DOI 10.1016/S0065-7743(08)00019-5

researching and developing new compounds faster. For example, combinatorial and other rapid chemistry approaches are being used to synthesize hundreds of compounds at a time. Such compounds are screened for potency and selectivity, employing automated high-throughput screening platforms. However, one cannot rely solely on greater potency and increased selectivity to advance compounds to the clinic. Attention must also be paid to the "developability" of drug candidates. Suboptimal drug metabolism (DM) and pharmacokinetic (PK) properties have long been recognized as major contributors to the failure of potential new therapies in early clinical trials. This has precipitated the realignment of DMPK scientists toward a focused discovery effort within the pharmaceutical industry. The aim of DMPK scientists has expanded well beyond simple measurement of PK parameters, such as absolute bioavailability and half-life, and now encompasses all aspects of absorption–distribution–metabolism–excretion (ADME) and pharmacodynamics (PD) *in vitro* and *in vivo*. The goal of ADME scientists is to view such data holistically and guide medicinal chemists toward compounds with the best balance of properties.

1.1 Why predict human exposure?

Theoretical principles dictate that the efficacy of a new molecular entity will be related to its concentration at the molecular target. In the absence of knowing the target concentrations (e.g., liver, brain or tumor concentrations), efficacy is often correlated to the plasma concentration vs. time profile and PK parameters derived from it (e.g., maximum vs. trough plasma concentrations (C_{max} vs. C_{min}), area under the curve (AUC)). Toxicity is often manifested as exaggerated pharmacology or off-target effects and is similarly related to plasma exposures. Novel targets with intricate underlying pharmacology often require optimal PK-ADME profiles for dissociating efficacy from toxicity. Projection of human exposure prior to human dosing is a major effort during the characterization of lead compounds in discovery. Although the no observable adverse effect level (NOAEL) is one of the important parameters used to guide dose selection in first-in-human studies, analysis of projected human exposure allows appropriate escalation of doses to reach the efficacious exposures quickly. This is especially important when treating patients in early clinical trials (e.g., in oncology), because exposure to subefficacious doses is considered unethical. Furthermore, prediction of human exposure allows one to model human PK at steady-state and set safety margins (relative to animal toxicokinetic data). Each of the exposure parameters (C_{max}, C_{min}, AUC, half-life) are useful in guiding selection of appropriate dosing regimens in clinical trials.

1.2 Why predict human efficacious dose?

The human efficacious dose is the dose in humans that leads to efficacious exposures. While achieving appropriate exposures enables successful clinical trials, the dose needed to get there is an important consideration for cost of goods. Prediction of efficacious doses focuses attention on the feasibility of various dosage forms and dosing regimens at the time of nomination. For example, compound solubility, permeability and dissolution rate will dictate the maximum absorbable dose from the gastrointestinal tract. A high predicted human dose is an important risk-benefit consideration as it will require considerable resources to develop an appropriate formulation to deliver the needed dose, as well as presenting a risk of off-target toxicity. Moreover, the dose projection provides an early estimate of the amount of compound needed to complete early clinical studies, and so provides process chemists the information they need.

1.3 Prospective vs. retrospective projections

The projection of human PK involves the generation and integration of preclinical datasets, and numerous approaches have been published over the years [1,2]. The literature is replete with examples of retrospective projections where availability of human data (exposures and efficacious doses) on diverse sets of compounds has enabled comparison of various methodologies for successful predictions [3–5]. For example, allometric scaling with a rational use of rule of exponents was shown to yield predictions of human clearance with better precision and accuracy than simple allometry [6]; or monkey tended to provide the most qualitatively and quantitatively accurate predictions of human clearance compared to other species [7]. Such information, while valuable in furthering our understanding of ADME properties in affecting human exposures, does not provide guidance in the *prospective* design of clinical trials. On the other hand, projections conducted in the preclinical discovery phase at the time of candidate nomination, without prior knowledge of the human PK behavior, are highly sought after in the pharmaceutical industry. Because of the prospective nature of these projections, one may not be able to compare different methodologies to see which ones offer the best predictions. However, when carried out prospectively, projections enable decision making; allocate resources to the most worthwhile series (based on preclinical data) and support the design of first-in-human studies.

2. THEORETICAL FRAMEWORK AND APPROACHES

There are many reports in the literature describing the prediction of human PK and some success has been reported. However, relatively fewer reports of prediction of human efficacious doses have been published perhaps due to a lack of appropriate PK/PD models, especially for newer therapeutic targets [8].

2.1 Methods for prediction of human PK parameters

2.1.1 Prediction of CL

Allometric scaling. The concept of allometric scaling is based on the assumption that the anatomical, physiological, and biochemical variables of mammals can be scaled across species based on the body weight [9,10]. The allometric approach is based on the power function, where the body weight of several species is plotted against the PK parameter of interest on a log–log scale. The log–log plot of clearance (CL) vs. body weight of the preclinical species then scales to human CL. Biliary clearance has been scaled with correction factors such as liver blood flow, bile flow, and UDP-glucuronosyl transferase (UDPGT) activity for biliary excreted drugs [6,11]. Similarly correction factors such as glomerular filtration rate and kidney blood flow are used to scale renal clearance for drugs exhibiting passive urinary excretion [2,12,13]. Other correction factors such as maximum life span potential, brain weight (rule of exponents) and protein binding (f_u corrected intercept method) have also been employed as a modification of simple allometry [6].

Scaling of in vitro intrinsic clearance. Because the liver is the major (metabolic) clearance organ for most drugs, subcellular fractions such as hepatic microsomes and hepatocytes have been used to project *in vivo* hepatic clearance [14–16]. The hepatic intrinsic clearance ($CL_{h,int}$) in various species is estimated from liver microsome or hepatocyte data [14,17,18]. Assuming linear kinetics and similar $CL_{h,int}$ of unbound drug *in vitro* and *in vivo*, the $CL_{h,int}$ is calculated from the disappearance of the parent drug after incubation with liver microsomes or hepatocytes. Correction for protein binding differences in microsomes (or hepatocytes) and blood is made with variable degrees of success. Assuming the well-stirred model of hepatic clearance, the hepatic blood clearance (CL_{hb}) is estimated by correcting for liver weight and microsomal (or hepatocyte) content [19]. The projected hepatic clearance is compared to the observed *in vivo* clearance in each animal species evaluated. If the two values are in reasonable agreement (e.g., within 3-fold), systemic clearance in humans is projected from the corresponding *in vitro* human system. If a correction factor is identified for projection of *in vivo* clearance across all the animal species evaluated, a similar correction factor is employed to project human *in vivo* clearance [14].

2.1.2 Prediction of volume of distribution

Allometric scaling. Similar to allometric scaling of CL, the log–log plot of volume of distribution at steady-state (V_{ss}) vs. body weight of the preclinical species scales to human V_{ss}. Plasma and tissue protein binding may play a role in the distribution of a compound and have been used as correction factors for scaling of V_{ss}. Scaling of volume of distribution during the elimination phase (V_β) and tissue distribution volume (V_t) are also reported [7,20].

Physiologically based pharmacokinetic (PBPK) modeling. Volume of distribution in human can also be predicted using PBPK modeling with compound specific (e.g., Log P, pK_a, protein binding) and physiological inputs (e.g., fractional tissue volume of water, neutral lipids, phospholipids, ratio of binding proteins in extracellular fluid to plasma) [20–23]. Mechanistic equations incorporating expressions of dissolution in water, lipid partition, and affinity for cellular constituents are employed to predict V_{ss}. Further improvement has been realized by incorporating binding to extracellular proteins and differential protein binding of weak bases, acids, and neutral compounds. For example, incorporating blood-brain barrier permeability and enterohepatic recirculation (recycling of drug between the intestine and liver) in a generic PBPK model has improved prediction for brain and gut distribution, respectively, of diazepam [24].

Computation modeling based on in vitro parameters. Based on *in vitro* physicochemical parameters (e.g., pK_a), V_{ss} in obese subjects has been predicted for aminoglycosides such as amikacin and gentamicin [25]. Lombardo et al. [26,27] utilized the Oie–Tozer equation, with incorporation of $E \log D(7.4)$ and fraction of compound ionized at pH 7.4, for the prediction of neutral and basic drug V_{ss} in humans.

Single preclinical species. Although this approach is not widely applied, Ward et al. [7] have shown that volume of distribution in humans is similar to that in monkeys. Furthermore, because of the similar anatomy, physiology, and body size between chimpanzees and humans, V_{ss} in humans may also be predicted from chimpanzees with or without correction for protein binding [28].

2.1.3 Prediction of half-life

Allometric scaling. Human half-life ($t_{1/2}$) is calculated from the allometrically scaled CL and V_{ss} assuming one compartment kinetics [21]. However, this approach may not be applicable when drugs follow multi-compartment kinetics in which case scaling of V_{β} is recommended for prediction of $t_{1/2}$. Similar to the allometric scaling for CL and V_{ss}, the log–log plot of half-life ($t_{1/2}$) vs. body weight of the preclinical species scales to human $t_{1/2}$ [12]. However, this approach is not recommended since $t_{1/2}$ is a PK parameter dependent on CL and V_{ss}.

PBPK modeling. Many commercial PBPK modeling programs have been developed in recent years. Plasma concentration profiles can be simulated based on physiological- and compound-specific parameters. Half-life is then determined from the simulated PK profiles [29,30].

2.1.4 Prediction of oral absorption and oral bioavailability

Hepatic CL and oral bioavailability in animals. Based on the hepatic clearance projected from *in vitro* data or observed in animals, as well as animal oral bioavailability, the fraction of oral dose absorbed into the portal vein ($f_a \times f_g$) can be estimated. In turn, such information can be used to project human oral bioavailability by considering first-pass liver metabolism estimated from *in vitro* human data. The average value of $f_a \times f_g$ from animal species is then used along with human *in vitro* data from liver microsomes (or hepatocytes) to project human oral bioavailability.

Computation techniques. Hou et al. [31,32] computed oral absorption of compounds absorbed by passive diffusion using several molecular descriptors (e.g., lipophilicity, topological polar surface area, square of the hydrogen-bond donor number). The model was able to predict the fraction absorbed with less than 10% prediction errors. More work is needed when absorption involves active transporters.

Compartmental absorption modeling. Compartmental absorption and transit models have been used to predict oral absorption in humans [33]. For example,

GastroPlus™ utilizes the Advanced Compartmental Absorption and Transit Model (ACAT model) to predict the extent of oral absorption through the gastrointestinal tract [34–36]. Several other commercial computer programs also employ such models to predict oral absorption.

Average from animal species. A conventional approach for the prediction of fraction absorbed and oral bioavailability in humans is using an average from animal species. Although these parameters are obtained from a solution formulation, it is important to assess the impact of aqueous solubility and dissolution rate by testing simple formulations such as an aqueous suspension or drug in capsule. Zhao et al. [37] have shown that rat oral bioavailability may be a good predictor of human oral bioavailability. Other scientists have reported better success with the monkey [4]. Understanding the physicochemical and metabolism properties of a compound is critical when selecting the best strategy for predicting human oral bioavailability.

2.2 Methods for prediction of human PK profiles

Species-invariant time method. Dedrick plots and related modified methods are based on the concept of physiological time, with the concentration vs. time profile in animals being transformed by correction factors to yield a human profile [38,39]. The species-invariant time method is a complex Dedrick plot, with exponents obtained from best fitting of the intravenous plasma concentration vs. time profiles in rats, dogs, and monkeys. The best fit of the exponents of body weight is then used to transform and project the plasma concentration vs. time profile in humans [39,40] from which CL, half-life, and other PK parameters can be calculated. Combined with oral absorption and bioavailability information, plasma concentration vs. time profile after oral administration can be predicted.

C_{ss}-MRT method. This method is based on the assumption that concentration vs. time profiles are linear and similar across species including humans, and that normalized curves derived from a variety of animal species can be superimposed [27]. The normalized curve is derived by dividing the concentration and time scales by C_{ss} (defined as dose/V_{ss}) and MRT, respectively. The plasma concentration vs. time profile in humans can be obtained by multiplying the concentration and time scales of the normalized curve obtained from the animal data by projected C_{ss} and MRT in humans, respectively. Together with the fraction absorbed and oral bioavailability, plasma concentration vs. time profile after oral administration can be predicted.

Chimpanzee as an animal model. Other than the typical laboratory animals (i.e., mouse, rat, dog, and cynomolgus monkey), the chimpanzee (*pan troglodytes*) has been used occasionally for projection of human PK [28,41,42]. The chimpanzee has 98.8% genetic similarity to human [43] and shows similar anatomy, physiology, and endocrinology compared to human. Although the hepatic blood flow is comparable between the chimpanzee and human, the hepatic microsomal cytochrome P450 activities in this species are not fully characterized. Therefore, its use for human PK projection should be examined carefully for oxidatively metabolized drugs.

2.3 Methods for prediction of human efficacious doses

Exposure required for efficacy in animal disease models. Animal models have been widely used in various therapeutic areas to mimic human diseases. By combining the projected human PK parameters, and the exposure required for efficacy in animal disease models, it is possible to project efficacious doses in humans. For example, if a minimum AUC is necessary for demonstrating efficacy in the animal model, that AUC along with the projected human CL and oral bioavailability are used to project human efficacious dose. A correction for species differences in potency against the target is applied whenever necessary.

A specialized approach is adopted to estimate human efficacious dose and exposure when the liver is the target organ. The compound is administered *via* intravenous (IV), intraportal, and oral routes to rats and guinea pigs and liver exposure, oral absorption, and hepatic extraction ratio (ER) are determined. Liver exposure is calculated as the maximum amount of the compound available to the liver (dose $\times f_a \times$ ER) and used to project human efficacious doses.

Exposure required to maintain adequate in vivo *concentrations based on* in vitro *potency.* In many disease areas (e.g., antiviral, ion channel inhibition, oncology, and thrombosis), a good correlation has been demonstrated between *in vitro* potency and *in vivo* activity [44]. For example, free plasma concentrations and the *in vitro* IC_{50} value for HERG-encoded potassium channel blockade are used for projection of concentrations at which QT prolongation may be a safety risk. For some targets, it is desirable to achieve trough concentrations that are 2–3-fold in excess of the *in vitro* IC_{50} or IC_{90} values. In certain therapeutic areas, an *in vivo* animal disease model representative of human disease is not available [45,46]. In such cases, it is necessary to project human efficacious dose by considering *in vitro* potency as the threshold concentration for the desired therapeutic effect. Human PK profiles may be projected using the species-invariant time method or the C_{ss}-MRT method and the efficacious dose is derived as the dose that maintains the required C_{min} for efficacy. When deemed important, *in vivo* potency may be corrected by differences in protein binding [47,48].

Experience with clinical leads. Prior experience garnered with drugs already in clinical trials, which have shown proof of concept, can provide significant insight into the efficacious doses of the new chemical entity in question. The feedback from such clinical leads is extremely important because back-up compounds are often from a similar chemotype or have similar physicochemical properties and mechanism of action. Therefore, lead compounds can be used to set up correction factors that can be applied when projecting efficacious dose.

2.4 Methods for evaluating success of predictions

Projection ratio. To assess the success of the projections, the projected parameters are compared to the observed clinical data. When a range of values is projected an arithmetic mean is calculated and used for the assessment. When the projection ratio (observed vs. projected or projected vs. observed) is within 2-fold, the projection is generally considered "accurate". A projection ratio of

Table 1 Overall success rate for the projection of human exposure and efficacious dose [1]

Assessment of the projection (projection ratio)[a]		Human exposure (AUC) (35 compounds)	Human efficacious dose (10 compounds)
Accurate	≤ 2	20 (57%)[b]	7 (70%)
Acceptable	> 2–4	11 (31%)	2 (20%)
Needs improvement	> 4	4 (11%)	1 (10%)

[a]For both AUC and efficacious dose, the projection ratio is defined as the ratio of observed vs. projected or projected vs. observed.
[b]Values in parentheses represent percentage of total number of compounds in the dataset.

greater than 2-fold and up to 4-fold is characterized "acceptable". A projection ratio of greater than 4-fold usually indicates that an improvement is needed and likely reasons for the discrepancy are investigated (Table 1).

Average fold error and root mean squared error. The success (bias) of the projections is assessed by calculation of the geometric mean of the ratio of projected and observed values (average fold error) for all compounds in the dataset [30,49]. An average fold error of ≤2 is considered to yield successful predictions.

$$\text{Average fold error} = 10^{\left[\frac{\Sigma \left| \log \frac{\text{Projected}}{\text{Observed}} \right|}{N}\right]}$$

The observed and predicted parameter values are used to calculate the root mean squared error as another measure of precision [2,36].

$$\text{RMSE} = \sqrt{\frac{\sum_{1}^{N}(\text{observed} - \text{predicted})^2}{N}}$$

Graphical plots of predicted vs. observed values. The predicted values are plotted vs. the observed values and compounds are categorized in various bins (e.g., low, medium, and high CL). Accurate categorization of the compounds is difficult when the predictions are close to the boundary of the bins. The predicted values can also be correlated to the observed values, calculating an R^2 value, and comparing with the line of identity [1,7,50].

3. DATA ANALYSIS AND APPLICATIONS

3.1 Collection of data

3.1.1 *In vivo* preclinical data
As part of lead characterization, *in vivo* PK studies are conducted in various animal species (mouse, rat, dog, monkey). Typically, the animals receive a single IV injection and single oral dose. The choice of vehicles depends on the compound, route of administration, and animal species. Serial blood samples are collected, plasma prepared, and the samples processed appropriately. Concentrations of each compound in plasma are determined using validated assays,

most typically involving liquid chromatography-tandem mass spectrometry (LC-MS/MS), and PK parameters are calculated. *In vivo* efficacy is assessed using various animal models (e.g., mice, rats, rabbits, or guinea pigs) representative of human diseases using single or repeated dosing. To assess the efficacious exposure, blood is sampled serially either from the same group of animals or a satellite group. Concentrations of each compound in plasma are determined similar to the PK studies and systemic exposure (e.g., C_{max}, AUC, and C_{min}) required for efficacy or PD response is determined. To assess the relevance of each of the PK parameters in determining observed response, efficacy studies are undertaken by altering the dosing regimen (e.g., QD vs. BID, bolus vs. infusion).

3.1.2 Clinical data

PK information in humans (e.g., AUC) is obtained typically from single ascending dose studies. The studies consist of normal healthy volunteers or patients depending on the therapeutic area. Concentrations of each compound in plasma are determined using validated LC-MS/MS assays. Exposure information is obtained at doses in the lower range of the dose-escalation scheme or from doses close to the projected efficacious doses. In the absence of definitive dose–response studies, efficacious dose information in humans is obtained from either biomarker response in Phase I studies or doses selected for Phase II studies.

3.1.3 *In vitro* metabolism data

The use of *in vitro* DM data to predict PK is now commonplace [10,21,51,52]. In this instance, metabolic turnover rates are calculated after incubation with liver microsomes or suspensions of primary hepatocytes. The $CL_{h,int}$ is then scaled to an *in vivo* intrinsic clearance by using appropriate scaling factors [53] and various models of liver clearance [54].

3.1.4 *In vitro* determination of potency

In vitro potencies of compounds, against the therapeutic target, are determined by optimized cellular assays and concentrations required for 50% or 90% inhibition or receptor occupancy (IC_{50} or IC_{90}) are calculated. The *in vitro* IC_{50} or IC_{90} values corrected for protein binding differences in the cell media (vs. serum) are used to project efficacious doses in certain therapeutic areas (e.g., antivirals).

3.1.5 Molecular descriptors

As described earlier, PBPK modeling is being recognized as a valuable methodology to understand a drug's PK in a complex biological system. The PBPK model input parameters include both drug-independent and drug-specific properties. The first set of properties comprises data underlying the physiological processes (e.g., blood flow and organ volumes). The second set comprises drug-specific biochemical parameters (e.g., intrinsic clearance of each organ involved in its elimination and tissue-to-plasma partition coefficients). Obtaining these partition coefficients (k_p) experimentally is resource intensive and a variety of *in vitro*-based prediction tools have been developed for the estimation of PBPK model input parameters [24,55]. Such prediction tools require commonly determined biochemical and physicochemical molecular descriptors such as molecular weight (MW), calculated partition coefficient ($c \log P$), ionization constant (pK_a), number of hydrogen-bond acceptors (HA), number of rotatable bonds, and polar surface area (PSA).

3.2 Integration of data

Preclinical data are obtained from a variety of sources such as *in vitro* metabolism studies, *in vivo* PK and efficacy studies and physicochemical properties. Proper integration of this information is critical to the success of both prospective and retrospective projections of human PK and efficacious doses.

Conventional allometric scaling has been shown to be predictive for drugs with passive renal excretion as well as for drugs rapidly metabolized where clearance can be related to hepatic blood flow. However, for metabolized compounds characterized by low or intermediate hepatic extraction ratio, the elimination is dependent on biochemical parameters such as $CL_{h,int}$ and protein binding. To improve the predictions of metabolic CL for this class of compounds, correction factors such as brain weight or maximum life span have been used with some success [11]. Increased accuracy in the prediction of metabolic clearance is also achieved by combining *in vitro* and *in vivo* metabolic data in animals and humans [56].

Interspecies differences in biliary excretion and bile flow rates make allometric scaling generally not applicable to drugs which are excreted in bile as a predominant pathway. For such compounds, various correction factors have been proposed such as bile flow rate, microsomal UDPGT activity, and liver blood flow [57,58]. In a study by Mahmood [11], the predictive performance of several methods was evaluated for the prediction of human clearance of eight drugs which are excreted in the bile. The criteria for a drug undergoing biliary excretion was based on the percentage of unchanged drug or conjugate excreted in the bile. The results indicated that the rule of exponents with a correction factor based on bile flow rate per kilogram liver weight was the best approach. The worst approach was the product of CL, bile flow and UDPGT activity.

To identify the most appropriate species for optimization of human clearance, Ward and Smith [7] utilized a large dataset consisting of 103 nonpeptide xenobiotics with IV PK data (rat, dog, monkey, and human) and both body weight- and hepatic blood flow-based methods for scaling of CL. Allometric scaling approaches, particularly those using data from only two of the preclinical species, were less successful than methods based on CL as a set fraction of liver blood flow from an individual species. The monkey tended to provide the most qualitatively and quantitatively accurate predictions of human clearance. Additionally, the availability of data from both commonly used nonrodent species (dog and monkey) did not ensure enhanced predictive quality compared with having only monkey data.

As oral exposure of drugs is governed by oral clearance (CL/F), several studies have evaluated the success of predicting oral clearance in humans. In a study by Wajima et al. [59], a partial least squares regression model, using animal oral clearance data from only two species and calculated molecular structural parameters (MW, $c \log P$, and HA) were found to predict human oral clearance better than the allometric approaches. In another study by Feng et al. [60], human CL/F was more accurately estimated using unbound drug concentration and incorporating brain weight in the allometric relationship.

De Buck et al. [2] reported retrospective predictions of human PK of 26 clinically tested drugs using PBPK modeling. Best V_{ss} predictions were obtained using combined predicted and experimentally determined rat tissue-to-plasma partitioning (k_p) data and correcting for interspecies differences in plasma protein binding. Best CL predictions were obtained as the sum of scaled rat renal clearance and hepatic clearance projected from *in vitro* metabolism data and disregarding both blood and microsomal or hepatocyte binding. Predictions of oral C_{max} and AUC with Gastroplus™ were acceptable with 65% and 74% of drugs within 2-fold.

Another study of retrospective predictions of human PK, exposures, and therapeutic dose of 63 marketed oral drugs was recently reported by McGinnity et al. [8]. Oral absorption predicted from oral rat PK studies was lower than the observed human absorption for most drugs, even when solubility and permeability appeared not to be limiting. Absorption in the dog appeared to be more representative of human for compounds absorbed *via* the transcellular pathway. Using predicted as well as observed PK and maintaining minimum efficacious concentrations (MEC) estimated from *in vitro* potency assays as minimum concentrations at steady-state ($C_{ss,min}$), the projected dose was equal to or greater than the actual clinical dose for 80% of the drugs. In the drug discovery environment, significant under- rather than over-prediction of human efficacious dose may be of most concern; hence targeting MEC as $C_{ss,min}$ appears to be a conservative approach acceptable to medicinal chemists.

An excellent example of integration of preclinical data to predict human exposures and efficacious doses was recently reported [61]. An exposure–response relationship for an immunoreactive compound was established in the cynomolgus monkey and dog, with *ex vivo* receptor occupancy used as a PD marker from prior experience with a clinical candidate. The *ex vivo* IC$_{50}$ was 1.5- to 3-fold higher than the *in vitro* IC$_{50}$ value. PK after IV and oral administration were available from rat, marmoset, and cynomolgus monkey. *In vitro* plasma protein binding and blood cell partitioning data from rat, dog, cynomolgus monkey, and human were available, as were *in vitro* microsomal CL in rat, monkey, and human. The human exposure–response relationship was assumed to follow a model similar to dog and monkey with human *in vivo* IC$_{50}$ assumed to be 2-fold higher than the *in vitro* IC$_{50}$. The *in vivo* CL was first predicted from microsomal data in rat and monkey, using physiological scaling parameters and well-stirred model of hepatic CL. Since a greater than 2-fold difference was observed between predicted and observed *in vivo* CL values, allometric scaling was applied based on rat, marmoset, and monkey IV clearances and also to predict V_{ss} from the same three species. Because oral bioavailability was moderate to high in three species, the human bioavailability was also assumed to be high with no absorption limitation. The bioavailability in human was predicted from hepatic CL and the oral PK profile in human was predicted assuming a nominal value for the absorption rate constant. Finally, the human predicted PD relationship was integrated with the human predicted PK parameters to provide simulations of response-time profile with different dosing regimens.

3.3 Use of multiple approaches

Several studies have reported the use of multiple approaches for predicting human PK parameters. Obach et al. [21] retrospectively examined 4 methods for predicting volume of distribution, 12 methods for predicting clearance, and 4 methods for predicting half-life and found many of them to provide successful prediction of human PK parameters. Ward et al. [7] examined methods for predicting human V_{ss} and MRT and concluded that monkey data provided the most accurate prediction. When comparing various projection methods, it is clear that no single method is more predictive than another. In fact, a combination of various methods yielding similar projections gives greater confidence at the time of compound nomination. Because prospective projections are difficult, especially for first-in-class compounds, the use of more than one method for consistency in projections is strongly recommended. It is also obvious that clinical data is extremely useful and needs to be made available to the discovery scientists through close communication between development and discovery scientists. Some prediction approaches are developed based on chemotype- or compound-specific requirements and require extensive validation to understand their limitations. Because marketed drugs have very different physicochemical properties compared to the majority of compounds in development, utilization of different methods integrating all available *in silico*, *in vitro*, and preclinical *in vivo* data is necessary to predict human PK and PD parameters.

3.4 Special cases and challenges

3.4.1 Lack of availability of final form and formulation

Very often in drug discovery, preclinical PK studies are conducted with less than the most optimal form of the drug substance by administering the compound(s) in a mixture of nonaqueous solvents. The former situation is likely to reduce the observed exposure while the latter is likely to increase the exposure observed in preclinical species. These uncertainties need to be considered when predicting oral bioavailability in humans. For example, oral exposure for a compound with dissolution-rate limited absorption is likely to decrease from a solid dosage form. Unfortunately, most formulation efforts only start after the compound is nominated for development. Such efforts lead to selection of the most optimal form (salt, free base, or specific polymorph) and enable optimization of formulations after selection of the most appropriate excipients. The selection of the best form and formulation at the development stage is likely to yield oral exposures at least similar to, or at times greater than, those obtained in a discovery setting.

3.4.2 Species differences in PK and PD

Although animal PK data are useful for predicting human PK, there is no general agreement as to which species best represents human. Occasionally it is possible to consider the similarity in the *in vitro* metabolic profile as a determinant of which species may (or may not) be more predictive of human. If a compound shows propensity for Phase II metabolism (e.g., direct glucuronidation or sulfation), primary hepatocytes are employed for CL prediction [62,63]. Human CL from liver microsomes and hepatocytes can be underestimated when a compound is metabolized by extra-hepatic routes. Unique metabolic pathways in man, enterohepatic recirculation, and species differences in the conversion of prodrugs to the active moiety, are additional complicating factors that need to be considered. Despite concerns related to species differences, several correction factors such as maximum life span potential and brain weight have been proposed for the projection of human clearance [64].

Projection of efficacious doses is heavily dependent on the animal disease models used in each of the therapeutic areas. Assumptions involved in projecting the efficacious dose go well beyond those used in the projection of human exposure. One not only needs to understand the PK/PD relationship associated with the drug action at the biological target, but one also needs to assume that the molecular basis of animal disease is similar to human disease, and that the drug molecules will act in a similar fashion in both species. Active metabolites further complicate efficacy projections, especially when their contribution varies from animal species to human [65,66].

3.4.3 Prodrugs

When the administered compound is a prodrug, projection of human exposure from preclinical species is quite challenging. Typically this approach is used to increase either the permeability or aqueous solubility of the parent compound. The PK and oral bioavailability of prodrugs are assessed in preclinical species, and *in vitro* in animal and human gut and liver subcellular fractions, in order to ensure conversion to the active moiety. Rapid and complete conversion of the prodrugs to the active moieties is assumed in the projection of the human exposure and efficacious doses. In general, liver has the highest enzyme activity for conversion of prodrugs to their active moieties [67]. Different species have shown different *in vitro* and *in vivo* enzymatic activity for bioconversion of prodrugs depending on their chemical structures. For example, an oxymethyl-modified coumarinic acid cyclic prodrug of an opioid peptide was relatively stable in human and canine plasma compared to mouse and rat plasma [68–70]. Cook et al. [71] demonstrated species differences in the hydrolysis rate for glycovir in the small intestinal mucosa but the hydrolysis rate correlated well with the observed oral bioavailability. Establishing such *in vitro–in vivo* correlation is necessary for successful projection of human exposures when species differences exist.

3.4.4 Nonlinear PK

An important assumption throughout the various projections methods is that each compound exhibits linear PK in animals and man. Linearity is also assumed when projections are based on scaled *in vitro* $CL_{h,int}$. As the dose increases, a compound may exhibit nonlinear PK because of saturation of metabolic and/or transporter pathways involved in its disposition. The extent of nonlinearity is difficult to predict from preclinical and *in vitro* human data. In addition, at higher doses, drugs may demonstrate flip-flop or absorption limited kinetics and oral exposure may decrease due to dissolution rate or solubility-limited absorption.

Another important point to consider is that all the AUC projections are typically based on observations after a single dose whereas most drugs require chronic dosing for achieving the desired efficacy. When linear PK prevails, the area under the plasma concentration vs. time curve from zero to infinity is the same as the area under the plasma concentration vs. time curve over a dosing interval at steady-state. However, if the steady-state human PK cannot be projected based on the single dose data, it is difficult to project steady-state human AUC based on preclinical data. This leads to erroneous projections of efficacious doses, especially when chronic dosing is required.

4. CONCLUSION AND FUTURE DIRECTIONS

4.1 Refining projections of human PK profile

Projections of the overall plasma concentration vs. time profile (e.g., C_{max}, C_{min}, or half-life) are valued indicators of safety, efficacy, and frequency of dosing. Multi-exponential PK in preclinical species can make projection of human PK more difficult and simple allometric scaling is generally not adequate. As discussed earlier, Dedrick plot, species-invariant time, and C_{ss}-MRT methods have increasingly been used to predict human PK profile. To date, minimal modification and correction based on *in vitro* data have been integrated into these methods. Further integration of all *in silico*, *in vitro*, and *in vivo* data should improve the accuracy of predicting human PK profile.

4.2 Impact of transporters on drug disposition

In recent years there has been an explosion in the knowledge related to drug transporters and their impact on drug disposition, tissue distribution, and PK. Going forward it may be possible to evaluate species differences in the transporter activity and incorporate such information in target exposure and dose projections [72,73]. Hepatic and renal transporters play a key role in DM and distribution. Currently, P-glycoprotein (P-gp) is the most common transporter considered in designing and optimizing a new series. Understanding what hepatic and renal transporters interact with a compound can help predict its uptake, metabolism, clearance, PK profile, and potential drug–drug interactions [74]. For example, integrating transporter information for

3-hydroxy-3-methyl-glutaryl-CoA (HMG-CoA) reductase inhibitors is critical for understanding uptake in the liver vs. muscle for differentiating efficacy from toxicity [75]. A recent publication demonstrated that peptide transporter 2 (PEPT2) is the major transporter involved in the cefadroxil renal tubular reabsorption and efflux from CSF to blood [76] and provided evidence for the importance of incorporating transporters in target exposure and efficacious dose projection.

4.3 Incorporating variability

Most published reports on human exposure and dose projection provide a summary of average projected parameters. To truly appreciate the success of human projections, the variability associated with predictions needs to be taken into consideration. Because of variability associated with the laboratory data, and the imperfect predictability of preclinical biological models and preclinical scaling methods, preclinical investigations can only provide inferences that in reality include considerable uncertainties for the expected human exposure and efficacious dose. It is expected that calculations based on point estimates will always represent a highly optimistic view of the state of our knowledge regarding developability of drug candidates. There are several examples in the literature where an attempt has been made to incorporate uncertainties associated with predicted plasma and tissue concentrations and hepatic CL predicted from hepatocytes [77–79]. In practice, predicting human attributes within a 2–4-fold range is considered acceptable. Although by itself, incorporating known uncertainty into these predictions would not change the outcome, it would be a more honest representation of the state of knowledge and, therefore, would facilitate an appraisal of the drug candidate for either advancement or discontinuation.

In conclusion, prospective predictions are difficult especially for first-in-class compounds. To increase the confidence and success rate, use of more than one method for consistency in predictions is strongly recommended. Clinical data on competitor compounds or internal lead compounds are extremely useful. This highlights the importance of close communication between the development and discovery organizations in the pharmaceutical industry.

ABBREVIATIONS

ADME	absorption, distribution, metabolism, and excretion
AUC	area under the concentration versus time curve
CL	clearance
CL_{hb}	hepatic blood clearance
$CL_{h,int}$	hepatic intrinsic clearance
C_{min}	trough concentration
C_{max}	maximum concentration
C_{ss}	dose divided by steady-state volume of distribution

ER	hepatic extraction ratio
f_a	fraction of the dose absorbed
f_g	fraction of the dose escaping first-pass metabolism by the gastrointestinal mucosa
f_u	fraction unbound
IC_{50}	concentration required for 50% inhibition
IC_{90}	concentration required for 90% inhibition
IV	intravenous
LC-MS/MS	liquid chromatography-tandem mass spectrometry
MRT	mean residence time
NOAEL	no observable adverse effect level
PBPK	physiologically based pharmacokinetic model
PK	pharmacokinetics
PD	pharmacodynamics
Q_{hb}	hepatic blood flow
$t_{1/2}$	half-life
V_{ss}	steady-state volume of distribution
V_β	volume of distribution during the elimination phase
V_t	tissue distribution volume

REFERENCES

[1] C. Huang, M. Zheng, Z. Yang, A. D. Rodrigues and P. Marathe, *Pharm. Res.*, 2007, **25**, 713–726.

[2] S. S. De Buck, V. K. Sinha, L. A. Fenu, M. J. Nijsen, C. E. Mackie and R. A. Gilissen, *Drug Metab. Dispos.*, 2007, **35**, 1766.

[3] Y. Sawada, M. Hanano, Y. Sugiyama and T. Iga, *J. Pharmacokinet. Biopharm.*, 1985, **13**, 477.

[4] K. W. Ward, R. Nagilla and L. J. Jolivette, *Xenobiotica*, 2005, **35**, 191.

[5] R. S. Obach, *Drug Metab. Dispos.*, 1999, **27**, 1350.

[6] I. Mahmood, *J. Pharm. Sci.*, 2006, **95**, 1810.

[7] K. W. Ward and B. R. Smith, *Drug Metab. Dispos.*, 2004, **32**, 612.

[8] D. F. McGinnity, J. Collington, R. P. Austin and R. J. Riley, *Curr. Drug Metab.*, 2007, **8**, 463.

[9] R. L. Dedrick, *J. Pharmacokinet. Biopharm.*, 1973, **1**, 435.

[10] J. B. Houston, *Biochem. Pharmacol.*, 1994, **47**, 1469.

[11] I. Mahmood, *J. Pharm. Sci.*, 2005, **94**, 883.

[12] I. Mahmood, *Life Sci.*, 1998, **63**, 2365.

[13] W. Y. Lin, S. P. Changlai and C. H. Kao, *Urol. Int.*, 1998, **60**, 11.

[14] J. B. Houston, *Biochem. Pharmacol.*, 1994, **47**, 1469.

[15] B. A. Hoener, *Biopharm. Drug Dispos.*, 1994, **15**, 295.

[16] K. Bachmann, J. Byers and R. Ghosh, *Xenobiotica*, 2003, **33**, 475.

[17] T. Iwatsubo, N. Hirota, T. Ooie, H. Suzuki and Y. Sugiyama, *Biopharm. Drug Dispos.*, 1996, **17**, 273.

[18] R. S. Obach, *Drug Metab. Dispos.*, 1997, **25**, 1359.

[19] B. Davies and T. Morris, *Pharm. Res.*, 1993, **10**, 1093.

[20] Y. Sawada, M. Hanano, Y. Sugiyama, H. Harashima and T. Iga, *J. Pharmacokinet. Biopharm.*, 1984, **12**, 587.

[21] R. S. Obach, J. G. Baxter, T. E. Liston, B. M. Silber, B. C. Jones, F. MacIntyre, D. J. Rance and P. Wastall, *J. Pharmacol. Exp. Ther.*, 1997, **283**, 46.

[22] G. E. Blakey, I. A. Nestorov, P. A. Arundel, L. J. Aarons and M. Rowland, *J. Pharmacokinet. Biopharm.*, 1997, **25**, 277.

[23] O. Luttringer, F. P. Theil, P. Poulin, A. H. Schmitt-Hoffmann, T. W. Guentert and T. Lave, *J. Pharm. Sci.*, 2003, **92**, 1990.

[24] P. Poulin and F. P. Theil, *J. Pharm. Sci.*, 2002, **91**, 1358.

[25] W. A. Ritschel and S. Kaul, *Methods Find. Exp. Clin. Pharmacol.*, 1986, **8**, 239.

[26] F. Lombardo, R. S. Obach, M. Y. Shalaeva and F. Gao, *J. Med. Chem.*, 2002, **45**, 2867.

[27] F. Lombardo, R. S. Obach, M. Y. Shalaeva and F. Gao, *J. Med. Chem.*, 2004, **47**, 1242.

[28] H. Wong, S. J. Grossman, S. A. Bai, S. Diamond, M. R. Wright, J. E. Grace, Jr., M. Qian, K. He, K. Yeleswaram and D. D. Christ, *Drug Metab. Dispos.*, 2004, **32**, 1359.

[29] F. P. Theil, T. W. Guentert, S. Haddad and P. Poulin, *Toxicol. Lett.*, 2003, **138**, 29.

[30] H. M. Jones, N. Parrott, K. Jorga and T. Lave, *Clin. Pharmacokinet.*, 2006, **45**, 511.

[31] T. Hou, J. Wang, W. Zhang and X. Xu, *J. Chem. Inf. Model.*, 2007, **47**, 460.

[32] T. Hou, J. Wang, W. Zhang and X. Xu, *J. Chem. Inf. Model.*, 2007, **47**, 208.

[33] S. Willmann, W. Schmitt, J. Keldenich, J. Lippert and J. B. Dressman, *J. Med. Chem.*, 2004, **47**, 4022.

[34] M. Kuentz, S. Nick, N. Parrott and D. Rothlisberger, *Eur. J. Pharm. Sci.*, 2006, **27**, 91.

[35] B. Agoram, W. S. Woltosz and M. B. Bolger, *Adv. Drug Deliv. Rev.*, 2001, **50**(Suppl. 1), S41.

[36] N. Parrott and T. Lave, *Eur. J. Pharm. Sci.*, 2002, **17**, 51.

[37] Y. H. Zhao, M. H. Abraham, J. Le, A. Hersey, C. N. Luscombe, G. Beck, B. Sherborne and I. Cooper, *Eur. J. Med. Chem.*, 2003, **38**, 233.

[38] R. L. Dedrick, *J. Pharmacokinet. Biopharm.*, 1973, **1**, 435.

[39] H. Boxenbaum and R. Ronfeld, *Am. J. Physiol.*, 1983, **245**, R768.

[40] I. Mahmood and R. Yuan, *Biopharm. Drug Dispos.*, 1999, **20**, 137.

[41] W. F. Mueller, F. Coulston and F. Korte, *Regul. Toxicol. Pharmacol.*, 1985, **5**, 182.

[42] B. M. Nath, K. E. Schumann and J. D. Boyer, *Trends Microbiol.*, 2000, **8**, 426.

[43] D. E. Wildman, *Bioessays*, 2002, **24**, 490.

[44] D. M. Jonker, L. A. Kenna, D. Leishman, R. Wallis, P. A. Milligan and E. N. Jonsson, *Clin. Pharmacol. Ther.*, 2005, **77**, 572.

[45] K. A. Walters, M. A. Joyce, J. C. Thompson, M. W. Smith, M. M. Yeh, S. Proll, L. F. Zhu, T. J. Gao, N. M. Kneteman, D. L. Tyrrell and M. G. Katze, *PLoS Pathog.*, 2006, **2**, e59.

[46] P. Turrini, R. Sasso, S. Germoni, I. Marcucci, A. Celluci, A. Di Marco, E. Marra, G. Paonessa, A. Eutropi, R. Laufer, G. Migliaccio and J. Padron, *Transplant Proc.*, 2006, **38**, 1181.

[47] M. van der Lee, G. Verweel, R. de Groot and D. Burger, *Antivir. Ther.*, 2006, **11**, 439.

[48] P. Yeni, *J. Acquir. Immune. Defic. Syndr.*, 2003, **34**, S91.

[49] R. S. Obach, J. G. Baxter, T. E. Liston, B. M. Silber, B. C. Jones, F. MacIntyre, D. J. Rance and P. Wastall, *J. Pharmacol. Exp. Ther.*, 1997, **283**, 46.

[50] T. Rodgers and M. Rowland, *J. Pharm. Sci.*, 2006, **95**, 1238.

[51] J. B. Houston and D. J. Carlile, *Drug Metab. Rev.*, 1997, **29**, 891.

[52] D. J. Carlile, K. Zomorodi and J. B. Houston, *Drug Metab. Dispos.*, 1997, **25**, 903.

[53] K. Ito and J. B. Houston, *Pharm. Res.*, 2005, **22**, 103.

[54] K. Ito and J. B. Houston, *Pharm. Res.*, 2004, **21**, 785.

[55] T. Rodgers, D. Leahy and M. Rowland, *J. Pharm. Sci.*, 2005, **94**, 1259.

[56] T. Lave, P. Coassolo and B. Reigner, *Clin. Pharmacokinet.*, 1999, **36**, 211.

[57] I. Mahmood and C. Sahajwalla, *J. Pharm. Sci.*, 2002, **91**, 1908.

[58] K. W. Ward, L. M. Azzarano, W. E. Bondinell, R. D. Cousins, W. F. Huffman, D. R. Jakas, R. M. Keenan, T. W. Ku, D. Lundberg, W. H. Miller, J. A. Mumaw, K. A. Newlander, J. L. Pirhalla, T. J. Roethke, K. L. Salyers, P. R. Souder, G. J. Stelman and B. R. Smith, *Drug Metab. Dispos.*, 1999, **27**, 1232.

[59] T. Wajima, K. Fukumura, Y. Yano and T. Oguma, *J. Pharm. Sci.*, 2003, **92**, 2427.

[60] M. R. Feng, X. Lou, R. R. Brown and A. Hutchaleelaha, *Pharm. Res.*, 2000, **17**, 410.

[61] P. J. Lowe, Y. Hijazi, O. Luttringer, H. Yin, R. Sarangapani and D. Howard, *Xenobiotica*, 2007, **37**, 1331.

[62] J. F. Levesque, M. Gaudreault, R. Houle and N. Chauret, *J. Chromatogr. B Analyt. Technol. Biomed. Life Sci.*, 2002, **780**, 145.

[63] Q. Wang, R. Jia, C. Ye, M. Garcia, J. Li and I. J. Hidalgo, *In Vitro Cell. Dev. Biol. Anim.*, 2005, **41**, 97.

[64] I. Mahmood and J. D. Balian, *Xenobiotica*, 1996, **26**, 887.

[65] A. Fura, *Drug Discov. Today*, 2006, **11**, 133.

[66] A. Fura, Y. Z. Shu, M. Zhu, R. L. Hanson, V. Roongta and W. G. Humphreys, *J. Med. Chem.*, 2004, **47**, 4339.

[67] B. M. Liederer and R. T. Borchardt, *J. Pharm. Sci.*, 2006, **95**, 1177.

[68] M. Hosokawa, T. Maki and T. Satoh, *Arch. Biochem. Biophys.*, 1990, **277**, 219.

[69] Y. Yoshigae, T. Imai, A. Horita, H. Matsukane and M. Otagiri, *Pharm. Res.*, 1998, **15**, 626.

[70] J. Z. Yang, W. Chen and R. T. Borchardt, *J. Pharmacol. Exp. Ther.*, 2002, **303**, 840.

[71] C. S. Cook, P. J. Karabatsos, G. L. Schoenhard and A. Karim, *Pharm. Res.*, 1995, **12**, 1158.

[72] K. Bleasby, J. C. Castle, C. J. Roberts, C. Cheng, W. J. Bailey, J. F. Sina, A. V. Kulkarni, M. J. Hafey, R. Evers, J. M. Johnson, R. G. Ulrich and J. G. Slatter, *Xenobiotica*, 2006, **36**, 963.

[73] Y. Shitara, T. Horie and Y. Sugiyama, *Eur. J. Pharm. Sci.*, 2006, **27**, 425.

[74] J. Sahi, *Expert. Opin. Drug Metab. Toxicol.*, 2005, **1**, 409.

[75] R. F. Reinoso, N. A. Sanchez, M. J. Garcia and J. R. Prous, *Methods Find. Exp. Clin. Pharmacol.*, 2001, **23**, 541.

[76] H. Shen, S. M. Ocheltree, Y. Hu, R. F. Keep and D. E. Smith, *Drug Metab. Dispos.*, 2007, **35**, 1209.

[77] I. Gueorguieva, II., I. A. Nestorov and M. Rowland, *J. Pharmacokinet. Pharmacodyn.*, 2004, **31**, 185.

[78] I. Nestorov, I. Gueorguieva, H. M. Jones, B. Houston and M. Rowland, *Drug Metab. Dispos.*, 2002, **30**, 276.

[79] N. Blanchard, N. J. Hewitt, P. Silber, H. Jones, P. Coassolo and T. Lave, *J. Pharm. Pharmacol.*, 2006, **58**, 633.

CHAPTER 20

Modeling Networks of Glycolysis, Overall Energy Metabolism and Drug Metabolism under a Systems Biology Approach

Zoltán Sarnyai* and **László G. Boros****

Contents			

* Department of Pharmacology, University of Cambridge, Tennis Court Road, Cambridge CB2 1PD, UK

** UCLA School of Medicine, Division of Pediatric Endocrinology, Los Angeles Biomedical Research Institute at the Harbor-UCLA Medical Center, 1124 West Carson Street, RB1, Torrance, CA 90502-2910, USA; SIDMAP, LLC, Los Angeles, CA 90064, USA

Annual Reports in Medicinal Chemistry, Volume 43
ISSN 0065-7743, DOI 10.1016/S0065-7743(08)00020-1

1. INTRODUCTION

The field of biology currently enjoys enormous opportunities arising from the availability of complete genomic sequences and the emerging technologies of proteomics and metabolomics, which inspire confidence that complex biological phenomena and systems can be understood completely [1,2]. This optimistic view is strengthened by the rapidly expanding technological capabilities to undertake global, whole cell-wide analyses of gene and protein expression in mammalian cells [3]. At the same time, there is an increasing gap between the amount of biochemical and molecular genetic information and our under-standing of the implication of these data for cell system function. In fact, life scientists have quickly realized that an encyclopedia of genes does not explain how the genome influences the phenome, nor does it provide a menu list for selecting drug targets [4]. Moreover, the genome alone (including even the proteins it encodes) does not determine the physiological state of any cell at a given moment [5]. Thus, whereas genes and proteins set the stage of what can happen in the cell, much of cell activity is actually triggered at the metabolite level: cell signaling, energy transfer and cell-to-cell communication are regulated by metabolites [6,7].

To understand how the parts (genes, proteins and metabolites) make up the whole organism, a systemic view is required, with genes and proteins seen more as parts of a highly interactive network with the potential to affect the network – and be affected by it – in many certain and sometimes unexpected ways, instead of as isolated entities entirely or largely determining cell function. Molecular function can then be more realistically recognized as very significantly influenced by cellular context rather than solely dependent on properties of the individual molecule. This change of perspective is accompanied by the recognition that bioinformatics plays an indispensable role in extracting the most relevant information from huge amounts of related data. Such a systemic view of cells demands the capacity to quantitatively predict, rather than simply qualitatively describe, cell behavior. In parallel with the data-driven research approach that focuses on speedy handling and analyzing of huge amounts of data, a new approach called Model-Driven Research (MDR) is gradually gaining influence and increased recognition [8–10].

MDR takes the approach that sets a biological model by combining the knowledge of the system with related data and simulates the behavior of the system in order to understand the biological mechanisms of the system. Typically, this is simply called systems biology [11] in which it is recognized that the living body is composed of numerous tissues, organized in the forms of various subsystems, large and small, by which the flows of energy, material and information are controlled. This is a hierarchical system involving subsystems such as metabolism, transcriptional control, signal transduction, the cell cycle and apoptosis at the cell level. All of these in turn are implemented as subsystems of interacting molecules. At a higher level, populations of cells are the subsystems of various physiological and pathological organ systems, up to and including the whole body level. Systems biology aims to model and simulate the various subsystems and the interactions that are established between and among them, for the better understanding of life system mechanisms.

The current efforts to provide analyses of protein–protein interactions can be considered as the first steps towards the construction of interaction networks. These qualitative networks provide the basis for the quantitative spatio–temporal simulation of the evolution of a protein network at the whole cell level [12]. It is foreseeable that systematic data will become available in the next few years to allow the design of a virtual cell *in silico*, an accomplishment that will have an enormous impact on biomedical and pharmaceutical areas [13,14]. With this perspective, large-scale initiatives have been launched in Japan, in the USA and to a smaller extent in Europe, aimed at the computational simulation of whole-cell functions, examples of which have been initiated at the Alliance for Cellular Signaling [15,16]. Apart from general-purpose mathematical software, there are tools specifically designed to handle problems such as modeling of metabolic and signal transduction that have been extensively used on small- and medium-sized subsystems. The next challenge is to increase the size of the systems that can be represented by combining models of subsystems into larger wholes.

A cell is classically seen as a physical entity with a definite volume and spatial architectural substructures such as nucleus, mitochondria, Golgi apparatus, ribosomes and so on. The availability of genomic and proteomic data as well as the growing successes in building up protein interaction maps, makes increasingly possible a "system view" of the cell. Moreover, the use of new technologies, such as DNA and protein microarrays, two-system hybrids and protein tagging techniques coupled with mass spectrometry (MS), is enabling a rapid increase in the amount of information concerning protein–protein interactions [17,18].

From a systems biology point of view, a cell can be considered as composed of a complex network of interacting macromolecules (proteins, nucleic acids) and small molecules (lipids, carbohydrates, ions, etc.). These entities form a number of distinct, highly connected networks that make the cell a dynamic ensemble of interconnected molecular networks. If we adopt a hierarchical picture of the network, we might attribute the most relevant role to the protein networks as they trigger the fundamental activities of the cell: all the biochemical reactions are catalyzed by proteins and proteins themselves interact with each other. This fact is consistent with the view of a "central integrated" protein network, spanning the whole cellular volume (a feature particularly relevant for cells with a complex morphology, such as neurons) to which a number of functional molecular subnetworks are interconnected. For a nerve cell, an additional level of complexity is constituted by its property of electrical excitability across the cell membrane.

This review looks at the newly emerging emphasis on systems biology as a tool to more fully understand perturbations in mammalian metabolism caused by disease and medicine. It examines as well the rapidly evolving other tools used within that context, new information being obtained as a result and, finally, what reasonably may be expected as these new models evolve into the future.

2. CHANGING ARCHITECTURE IN NEWER METABOLIC MODELS

The first question to answer before beginning to design a model is always the same: how many details of its components must we include? Must we include all the individual molecules and exhaustively compute their interactions? As D. Noble pointed out in an earlier review on this topic, there are problems in trying to reproduce nature in a complete way, and even if we were successful in that task, the result would not be a model in the strict sense, as models are by definition partial representations [19]. In fact, he noted, the power of a model lies in determining what is essential, which allows identification of the controlling steps of the modeled system. A complete representation of that system would leave us just as wise, or as ignorant, as before.

A second question is how to define the boundaries of a model. Traditionally, the accepted boundaries for a metabolic model have been the metabolic pathways. Models of traditional pathways (like glycolysis, the pentose phosphate pathway and the tricarboxylic acid (TCA) cycle and purine metabolism) have been developed for different cell lines and tissues [20,21]. Although useful to analyze metabolism in cells in which they represent significant metabolic activities (e.g. glycolysis in yeast or the TCA cycle in anaerobic muscle), these metabolic pathway models do not provide for quantitative, systemic evaluations of full cell metabolic reaction networks because they focus on subsets of networks without considering network-wide interactions [22].

The availability of full sets of data at the genomic level and the emerging technologies of proteomics and metabolomics show that an attempt to fully and usefully understand metabolism in any cell or tissue requires a network approach. Thus, for instance, knock-outs of one or more steps in a metabolic pathway often result in a rearrangement not only of that pathway alone, but also cause a full reorganization of the network such that enzymes from other pathways are recruited to create alternative pathways that bypass the deleted step. These network rearrangements and network interactions explain failures in metabolic engineering caused by attempting to achieve significant metabolic phenotype changes solely through genetic tools, for example, the failure to enhance ethanol production in yeast by over-expressing each enzyme in the glycolytic pathway, or the failure to increase penicillin production by over-expressing all the genes of the biosynthetic pathway in an industrial strain of the mold [23].

Modeling metabolic networks such that the main pathways are integrated in a single model and interactions between them are taken into account, requires experimental approaches that simultaneously provide information of the different metabolic network components and the measure of metabolic fluxes through the different branches of the network.

Throughout recent decades, an extremely large, ongoing effort has continued to add to the collection of information about the characteristics of the components of metabolic networks and their interactions. This information now comprises very large databases that are updated continuously. Examples of such databases are those containing information on genes, proteins, small molecules and interactions among proteins such as those housed and maintained at the USA National Center for Biotechnology Information (http://www.ncbi.nlm.nih.gov); the Protein Data Bank (http://www.rcb.org/pdb); the Kyoto Encyclopedia of Genes and Genomes (http://www.genome.jp/kegg); and The Biomolecular Interaction Network Database (http://bind.ca). However, these databases depict only modular or static biochemical states. To be more meaningful, this information needs to be complemented with data from experiments using different substrates such as limiting substrates or pulse substrates (e.g. glucose and glycerol). The resulting plots of intracellular metabolite concentrations against time, when such measurements can be obtained, allow the useful estimation of fluxes throughout the different branches of the metabolic network.

Functional characteristics of a cellular metabolic network represent the metabolic phenotype of the cell. The characterization of metabolic functions requires the simultaneous measurement of substrate fluxes both within and among the interconnecting pathways. The simultaneous assessment of substrate flux within and among major metabolic branches inside the cellular metabolic network gives us the increasingly useful metabolic profile of a given cell.

2.1 Stable isotope labeling tracks glucose pathway footprints

In general, the degree of difficulty in measuring metabolic fluxes depends on the flux we want to calculate [24,25]. The easy way to measure is the uptake of external substrates or the output of end products fluxes, which can be calculated from the rates of disappearance or appearance of the relevant compounds. More difficult is to determine metabolic fluxes within each internal branch of the network. The most common experimental method to determine fluxes is the utilization of isotope-labeled molecules bearing either stable or radioactive isotopes. Isotope labeling studies allow the introduction of the dimension of time and the quantitative measurement of specific kinetic processes occurring in any given time period. Many classical metabolic studies in recent decades have been carried out by measuring radioactivity associated with unstable isotopes such as 3H or 14C. However, relatively recent advances in MS and nuclear magnetic resonance (NMR) technologies have resulted in an increase in the application of stable isotope-based tracer methodologies to characterize the metabolic profile in different cells, tissues and even whole organisms [26,27].

Currently, the utilization of stable isotopes has almost completely replaced use of radioisotopes in pathway elucidation and flux analyses [28]. Partly, this is occurring because the stable isotopes allow simultaneous quantification in one experiment of multiple different specific-position and mass isotopomer molecules, facilitating identification in one experiment of cell use of different metabolic pathways [29]. In addition, the stable isotope methods are easily and safely translatable from animal to human studies [30]. Still, it has to be taken into account that tracer choice is determined by the fluxes of interest in the experiment at hand.

Stable isotope-based dynamic metabolic profiling (SIDMAP) technology, combined with detailed kinetic models of the metabolic networks and appropriate software permit quantitative and highly detailed characterization of the specific steps and routes taken by glucose carbons in the macromolecule and energy-producing metabolic reactions [31]. In particular, incubation with [1,2-^{13}C]-glucose or uniformly labeled glucose and the use of MS enables the measurement of glucose redistribution among glycolysis and other major metabolic pathways (see Figure 1). This methodology has been applied to various different cell types to obtain new information on molecular-level metabolic pathway interactions in disease conditions with strong metabolic components [32,33]. Applied to different types of cancer cells, SIDMAP studies demonstrated that tumor growth is primarily determined by specific glucose

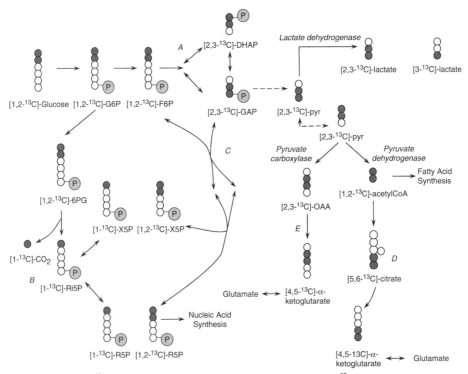

Figure 1 Possible ^{13}C isotopomers formed after incubation with [1,2-^{13}C]-glucose as a single tracer. Labeled glucose can enter the cell and be metabolized through glycolysis to form lactate with two labeled carbons (a). Glucose can also enter the oxidative pentose phosphate pathway (b) and be recycled back to glycolysis through the non-oxidative pentose phosphate pathway (c), forming lactate with a single carbon label. Analysis of carbon label in ribose from nucleic acids will indicate if the ribose is produced through the oxidative pathway, in which case ribose with one carbon label will be obtained, or through the non-oxidative pathway, in which case ribose with two carbon labels will be obtained. Analysis of the isotopomers from fatty acids can indicate if glucose is utilized to synthesize them. Glutamate isotopomer analysis will indicate if glucose enters the TCA cycle through the pyruvate dehydrogenase (d), or through the pyruvate carboxylase (e) enzyme, demonstrated by the position of the label in glutamate carbons.

metabolic reactions responsible for nucleic acid synthesis, lipid production and TCA cycle anaplerotic flux. Significantly, these closely measured metabolic reactions showed diametrically opposite changes in response to tumor growth-promoting or growth-inhibiting treatments. Among these pathways, glucose intake and the non-oxidative steps of the pentose cycle, which provide the central reactions for nucleic acid ribose synthesis from glucose, are the most important pathways that directly determine tumor growth and cell transformation [34,35].

2.2 Quantitative tools to identify which components control the metabolome and metabolic flux distribution within the network

A classical question when analyzing a metabolic network such as glucose metabolism is whether a rate-limiting reaction exists. If a rate-limiting step were to exist in a pathway, varying the activity of that step alone would change the flux in that pathway. However, as reviewed by Fell, there are few definitive experimental observations of such a phenomenon. Instead the existing evidence suggests that most pathways are significantly affected by the activities of several steps [24]. Furthermore, the large number of failures reported in the field of metabolic engineering, for example, when trying to increase the rate of a pathway by over-expressing the predicted rate-limiting step, reinforces the concept that the control of the metabolic network is often shared among several steps [36].

The concept that several enzymes might affect the flux in a pathway and that the effect of each enzyme on metabolite concentrations and metabolic fluxes can be quantified are the basis of two main theoretical frameworks that have been developed during the last two decades for studying the genetic, enzymatic and substrate level control mechanisms in metabolic networks: the biochemical systems theory (BST) [37,38] and the metabolic control analysis (MCA) [24,39]. Both approaches are equivalent in essential parts as they are based on general systems sensitivity theory [40]. Sensitivity theory addresses the impact of changes in parameter values (or independent variables like enzyme concentrations, that can be independently manipulated by genetic engineering) on responses of the integrated system. Thus, it relates properties of individual system components (local) to properties of the intact system (global).

Both theories aim to quantify how sensitive the steady-state variables (the fluxes and concentrations) are to variations in the parameters (or independent variables as, for instance, the component enzymes). These sensitivities are global or systemic properties as they depend on all the interactions within a cellular system. Moreover, both theories seek to relate these global sensitivities to local properties of each individual metabolic reaction (enzyme reactions).

These theories have been developed almost independently of each other during the past three decades and use different nomenclatures to describe local and global properties. In the following paragraphs, we explain how these local and global properties are described using MCA and BST nomenclature.

In MCA, control coefficients are defined that describe quantitatively the sensitivity of systemic variables (metabolic fluxes, intermediary metabolite concentrations, hormone secretion, brain electrical activity, etc.) to variations in enzyme activities or to any other parameters or independent variables of the system [24,41]. Below, we illustrate how control coefficients are defined using a simple two-step exemplary pathway

$$\text{So} \xrightarrow[v_1]{E_1} \text{S} \xrightarrow[v_2]{E_2} \text{P}$$

Steady state flux:J

where So is the initial substrate, S the intermediary metabolite, P the final product, E_1 and E_2 are the enzymes that catalyze the two metabolic reactions respectively; v_1 and v_2 are the corresponding local rates and J is the steady-state pathway flux. Let us suppose first that a small change, δE_1, is made in the amount of enzyme E_1, and that this change produces a small change, δJ, in the steady-state pathway flux, J. If the change made is small enough, then the ratio $\delta J / \delta E_1$ becomes equal to the slope of the tangent to the curve of J against E_1.

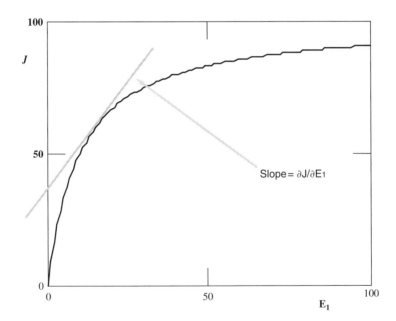

This slope is a measure of how sensitive the flux is to variations in E_1, but it has the disadvantage that it will depend on the units used to measure the flux and the enzyme. This slope can be made dimensionless by dividing ∂J and ∂E_1 by the steady-state values of J and E_1 respectively. In MCA this scaled slope is called "flux control coefficient" because it is a measure of the "control" of one enzyme on the pathway flux

$$C^J_{E_1} = \frac{\partial J/J}{\partial E_1/E_1}$$

Mathematically, this is equivalent to

$$C^J_{E_1} = \frac{\partial \ln J}{\partial \ln E_1}$$

and graphically it can be visualized as the slope of the tangent to the curve of $\ln J$ against $\ln E_1$

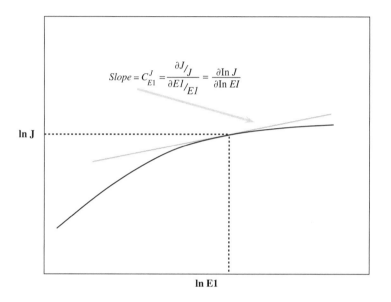

In MCA, the meaning of "to control" is understood as "to be capable of influencing something". In general, for metabolic fluxes or other systemic variables (Y), control coefficients of each enzyme or parameter (X) have been defined as the fractional change in the system variable (Y) over the fractional change in the enzyme activity or parameter (X)

$$C^Y_X = \frac{\partial Y/Y}{\partial X/X}$$

In BST the same magnitude is defined as logarithmic gains

$$L(X, Y) = \frac{\partial \ln Y}{\partial \ln X} = C^Y_X$$

Thus, logarithmic gains in BST are equal to control coefficients in MCA (see Refs 53 and 54 for a more complete comparison).

Flux control coefficients are often determined by measuring these fractional changes after applying specific enzyme inhibitors. A high-control coefficient indicates that the flux of the system is highly sensitive to changes in the concentration of this enzyme. For instance, control coefficients have been measured in rat hepatocytes using adenovirus-mediated enzyme over-expression. The control coefficient of glucokinase on hepatic glycogen synthesis has been reported as being close to one, indicating a true rate-limiting step [42]. This high-control coefficient can explain the abnormal hepatic glycogen synthesis in individuals with maturity-onset diabetes of the young type-2 who have just a single mutant allele [43]. On the other hand, a low control coefficient indicates that the flux of the system is low in sensitivity to changes in the concentration of this enzyme. An illustrative example is the low control coefficient (lower than 0.1) reported for triose phosphate isomerase in erythrocytes [44]. From this low control coefficient it can be correctly predicted that individuals heterozygous for the mutated allele do not show clinical symptoms. Moreover, it has been reported that homozygous patients suffer clinical symptoms only when they present with a very low activity of this enzyme, which is in accordance with the MCA prediction for an enzyme with a very low degree of control on the pathway flux [45,46]. The complete set of control coefficients allows us to predict the response of the system to any perturbation. In an analogous way, control coefficients of intermediate metabolite concentrations with regard to enzymes have been defined with the same finality, which is to quantify the variation of metabolite intermediates with regard to the enzymes

Thus, knowledge of control coefficient distribution among the different steps of a pathway gives us a useful orientation from which to manipulate a metabolic network (genetically or with drugs) to restore anomalous metabolic profiles accompanying several diseases, such as atherosclerosis or Alzheimer's disease.

Moreover, control coefficients are correlated with the kinetic properties of the pathway's enzymes. To correlate control coefficients with the kinetic properties of individual pathway enzymes, we first must define the kinetic properties of enzymes in terms of the sensitivity of the local rate to enzyme effects with variations in concentrations of its directly connected metabolites (substrates, products, effectors, etc.). In MCA, the effect of a metabolite (S) on the velocity (vi) of the enzyme Ei is quantified by means of the elasticity coefficients, defined as the fractional change in rate of the isolated enzyme for a fractional change in substrate S, with all other effectors of the enzyme held constant at the value they have in the metabolic pathway

$$\varepsilon_S^{vi} = \frac{\partial vi/vi}{\partial S/S} = \frac{\partial \ln vi}{\partial \ln S}$$

Elasticities have positive values for substrates and metabolites (activators) that stimulate the rate of a reaction and negative values for products or inhibitors that decrease the reaction rate. They can be calculated from a rate equation using differential calculus. For example, if an enzyme obeys the Michaelis–Menten equation

$$v = \frac{V_{\text{MAX}} S}{S + K_{\text{m}}}$$

then the elasticity coefficients for each concentration of S can be calculated by differential calculus as

$$\varepsilon_S^v = \frac{K_{\text{m}}}{S + K_{\text{m}}}$$

This formula facilitates observation of the link between elasticity coefficients and kinetic parameters and the typical range of values that elasticity can achieve. Elasticity coefficient values go from 1, when S is much below K_{m}, to 0, when S is saturated by its substrate ($S \gg K_{\text{m}}$). The function of elasticity in MCA is to describe quantitatively the responsiveness of an enzyme rate to a metabolite. A high-elasticity coefficient of an enzyme rate to a metabolite, S, means that the rate through the enzyme is very sensitive to a perturbation in this substrate concentration.

The elasticity coefficient of an enzyme to a metabolite can be obtained directly as the slope of the logarithm of the rate plotted against the logarithm of the metabolite concentration

$$\text{slope} = \varepsilon_S^{vi} = \frac{\partial vi / vi}{\partial S / S} = \frac{\partial \ln vi}{\partial \ln S} \approx 0$$

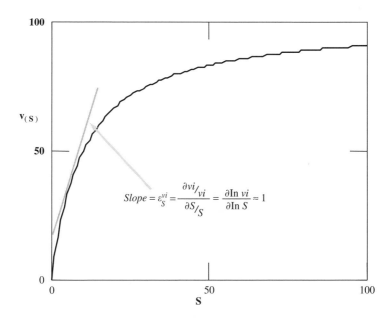

In BST, the relative derivative of rate versus metabolite concentration is named the kinetic order (g_{ij}) of the reaction, following the classical chemical kinetics nomenclature. In fact, kinetic orders are equivalent to elasticity coefficients

$$g_{ij} = \frac{\partial vi/vi}{\partial Sj/Sj} = \frac{\partial \ln vi}{\partial \ln Sj} = \varepsilon_{Sj}^{vi}$$

Moreover, in BST individual enzyme rates, and in some cases also the net rates of synthesis or degradation of a metabolite, are described using a power-law approximation

$$v_i = \alpha \prod_{j=1}^{n} X_j^{g_{ij}}$$

where X_j includes all the vi directly connected metabolites (substrates, products, effectors, etc.).

Elasticity coefficients and control coefficients are related by the so-called connectivity theorem [47,48]. For a simple two-step metabolic pathway with a single intermediary metabolite, S, the connectivity theorem is written as

$$C_{E_1}^J \varepsilon_s^{v_1} + C_{E_2}^J \varepsilon_s^{v_2} = 0$$

From the connectivity theorem we obtain

$$\frac{C_{E_1}^J}{C_{E_2}^J} = \frac{\varepsilon_s^{v_2}}{\varepsilon_s^{v_1}}$$

This ratio shows the tendency of large elasticities to be associated with small flux control coefficients, and vice versa. In general, matrix equations can be written, allowing easy computation of all control coefficients in terms of the elasticity coefficients equations [24, 49–52]. Connectivity theorem allows us to easily observe that although the control coefficients are system properties of the pathway, they are explainable in terms of the kinetic properties of the constituent enzymes.

The complete set of elasticities and control coefficients allows us to predict the response of the system to any perturbation and to easily link the control properties of one enzyme with its kinetic properties.

3. ANALYSIS OF METABOLIC NETWORK CONTROL DISTRIBUTION IN DRUG DISCOVERY AND DISEASE

MCA has been applied in biomedicine to diagnose and understand several genetic disorders. For example, Mazat showed that control coefficient values can explain the threshold effect observed in mitochondrial diseases and used MCA to provide a possible explanation of the heterogeneous phenotypes of these pathologies [53]. Another example was provided by Agius, which showed that the high-control coefficient of glucokinase on hepatic glycogen synthesis explains the abnormalities observed in this process for patients with a single mutant allele of the glucokinase gene suffering maturity-onset diabetes of the young, type II [54]. Moreover, it was demonstrated that mutations of glycolytic enzymes displaying a low glycolytic flux control coefficient are recessive.

Another potentially promising area of application of MCA is the rational design of combined drug therapies, identifying particularly suitable sites for the manipulation of the metabolism with drugs. This is very useful because, as was pointed out by Salter, every enzyme in a sequence is essential for the metabolic process to work, so the effects on metabolism are likely to be obtained with lower concentrations of drug if an enzyme with a higher flux control coefficient is inhibited rather than one with a lower coefficient [55].

An example of the application of MCA and BST to determine the cause that led to neurological dysfunctions and mental retardation related to purine metabolism was provided by Curto [56]. They modeled purine metabolism in humans utilizing different variables and looked at metabolic and flux changes when some of these variables were altered. They observed that hypoxanthine:guanine phosphoribosyltransferase deficiency affected all the variables of the system and that the guanylate pool might be the cause of the observed neurological dysfunctions.

MCA can also provide a theoretical framework to identify key targets in disease pathways. MCA approaches the problem of drug targeting by examining the contribution of individual components within a metabolic network and quantifying it by means of control coefficients [57]. For instance, a high-control coefficient of a protein on an enhanced metabolite concentration associated with a pathological condition indicates that targeting this protein will result in an efficient inhibition of this metabolite accumulation. Thus, once metabolic profile differences between normal and pathological subjects are established and the steps with higher control on the relevant pathological metabolic properties are identified (steps with higher control coefficients), a rational drug design strategy can be carried out to correct the pathological situation [58]. A specific metabolic target to inhibit tumor cell growth that has been identified following such an approach is the enzyme transketolase. The control coefficient of transketolase on pentose phosphate synthesis has been reported as exceptionally high [59]. Accordingly, it has been predicted that inhibition of transketolase will result in tumor proliferation inhibition.

4. MODELING METABOLISM WITHIN THE BRAIN

An examination of information available from studies of brain metabolism provides an excellent example of the versatility and utility of metabolic assessment tools that can be applied to a wide variety of cell, organ and larger systems.

To more fully understand and better model metabolism of energy and of drugs within the brain, it is necessary to integrate key information on characteristics of the metabolic substrates glucose, ketone bodies and fatty acids with information on important differences in both function and function-specific metabolism among different types of brain cells.

4.1 Neuronal activation and energy metabolism

The nerve impulse is initiated by the depolarization of the neuronal plasma membrane. This depolarization releases potential energy earlier established through the formation of transmembrane ion gradients. The amount of energy released by depolarization is fixed, so to obtain an increase in signal intensity, an increase in the frequency sufficient to re-establish the ion gradient is necessary. For this purpose, chemical energy in the form of adenosine-5'-triphosphate (ATP) is needed, in an amount at least equal to the potential energy that will be consumed in the next depolarization [60].

The increased neuronal ATP generation creates a demand for increased availability of oxidizable substrates and increased activity of cellular mechanisms to enable its utilization. Several substrates, such as glucose, ketone bodies (β-hydroxybutyrate and acetoacetate) or free fatty acids (FFA), can be used to produce the necessary ATP for neuronal activation depending on substrate availability and the kinds of cells involved.

Another important metabolic aspect of neuronal cell function is the storage and timely release of neurotransmitters upon receipt of signals, functions which require intense membrane fatty acid synthesis and turnover. Insulin production exemplifies the importance of neuropeptides in regulating *de novo* fatty acid synthesis and turnover in differentiating islet beta cells as outlined later in this review. Glucagon-like peptide-l (GLP-1) is a powerful neuropeptide regulating beta cell *de novo* fatty acid synthesis and turnover with a primary effect of increasing glucose carbon channeling towards fatty acid synthesis, chain elongation and desaturation of the long chains. Further explorations of these unique metabolic effects of hormones acting in the brain, for example, GLP-1, improve the understanding of neuronal response and neurological and psychiatric diseases and lead to the discovery of new target sites for drug interactions with molecular mechanisms responsible for defective neurotrans-mitter synthesis and responses.

4.2 Metabolic substrates

4.2.1 Glucose

Glucose uptake by the cell is mediated by members of the facilitative glucose transporters. Glucose transporter class I molecules (GLUT 1) are found in the blood–brain barrier and in astrocytes, whereas GLUT 3 is found in neurons. GLUT 3 is kinetically more efficient than GLUT 1, but this efficiency is affected *in vivo* by anatomic factors and by the transporter abundance. The final end products of the complete oxidation of glucose by glycolysis and the TCA cycle are CO_2 and H_2O. This oxidation yields approximately 30 ATP molecules and consumes 6 O_2, being the more efficient substrate. In anaerobic conditions (anaerobic glycolysis) one glucose yields 2 ATP and 2 lactates. The obtained lactate can be exported to other cells, which can oxidize it, or it can be reconverted to glucose in hepatocytes by gluconeogenesis. Lactate release seems to be important to support the metabolic and energetic requirements of neurons, but controversy exists about its main utilization during neuron activity [61].

Aerobic glycolysis is highly efficient, partly because of characteristics such as the presence of hexokinase type II, which is associated with the mitochondria. This association appears to enhance the rate and efficiency of glucose phosphorylation and ATP synthase-mediated oxidative phosphorylation [62].

The utilization of the pyruvate produced by aerobic glycolysis in the TCA cycle is highly efficient. This results from the fact that pyruvate dehydrogenase is subject to kinase-mediated negative feedback control by high ratios of reduced nicotinamide adenine dinucleotide/nicotinamide adenine dinucleotide (NADH/NAD), acetylCoA/CoA and ATP/adenosine-5′-diphosphate (ADP) in the mitochondrial matrix, preventing excessive oxidative decarboxylation of pyruvate to acetylCoA and controlling the flow of electrons into the electron transport chain. This control avoids loss of energy by disruption of the flow of electrons and prevents oxidative injury in the cell.

4.2.2 Ketone bodies

Ketone bodies (β-hydroxybutyrate and acetoacetate) can be used by brain cells instead of glucose, and are quickly transferred to neurons and astrocytes. Moreover, utilization of ketone bodies suppresses glucose utilization. This seems to be due to the transfer of ketone bodies from plasma to brain to neuron to mitochondria, where they generate ATP. The flow from plasma into the brain is proportional to plasma concentration as they share the transport mechanism in the blood–brain barrier with other low-weight monocarboxylates. The ability of ketone bodies to suppress and substitute for glucose utilization in the brain can be understood via its intramitochondrial metabolism: β-hydroxybutyrate is oxidized by β-hydroxybutyrate dehydrogenase to acetoacetate, which is converted to its acylCoA thioester directly or by the succinylCoA acyltransferase entering into the oxidative pathways in the mitochondrial matrix. This metabolism will affect pyruvate dehydrogenase regulation, avoiding glucose oxidation and inhibiting pyruvate utilization, glycolysis and glucose uptake. Because of the lack of a regulatory mechanism to prevent the increased oxidation of ketone bodies, there is an uncoupling of the oxidative phosphorylation, affecting the efficiency of this process, and an increase of the redox potential and generation of reactive oxygen species (ROS), promoting oxidative stress and cell injury.

4.2.3 Fatty acids

Within the brain, fatty acids are utilized only in astrocytes and other glial cells, not in neurons. These long chain fatty acids are converted to their acylCoA form and then an acylcarnitine derivative is obtained. The entrance of this derivative into the mitochondria is through carnitine palmitoyltransferase-I (CPT-I).

Once inside the mitochondria, the fatty acid can be β-oxidized, generating acetylCoA, which through the TCA cycle will produce ATP. These fatty acids can also produce ketone bodies that can be exported from the astrocytes (in a manner similar to what hepatocytes do) and which can then be used directly by nearby neurons in preference to glucose utilization.

Mitochondrial oxidation of fatty acids inhibits several steps of glucose uptake and glycolysis and in astrocytes it feeds gluconeogenesis. The main control point of long chain fatty acids utilization is CPT-I, but this control is weaker than the one exerted by pyruvate dehydrogenase in glycolysis.

To summarize, the metabolic behavior of brain cells depends on the cell type: neurons can oxidize glucose, lactate, ketone bodies and TCA cycle intermediates to produce ATP, but they cannot oxidize fatty acids. On the other hand, astrocytes are more versatile, as they can also oxidize fatty acids and generate *de novo* glucose by gluconeogenesis (but they cannot export it) and export lactate and/or ketone bodies to the surrounding cells. In this context it is important to model glucose metabolism both by itself and also when there are other substrates available, as it will indicate in each situation which are the limiting and controlling steps of the brain cells' metabolism.

5. MODELING BRAIN METABOLIC NETWORKS

For the most comprehensive understanding of brain metabolism it is necessary to integrate the existing data on the working of the different metabolic pathways in neurons and astrocytes, and particularly how glucose is metabolized depending on the availability of other substrates. This integration of the data obtained in metabolic network models will be necessary to understand the relationship between the main metabolic pathways inside the cell and to comprehend the metabolic astrocyte-neuron interactions.

Any model of brain metabolism needs to take into account the differences in the organization of metabolic networks as well as the metabolic interactions established between these two types of brain cells.

Metabolites are the end products of cellular regulatory processes and their levels can be regarded as the ultimate responses of biological systems to genetic and environmental changes. The metabolic phenotype is a manifestation of gene expression and it produces a metabolic fingerprint unique to its phenotype that can be related to physiological characteristics. In fact, the phenotype and function of mammalian cells greatly depends on metabolic adaptation. Thus, the metabolic profile of a given cell represents the integrated end-point of many growth modifying signaling events.

Metabolic adaptations of different cell types are not independent of their neighbors, as the excreted products of one cell type can be used as energy supplies for others, establishing a connection of the metabolic networks from one cell to another. The high complexity and branching nature of the neuronal glucose metabolic network permits multiple metabolic endpoints (results) in response to changes in environmental factors for a given genomic or proteomic expression. Thus, the metabolic phenotype expression or the physiological state of neurons at a given moment cannot be simply predicted by the genome alone (or even by the proteins it encodes).

Some studies of metabolic profiling using stable isotope tracer technology have been performed using labeled acetate to determine glutamine synthesis and oxidation in astrocytes, looking at the different mass and positional isotopomers of glutamate [63]. This study determined that glutamine was not the main energetic substrate in astrocytes *in vivo* or *in vitro*, and that the anaplerotic flux responsible for glutamine synthesis was about 1.5 times that of the TCA cycle flux when glutamine or glucose was supplied in the medium.

A similar approach using stable isotope tracer technology and NMR instead of MS has been used to characterize the TCA cycle in cerebellar astrocytes. In this case, a mathematical model was developed to determine the relative flux of molecules through the anaplerotic versus oxidative pathways to analyze the experimental data. The model was also applied to data from granule cells and permitted identification of important differences in carbon metabolism between cerebellar astrocytes and granule cells [64].

6. CAN STABLE ISOTOPE METHODS PROVIDE NEW INSIGHTS INTO OPEN QUESTIONS IN BRAIN METABOLIC REGULATION AND NEUROLOGICAL DISORDERS?

Despite the fact that increasing evidence exists of the utility of stable isotope-based dynamic metabolic profiling (SIDMAP) studies in elucidating metabolic adaptations to different treatments or diseases in cells and tissues, these techniques have not been applied systematically to achieve a better under-standing of multifactorial brain diseases and drug discovery strategies.

An open question in which SIDMAP can provide new insights is in the understanding of the effect of GLP-1 in the brain. GLP-1 is produced in the brain and transported along axonal networks to diverse central nervous system regions [65]. It has been suggested that GLP-1 could be downstream of leptin action in the brain whereas it is still an open question whether GLP-1 needs to enter the brain to affect gut motility and food intake [66]. Taking into account that increased evidence has been provided recently on the relationship between enhanced glucose metabolism and resistance to apoptosis [67], and that fatty acids induce apoptosis in β-cells [68] we hypothesize that SIDMAP characterization of GLP-1 effect on the neuronal metabolic network can help to illuminate the molecular basis of the described neuroprotective effect of GLP-1, which activates different anti-apoptotic signaling pathways in specific neurons [69]. This hypothesis is supported by the recent evidence provided on the effect of GLP-1 in the glucose metabolic network in pancreatic cells. Those studies showed that cell differentiation and regulation of insulin release after GLP-1 treatment were accompanied by important metabolic adaptive changes that primarily affected the contribution of glucose to *de novo* fatty acid synthesis and chain elongation of the saturated long chain species primarily utilized for triglyceride and membrane synthesis [70]. These results indicate that a selective signaling mechanism was responsible for the differentiation-inducing metabolic effects of this hormone, being a peroxisome proliferator activated receptor-related (PPAR-related) nuclear signaling mechanism, which is a strong candidate for this action. This metabolic effect on glucose metabolic network induced by GLP-1 is totally different from the effect reported for the anticancer drug imatinib (Gleevec), which acts through a mechanism involving tyrosine kinase inhibition. In this case, the effect on glucose metabolism was on glycolysis whereas fatty acid metabolism was unaffected [71]. In light of these results it could be expected that SIDMAP characterization of GLP-1 effects on neuronal cells can aid in the identification of the signaling cascades and metabolic adaptations involved in the GLP-1 neuroprotective effect.

7. FUTURE DIRECTIONS IN METABOLIC MODELING

Because it is increasingly acknowledged that the complexity of interconnected networks must be considered in constructing the most useful models of metabolism, there are those who argue that a truly accurate model can be had only *in vivo* [72]. However, enormous progress has been made in recent years and is continuing today at a more rapid pace, in bioinformatics computing that allows incorporation of information obtained from both *in vivo* and *ex vivo* models into *in silico* models in which almost infinite numbers of variables can be both considered and manipulated via advanced, dedicated software programs [73]. While proponents of each model sometimes focus entirely on the option each knows best, the fact is that these options increasingly are seen in a similar light as genomics, proteomics and metabolic modeling: not as single answers to simple questions but as complementary approaches to be integrated into an ever-more-successful effort to accurately reflect — without prohibitive cost — the wide variety of important, causal and controlling influences on organisms that can be successfully modeled with a combination of the best tools available. Thus, the answer to the question of which model — *in vivo* or *ex vivo* — will become more widely used in the future is likely not to be one or the other, but both. Already the inevitable hybrid models are developing, in which *in vivo* and *ex vivo* models each inform the other, and hence progress in tandem towards a more perfect utility [74].

8. CONCLUSION

Like the empirical and sometimes serendipitous methods of drug discovery and evaluation that came before it, the molecular-target approach of modeling metabolism of energy and drugs is increasingly recognized as only a partial – and often too simple – answer to biological questions that are more complex than was fully appreciated until relatively recently. The realization that all biochemical phenomena within an organism have important and varying influence on each other depending of many variables has, some would say, been late to be applied to metabolic modeling and drug discovery and development. In response, the systems biology approach is rapidly becoming the bedrock of new progress in metabolic and medicinal research. A systems biology approach incorporates both the micro- and macro- views of organism function in health and disease. Improved models of such function are most likely to help break the logjam of stalled progress in developing more effective treatments in a wide variety of diseases. Evolving are more accurate models *in vivo*, *ex vivo* and *in silico*, and combinations as interactive and dynamic in change as the biologic systems they represent.

REFERENCES

[1] M. E. Frazier, G. M. Johnson, D. G. Thomassen, C. E. Oliver and A. Patrinos, *Science*, 2003, **300**, 290.

[2] Z. N. Oltvai and A. L. Barabasi, *Science*, 2002, **298**, 763.

[3] T. J. Phelps, A. V. Palumbo and A. S. Beliaev, *Curr. Opin. Biotechnol.*, 2002, **13**, 20.

[4] S. Huang, *Drug Discov. Today*, 2002, **7**, S163.

[5] M. E. Frazier, G. M. Johnson, D. G. Thomassen, E. E. Oliver and A. Patrinos, *Science*, 2003, **300**, 290.

[6] C. W. Schmidt, *Environ. Health Perspect.*, 2004, **112**, A410.

[7] M. Hellerstein, *J. Pharmacol. Exp. Ther.*, 2008, **325**(1), 1.

[8] O. Alter, *Methods Mol. Biol.*, 2007, **377**, 17.

[9] D. S. Morris, S. A. Tomlins, D. R. Rhodes, R. Mehra R, R. B. Shah and A. M. Chinnaiyan, *Cell Cycle*, 2007, **6**(10), 1177.

[10] K. Aoki, *Plant Cell Physiol.*, 2007, **48**(3), 381.

[11] T. Yao, *Prog. Biophys. Mol. Biol.*, 2002, **80**, 23.

[12] D. W. Ritchie, *Curr. Protein Pept. Sci.*, 2008, **9**(1), 1.

[13] T. W. Huang, A. C. Tien, W. S. Huang, Y. C. Lee, C. L. Peng, H. H. Tseng, C. Y. Kao and C. Y. Huang, *Bioinformatics*, 2004, **20**, 3273.

[14] S. Ekins, *Br. J. Pharmacol.*, 2007, **152**(1), 217.

[15] J. Li, Y. Ning, W. Hedley, B. Saunders, Y. Chen, N. Tindill, T. Hannay and S. Subramaniam, *Nature*, 2002, **420**, 716.

[16] J. R. Zavzavadjian, *Mol. Cell Proteomics*, 2007, **6**(3), 413.

[17] A. C. Gavin, M. Bosche, R. Krause, P. Grandi, M. Marzioch, A. Bauer, J. Schultz, J. M. Rick, A. M. Michon, C. M. Cruciat, M. Remor, C. Hofert, M. Schelder, M. Brajenovic, H. Ruffner, K. Merino, M. Klein, D. Hudak, T. Dickson, V. Rudi, A. Gnau, S. Bauch, S. Bastuck, B. Huhse, C. Leutwein, M. A. Heurtier, R. R. Copley, A. Edelmann, E. Querfurth, V. Rybin, G. Drewes, M. Raida, T. Bouwmeester, P. Bork, B. Seraphin, B. Kuster, G. Neubauer and G. Superti-Furga, *Nature*, 2002, **415**, 141.

[18] D. Bonatto, A systems biology analysis of protein-protein interactions between yeast superoxide dismutases and DNA repair pathways. *Free Radic. Biol. Med.*, 2007, 43(4), 557–567.

[19] D. Noble, *Nat. Rev.*, 2002, **3**, 460–463.

[20] L. Sabaté, R. Franco, E. I. Canela, J. J. Centelles and M. Cascante, *Mol. Cell. Biochem.*, 1995, **142**, 9.

[21] R. Curto, E. O. Voit, A. Sorribas and M. Cascante, *Biochem. J.*, 1997, **324**, 761.

[22] J. A. Papin, N. D. Price, S. J. Wiback, D. A. Fell and B. O. Palsson, *Trends Biochem. Sci.*, 2003, **28**, 250.

[23] J. E. Bailey, *Nat. Biotechnol.*, 1999, **17**, 616.

[24] D. Fell, *Understanding the Control of Metabolism*, Portland Press, Oxford, 1997.

[25] M. K. Hellerstein, *Annu. Rev. Nutr.*, 2003, **23**, 379.

[26] W. van Winden, P. Verheijen and S. Heijnen, *Metab. Eng.*, 2001, **3**, 151.

[27] T. Kuhara, *J. Chromatogr. B Analyt. Technol. Biomed. Life Sci.*, 2007, **855**(1), 42.

[28] U. Sauer, *Mol. Syst. Biol.*, 2006, **2**, 62.
[29] J. Boren, W.-N. P. Lee, S. Bassilian, J. J. Centelles, S. Lim, S. Ahmed, L. G. Boros and M. Cascante, *J. Biol. Chem.*, 2003, **278**, 28395.
[30] M. Hellerstein, *J. Pharmacol. Exp. Ther.*, 2008, **325**, 1.
[31] V. A. Selivanov, J. Puigjaner, A. Sillero, J. J. Centelles, A. Ramos-Montoya, P. W. Lee and M. Cascante, *Bioinformatics*, 2004, **20**, 3387.
[32] L. G. Boros, M. Cascante and W. N. Lee, *Drug Discov. Today*, 2002, **7**, 364.
[33] W. N. Lee, L. G. Boros, J. Puigjaner, S. Bassilian, S. Lim and M. Cascante, *Am. J. Physiol.*, 1998, **274**, E843.
[34] J. Boren, M. Cascante, S. Marin, B. Comin-Anduix, J. J. Centelles, S. Lim, S. Bassilian, S. Ahmed, W. P. Lee and L. G. Boros, *J. Biol. Chem.*, 2001, **276**, 37747.
[35] W. N. Lee, L. G. Boros, J. Puigjaner, S. Bassilian, S. Lim and M. Cascante, *Am. J. Physiol.*, 1998, **274**, E843.
[36] J. E. Bailey, *Nat. Biotechnol.*, 1999, **17**, 618.
[37] M. Savageau, *Biochemical System Analysis. A Study of Function and Design in Molecular Biology*, Adison-Wesley, Reading, 1976.
[38] E. O. Voit, *Computational Analysis of Biochemical Systems*, Cambridge University Press, Cambridge, 2000.
[39] M. Cascante, L. G. Boros, B. Comin-Anduix, P. de Atauri, J. J. Centelles and P. W. Lee, *Nat. Biotechnol.*, 2002, **20**, 243.
[40] P. M. Frank, *Introduction to System Sensitivity Theory*, Academic Press, New York, 1978.
[41] M. Cascante, L. G. Boros, B. Comin-Anduix, P. de Atauri, J. J. Centelles, P. W. Lee and P. W. Nat, *Biotechnol.*, 2002, **20**, 243.
[42] L. Agius, M. Peak, C. B. Newgard, A. M. Gomez-Foix and J. J. Guinovart, *J. Biol. Chem.*, 1996, **271**, 30479.
[43] G. Velho, K. F. Petersen, G. Perseghin, J. H. Hwang, D. L. Rothman, M. E. Pueyo, G. W. Cline, P. Froguel and G. I. Shulman, *J. Clin. Invest.*, 1996, **98**, 1755.
[44] R. Schuster and H. G. Holzhutter, *Eur. J. Biochem.*, 1995, **229**, 403.
[45] F. Orosz, B. G. Vertessy, S. Hollan, M. Horanyi and J. Ovadi, *J. Theor. Biol.*, 1996, **182**, 437.
[46] A. Repiso, J. Boren, F. Ortega, A. Pujades, J. Centelles, J. L. Vives-Corrons, F. Climent, M. Cascante and J. Carreras, *Haematologica*, 2002, **87**, ECR12.
[47] H. Kacser and J. A. Burns, *Symp. Soc. Exp. Biol.*, 1973, **27**, 65.
[48] H. Kacser and J. A. Burns, *Biochem. Soc. Trans.*, 1995, **23**, 341.
[49] M. Cascante, R. Franco and E. I. Canela, *Math. Biosci.*, 1989, **94**, 271–288.
[50] M. Cascante, R. Franco and E. I. Canela, *Math. Biosci.*, 1989, **94**, 289.
[51] C. Reder, *J. Theor. Biol.*, 1988, **135**, 175–201.
[52] C. Giersch, *Eur. J. Biochem.*, 1988, **174**, 509.
[53] J. P. Mazat, R. Rossignol, M. Malgat, C. Rocher, B. Faustin and T. Letellier, *Biochim. Biophys. Acta*, 2001, **1504**, 20.
[54] L. Agius, *Adv. Enzyme Regul.*, 1998, **38**, 303.
[55] M. Salter, R. G. Knowles and C. I. Pogson, *Essays Biochem.*, 1994, **28**, 1.
[56] R. Curto, E. O. Voit and M. Cascante, *Biochem. J.*, 1998, **329**, 477–487.
[57] L. G. Boros, M. Cascante and W. N. Lee, *Drug Discov. Today*, 2002, **7**, 364.
[58] M. Cascante, L. G. Boros, B. Comin-Anduix, P. de Atauri, J. J. Centelles and P. W. Lee, *Nat. Biotechnol.*, 2002, **20**, 243.
[59] L. Sabaté, L. R. Franco, E. I. Canela, J. J. Centelles and M. Cascante, *Mol. Cell. Biochem.*, 1995, **142**, 9.
[60] R. K. Ockner, *Integration of Metabolism, Energetics, and Signal Transduction*, Kluwer Academic/Plenum Publishers, New York, 2004.
[61] C. P. Chih, P. Lipton and E. L. Roberts, *Trends Neurosci.*, 2001, **24**, 573.
[62] V. V. Lemeshko, *Biophys. J.*, 2002, **82**, 684.
[63] W. N. Lee, L. G. Boros, J. Puigjaner, S. Bassilian, S. Lim and M. Cascante, *Am. J. Physiol.*, 1998, **274**, E843.
[64] M. Merle, M. Martin, A. Villegier and P. Canioni, *Eur. J. Biochem.*, 1996, **239**, 742.

[65] M. D. Turton, D. O'Shea, I. Gunn, S. A. Beak, C. M. Edwards, K. Meeran, S. J. Choi, G. M. Taylor, M. M. Heath, P. D. Lambert, J. P. Wilding, D. M. Smith, M. A. Ghatei, J. Herbert and S. R. Bloom, *Nature*, 1996, **379**, 69.

[66] C. F. Elias, J. F. Kelly, C. E. Lee, R. S. Ahima, D. J. Drucker, C. B. Saper and J. K. Elmquist, *J. Comp. Neurol.*, 2000, **423**, 261.

[67] N. N. Danial, C. F. Gramm, L. Scorrano, C. Y. Zhang, S. Krauss, A. M. Ranger, S. R. Datta, M. E. Greenberg, L. J. Licklider, B. B. Lowell, S. P. Gygi and S. Korsmeyer, *Nature*, 2003, **424**, 952.

[68] M. Shimabukuro, M. Y. Wang, Y. T. Zhou, C. B. Newgard and R. H. Unger, *Proc. Natl. Acad. Sci.*, 1998, **95**, 9558.

[69] T. Perry, N. J. Haughey, M. P. Mattson, J. M. Egan and N. H. Greig, *Pharmacol. Exp. Ther.*, 2002, **302**, 881.

[70] A. Bulotta, R. Perfetti, H. Hui and L. G. Boros, *J. Lipid Res.*, 2003, **44**, 1559.

[71] J. Boren, M. Cascante, S. Marin, B. Comin-Anduix, B. J. J. Centelles, S. Lim, S. Bassilian, S. Ahmed, W. P. Lee and L. G. Boros, *J. Biol. Chem.*, 2001, **276**, 37747.

[72] M. K. Hellerstein, *Annu. Rev. Nutr.*, 2003, **23**, 379.

[73] T. J. Carlson and M. B. Fisher, *Comb. Chem. High Throughput Screen.*, 2008, **11**(3), 258.

[74] M. Canovas, V. Bernal, A. Sevilla and J. L. Iborra, *In Silico Biol.*, 2007, **7**(Suppl. 2), S3.

PART VII:
Topics in Drug Design and Discovery

Editor: Manoj C. Desai
Gilead Sciences, Inc.
Foster City
California

CHAPTER **21**

Contrasting the Gastrointestinal Tracts of Mammals: Factors that Influence Absorption

John M. DeSesso and **Amy Lavin Williams**

1. INTRODUCTION

The rate and extent of absorption of orally ingested compounds are influenced by properties that are intrinsic to ingested substances themselves as well as factors that are associated with the milieu of the alimentary canal and its absorptive surface. The physicochemical properties of drugs and chemicals (e.g., molecular size, aqueous solubility, lipophilicity) are relatively invariable in different test systems, whereas factors such as the absorptive surfaces and luminal milieus are greatly influenced by the gastrointestinal anatomy and physiology of the species under consideration. Previous reports from this group have described the interspecies similarities and differences of the gastrointestinal tracts of rats and

Noblis, 3150 Fairview Park Drive, Falls Church, Virginia 22042, USA

Annual Reports in Medicinal Chemistry, Volume 43
ISSN 0065-7743, DOI 10.1016/S0065-7743(08)00021-3

humans [1,2]. Those reports identified several anatomical and physiological parameters that affect gastrointestinal absorption. It is the purpose of the present report to expand upon those papers by presenting comparative, quantitative data for the previously identified important anatomical and physiological parameters of the gastrointestinal tract for several additional mammalian species. Data for the dog, rabbit, mouse, pig, and monkey will be presented herein, as well as updated information for rats and humans.

It is emphasized that the parameters described herein relate to the crossing of materials from the gastrointestinal lumen into the bloodstream. Blood from the major absorptive areas of the gastrointestinal tract passes through the liver by means of the hepatic portal system and the absorbed materials carried therein are subject to possible metabolism or storage. This process is termed the "first pass effect" because it effectively reduces the concentration or amount of material absorbed from the gastrointestinal tract that enters the systemic circulation. Thus, while information about absorption of materials from the gastrointestinal tract is the first step in determining potential biological effects, this does not always provide accurate measures of concentrations found at target tissues that are distant from the gastrointestinal tract (see discussion in Ref. [3]).

2. OVERVIEW OF THE ALIMENTARY CANAL

The alimentary canal is a tube that extends from an entryway (the mouth) through the body to an exit point (the anus). Because the lumen of the alimentary canal connects to the environment at both ends and is lined by a continuous epithelium, the lumen of the alimentary canal is essentially an extension of the environment that passes through the body. Materials that travel through the alimentary canal must traverse its walls to gain access to the internal environment of the body and be absorbed into the bloodstream. In mammals, the alimentary canal is subdivided into regions with distinct roles. Ingested material enters the mouth, where it is broken up through the process of mastication (chewing) and made somewhat fluid by the secretion of saliva. When the bolus of material attains an appropriate consistency, it is propelled through swallowing into the muscular esophageal tube, which conducts it to the stomach. Little, if any, absorption occurs in the esophagus.

The stomach is a capacious muscular organ that mixes the ingested material with additional secretions to facilitate digestion. The stomachs of rodents and lagomorphs have two distinct regions (forestomach and glandular stomach) that are separated by a limiting ridge, which prevents these species from vomiting. The forestomach of rats and mice is populated by numerous bacteria that are important in their digestive process. It is lined by epithelium that resembles the mucous membrane of the oral cavity. The glandular region contains the acid-secreting glands. In dogs, pigs, and primates, the stomach has a single chamber that is lined by mucus- and acid-secreting cells. The predominant gastric secretion in all species considered herein is rich in hydrochloric acid. In humans, the daily volume of gastric secretions can reach 1.5–21 [2]. Gastric secretions help to denature proteins and to further break up the food particles. After processing in the stomach, which may take several hours, small boli of the ingested material (now termed chyme) pass through the pyloric sphincter to the small intestine, where digestion is essentially completed and the majority of absorption takes place.

The proximal part of the small intestine, the duodenum, receives secretions from numerous sources, including the pancreas (high in volume and rich in digestive enzymes and bicarbonate), liver (bile), and enteric glands (located in the wall of the small intestine, their secretions are rich in bicarbonate that neutralizes the acid from the stomach). The volumes of these secretions in primates, dogs, and pigs more than double the secretions from the salivary glands and stomach and serve to make the chyme watery; in rodents, the chyme remains rather pasty (cf. [2]). As the chyme moves through the duodenum to the jejunum and ileum, absorption reaches its zenith. At the distal end of the ileum, chyme passes into the large intestine. The first portion of the large intestine is the cecum, which has disparate functions among various species. In primates, fluids and salts that remain after completion of intestinal digestion are absorbed in the cecum. Rabbits, which are herbivores, have a much larger cecum that contains bacteria which help digest cellulose. The ceca of carnivores such as dogs use strong muscle contractions to reverse the propulsion of the chyme and bring it back from the next region of the large intestine to mix it with newly arrived chyme from the ileum. As the chyme moves through the cecum and the remainder of the colon, it is mixed with intestinal bacterial flora. Additionally, water is progressively absorbed from the chyme–bacterial matrix, which then forms solid feces. Fecal material is eventually propelled to the rectum, where it is stored until its expulsion during defecation.

Throughout the length of the alimentary canal, the lumen is bounded by a surface characterized as both a barrier to the entry of undesirable luminal contents, such as bacteria and undigested materials like cellulose, and a structure comprising various mechanisms that facilitate the absorption of preferred contents at specific locations. Detailed discussions of these topics can be found in other publications and textbooks (e.g., [2,4–7]). Briefly, the impediments to transfer of materials from the lumen to the bloodstream include a continuous layer of epithelium that is interconnected by tight junctions and desmosomes, plus a subjacent basal lamina. Mechanisms that provide passage of digested materials from the lumen to the bloodstream include passive diffusion across the epithelial membrane, facilitated (carrier-mediated) diffusion, active transport, aqueous pores within the cell membranes or tight junctions, and endocytosis (transport of material into the cell by folding of the cellular plasma membrane inwards to form a vacuole). These mechanisms of transfer are localized in select portions of the alimentary canal, predominantly the small intestine. Perhaps the most important factor that affects the rate and extent of absorption relates to alterations in the anatomical structure of the epithelial boundary that greatly increase the surface area available for absorption. These modifications of the epithelial surface (described later) include the presence of finger-like projections of epithelium (villi) and of a brush border (microvilli), as well as grossly observable folds of epithelium into the lumen as seen in some species, including primates and humans.

While the gross organization of the individual segments of the alimentary canal of all mammals is essentially similar, there are wide differences in the dimensions of some gastrointestinal structures, as well as in the lengths and absorptive surface areas of the various subdivisions. In addition, the "environmental" conditions (e.g., pH and fluidity of the chyme, number and type of bacterial flora, transit time) within the lumen of the various subdivisions differ greatly among species. These differences, and the comparison of such to the human condition, must be taken into account when selecting an animal model for drug disposition studies and when extrapolating the results of studies from animal to human. The purpose of the present chapter is to discuss and contrast these interspecies differences to assist in this process.

3. SITES OF AND BARRIERS TO ABSORPTION

Detailed consideration of the nature of the protective barrier in the gastrointestinal system can be found in our previous reports (see Refs [1,2]), in textbooks of histology (e.g., [8–10]), and in articles by others (e.g., [11]). Briefly, the gastrointestinal system can be divided into those regions which serve as conduits to propel ingested material through the alimentary canal (oral cavity, pharynx, esophagus, lower rectum, and anal canal), and those regions involved in digestion and absorption of ingested substances (stomach, small and large intestine). Regardless of its function, every region in the alimentary canal is lined by a mucous membrane composed of an epithelium of cells joined by tight junctions underlain by a basement membrane and loose connective tissue (lamina propria), which is richly vascularized in those areas of the alimentary canal involved in absorption. Absorptive areas of the alimentary canal are further characterized by modifications of the mucous membrane that increase the absorptive surface area (Figure 1). These modifications include large folds of the mucous membrane (plicae circulares, also known as valves of Kerckring) that are present in some species, finger-like projections of mucous membrane into villi, and numerous projections from the luminal surface of the epithelial cell membrane to form a brush border (microvilli).

Many substances that are absorbed in the alimentary canal diffuse across the luminal membrane into the vasculature of the lamina propria. Diffusion requires that the substance to be absorbed comes into intimate contact with the absorptive surface. As discussed later, factors that affect diffusion include the expanse of absorptive surface area and the time of contact with that surface area. Thus, to be able to relate gastrointestinal absorption in various animal species to humans, it is important to have an understanding of the dimensions of the absorptive regions of the alimentary canal, the available surface areas of those regions, and the transit time of ingesta through each subdivision of the alimentary across species. Such information is sparse for some species, but the authors have attempted to be as complete as possible in the tables and discussion later.

Nearly all substances taken up from the intestinal lumen must traverse the cytoplasm of an enterocyte (Figure 2). Amino acids, triglycerides, and

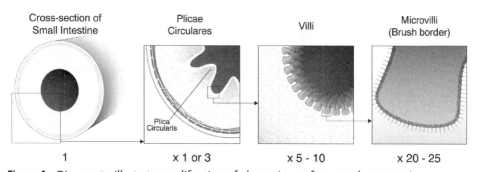

Cross-section of Small Intestine	Plicae Circulares	Villi	Microvilli (Brush border)
1	x 1 or 3	x 5 - 10	x 20 - 25

Figure 1   Diagram to illustrate amplification of absorptive surface area by anatomic modifications of the intestinal wall. At the left of the diagram, the absorptive surface area of the intestine is depicted as that of a cylinder having a nominal value of 1. Modifications of the surface are depicted as one moves to the right. In the second diagram, folds of mucosa (plicae circulares) are represented. In those species that have them (monkey and human, of the seven species considered here), the surface area is increased 3-fold. The next diagram portrays the fingerlike villi that project from the surface of the intestinal mucosa throughout the length of the intestine. The dimensions of the villi vary for each species depending on length, width, and density of villi per square millimeter of mucosal surface. The villi increase the absorptive surface by 5–10-fold. The last diagram illustrates the surface of individual enterocytes. Note the brush border, which is made up of microvilli that increase the absorptive surface area by 20–25-fold.

carbohydrates are absorbed along the length of the microvilli, traverse the enterocyte cytoplasm, and exit the cell at its base. Under certain conditions, solvent drag (or convection) can occur as a parallel route of absorption [12,13]. This condition results when there is a partial separation of the tight junctions between enterocytes that allows the passage of large amounts of water and small solute molecules. These molecules pass into the capillary plexus of the villi. In contrast, lipids and fats are absorbed at the base of the microvilli, traverse the enterocyte cytoplasm, and exit the enterocyte at its side, beneath the tight junctions.

The deep structure of each intestinal villus contains a large, blind-ending lymphatic vessel (the lacteal) which is responsible for absorbing higher molecular weight substances, including fats (Figure 3). The fatty fluid contained therein is termed "chyle". The lacteals coalesce to form the intestinal lymphatic channels that converge at the cisterna chyli on the posterior abdominal wall and empty into the left subclavian vein via the thoracic duct. This means that materials transported in the chyle bypass the liver and avoid the first pass effect.

4. DIMENSIONS OF REGIONS OF THE GASTROINTESTINAL TRACT

The portions of the gastrointestinal tract considered in the remainder of this document are those involved in absorption of ingested materials including the stomach, small intestine (duodenum, jejunum, and ileum), and the large intestine

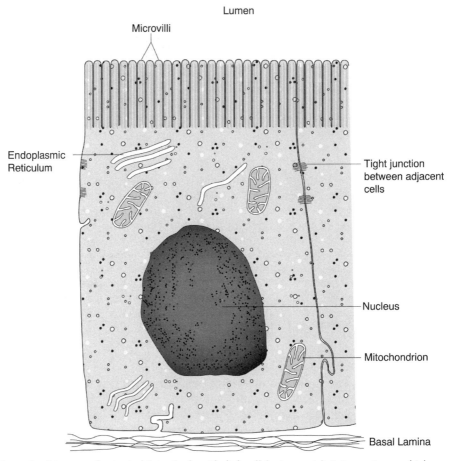

Figure 2 Diagram of a typical intestinal epithelial cell (enterocyte). Enterocytes are high cuboidal to low columnar epithelial cells that rest on a basement membrane. Enterocytes are joined firmly to adjacent cells by tight junctions. The nucleus resides in the basal portion of the cell. The apical surface of the cell exhibits numerous microscopic projections (microvilli) that give the appearance of a "brush border" when seen with a light microscope. Microvilli greatly increase the available absorptive surface area for the enterocyte. Nearly all substances taken up from the intestinal lumen must traverse the cytoplasm of an enterocyte. Lipids and fats are absorbed at the base of the microvilli, traverse the enterocyte cytoplasm, and exit the enterocyte at its side, beneath the tight junctions. In contrast, amino acids, triglycerides, and carbohydrates are absorbed along the length of the microvilli, traverse the enterocyte cytoplasm, and exit the cell at its base. Reproduced from DeSesso and Jacobson [2], with permission.

(cecum, ascending colon, transverse colon, descending colon, and rectum). The lengths and percentages of total length of the small and large intestinal regions of the gastrointestinal tract for seven species (mouse, rat, rabbit, dog, pig, monkey, and human) are presented in Table 1. Where available, the lengths for

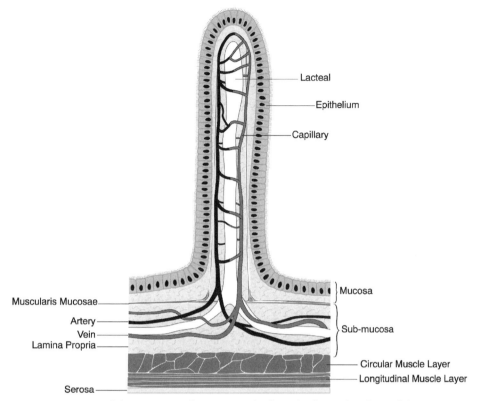

Figure 3 Diagram of the structure of an intestinal villus. The luminal surface of the entire intestinal tract, including each villus, is lined by enterocytes. Note the fuzzy appearance of the brush border on the surface of the villus. The deep structure of the villus is characterized by a large, blind-ending lymphatic vessel (the lacteal) which absorbs higher molecular weight substances, including fats. The lacteals are tributaries to the intestinal lymphatic channels that coalesce at the cisterna chyli, the origin of the thoracic duct. Materials transported in the chyle bypass the liver and avoid the first pass effect. Note the small plexus of blood capillaries that surround the lacteal. In addition to the vascular structures, the villus contains loose connective tissue and small amounts of smooth muscle (muscularis mucosae). The muscularis mucosae wiggle the villi back and forth in the liquid layer in the lumen adjacent to the intestinal cells, thereby increasing the efficiency of absorption. Reproduced from DeSesso and Jacobson [2], with permission.

subdivisions of each region are reported. The data have been collected from multiple sources and, in some cases, they are presented as ranges.

Inspection of Table 1 reveals that the rabbit has the longest cecum of the species considered in this chapter. Microbial fermentation of dietary fiber in the rabbit cecum is the source of rapidly absorbed volatile fatty acids [14]. We also note that there are wide ranges in the reported lengths of various portions of the gastrointestinal tract, which may be due to differences in methods of measurement as noted by Snyder et al. [15] and Choi and Chiou [16].

Table 1 Comparison of lengths of the intestinal tract and its major subdivisions in seven different species

Region of intestinal tract	Mouse			Rat			Rabbit			Dog			Pig			Monkey			Human		
	Length (cm)	% of total	% of subd	Length (cm)	% of total	% of subd	Length (cm)	% of total	% of subd	Length (cm)	% of total	% of subd	Length (cm)	% of total	% of subd	Length (cm)	% of total	% of subd	Length (cm)	% of total	% of subd
Duodenum	5.3[a]	—	—	9.5–10[b]	—	8	—	—	—	25[c,d]	—	6	—	—	—	—	—	—	25[e]	—	4
Jejunum	46.2[a]	—	—	90–135[b]	—	90	—	—	—	360	—	90	—	—	—	—	—	—	260[e]	—	38
Ileum	—	—	—	2.5–3.5[b]	—	2	—	—	—	15[c,d]	—	4	—	—	—	—	—	—	395[e]	—	58
Total small intestine	45	~75	—	125[b]	83	—	339–398[f,c]	61	—	~400[c,g,h]	83	—	1,500–2,000[c,g,h]	70–85	—	—	—	—	680[e]	81	—
Cecum	3.4[a]	—	—	5–7[b]	—	26	41–61[c,g,i]	—	28	5–8[c,d,g]	—	~11	20–30[c,g,h]	—	—	—	—	—	7[e]	—	5
Colon	10.0[a]	—	—	9–11[b]	—	42	114–165[c,g,i,j]	—	72	25–60[c,d, g,h]	—	~81	400–500[c,g,h]	—	—	—	—	—	93[e]	—	60
Rectum	—	—	—	8[b]	—	32	—	—	—	5[d]	—	~8	—	—	—	—	—	—	55[e]	—	35
Total large intestine	—	~25	—	25[b]	17	—	—	39	—	~82	17	—	350–850	15–30	—	—	—	—	155[e]	19	—
Total intestinal tract	—	—	—	150[b]	—	—	582[c,g]	—	—	482[c,g]	—	—	2,350[c,g]	—	—	—	—	—	835[e]	—	—

[a] [45]; intestinal segments unfixed.
[b] [46].
[c] [47].
[d] [51].
[e] [15].
[f] [31].
[g] [48].
[h] [52].
[i] [49]; formaldehyde fixed segments.
[j] [50]; Bouins solution fixed segments.

For example, measurements made using intubation procedures may be too short because the intestines tend to gather on the tube and due to the penchant for the tube to stray from the center of the lumen by cutting corners. If measurements are made post-mortem, the loss of smooth muscle tone in the intestine can lead to lengths that are too long. In cases where the intestines had been placed in fixative, the measurements are likely to be distorted by the shrinkage that is attendant to the fixation process.

5. SURFACE AREAS OF THE GASTROINTESTINAL TRACT

The vast majority of substances that are absorbed in the gastrointestinal tract gain access to the bloodstream by means of passive diffusion, which occurs predominantly in the small intestine. Passage of material by passive diffusion follows Fick's Law of diffusion ([17]; for discussion of simple diffusion in physiology texts see Guyton and Hall [18]). Because the details of gastrointestinal absorption are quite complex, Fick's Law is really an approximation [3]. Nevertheless, there are several characteristics of the luminal surface of the intestine that can be explained by considering the impact of morphology on this approximation of intestinal absorption.

Basically, Fick's law states that bulk diffusion at steady state is proportional to the efficiency of exchange (membrane characteristics) and available surface area. Because the amount of available surface area plays a dominant role in passive diffusion, it is important to know this value for the species of interest. Table 2 provides the smooth luminal surface areas of the small intestine for seven species, including humans. The smooth luminal values were calculated in various reports by conceiving the small intestine as a cylinder and using the length and diameter measurements. As noted in Table 1, the values provided should be considered as estimates rather than absolute, true values. As previously mentioned, in organs like the small intestine that are specialized for absorption, the rate and extent of absorption is greatly increased by expansion of the absorptive surface area. Each of the seven species examined in this report exhibits modifications to the absorptive surface that lines the lumen of the small intestine, which increase the surface area of the small intestinal epithelium by a combined factor of 100–750, as shown in Table 2. Among the species listed in Table 2, only two possess plicae circulares: monkeys and humans. The plicae circulares increase the luminal surface area approximately 3-fold. The presence of villi increases the luminal surface area by a factor ranging from 5 to 10 depending upon the species. The range in values is related to differences in the dimensions of the villi for each species (e.g., length, width, density of microvilli), and within the different regions of the small intestine. Microvilli, which are the major component of the brush border, further increase the absorptive surface area by a factor of 20–25.

It should be noted that within a given species, the architecture of villi change as one progresses distally through the intestine. Thus, villi in the duodenum tend to be broader than those of the ileum [19,20], resulting in a gradient of increased

Table 2 Calculated total small intestinal surface areas for seven species

	Body weights (kg)	Smooth lumenal surface area (m²)	Fold-increase factors			Combined multiplication factors	Estimated total surface area (m²)
			Plicae	Villi	Micro-villi		
Mouse	0.032	0.004[a]	(1)[b]	(5)[b]	(20)[b]	100	(0.4)
Rat	0.3	0.016[a]	1[c]	5[c]	20[c]	100	1.6
Rabbit	2.6–3.9	0.087–0.096[a]	1[d]	5.7[e]	24[f]	136.8	11.9–13.1
Dog	10.7–12.6	0.099–0.14[a,g]	1[d]	10[h]	25[i]	250	24.75–35
Pig	47	1.4[g]	1[d]	6[j]	(20–25)[k]	120	(168–210)
Monkey	2.7–3.0	0.091–0.11[g,l]	(3)[m]	(10)[m]	(20)[m]	600	(54.6–66)
Human	70	0.42[a]	3[c]	10[c]	20[c]	600	252

[a]Values as reported in Ref. [31].
[b]Extrapolated from values for rat.
[c]From Ref. [2].
[d]Value based on the absence of plicae in the small intestine.
[e]Value based on comparison of villus lengths between rabbits and humans as reported in Refs [79] and [15].
[f]As cited in Ref. [54].
[g]Values as reported in Ref. [53].
[h]Based on same size villi as humans as reported in Refs [15,55].
[i]As reported in Ref. [47].
[j]Based on comparison of villi lengths between pigs and humans as reported in Refs [15,56].
[k]Extrapolated from values reported for other animal species.
[l]Values as reported for Cynomolgus monkeys.
[m]Extrapolated from values for human.

surface area in the proximal small intestine [21]. This gradient is even greater in those species (monkey and human) that possess plicae circulares (which tend to be larger and more numerous in the proximal intestine).

Because animals of dissimilar sizes have different basal metabolic rates, they have distinctive requirements for caloric intake. This makes direct comparisons of the absolute surface areas of the gastrointestinal tracts of various species difficult to interpret. In an attempt to normalize across species the surface areas of the stomach, small intestine, and large intestine, the surface areas of these regions of the gastrointestinal tract are presented in terms of both absolute surface area and a relative surface area in Table 3. (The relative surface area is herein defined as the absolute surface area of the gastrointestinal tract, divided by the body surface area.) The metabolic rate of most mammalian species is proportional to its body surface area [22,23]. Consequently, the relative surface area serves to normalize the effective absorptive surface based on a species' metabolic requirements.

Humans are the largest species analyzed in Table 3; thus, it is not surprising that they have the largest absolute small intestinal surface area. It is noteworthy, however, that when the basal metabolic rates of the species are taken into consideration, humans still have the largest relative surface area.

It should be noted that the values in Table 3 can be altered by physiological changes that occur in response to nutritional challenges. For example, during times of starvation or disease, such as diabetes, [24] or during normal physiologic

Table 3 Comparison of absolute surface areas and surface areas relative to resting metabolic rates of the gastrointestinal tract and its major subdivisions in seven different species

	Mouse		Rat		Rabbit		Dog		Pig		Monkey		Human	
Body weight (kg)	0.032		0.3		2.6–3.9		10.7–12.6		47		2.7–3.0		70	
Body surface area (BSA; m²)[a]	0.0066		0.03–0.06		0.23		0.39–0.78		—		0.3		1.8	
Region	Absolute SA (m²)	SA relative to BSA (% total)	Absolute SA (m²)	SA relative to BSA (% total)	Absolute SA (m²)	SA relative to BSA	Absolute SA (m²)	SA relative to BSA (% total)	Absolute SA (m²)	SA relative to BSA (% total)	Absolute SA (m²)	SA relative to BSA (% total)	Absolute SA (m²)	SA relative to BSA (% total)
Stomach	0.00035[b]	0.05 (0.1%)	0.00062[c]	0.01–0.02 (0.0%)	—	—	0.0344–0.0426[d]	0.04–0.11 (0.1%)	0.016[e]	—	0.0143–0.0306[f]	0.048–0.10 (0.0%)	0.053[g]	0.029 (0.0%)
Total small intestine[h]	0.4	60.61 (99.0%)	1.6	26.67–53.33 (97.9%)	11.9–13.1	51.74–56.96	24.75–35	31.73–89.74 (99.3%)	168–210	—	54.6–66	182–220 (99.8%)	252	140 (99.8%)
Total large intestine	0.0036[i]	0.55 (0.9%)	0.034[j]	0.57–1.13 (2.1%)	0.1606[k]	0.7	0.023–0.245[d,l]	0.029–0.63 (0.5%)	0.5142[e]	—	0.0687–0.0933[f]	0.23–0.31 (0.1%)	0.35[g]	0.19 (0.1%)

[a] As reported in Ref. [57].
[b] [45]; mean bw=~43 g: unfixed tissue.
[c] [58].
[d] [53]; values from 2 dogs (bw = 10.68 and 12.55 kg).
[e] [53]; values based on single pig (bw = 47.98 kg); formaldehyde fix segments; large intestinal value based on surface areas of cecum and colon combined.
[f] Derived from Ref. [53]; values for Cynomolgus monkeys (bw = 2.7–3.1 kg); unfixed tissues.
[g] [15] as reported in [2].
[h] Small intestinal surface areas reported as calculated in Table 2.
[i] Derived from Ref. [49]; mean bw = 47.5 g; formaldehyde fixed segments; large intestinal value based on surface areas of cecum and colon combined.
[j] [59].
[k] Derived from Ref. [49]; mean bw = 3.6 kg; formaldehyde fixed segments; large intestinal value based on surface areas of cecum and colon combined.
[l] Derived from Ref. [49]; mean bw = 13.75 kg; formaldehyde fixed segments; large intestinal value based on surface areas of cecum and colon combined.

stress brought on by pregnancy, the mucosa of the stomach undergoes hyperplasia in rats [25], and the mucosal surface area of the small intestine increases as the height of villi increase temporarily during gestation [26]. This condition of increased absorptive surface area is maintained throughout the period of lactation [27]. Similar changes occur in all species.

Changes in effective surface areas occur as animals age into senescence. The villi present in the small intestine decrease in height during old age [28–30]. This results in a decreased surface area for absorption and, consequently, in a reduced efficiency of the absorption of nutrients and other materials from chyme.

It is of note that Pappenheimer [31] reported that the mucosal surface area of the small intestines among nonruminant eutherian mammals increases approximately in proportion to the 0.6 power of body mass.

6. TRANSIT TIMES

Absorption of materials from the gastrointestinal tract is influenced not only by the total available surface area but also by the amount of time that the substance to be absorbed is in contact with the absorptive epithelium. Because chyme is propelled through the gastrointestinal tract, it is in contact with the absorptive epithelium for a finite period. The duration of time that chyme is in contact with the epithelium of a particular subdivision of the gastrointestinal tract is dictated by the rate of propulsion (which is specific to each species) and the length of that particular region of the gastrointestinal tract. These periods are described as transit times or as emptying half-times.

The half-time for emptying of stomach contents after a meal for each of the species is presented in Table 4. Inspection of the table reveals that, for the most part, liquids typically pass through the stomach at a faster rate than solids. It should be noted, however, that ingestion of a fatty meal will greatly slow the

Table 4 Range of gastric emptying half-times (in hours) as reported for seven species of animals

Species	Liquids	Particulates
Mouse	1.23[a]	
Rat	—	—
Dog	1.5[b]	1.5[b]
Rabbit	1.3[b]	12[b]
Pig	2[b]	10[b]
Monkey	0.4–0.5[c]/2.4[d]	—
Human	0.2[e]	0.5–3.4[a,f]

[a][60].
[b]Values reported in Refs [47,48].
[c][61]; measurements from fasted cynomolgus monkeys.
[d][62]; measurements from fed cynomolgus monkeys.
[e][4].
[f][63]; measurements for males and females, 19–45 years of age.

passage of stomach contents to the duodenum [4,7]. This is illustrated by the significantly different gastric emptying half-times for liquids given to monkeys depending on whether or not they were fasted or fed.

The range of transit times through various portions of the gastrointestinal tract for each of the species is presented in Table 5. In general, transit times are shortest in the stomach and longest in the large intestine. Again, it should be noted that the fasting state of the individual can affect transit time. For instance, prolonged fasting slows the oro-cecal transit time, likely because the body needs to absorb more calories from the ingested material (discussed in Ref. [32]). The amount of fiber in the diet affects both transit time and retards the access of some substances, such as cholesterol, to the absorptive surface [33,34].

Physiological status can also influence transit time. For example, aging has been shown to affect transit time in dogs and humans. The mean oro-cecal transit time decreased with age in large breed dogs (Great Dane and Giant Schnauzer). This relationship did not hold for smaller breeds (Miniature Poodle and Standard Schnauzer), however. It is believed that the decrease is due to the increase in body size associated with growth as those breeds matured to final size (resulting in a longer gastrointestinal tube) rather than to a slower rate of transport [35,36].

In humans, transit times through the stomach and small intestine are significantly longer among older people compared to young adults, but the large intestinal transit time increased significantly only among women [37]. Recall that the dimensions of intestinal villi change such that the available absorptive surface area decreases with age [38]. The slowing in transit time may be a compensation that allows sufficient nutritional absorption. This same group of investigators reported that the overall gastrointestinal transit time was significantly longer in women than in men at all ages.

7. NATURE OF LUMINAL CONTENTS

As discussed in our previous report [2], there is a dramatic difference in the nature of the chyme that enters the duodenum of rats compared to that of humans. Rat chyme contains many bacteria and has the consistency of a moist paste when entering the duodenum, whereas human chyme is watery and its bacterial content is negligible. The chyme of mice and rabbits is quite similar to that of the rat. The chyme of dogs, pigs, and monkeys is more similar to human chyme than to that of the rat. As chyme moves through the distal portion of the small intestine in all species, it begins to acquire a more viscous consistency, especially at the distal end of the ileum.

The pH of the luminal contents also varies widely among species (Table 6). As a general rule, the pH of contents is lowest in the stomach and increases as the chyme progresses distally through the gastrointestinal tract, its pH approaches neutrality. This accomplished by the secretion of bicarbonate from the intestinal glands of Brunner and, predominantly, from the mucosa [39]. In carnivores (including humans, pigs, and dogs) as well as the rabbit (an herbivore), the pH of the stomach is quite low, whereas the gastric pH of rodents (rats and mice) is only

Table 5 Range of Gastrointestinal Transit Times (Hrs) Through Different Portions of the Gastrointestinal Tract as Reported for Seven Species

Species	Oro-Cecal — Stomach Liquid	Stomach Solid	Small intestine Liquid	Small intestine Solid	Large intestine Liquid	Large intestine Solid	Total Liquid	Total Solid
Mouse	—		—	—		—		—
Rat	0.7–2.1[a]			2.6–3.3[a]		15.5[b]		38.9[c]
Dog	1.7[d]		14.3[e]			18.5[e]	27.0–37.4[e,f,g]	
Rabbit			9.1[h]			9.6[h]	18.1–20.6[h]	
Pig	0.8–0.9[i]	1.0–1.3[i]	3.9–4.4[i]	3.7–4.3[i]	24.9–41.3[i]	35.6–44.4[i]	29.1–46.6[i]	40.5–49.7[i]
Monkey	1.1[l]	1.8/2.3–2.5[k]	7.5[l]		17.2–40[l,m]	17.5[l]		
Human — Child			1.8–8.0[l,o,q]			35.0[n]		
Human — Young adult	1.0–1.6[d,n] 1.1[o,p]	1.9–2.3[n]	3.8–3.9[n]	3.4–3.5[n]				
Human — Mature adult	0.8–1.1[n]	1.3–1.8[n]	2.0–4.2[n]	3.0–3.8[n]	33.5–61.5[n]		57.3–58.2[r]	

a[64]; measurements from fed rats.
b[65]; measured by the insertion of dye in the cecum.
c[66]
d[40]; liquid transit in fasted Beagle dogs and young adults (21–29 years of age).
e[77]; measured in standard Schnauzers.
f[76]; measured in Beagle dogs.
g[67]; measured in French Bulldogs, English Cocker Spaniels, and standard Schnauzers.
h[68].
i[69]; measurements derived from pigs with average weight of 33 kg.
j[61]; measurements from fed cynomolgus monkeys.
k[62]; measurements from fasted cynomolgus monkeys.
l[70]; determined in fasted cynomolgus monkeys.
m[71]; determined in children 8–14 years of age.
n[37]; determined in adults 20–30 years of age (young) and 38–53 years of age (mature).
o[72]
p[73]
q[43]; measurements for males and females, 19–45 years of age.
r[74]; males only.

Table 6 Range of pH values reported for different portions of the gastrointestinal tract for seven species of animals[a]

Species	Stomach		Small intestine[b]			Large intestine		
	Anterior	Posterior	Duodenum	Jejunum	Ileum	Cecum	Colon	Rectum/Feces
Mouse	4.5	3.1	—	—	—	—	—	—
Rat	4.3–5.1	2.3–4.0	6.5–7.1	6.7–6.8	7.1–8.0	6.4–7.2	6.6–7.6	6.9
Dog	1.5–5.5	1.5–3.4	6.2	6.2–7.3	7.5	6.4	6.5	6.2
Rabbit	1.9–2.2	1.9–2.2	6.0–6.1	6.8–7.5	7.2–8.0	5.7–6.6	6.1–7.2	7.2
Pig	1.6–4.3	6.0	6.2–6.9	7.5	6.3	6.8	7.1	—
Monkey	4.7–5.0	2.3–2.8	5.6–6.0	5.8–6.0	6.0–6.7	4.9–5.1	5.0–5.9	5.5
Human	1.5–5.0	5.0–7.0	6.0–7.0	7.0–7.4	5.7–5.9	5.5–7.5	6.5–7.0	—

[a]The range of values reported here are from the following references: [14,40,41,47,73,75,78].
[b]Some studies reported pH values for segments 1, 3, 5, and 7 of the small intestine. For the purposes of this table, segment 1 was designated the duodenum, segments 3 and 5 were designated the jejunum and segment 7 was designated the ileum.

slightly more acidic than that of the skin. The moderate pH in the stomach of rodents is a consequence of the requirement for microfloral digestion of ingested material that begins before entry to the small intestine. Animals with very acidic stomach contents begin digestion of proteins therein and have no requirement for a large bacterial colony in their stomach. Despite its status as a carnivore, the pH of the gastrointestinal tract of the dog presents some unique characteristics compared to humans. For instance, the fasting intestinal pH in dogs is higher than that of humans (7.3 in the dog versus 6.0 in humans; [40]).

Most elderly humans exhibit the same pH behavior as younger humans, but there are notable exceptions. Some seniors exhibit elevated gastric pH or even achlorhydria (the inability to secrete gastric acid; [41]). Achlorhydria is often accompanied by delayed gastric emptying times.

In addition to the pH gradient that exists distally along the gastrointestinal tract, a pH gradient also exists from the center of the lumen moving radially towards the epithelial surface. The pH of the lumen is more acidic than the pH of contents at the epithelial surface. This is a function of the unstirred layer of water that is associated with the brush border (see discussion in Ref. [2]) and the alkaline secretions of the intestinal epithelia [42]. Furthermore, this gradient may affect the rate of uptake of various substances, especially xenobiotics.

8. BILE AND BILE SALTS

Among the species described in this chapter, the rat is the only one that lacks a gallbladder. While this makes the rat unique among commonly used experimental animals, other familiar vertebrates, including the horse and the pigeon, also lack gallbladders. In species that lack a gallbladder, bile is not concentrated but rather is added to the luminal contents of the duodenum on a continuous basis.

Table 7 Comparison of flow rates and composition of bile from seven different species

Species	Bile flow rate (ml/day/kg)	Total bile Salts (mmol/l)
Mouse	—	—
Rat	48–92	17–18
Rabbit	130	6–24
Dog	19–36	40–90
Pig	—	—
Monkey	19–32	22
Human	2.2–22.2	3–45

*Data from Ref. [47].

Table 7 presents the daily flow of bile into the duodenum (normalized by body weight in kilograms) and the concentration of bile salts within the bile for five of the seven species considered in this chapter. There is no clear relationship between the size of the species and its bile flow rate. Interestingly, the highest rate is found in the rabbit, the only herbivore among the set.

9. CONCLUSIONS ABOUT ABSORPTION

By far, the greatest amount of experimental information regarding gastrointestinal absorption is available for the rat. Due to similar small intestinal transit times, it is likely that the total amount of material absorbed by the small intestine after oral administration is similar between humans and rats, as has been concluded by others [43]. The difference between the two species, however, is the rate at which ingested materials are absorbed. This is largely a function of the great amount of surface area found in the proximal portion of the small intestine of humans. This led us to conclude in our previous paper [2], that humans are approximately five times more efficient at absorbing ingested materials than rats. Our conclusions compare favorably to the report of Pappenheimer [31], who found that perfused jejunal segments of normal human subjects absorb fluids at a rate that is 5–10 times greater per unit area of mucosa than that of laboratory rats.

It is also of interest that the oral absorption of a series of 43 drugs in dogs, which share many of the same physiological characteristics as humans, was found to differ significantly from the human [44], whereas the absorption of rats was much closer to that of humans [43].

While the information provided in this chapter is not able to answer the question of which species is the most suitable model to study gastrointestinal absorption, it is hoped that it will be useful for investigators who need to determine experimental conditions that test the pharmacological and toxicological effects of compounds in various species for the purpose of extrapolating the findings to humans.

ACKNOWLEDGEMENTS

The authors are grateful to Ms. Elaine Mullen and Mr. Ken Arevalo for their patience and attention to detail in creating the figures that illustrate this manuscript.

Funding to support preparation of this manuscript came, in part, from the Noblis Research Program.

REFERENCES

[1] J. M. DeSesso and R. D. Mavis, Identification of Critical Biological Parameters Affecting Gastrointestinal Absorption, MITRE Technical Report (MTR 89W223), 120pp, 1989.

[2] J. M. DeSesso and C. F. Jacobson, *Fd. Chem. Toxicol.*, 2001, **39**, 209–228.

[3] W. L. Chiou, *Pharmacokinet. Biopharm.*, 1996, **24**, 433.

[4] D. N. Granger, J. A. Barrowman and P. R. Kvietys, *Clinical Gastrointestinal Physiology*, WB Saunders, Philadelphia, 1985.

[5] T. T. Kararli, *Crit. Rev. Ther. Drug Carrier Syst.*, 1989, **6**, 39–86.

[6] K. K. Rozman and C. D. Klaassen, in *Casarett and Doull's Toxicology: The Basic Science of Poisons* (ed. C. D. Klaassen), 5th edition, McGraw-Hill, New York, 1996, pp. 91–112.

[7] L. R. Johnson, *Gastrointestinal Physiology*, C. V. Mosby, St. Louis, 2007, 7th edition.

[8] L. Weiss, *Cell and Tissue Biology: A Textbook of Histology*, Urban and Schwartzenberg, Baltimore, 1988, 6th edition.

[9] H. G. Burkitt, B. Young and J. W. Heath, *Wheater's Functional Histology*, Churchill Livingstone, Edinburgh, 1993, 3rd edition.

[10] L. C. Junqueira and J. Carneiro, *Basic Histology: Text & Atlas*, McGraw-Hill, New York, 2005, 11th edition.

[11] J. L. Werther, *Mt. Sinai J. Med.*, 2000, **67**, 41.

[12] J. R. Pappenheimer and K. Z. Reiss, *J. Mem. Biol.*, 1987, **100**, 123.

[13] J. R. Pappenheimer, *Am. J. Physiol.*, 1990, **259**, G290.

[14] S. Fekete, *Acta Vet. Hung.*, 1989, **37**, 265.

[15] W. S. Snyder,, M. J. Cook,, E. S. Nasset,, L. R. Karhausen,, G. P. Howells, and I. H. Tipton (eds), *Report of the Task Group on Reference Man*, Pergamon, New York, 1975.

[16] Y. M. Choi and W. L. Chiou, *Biopharm. Drug Dispos.*, 1997, **18**, 271.

[17] H. Schaefer, A. Zesch and G. Stuttgen, *Skin Permeability*, Springer-Verlag, Berlin, 1982, pp. 588–590.

[18] A. C. Guyton and J. E. Hall, *Textbook of Medical Physiology*, WB Saunders, Philadelphia, 1996, 9th edition.

[19] W. Bloom and D. W. Fawcett, *A Textbook of Histology*, WB Saunders, Philadelphia, 1968, 9th edition.

[20] W. M. Copenhaver, R. P. Bunge and M. B. Bunge, *Bailey's Textbook of Histology*, The Williams and Wilkins Company, Baltimore, 1971, 16th edition.

[21] R. B. Fisher and D. S. Parsons, *J. Anat.*, 1950, **84**, 271.

[22] M. Kleiber, *Hilgardia*, 1932, **6**, 315.

[23] M. Kleiber, *Physiol. Rev.*, 1947, **27**, 511.

[24] T. M. Mayhew and F. L. Carson, *J. Anat.*, 1989, **164**, 189.

[25] G. P. Crean and R. D. E. Rumsey, *J. Physiol.*, 1971, **215**, 181.

[26] L. Penzes and O. Regius, *J. Anat.*, 1985, **140**, 389.

[27] R. Boyne, B. F. Fell and I. Robb, *J. Physiol.*, 1966, **183**, 570.

[28] L. Penzes and I. Skala, *J. Anat.*, 1977, **124**, 217.

[29] L. Penzes, D. Kranz, K. Kretschmar, V. Rosenthal, D. Januschkewitz, M. Kramer and O. Regius, *Z. Alternsforsch.*, 1988, **43**, 251.

[30] T. S. Chen, G. J. Currier and C. L. Wabner, *J. Gerontol.*, 1990, **45**, B129.

[31] J. R. Pappenheimer, *Comp. Biochem. Physiol. Part A*, 1998, **121**, 45.

[32] W. H. Karasov and J. M. Diamond, *Am. J. Physiol.*, 1983, **245**, G443.

[33] J. H. Cummings and A. M. Stephen, *Can. Med. Assoc. J.*, 1980, **123**, 1109.

[34] A. B. R. Thomson, *J. Lipid. Res.*, 1980, **21**, 1097.
[35] M. P. Weber, L. J. Martin, V. C. Bourge, P. G. Nguyen and H. J. Dumon, *Am. J. Vet. Res.*, 2001, **64**, 1105.
[36] M. Weber, F. Stambouli, L. Martin, H. Dumon, V. Bourge and P. Nguyen, *J. Anim. Physiol. Anim. Nutr.*, 2003, **85**, 242.
[37] J. Graff, K. Brinch and J. L. Madsen, *Clin. Physiol.*, 2001, **21**, 253.
[38] D. Hollander, V. D. Dadufalza and E. G. Sletten, *J. Lipid. Res.*, 1984, **25**, 129.
[39] M. A. Ainsworth, M. A. Koss, D. L. Hogan and J. I. Isenberg, *Gastroenterology*, 1995, **109**, 1160.
[40] C. Y. Lui, G. L. Amidon, R. R. Berardi, D. Fleisher, C. Youngberg and J. B. Dressman, *J. Pharm. Sci.*, 1986, **75**, 271.
[41] J. B. Dressman, P. Bass, W. A. Ritschel, D. R. Friend, A. Rubinstein and E. Ziv, *J. Pharm. Sci.*, 1993, **82**, 857.
[42] G. Flemstrom and E. Kivilaasko, *Gastroenterology*, 1983, **84**, 787.
[43] W. L. Chiou, C. Ma, S. M. Chung, T. C. Wu and H. Y. Jeong, *Int. J. Clin. Pharmacol. Ther.*, 2000, **38**, 532.
[44] W. L. Chiou, H. Y. Jeong, S. M. Chung and T. C. Wu, *Pharm. Res.*, 2000, **17**, 135.
[45] L. Ogiolda, R. Wanke, O. Rottmann, W. Hermanns and E. Wolf, *Anat. Rec.*, 1998, **250**, 292.
[46] R. Hebel and M. W. Stromberg, *Anatomy and Embryology of the Laboratory Rat*, BioMed Verlag, Germany, 1986.
[47] T. T. Kararli, *Biopharm. Drug Dispos.*, 1995, **16**, 351.
[48] C. E. Stevens, in *Dukes' Physiology of Domestic Animals* (ed. M. J. Swenson), 9th edition, Comstock Publishing Associates, Ithaca, NY, 1977.
[49] R. L. Snipes, *Adv. Anat. Embryol. Cell Biol.*, 1997, **138**, 1.
[50] R. L. Snipes, W. Clauss, A. Weber and H. Hörnicke, *Cell Tiss. Res.*, 1982, **225**, 331.
[51] H. E. Evans, in *Miller's Anatomy of the Dog* (ed. H. E. Evans), 3rd edition, WB Saunders, Philadelphia, PA, 1993, pp. 385–462.
[52] J. B. Dressman and K. Yamada, in *Pharmaceutical Bioequivalence* (eds P. G. Welling, F. L. S. Tse, and S. V. Dighe), Marcel Dekker, New York, 1991, pp. 235–266.
[53] D. J. Chivers and C. M. Hladik, *J. Morphol.*, 1980, **166**, 337.
[54] H. Westergaard and J. M. Dietschy, *J. Clin. Invest.*, 1974, **54**, 718.
[55] D. B. Paulsen, K. K. Buddington and R. K. Buddington, *Am. J. Vet. Res.*, 2003, **64**, 618.
[56] T. W. Shirkey, R. H. Siggers, B. G. Goldade, J. K. Marshall, M. D. Drew, B. Laarveld and A. G. Van Kessel, *Exper. Biol. Med. (Maywood, NJ)*, 2006, **231**, 1333.
[57] M. J. Derelanko, *Toxicologists Pocket Handbook*, CRC Press LLC, Boca Raton, FL, 2000.
[58] L. R. Jarvis and R. Whitehead, *Gastroenterology*, 1980, **78**, 1488.
[59] J. A. Young, D. I. Cook, A. D. Conigrave and C. R. Murphy, *Gastrointestinal Physiology*, Globe, Melbourne, 1991.
[60] R. Schwarz, A. Kaspar, J. Seelig and B. Kunnecke, *Mag. Res. Med.*, 2002, **48**, 255.
[61] H. Kondo, T. Watanabe, S. Yokohama and J. Watanabe, *Biopharm. Drug Dispos.*, 2003, **24**, 141.
[62] H. Kondo, Y. Takahashi, T. Watanabe, S. Yokohama and J. Watanabe, *Biopharm. Drug Dispos.*, 2003, **24**, 131.
[63] L. P. Degen and S. F. Phillips, *Gut*, 1996, **39**, 299.
[64] C. Tuleu, C. Andrieux, P. Boy and J. C. Chaumeil, *Int. J. Pharm.*, 1999, **180**, 123.
[65] P. Enck, V. Merlin, J. F. Erckenbrecht and M. Wienbeck, *Gut*, 1989, **30**, 455.
[66] N. Sastry and B. S. Rao, *Ind. J. Physiol. Pharmacol.*, 1991, **35**, 183.
[67] D. C. Hernot, V. C. Biourge, L. J. Martin, H. J. Dumon and P. G. Nguyen, *J. Anim. Physiol. Anim. Nutr. (Berlin)*, 2005, **89**, 189.
[68] T. Gidenne and Y. Ruckebusch, *Reprod. Nutr. Dev.*, 1989, **29**, 403.
[69] A. Wilfart, L. Montagne, H. Simmins, J. Noblet and J. Milgen, *Br. J. Nutr.*, 2007, **98**, 54.
[70] J. Fallingborg, L. A. Christensen, M. Ingeman-Nielsen, B. A. Jacobsen, K. Abildgaard, H. H. Rasmussen and S. N. Rasmussen, *J. Pediatr. Gastroenterol. Nutr.*, 1990, **11**, 211.
[71] S. Wagener, K. R. Shankar, R. R. Turnock, G. L. Lamont and C. T. Baillie, *J. Pediatr. Surg.*, 2004, **39**, 166.
[72] K. Ewe, A. G. Press and W. Dederer, *Eur. J. Clin. Invest.*, 1989, **19**, 291.
[73] J. Fallingborg, L. A. Christensen, M. Ingeman-Nielsen, B. A. Jacobsen, K. Abildgaard and H. H. Rasmussen, *Aliment. Pharmacol. Ther.*, 1989, **3**, 605.

[74] J. H. Cummins, H. S. Wiggins, D. J. A. Jenkins, H. Houston, T. Jivraj, B. S. Drasar and M. J. Hill, *J. Clin. Invest.*, 1978, **61**, 953.

[75] M. R. Bruorton, C. L. Davis and M. R. Perrin, *Appl. Environ. Microbiol.*, 1991, **57**, 573.

[76] C. F. Burrows, D. S. Kronfeld, C. A. Banta and A. M. Merritt, *J. Nutr.*, 1982, **112**, 1726.

[77] D. C. Hernot, H. J. Dumon, V. C. Biourge, L. J. Martin and P. G. Nguyen, *Am. J. Vet. Res.*, 2006, **67**, 342.

[78] E. J. Calabrese, *Principles of Animal Extrapolation*, Wiley, New York, 1983.

[79] A. B. R. Thomson, *Q. J. Exp. Physiol.*, 1986, **71**, 29.

CHAPTER 22

The Emerging Utility of Co-Crystals in Drug Discovery and Development

Nicholas A. Meanwell

Contents

1. INTRODUCTION

It has been proposed that co-crystals be defined as mixed crystals containing two components that individually are solids at room temperature and which do not interact to form a salt, a description designed to discriminate these interesting molecular architectures from the more commonly encountered solvates [1–3]. The first examples of co-crystals were prepared in the 19th century but detailed insights into the nature and scope of the intermolecular interactions involved emerged only in the latter part of the 20th century, coincident with the

Bristol-Myers Squibb Research and Development, 5 Research Parkway, Wallingford, CT 06492, USA

Annual Reports in Medicinal Chemistry, Volume 43
ISSN 0065-7743, DOI 10.1016/S0065-7743(08)00022-5

widespread application of single crystal X-ray diffraction analysis [4–7]. Early studies of co-crystals were focused mainly on developing an understanding of the nature and patterns of intermolecular interactions, with the result that interesting combinations of molecules were prepared and characterized as part of an initiative that has contributed significantly to the development of the field of crystal engineering [7–9]. However, an appreciation has begun to emerge recently that co-crystals hold considerable promise to address a number of problems being encountered more frequently in contemporary drug discovery campaigns, including modulating dissolution, conferring crystallinity, reducing hygroscopicity and controlling polymorphism of candidate compounds [4,10–14]. This development is a consequence, in part, of the evolution of candidate drugs over the last decade toward molecules with higher molecular weight and increased lipophilicity [13,15–17], parameters that confer the kind of physical properties that present a considerable challenge to the optimization of a successful formulation [18–20]. Much of the initial effort has been directed toward understanding the potential of co-crystals in the context of relatively simple, well-established drug and drug-like molecules that present specific formulation issues [4,10–14]. However, the descriptions of successful applications of this technology in the preclinical and clinical settings are stimulating increased experimentation with structurally more sophisticated and demanding molecules that are, in turn, promoting both further advances and a greater awareness. Examples of pharmaceutical co-crystals prepared on multi-kilogram scale are described in the recent patent literature, and select compounds formulated in this fashion have been advanced into clinical studies. Although drug formulation is a recently emerging emphasis, co-crystals have also demonstrated utility in the resolution of racemic compounds, a practical approach to enantiomer separation in circumstances where chiral salt formation may be precluded. In this review, we will provide a synopsis of the potential of pharmaceutical co-crystals to address issues associated with drug formulation by presenting illustrative examples reported in both the primary and patent literature and summarize applications of co-crystals as an approach to the resolution of racemic mixtures. It should be noted that the strict definition of co-crystals proposed in the literature will not be adhered to since examples that fall under the broader umbrella of crystalline complexes offering advantageous physical properties for drug formulation or resolution are included in the discussion [1,2].

2. FUNDAMENTALS OF CO-CRYSTALS: NOMENCLATURE AND INTERACTIONS

The key elements involved in co-crystal formation are typically strong hydrogen bonds that are frequently augmented by additional weaker hydrogen bonds [5,21], halogen bonds (non-covalent interactions between halogens that act as Lewis acids, complementing neutral or anionic Lewis bases), π-stacking interactions or van der Waals forces, although a combination of the weaker forces can, on occasion, be sufficient. These are intermolecular interactions available to a diverse range of functionalities and topologies that provides for a broad scope of opportunity, readily encompassing compounds that are devoid of the kind of overtly acidic or basic moieties that facilitate salt formation [4,10–12].

2.1 Nomenclature and definitions of co-crystals

Compounds capable of establishing a complementary pattern of non-bonded interactions with partner molecules to produce stable, crystalline complexes are described as *co-crystal formers*. The structural motifs that productively engage in such interactions are designated as *supramolecular synthons*, with *supramolecular homosynthons* representing identical interacting functionality and *supramolecular heterosynthons* describing co-crystals based on distinctly different interacting structural elements. The examples depicted in Figure 1 provide representatives of two supramolecular motifs in the context of some of the more common functionalities that are established co-crystal formers.

2.2 Intermolecular interactions in co-crystals

To provide insight into the potential for the formation of a particular bimolecular interaction, a statistical analysis was conducted of 75 motifs that appeared more than 12 times in the Cambridge Structural Database (CSD) prior to October 1996, providing a ranking of the examples according to the frequency with which they occurred that was corrected by an assessment of the potential for their formation [5]. This survey allowed a ranking of the capacity for supramolecular homo- and heterosynthon formation based on the data available at the time the analysis was conducted. However, it should be emphasized that in molecules with multiple possibilities, crystal-packing interactions may dictate the experimental outcome. A synopsis of some of the more interesting motifs that may have utility in active pharmaceutical ingredient (API) formulation is presented in Figure 2, which is organized based on the index of probability of formation, designated as Pm, with the preceding number referring to the hierarchical ranking reported for the 75 motifs [5]. As might be anticipated, an interaction involving 3 complementary hydrogen bonds provides the highest probability of formation but the lactam/ 2-amino pyridine interaction is also highly likely to form, preferred over the carboxylic acid/2-amino pyridine motif, and both are significantly more

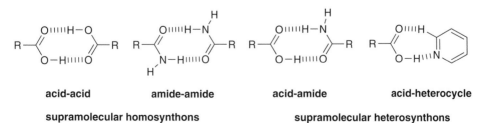

acid-acid   **amide-amide**   **acid-amide**   **acid-heterocycle**

supramolecular homosynthons   **supramolecular heterosynthons**

Figure 1 Examples of supramolecular synthons.

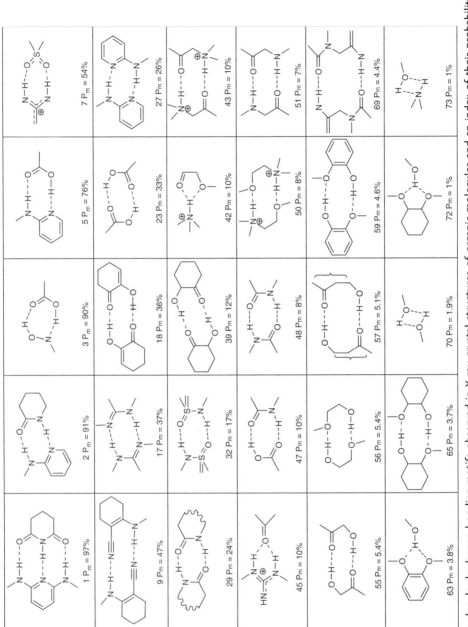

Figure 2 Bimolecular hydrogen bonding motifs observed in X-ray crystal structures of organic molecules and an index of their probability of formation.

Table 1 Hierarchy of the formation of successful hydrogen bonding interactions between acceptors and the carboxylic acid donor.

Acceptor	Success index
Fluoride, F^-	1.0
Chloride, Cl^-	0.971
RCO_2^-	0.966
Pyridine, N	0.91
H_2O	0.84
Bromide, Br^-	0.81
C=N–O	0.8
Urea carbonyl	0.78
Amide carbonyl	0.78
Sulfoxide S=O	0.78
sp^3 C–OH	0.78
Ketone carbonyl	0.55
RCO_2H	0.5
Carboxylic ester	0.28
C=S	0.27
R–C≡N	0.26
PhOH	0.20
C–O–C	0.13
C–F	0.00
C–Cl	0.00
C–S–C	0.00

prevalent than the 2-amino pyridine, carboxylic acid or lactam homosupramolecular synthons. Not surprisingly, intermolecular interactions are facilitated by ring scaffolds that provide the appropriate preorganization of the interacting elements.

A subsequent analysis focused specifically on the carboxylic acid moiety, a strong hydrogen bond donor with high prevalence and versatility as a co-crystal former and which partners readily with a wide range of the functionalities typically being encountered in contemporary drug design. The assessment of the potential for successful interaction of carboxylic acids with a series of acceptors is summarized in Table 1 and was found to correlate with hydrogen bond length [22]. Interestingly, fluoride ion showed the highest success index, reflecting its potential as a hydrogen bond acceptor with practical utility as a base [23], followed closely by chloride ion, a common salt form of APIs and an interaction that has been utilized in co-crystal formation (*vide infra*).

2.3 Topological and topographical aspects of co-crystals

Crystal lattice variation induced by co-crystal formation can be categorized into two fundamental topological motifs — one that preserves the homosynthon in a molecule of interest and a second that replaces it with a heterosynthon. This is most effectively illustrated by experimental observations with the anticonvulsant carbamazepine, a compound sparingly soluble in water that demonstrates poor dissolution properties. Carbamazepine exhibits considerable diversity in the crystalline state with 4 anhydrous polymorphs, several solvates and a stable hydrate known, all of which have been characterized by X-ray crystallography and found to be based on a carboxamide–carboxamide supramolecular homosynthon, as depicted in Figure 3. The dibenzazepine ring system of carbamazepine dominates intermolecular packing interactions and provides steric shielding of the anti-disposed carboxamide NHs that prevents them from engaging in additional intermolecular interactions. This contrasts with smaller primary carboxamides where the anti-disposed hydrogen atoms can act as donors to the carbonyl moieties of adjacent molecules, forming the ribbon-like structures typical of benzamide, as depicted in Figure 4 [24].

This topography creates a cavity in the crystal lattice of carbamazepine that can readily be filled by small molecule hydrogen bond acceptors capable of interacting with the available carboxamide hydrogen atoms, providing an understanding of the propensity of this drug to hydrate or to form solvates with small molecules like acetone and DMSO [25,26]. Large molecules that are weaker co-crystal heterosynthons than the carboxamide moiety are also able to interact in a similar fashion, preserving the intrinsic homosynthon interactions but disrupting crystal packing by virtue of their increased size, which leads to modified solid-state structures and properties. This is exemplified by the saccharin-carbamazepine co-crystal, depicted in Figure 5, which exhibits a melting point ∼20°C lower than the parent and a markedly improved dissolution profile that translates into improved exposure *in vivo* [25,26]. The alternate approach relies upon co-crystallizing carbamazepine with a carboxylic acid, which disrupts the carboxamide–carboxamide homosynthon, replacing it with the carboxamide-carboxylic acid heterosynthon [25]. By necessity, this provides a markedly different crystal form with intermolecular interactions in the crystal lattice dependent upon the identity of the acid, an example of which is depicted in Figure 6 [25].

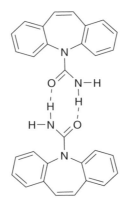

Figure 3 Supramolecular homosynthon interaction observed for carbamazepine in the solid state.

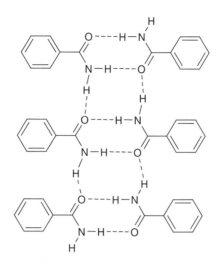

Figure 4 Ribbon-like structure adopted by benzamide based on intermolecular interactions of the supramolecular homosynthon.

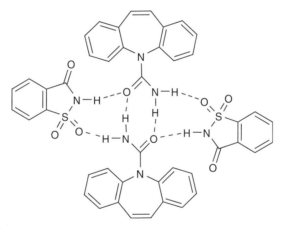

Figure 5 Intra- and inter-molecular interactions observed in the 1:1 carbamazepine-saccharin co-crystal.

3. A BRIEF HISTORY OF CO-CRYSTALS

The first characterization of a co-crystal is attributed to a complex of quinone and hydroquinone, isolated in the 19th century [27,28] but not revealed at the atomic level until late in the 20th century [29,30]. The first pharmaceutical application of a co-crystal is considered to be the 1:1 complex of sulfathiazole (**1**) and 3,6-diaminoacridine (proflavine, **2**) prepared by the Boots Pure Drug Company and used as an antibacterial agent during the Second World War [31]. The identity of this complex has not been established but sulfa drugs have proven to

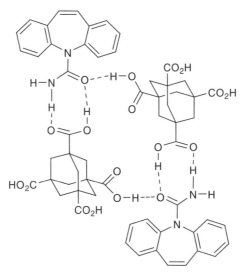

Figure 6 Intra- and inter-molecular interactions observed in the 1:1 carbamzaepine-adamantane-tetracarboxylic acid co-crystal.

be versatile co-crystal partners based on the constellation of H-bond donors and acceptors inherent in these molecules and a wide range of complexes have been isolated and characterized [32].

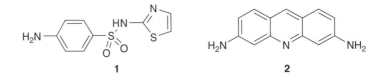

In the late 1940s, a complex of theophylline and glycine was prepared and evaluated clinically, with the complex dosed to more than 300 patients for a period of 18 months [33]. At doses up to 4 g (2 g of theophylline) in 24 h, this complex was considered to be well tolerated, producing the typical response to theophylline but without the nausea or vomiting more commonly associated with the theophylline-ethylenediamine formulation prevalent at that time [33].

The emergence of crystal engineering as a defined discipline produced significant advances in materials research that sought to understand the underlying interactions with the ultimate objective of designing and engineering supramolecular structures [7–9,34,35]. However, predicting the solid-state structures of even relatively simple molecules has proven to be a significant challenge [36–38], although some progress has been made based on the most recent of the CSD challenges, where 4 structures, including a co-crystal of 2-amino-4-methylpyrimidine and 2-methylbenzoic acid, were successfully predicted for the first time [39].

4. THE PREPARATION AND CHARACTERIZATION OF CO-CRYSTALS

Although the current understanding of the interactions between supramolecular synthons is well developed and hierarchical analyses have provided useful guides to the selection of potential partner molecules, the preparation of co-crystals remains fundamentally an experimental science. The wide range of structural diversity afforded by potential combinations of co-crystal partners and their unpredictable architecture lends itself to a high-throughput screening-based approach as the most effective method of identifying co-crystals of novel compounds [40]. Co-crystals can be prepared by the standard crystallization process of concentrating the partners in a hot solvent and cooling, promoted as appropriate by an antisolvent, [4,12], while a suspension/slurry technique takes advantage of a solution-mediated phase transformation [41]. Grinding partner molecules together is a useful and green process that can be facilitated by the addition of a drop of solvent (solvent-drop grinding) [42–49] or promoted by sonification of a paste [50,51], a process that allows some influence over polymorph production by modulation of the polarity of the solvent [52]. Co-crystals can also be produced by melting two solids in close proximity using a hot stage microscopy apparatus, a technique pioneered by Kofler in which co-crystals form at the interface [4,11,12,53]. Co-crystal formation under melt conditions can often be established by visual inspection but typical procedures for confirming co-crystal formation rely upon Raman spectroscopy, powder X-ray diffraction pattern analysis or, for a definitive understanding, single crystal X-ray diffraction analysis of the product. However, some care is required in the interpretation of X-ray crystallographic data in order to definitively discriminate between hydrogen-bonded complexes and salts, as exemplified by the refinement of the trimethoprim-sulfathiazole structure [32,54,55].

Co-crystals can also be obtained by exchanging synthons under grinding conditions. For example, grinding a sulfadimidine-salicylic acid co-crystal with an equivalent of 2-aminobenzoic acid produces the sulfadimidine-2-aminobenzoic acid co-crystal, which is also produced when sulfadimidine is ground with a mixture of benzoic acid and 2-aminobenzoic acid [56]. However, grinding a mixture of the sulfadimidine-benzoic acid co-crystal with salicylic acid does not lead to an exchange, indicative of the unique properties of 2-aminobenzoic acid. This result has been attributed to 2-aminobenzoic acid not forming stable dimers in the native state, which is comprised of one neutral and one zwitterionic molecule in the asymmetric unit, with the result that 2-aminobenzoic acid effectively has increased freedom to interact with the sulfa drug [55].

The propensity of drug-like molecules to form co-crystals was examined by scientists at SSCI in an experiment in which 53 structurally diverse APIs representing a total of 64 different compounds that included salt forms were studied [56]. In this experiment, 24 non-salts and 15 salts afforded a total of 192 co-crystals, with 32 molecules co-crystallizing with multiple partners, and salts and non-salts showing no significant difference in their propensity to form co-crystals [56]. Typically, the different preparatory techniques afford the same co-crystals but care has to be exercised with partners in which the difference in pKa approaches the 2–4 unit magnitude that is usually required for salt formation. For example, combining sulfamethoxypyridazine and trimethoprim in H_2O produces the hydrated salt **3**, in which the 2,4-diaminopyrimidine moiety of trimethoprim is protonated. However, when MeOH is used as the solvent, a co-crystal is isolated in which the sulfamethoxypyridazine has tautomerized to the exocyclic imine form, providing a scaffold for interaction with trimethoprim based on complementary hydrogen bonds, as depicted in **4** [58].

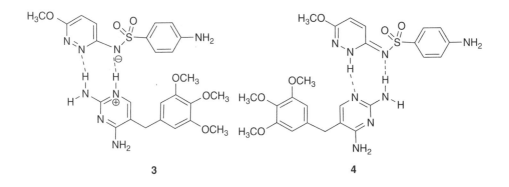

3   **4**

5. APPLICATIONS OF CO-CRYSTALS IN DRUG FORMULATION

The advantageous properties offered by co-crystals have begun to be appreciated by the pharmaceutical industry and examples of utility are beginning to be documented in both the patent and primary literature. The formation of co-crystals can reduce the tendency of an API to hydrate or form solvates, largely a function of satisfied hydrogen bonding. In addition, because of the lower vapor pressure of the solids used as partners, co-crystals are typically more stable than solvates, which can degrade by evaporation of the solvent from the crystal lattice. Co-crystals can also exhibit polymorphism, although the experience to date indicates a generally low propensity. In addition, by altering the arrangement within a lattice, co-crystals can increase chemical stability in the solid state, particularly for those compounds subject to photochemical-mediated reactions [59], with carbamazepine a prominent example [12,60]. Co-crystal formation has also been shown to confer crystallinity to syrups and waxes and to modulate drug dissolution, both enhancing and delaying, depending on the properties of the co-crystal former, which is probably one of the most prominent of the emerging utilities. Co-crystals can also be formulated with excipients and subjected to particle engineering as a means of further enhancing performance *in vivo* [26,61].

5.1 Co-crystals to modulate dissolution of APIs

The physical properties of a drug candidate can be a critical determinant of its potential to be successfully developed and poor dissolution can limit the oral bioavailability of compounds exhibiting low intrinsic solubility, a function of high crystal lattice energies in the solid-state or inadequate solvation properties [62–65]. Co-crystals can offer a useful formulation option that specifically addresses both of these problems, particularly for non-ionizable compounds. Co-crystals complement other formulation approaches, including nanoparticles, inclusion complexes and amorphous dispersions, with the improved reproducibility and stability of co-crystals presenting a significant advantage over the latter [14]. The first defined study of co-crystal formation as a strategy for influencing drug dissolution was described for itraconazole, an anti-fungal agent that is a mixture of 4 diastereomers [66]. This compound exhibits poor aqueous solubility and is marketed as an amorphous dispersion, designed to facilitate dissolution in the gut. Itraconazole possesses two weakly basic functionalities, the piperazine element with a pKa of $\sim$3.7, and the triazole moiety, pKa $\sim$2.5, that would be anticipated to form salts only with strong acids with a pKa of <1.7 based on the pKa difference of 2–4 that is usually required for reliable salt formation [67]. However, crystalline salts of itraconazole have not been described, presumably a challenging enterprise because of the diastereomeric complexity of the drug. A high-throughput crystallization screen in which itraconazole was combined with a range of pharmaceutically acceptable carboxylic acid derivatives that sampled variations in dissociation constant, shape, size and number of acid moieties conducted in both polar protic and aprotic solvent systems, provided an initial example of a co-crystal of the drug [66]. Wheras polar protic solvents, water and lower alcohols simply returned crystalline itraconazole, polar aprotic solvents, with and without added hydrocarbons as antisolvents, provided co-crystals of itraconazole, uniquely with 1,4-dicarboxylic acids. The initial screen identified co-crystals with fumaric acid and DL-tartaric acid, both 2:1 mixtures of the API to acid according to ^{1}H-NMR analysis, while co-crystals derived from monocarboxylic acids were not produced. A second round of screening focused specifically on dicarboxylic acids, affording additional complexes with D- and L-tartaric, succinic and malic acids, with only maleic acid from this series failing to provide a co-crystal. The homologous malonic, glutaric and adipic acids did not form co-crystals, indicating a precise requirement for the 1,4 spatial disposition of the two carboxylic acid moieties. The underlying premise for this observation was revealed by single crystal X-ray analysis of the co-crystal with succinic acid which demonstrated H-bonds from both CO_2H moieties to the triazole rings of different itraconazole molecules that were oriented in an anti-parallel fashion around the diacid, as depicted in Figure 7 [66]. Interestingly, the acid moieties complexed with the weaker base, presumably governed by the crystalline environment [67], and only 2 of the 4 diastereomers were found in the co-crystal [66].

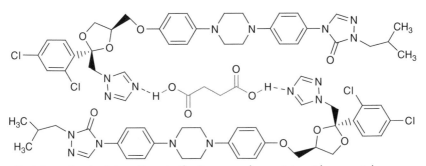

Figure 7 Intermolecular interactions in the itraconazole-succinic acid co-crystal.

Evaluation of dissolution rates of the itraconazole co-crystals formed with succinic, L-malic and L-tartaric acids in 0.1N HCl solution at 25°C revealed that all 3 dissolved faster than crystalline intraconazole. The malic acid co-crystal performed optimally, dissolving at a rate comparable to that of the commercial amorphous dispersion formulation of itraconazole, while the L-tartaric acid complex was somewhat less effective and the succinic acid co-crystal dissolved markedly more slowly, although still faster than crystalline parent drug. The dissolution rates of these 3 co-crystals correlated with the polarity of the partner, with the more hydrophilic tartaric and malic acid co-crystals dissolving faster than the succinic acid complex.

The hydrochloride salt of the antidepressant fluoxetine, a selective serotonin reuptake inhibitor (SSRI), provides an interesting substrate for co-crystal formation [68] in which the heterosynthons are the chloride ion of the salt and strong hydrogen bond donors, an observation anticipated based on a prior analysis of chloride interactions and which directed the selection of potential co-crystal partners [22,69–71]. The experimental exercise produced co-crystals with benzoic, succinic and fumaric acids, all analyzed by X-ray crystallography, which confirmed coordination of the carboxylic acid moiety with the chloride ion. Benzoic acid formed the 1:1 complex **5** while the two diacids produced complexes which were a 2:1 ratio of fluoxetine•HCl to the co-crystal former, a finding readily understood in the context of the succinic acid co-crystal structure **6**. The melting points of the benzoic and succinic acid co-crystals were over 20°C lower than that of fluoxetine•HCl, 157.2°C, while the fumaric acid complex melted almost 4°C higher at 161.1°C [68]. This is a particularly interesting observation given that the fumaric and succinic acid co-crystals are isostructural, a result that substantiates some of the difficulties associated with predicting the structures of complex crystals [36–39].

Powder dissolution experiments revealed that the co-crystals formed between fluoxetine•HCl and both benzoic acid and fumaric acid were stable in H_2O but the succinic acid complex disassociated readily, leading to crystallization of fluoxetine•HCl from the mixture. The rates of dissolution of these complexes in H_2O at 10°C were: succinic acid > fumaric acid = fluoxetine•HCl > benzoic acid, an order that correlates with the aqueous solubility of the heterosynthon since succinic acid is 10-fold more soluble than fumaric acid which, in turn, is 2-fold more soluble than benzoic acid [68].

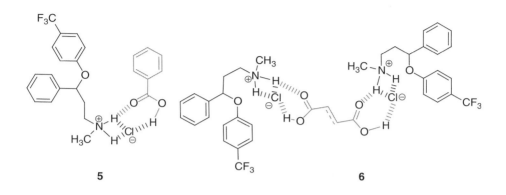

5   6

The *in vitro* dissolution rates and/or solubilities of co-crystals of a number of marketed and experimental drugs have been shown or claimed to offer improvement over existing formulations. The cyclic GMP phosphodiesterase (PDE) 5 inhibitor sildenafil is a strong enough base, pKa = 6.78, to form a salt with citric acid, pKa = 2.96, and it is this formulation that is marketed [72,73]. However, heating sildenafil in toluene/CH_3CN or iPrOH/CH_3CN with the less acidic acetylsalicylic acid, pKa = 3.5, followed by cooling and filtration, produces a co-crystal stabilized by a hydrogen-bonding interaction between the carboxylic acid and amine moieties, as determined by single crystal X-ray crystallographic analysis [74]. The sildenafil-acetylsalicylic acid co-crystal is claimed to offer a 2-fold increase in intrinsic dissolution rate at low pH compared to the citrate salt [74].

The quinolone antibacterial agent norfloxacin co-crystallizes with isonicotinamide, one of a series of useful amide-based co-crystal formers [75], to afford a complex that melts at a markedly lower temperature than the anhydrous parent [76]. This co-crystal assembles with the zwitterionic quinolone carboxylate anion accepting a hydrogen bond from the isonicotinamide homosupramolecular synthon, as depicted in Figure 8. The co-crystal shows an almost 3-fold improved apparent solubility after 72 h of equilibration compared to the anhydrous zwitterionic form, attributed to reduced crystal lattice energy, although salts derived from succinic, malic and maleic acids offer superior performance [76].

Formulation of the tetrapeptidic hepatitis C virus NS3 protease inhibitor telaprevir (VX-950, **7**) presents a considerable challenge given the inherent physical properties and high dosing requirements of 750 mg TID [77–79]. Solvent-drop grinding, melting and solvent evaporation of telaprevir (**7**) with salicylic acid and 4-aminosalicylic acid afforded co-crystals with significantly reduced melting points compared to crystalline parent and improved solid-state stability compared to amorphous forms of the drug [79]. The co-crystals, which have not been defined by crystallographic analysis, are claimed to show increased dissolution and higher aqueous solubility than amorphous forms, with the

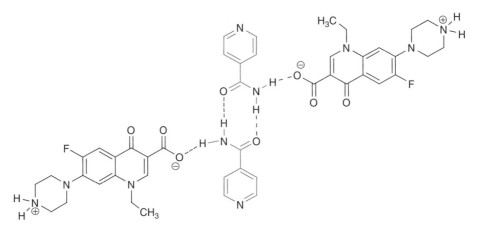

Figure 8 Intermolecular interactions in the norfloxacin-isonicotinamide co-crystal.

4-aminosalicylic acid complex stable in 1% hydroxypropylmethylcellulose (HPMC)/H_2O for 24 h, although not in H_2O alone [79]. In simulated gastrointestinal fluid at pH = 6.8, the amorphous form of telaprevir (**7**) was initially soluble at 25 µg/mL at room temperature but began to crystallize out of solution by 4 h, a process complete by 24 h. In contrast, the telaprevir-acetylsalicylic acid co-crystal remained dissolved at 24 hours with a reported solubility of 148 µg/mL [79].

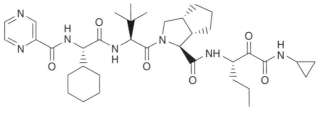

7 (telaprevir, VX-950)

The performance of several co-crystal formulations have been evaluated *in vivo* where improved dissolution properties are anticipated to lead to enhanced bioavailability. This is an approach to drug formulation that will be most successful with compounds designated as Biopharmaceutics Classifications System (BCS) class 2, those that exhibit low aqueous solubility and high membrane permeability [80]. The anti-epileptic drug carbamazepine falls into this category [81] and the highly polymorphic nature of this compound, in combination with its tendency to form hydrates and solvates, makes it an interesting prototype for crystal engineering designed to optimize formulation properties [25,26,82]. The carbamazepine-saccharin co-crystal depicted in Figure 5, readily prepared in good yield on a 30-g scale under conventional crystallization conditions from a ~2:1 mixture of ethanol and methanol, exhibits comparable stability to the marketed anhydrous polymorph form [26]. More importantly, this carbamazepine-saccharin co-crystal shows a remarkably reduced tendency for polymorphism, with 550 individual experiments encompassing 480 crystallization trials, mechanical grinding with 24 different solvents and slurry conversion experiments with 7 solvents failing to produce polymorphs of the co-crystal [26]. The initial rate of dissolution of the co-crystal in simulated gastric fluid revealed a dependence on particle size, with particles of less than 150 μm demonstrating the most rapid dissolution. Moreover, the co-crystal exhibited resistance toward forming the dihydrate form of carbamazepine in aqueous suspension on a timescale compatible with drug absorption. Bioavailability studies were conducted in beagle dogs to compare the carbamazepine-saccharin co-crystal, formulated at a particle size of <53 μm in capsules with lactose in a dry blending step, with the marketed Tegretol® immediate release tablet. The performance of the co-crystal was similar to the marketed formulation, producing a statistically insignificant higher AUC and C_{max} with a similar T_{max} [26].

Recently, a second polymorph of the carbamazepine-saccharin co-crystal presented in Figure 5 has been isolated after crystallization in the presence of functionalized cross-linked polymers, which function as heteronuclei for crystal growth, a useful technique for polymorph identification [83]. This complex, designated as form II, involves the heterosynthon arrangement depicted in Figure 9 and the co-crystal melts at 166.8°C, 7°C lower than form I (Figure 5). However, form II is a higher energy form of the carbamazepine-saccharin co-crystal since it converts to form I under slurry conditions at room temperature [83].

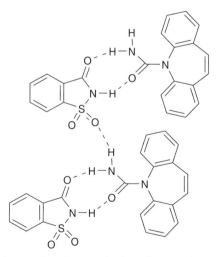

Figure 9 Intermolecular interactions in a saccharin-carbamazepine co-crystal based on a supramolecular heterosynthon motif.

The Na⁺ channel blocker **8**, a potential treatment for neuropathic pain [84], provides a particularly interesting example of a co-crystal formulation offering improved *in vivo* properties [53]. This compound presents a significant formulation challenge based on an intrinsic solubility of less than $0.1\,\mu g/mL$ and the absence of functionality that would support salt formation, since the pKa of the pyrimidine is estimated to be -0.7 [53].

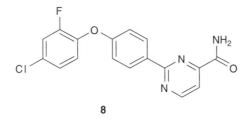

8

This compound demonstrated poor dissolution properties *in vitro* and oral administration of a crystalline suspension to dogs resulted in low systemic exposure [53]. Since an amorphous form of **8** could not be identified, the potential for co-crystal formation was explored by screening melts with 26 pharmaceutically acceptable carboxylic acid derivatives on microscope slides. This experiment identified 5 unique co-crystals of **8**, with a glutaric acid complex selected for further evaluation based on its physical properties [53]. The co-crystal melted at 142°C, lower than the parent drug, mp 206°C, and single crystal X-ray diffraction analysis revealed the intermolecular interactions depicted in Figure 10. The carboxamide moiety of **8** forms a heterosupramolecular synthon with one CO_2H moiety of glutaric acid while the other CO_2H donates a hydrogen bond to the pyrimidine N atom of an adjacent molecule. The co-crystal dissolves

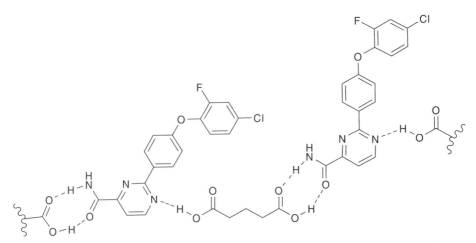

Figure 10 Intermolecular interactions in a co-crystal of the Na⁺ channel blocker 8 with glutaric acid.

Table 2 Pharmacokinetic parameters of the Na$^+$ channel blocker 8 in rats following dosing of parent and the glutaric acid co-crystal.

Dose	C_{max} (ng/mL)	AUC (ng h/mL)	T_{max} (h)
5 mpk parent	25.4±11.4	374±192	13±12
5 mpk co-crystal	89.2±57.7	1,234±613	6±9
50 mpk parent	89.2±68.7	889±740	13±14
50 mpk co-crystal	278±70.5	2,230±824	2±0

18-fold faster in water at 37°C than the crystalline parent and remains largely intact after 90 min, although complete dissociation occurs after 24 h. The enhanced dissolution properties translated into increased exposure in fasted dogs following oral dosing of the neat solid forms of the drug in gelatin capsules, results summarized in Table 2. The co-crystal complex provided a 2–3-fold increase in AUC and C_{max} compared to the unformulated drug at both doses, and exposure increased with dose, although not proportionally [53].

The PDE-4 inhibitor L-883555 (**9**) provides another example where co-crystal formation provided a solution to a difficult formulation problem [85]. This compound is weakly basic, with the pKa of the pyridine N calculated to be 4.21 and the other nitrogen atoms <1, and poorly soluble in water as the free base, 7.5 μg/mL, leading to poor bioavailability in rats and monkeys. Attempts to form salts of L-883555 (**9**) using strong acids failed but tartaric acid provided several crystalline forms with varying stoichiometric ratios, from which the 2:1 L-883555:tartaric acid complex offered the greater thermal stability [85]. The precise structure of the co-crystal was not determined but cross-polarized magic angle spinning (CPMAS) ^{15}N-NMR analysis indicated an interaction between tartaric acid and the pyridine N-oxide moiety, further supported by pertubations in the pyridine ring carbon atoms in the ^{13}C-NMR. The solubility of the co-crystal was 23.7 μg/mL, a 3-fold improvement compared to the crystalline free base, and oral administration of a methocel solution to rhesus monkeys at a dose of 3 mpk resulted in a 23-fold increased AUC and 14-fold higher C_{max} compared to the parent compound, providing a formulation with exposure suitable for safety evaluation [85].

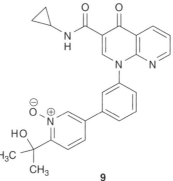

9

The potent vanilloid receptor (VR) 1 antagonist AMG-517 (**10**) demonstrated good bioavailability as a suspension of the free base in 10% (w/v) Pluronic F108® in OraPlus®, an oral suspending agent [86,87]. However, at higher doses, solubility-limited absorption intervened, compromising drug exposure.

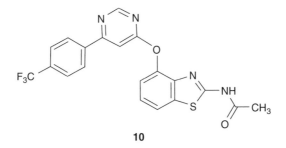

10

Close inspection of the formulation revealed that AMG-517 (**10**) formed a co-crystal with the sorbic acid included as a preservative in the OraPlus® suspending agent, a complex that could be isolated in discrete form from a slurry of the components in 2-butanol at 50°C or ethanol at room temperature [87]. Single crystal X-ray analysis revealed the intermolecular interactions depicted in Figure 11, in which the amide NH and benzothiazole N atom interact with the carboxylic acid moiety of sorbic acid via complementary hydrogen bonds, an arrangement that disrupts the intermolecular H-bonds seen in the free base. These interactions are similar to those observed in a complex of the anthelmintic agent mebendazole with propionic acid [88]. The AMG-517-sorbic acid co-crystal melts at 150°C, which compares with 230°C for form A of the free base, and demonstrates good stability at 90% relative humidity. This form offered advantage over crystalline salts which were found

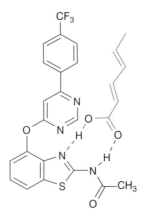

Figure 11 Intermolecular interactions in a 1:1 co-crystal of the vanilloid receptor antagonist AMG-517 with sorbic acid.

to dissociate when slurried in water, producing aqueous media with a low pH, problematic because the pyrimidine ether moiety is sensitive to hydrolysis under acidic conditions [87]. The co-crystal dissolved more readily in fasted simulated intestinal fluid at pH = 6.8 than the free base, providing a 7–9-fold solubility advantage (21–29 µg/mL compared to 3 µg/mL) between 0.2 and 1.1 h but dissociation occurred by 4 h. A pharmacokinetic study in rats compared a micronized form of the AMG-517-sorbic acid co-crystal at doses of 10, 30, 100 and 500 mpk to the free base at a dose of 500 mpk. AMG-517 (**10**) was absorbed more rapidly from the co-crystal at doses of 10, 30 and 100 mpk with a median T_{max} of 1–2 h, which compares with 8 h for the free base. The C_{max} of AMG-517 (**10**) from the co-crystal at 10 mpk was comparable to the 500 mpk dose of free base while the AUC was approximately half [87]. At a dose of 500 mpk, the co-crystal provided an 8-fold increase in C_{max} and enhanced the AUC by 9-fold.

CP-724714 (**11**), a potent and selective ErbB2 kinase inhibitor, is sparingly soluble in water as the free base and suffers from dissolution-limited absorption, compromising therapeutic exposure [89–92]. The pyridine nitrogen, pKa = 4.9, and quinazoline N1 atom, pKa = 6.1, of CP-724714 (**11**) are sufficiently basic to allow salt formation with strong acids, including HCl and CH_3SO_3H [92]. A salt screen identified 20 crystalline complexes from which the sesquisuccinate, dimalonate and dimaleate were profiled as possessing physico-chemical properties suitable for development [89,90,92]. The acidity of these organic acids, pKa = 1.91 for maleic acid, 2.85 for malonic acid and 4.21 for succinic acid, spans a range that would suggest salt formation with the more acidic partners and co-crystal formation for the weaker acid based on the differences in pKa. The three acid complexes were characterized by single crystal X-ray analysis and revealed that maleic acid formed an ionic salt at both basic N atoms while the malonate existed as a mixed ionic and zwitterionic complex in which the quinazoline N1 atom was partially protonated while the pyridine N atom engaged in a hydrogen-bonding interaction with the acid [92]. The weaker succinic acid formed a neutral co-crystal in which both basic N atoms accepted hydrogen bonds from the partner, as might be anticipated based on the relative pKa's [92]. Analysis of the complexes by solid-state ^{15}N-NMR established a correlation between resonance frequency and protonation state, with both the quinazoline N1 and pyridine N atoms shifted upfield by more than 90 ppm in the maleate salt. In the malonate complex, the quinazoline N1 atom resonated 85.2 ppm upfield of the signal for the free base while the pyridine N atom moved upfield by 40.4 ppm. For the succinate, the quinazoline N1 and pyridine N resonances were shifted upfield by only 31.2 and 22.6 ppm, respectively, leading to the conclusion that an upfield shift of >80 ppm reflects proton transfer while a 20–40 ppm shift is indicative of H bonding [92].

Each of these complexes demonstrated improved bioavailability and increased aqueous solubility compared to the free base, 0.372 µg/mL. The dimaleate salt was the most soluble, at 4.8 mg/mL, with the dimalonate 2-fold less soluble, 1.8 mg/mL. The sesquisuccinate was 1,000-fold more soluble than the free base and no polymorphs were found by screening, leading to this form being advanced into clinical studies [91].

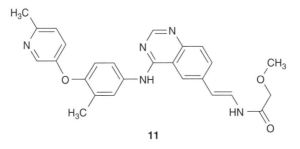

11

Co-crystals have also been examined as a potential formulation strategy specifically designed to slow the dissolution of an API under circumstances where this would offer advantage. The study of co-crystals of itraconazole and fluoxetine • HCl revealed a dependence of dissolution rate on the hydrophobicity of the partners, with the more lipophilic acids exhibiting reduced dissolution rates *in vitro* [66,68]. The caffeine-gentisic acid co-crystal was recognized as an early example of this phenomenon based on the observation that both the 1:1 and 1:2 complexes exhibited lower solubility and reduced dissolution rates in 0.1N HCl [93]. This was hypothesized as a potentially useful formulation approach to limit the release of caffeine from chewable tablets in the mouth, masking the inherently bitter taste of the drug.

A recent example is provided by the quinolone antibiotic gatifloxacin (**12**), the bitter taste of which required masking for use in a liquid formulation designed for the pediatric population [94]. Gatifloxacin forms a 1:2.1 complex with stearic acid when heated with excess acid in hot ethanol, a complex that might be anticipated to be a co-crystal based on the low pKa difference of 1.77 between the two partners. Aqueous suspensions of this complex are stable for 14 days without dissolution but dissolve in 0.1N HCl within 10 min to completely release the drug, reflected in the 99% bioavailability relative to the tablet form in humans [94].

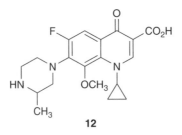

12

5.2 Co-crystals to confer crystallinity

There have been a number of examples where co-crystals of an API provide crystalline forms where other methods have been less successful. The free base of the MCR4 antagonist **13** could not be obtained in crystalline form and an amorphous dihydrochloride salt used for initial development work was both hygroscopic and unstable [95].

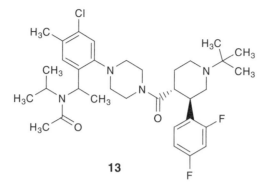

13

A high-throughput salt screen identified a single crystalline form, a phosphoric acid complex that was determined to contain 2 moles of H_3PO_4 per API molecule and which was not hydrated. The complex was anticipated to be a co-crystal after comparison of the basicity of **13**, pKa = 8.3 for the piperidine moiety and 2.7 for the piperazine heterocycle, with the acidity of phosphoric acid, lowest pKa = 2.15. This analysis suggested that salt formation with the piperazine would be unlikely [95]. Single crystal X-ray diffraction analysis confirmed the suspicion and revealed a protonated piperidine in which a hydrogen atom of the phosphate counter ion donates a hydrogen bond to the pendant amide carbonyl moiety. The second phosphoric acid molecule was observed to establish hydrogen bonds with the phosphate element of one molecule and the isopropyl amide carbonyl of a second molecule, as depicted in Figure 12 [95]. The co-crystal exhibited high solubility in water, >250 mg/mL, demonstrated good flow properties and was chemically stable, showing no change at 40°C and 75% relative humidity for 8 weeks. Moreover, high-throughput screening failed to identify additional crystalline forms.

Formulation issues associated with the polyether ester anti-inflammatory agent **14**, a waxy solid, were solved by co-precipitation of this compound with

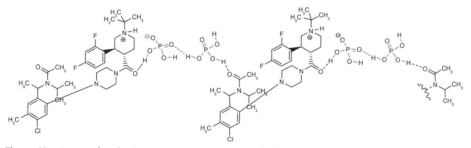

Figure 12   Intermolecular interactions in a co-crystal of the MCR4 antagonist 13 with phosphoric acid.

sorbitol from a mixture of acetic acid and iso-octane, which afforded a 2:1 complex of sugar to API [96]. The complex is claimed to be a free flowing, particulate solid that has advantageous pharmaceutical properties compared to the parent.

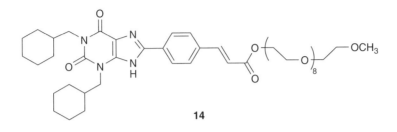

14

Inhibitors of sodium-glucose co-transporters, particularly SGLT2, are an emerging class of anti-hyperglycemic agents which act by blocking glucose reabsorption in the kidney [97,98]. Carbohydrate-based inhibitors of SGLT2 are typically isolated as viscous oils or amorphous, glassy solids that are difficult to formulate. These molecules possess an array of hydrogen bond donors and acceptors that might be anticipated to engage in the type interactions found in co-crystals. Co-precipitation of several SGLT2 inhibitors with the natural amino acids phenylalanine and proline provided 1:1 or 1:2 complexes that were isolated as white needles with sharp melting points [99–101]. The nature of the interactions in the co-crystal was revealed by single crystal X-ray analysis, with the structure of a representative example depicted in Figure 13 [99,100]. The phenylalanine co-crystal of the SGLT2 inhibitor shown in Figure 13 has been prepared on a 50-Kg scale [101].

5.3 Co-crystals that reduce hygroscopicity

Several co-crystals have been described in this review as displaying limited hygroscopicity, a physicochemical profile typically sought when formulating drug candidates. Caffeine has been studied specifically in this context as a model pharmaceutical since it forms a crystalline, non-stoichiometric hydrate that

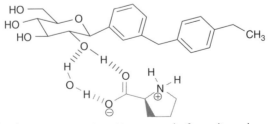

Figure 13 Intermolecular interactions in a 1:1 co-crystal of a sodium-glucose transporter inhibitor with proline.

incorporates 0.8 moles of water but readily dehydrates at low relative humidity to afford anhydrous β-caffeine [102]. The weakly basic nature of caffeine, pKa = 3.6, limits salt formation to only the strongest of acid partners, but this purine derivative readily forms co-crystals with a range of organic acids [102–104]. Oxalic, malonic and maleic acids combine with caffeine to produce 2:1 caffeine:acid complexes while glutaric, adipic and maleic acids form 1:1 co-crystals [102–104]. The supramolecular heterosynthon formed between caffeine and the carboxylic acid moiety in each of these structures comprises a hydrogen bond donated by the acid to the imidazole nitrogen atom complemented by a C–H hydrogen bond from the 2-postion of the imidazole to the acid carbonyl. The differences in stoichiometry reflect a dependence of the crystal packing on the structure of the co-crystal former, exemplified by comparing the interactions depicted for the oxalic acid complex in Figure 14 with that of the glutaric acid co-crystal shown in Figure 15. To further illustrate the dependence of co-crystal structure on the identity of the partner, co-crystals of caffeine with both 1-hydroxy-2-naphthoic acid and 3-hydroxy-2-naphthoic acid rely upon the same supramolecular heterosynthon observed with the simple dicarboxylic acids depicted in Figures 14 and 15. However, in co-crystals of caffeine and 6-hydroxy-2-naphthoic acid, the co-crystal former establishes the carboxylic acid dimer,

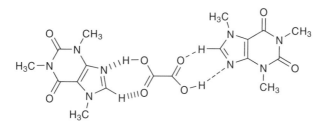

Figure 14   Intermolecular interactions in a 2:1 co-crystal of caffeine and oxalic acid.

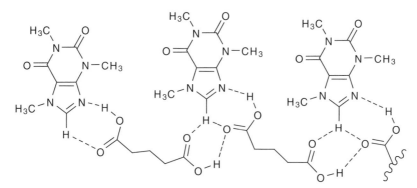

Figure 15   Intermolecular interactions in a co-crystal of caffeine and glutaric acid.

which allows each of the phenols to donate a hydrogen bond to the urea carbonyl of the purine, affording the 1:1 complex depicted in Figure 16 [105].

None of the caffeine:dicarboxylic acid co-crystals formed hydrates, even when prepared from hydrated drug, reflecting the satisfied hydrogen bonding [102]. The caffeine-oxalic acid co-crystal was non-hygroscopic and stable over several weeks at 43–98% relative humidity while the other co-crystals dissociated into the individual components under similar conditions [102].

The free NH in the structurally homologous theophylline provides an opportunity for alternative modes of intermolecular interaction [106]. Theophylline forms 2:1 co-crystals with oxalic acid in which hydrogen-bonded theophylline dimers are bridged in the crystal lattice by the co-crystal former, stabilized by hydrogen bonding to the imidazole moiety, as illustrated in Figure 17 [106]. However, with malonic, maleic and glutaric acids, stoichiometric co-crystals formed, also based on the theophylline dimer but with different, more complex arrangements of the diacids that retain a hydrogen-bonding interaction with the imidazole [106]. None of these co-crystals showed a tendency to hydrate at high relative humidity, with the oxalic acid co-crystal superior to anhydrous theophylline [106].

Interestingly, theophylline forms a different kind of co-crystal with 2,4-dihydroxybenzoic acid in which the theophylline supramolecular homosynthon is replaced by the heterosynthon shown in Figure 18 [107]. This complex is isolated as a hydrate, with the water molecule accepting a hydrogen bond from the isolated phenol moiety.

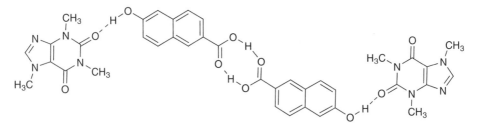

Figure 16   Intermolecular interactions in a co-crystal of caffeine and 6-hydroxy-2-naphthoic acid.

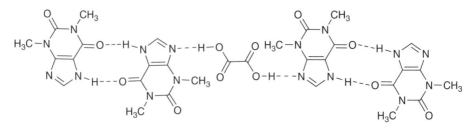

Figure 17   Intermolecular interactions in a co-crystal of theophylline and oxalic acid.

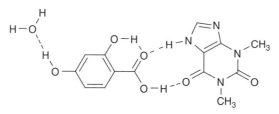

Figure 18 Intermolecular interactions in a co-crystal of theophylline and 2,4-dihydroxybenzoic acid.

A recent example of a more complex molecule demonstrating hygroscopicity that was solved by co-crystal formation is provided by AZD-1152 (**15**), the prodrug of a potent aurora kinase A, B and C inhibitor [108]. The free base of AZD-1152 (**15**) associates with 3 or 4 molecules of water, the stoichiometry dependent upon relative humidity and temperature. However, a co-crystal with maleic acid, formed by combining the components in methanol, DMSO/methanol or DMSO at 60°C and precipitating by adding CH₃CN, is claimed to be both anhydrous and non-hygroscopic [108].

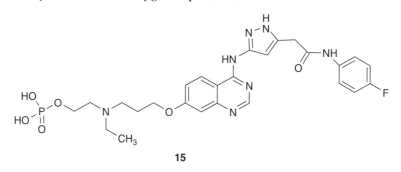

15

6. OPTICAL RESOLUTION BY CO-CRYSTAL FORMATION

Approximately 90% of racemic compounds crystallize as mixtures in which the (*R*)- and (*S*)-isomers interact, with the result that most racemates of solid compounds are, in essence, co-crystals [57]. In addition to the well-established potential for enantiomers to demonstrate different biological properties, they also possess physical properties that are distinct from the racemic mixture. Ibuprofen has been highlighted as a prototypical example in which the racemate melts at 71°C but the pure (*S*)-enantiomer melts at a much lower temperature, 46°C, exhibits a lower heat of fusion and shows almost 2-fold greater solubility in KCl–HCl at 25°C, reflecting differences in crystal packing [57]. The lactam 4-(4-chlorophenyl)pyrrolidin-2-one behaves similarly to ibuprofen, crystallizing such that both enantiomers are present in the single crystal, interacting via the amide-amide homosynthon, as illustrated in Figure 19 [109]. This observation indicates

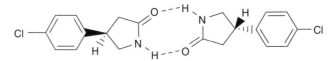

Figure 19 Intermolecular interactions in a crystal of racemic 4-(4-chlorophenyl)pyrrolidin-2-one.

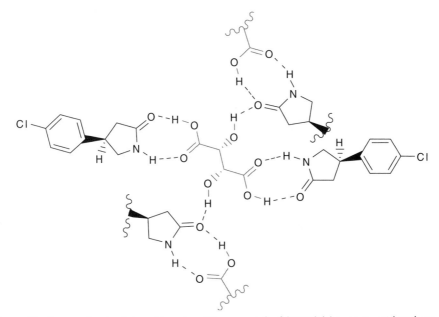

Figure 20 Intermolecular interactions in a 2:1 co-crystal of (2R,3R)-(+)-tartaric acid and one enantiomer of 4-(4-chlorophenyl)pyrrolidin-2-one.

that resolution by seeding, entrainment or the method that Pasteur utilized in his classic resolution of tartaric acid will not be successful [109,110].

However, co-crystallization of 4-(4-chlorophenyl)pyrrolidin-2-one with (2R,3R)-(+)-tartaric acid gave a 2:1 acid:lactam co-crystal in which only one enantiomer of the lactam was present, incorporated by the amide/carboxylic acid heterosynthon depicted in Figure 20 [109]. Since the chiral center of the lactam is quite remote from the asymmetry associated with tartaric acid, resolution appears to depend upon the acid acting as a scaffold to project the lactam molecules in a fashion that generates a chiral environment for molecular association and crystal growth [109].

There are many other examples of co-crystals being used to resolve racemic mixtures with 2,2'-dihydroxy-1,1'-binaphthyl (BINOL) a particularly useful reagent capable of resolving phosphinates, phosphine oxides [111], sulfoxides, including the proton pump inhibitors omeprazole (**16**) and esomeprazole (**17**) [112–115], selenoxides [116], amines [117,118] and ammonium salts [119,120].

The reciprocal process, in which co-crystals of racemic BINOL prepared from optically pure partners are used to resolve BINOL, is also successful [121–123].

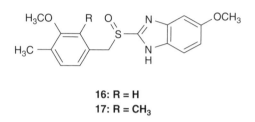

16: R = H
17: R = CH₃

A prominent example of optical resolution on a commercial scale that appears to rely on co-crystal complex formation is pregabalin (**17**), a successful drug marketed by Pfizer as Lyrica® for the treatment of fibromyalgia [124,125]. A screening campaign identified (*S*)-malic acid as a useful resolving agent that, when present in 50–100% excess, formed a complex with pregabalin (**17**) in aqueous *iso*-propanol [126]. Two recrystallizations in the presence of additional (*S*)-malic acid produced material with 99:1 diastereomeric purity in 70% yield. Recrystallization of the complex from THF and water took advantage of the fact that in the absence of excess (*S*)-malic acid, pregabalin (**17**) crystallizes as the free form. A final crystallization afforded 100% (*S*)-pregabalin (**17**) in 25–29% overall yield [126]. The single crystal X-ray structure of the pregabalin-mandelic acid complex reveals interesting and complex interactions in the solid state in which each carboxylate anion accepts 3 hydrogen bonds, an unusual arrangement [127]. These hydrogen bonds are donated by the hydroxyl and carboxylic acid moieties of mandelic acid and the ammonium moiety of an adjacent pregabalin molecule in the crystal lattice.

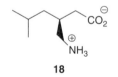

18

7. CO-CRYSTALS AND TOXICITY

A potential issue with the application co-crystals to formulate APIs is ensuring that the partner molecule is not associated with toxicity and meets the 'generally recognized as safe' (GRAS) standard [128]. The principles typically used to select salt forms are applicable and a wide range of carboxylic acids, the most versatile class of co-crystal formers, is available for this purpose [129,130]. However, APIs

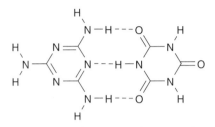

Figure 21 Hydrogen bonding interactions between melamine and cyanuric acid in a co-crystal.

not susceptible to co-crystal formation with carboxylic acids will require partners with alternative functionality, a potentially more challenging proposition [131].

An interesting toxicity that appears to be attributable to co-crystal formation *in vivo* occurred in the Spring of 2007 when contaminated pet food sold in the United States caused renal failure and death in domestic cats and dogs [132]. Melamine was shown to be a contaminant in the recalled food, postulated to have been added to increase the nitrogen content of the food since this was the simple assay used to reflect amino acid content. However, it was subsequently shown that cyanuric acid was also a significant contaminant. Individual administration of either melamine or cyanuric acid to cats as part of their food failed to cause toxicity at doses up to 120 mpk. However, combining the 2 compounds in the cats' diet produced renal failure within 48 h and autopsy revealed fan-shaped, birefringent crystals in the kidney tubules and distal nephron. Analysis indicated that melamine and cyanuric acid were present in the kidneys in a 1:1 ratio and mixing the 2 chemicals in cat urine afforded identical crystals [132]. Whereas the crystals have not been definitively identified, the data are consistent with co-crystals of melamine and cyanuric acid, strong co-crystal formers known to form stable complexes in which each molecule interacts via 3 complementary hydrogen bonds in the arrangement depicted in Figure 21 [133,134]. In the crystal lattice, these supramolecular synthons assemble in planar sheets of hexameric rosettes, with each molecule stabilized by engaging in a total of 9 N–H$\cdots$O and N–H$\cdots$N hydrogen bonds [134].

8. SUMMARY

The formation of co-crystals is emerging as a useful methodology with considerable potential for productively modulating the physical properties of drug candidates. The properties of co-crystals can be enhanced by structural variation of the co-crystal formers, an enterprise that is beginning to attract attention [135], and the application of conventional formulation approaches, including particle engineering and combination with excipients. The first APIs formulated as co-crystals have been advanced into clinical trials and it may be anticipated that others will follow as the potential of the technology is more fully explored and exploited.

REFERENCES

[1] F. Lara-Ochoa and G. Espinosa-Pérez, *Supramol. Chem.*, 2007, **19**, 553.

[2] N. J. Babu, L. S. Reddy and A. Nangia, *Mol. Pharmacol.*, 2007, **4**, 417.

[3] A. D. Bond, *Cryst. Eng. Commun.*, 2007, **9**, 833.

[4] P. Vishweshwar, J. A. McMahon, J. A. Bis and M. J. Zaworotko, *J. Pharm. Sci.*, 2006, **95**, 499.

[5] F. H. Allen, W. D. S. Motherwell, P. R. Raithby, G. P. Shields and R. Taylor, *N. J. Chem.*, 1999, **23**, 25.

[6] G. R. Desiraju, *Curr. Sci.*, 2001, **81**, 1038.

[7] G. R. Desiraju, *Angew. Chem. Int. Ed. Engl.*, 1995, **34**, 2311.

[8] C. B. Aakeröy, *Acta Crystallogr. Sect. B: Struct. Sci. B*, 1997, **53**, 569.

[9] G. R. Desiraju, *Angew. Chem. Int. Ed. Engl.*, 2007, **46**, 2.

[10] Ö. Almarsson and M. J. Zaworotko, *Chem. Commun.*, 2004, **1889**, .

[11] A. Park, L. J. Chyall, J. Dunlap, C. Schertz, D. Jonaitis, B. C. Stahly, S. Bates, R. Shipplett and S. Childs, *Expert Opin. Drug Disc.*, 2007, **2**, 145.

[12] N. Rodríguez-Hornedo, S. J. Nehm and A. Jayasankar, *Encycl. Pharm. Technol.*, 2007, **1**, 615.

[13] M. W. Cooke, M. Stanton, R. Shimanovich and A. Bak, *Am. Pharm. Rev.*, 2007, **10**, 54.

[14] N. Blagden, M. de Matas, P. T. Gavan and P. York, *Adv. Drug Deliv. Rev.*, 2007, **59**, 617.

[15] P. D. Leeson and A. M. Davis, *J. Med. Chem.*, 2004, **47**, 6338.

[16] M. C. Wenlock, R. P. Austin, P. Barton, A. M. Davis and P. D. Leeson, *J. Med. Chem.*, 2003, **46**, 1250.

[17] J. R. Proudfoot, *Bioorg. Med. Chem. Lett.*, 2005, **15**, 1087.

[18] L.-F. Huang and W.-Q. Tong, *Adv. Drug Deliv. Rev.*, 2004, **56**, 321.

[19] S. Datta and D. J. W. Grant, *Nat. Rev. Drug Discov.*, 2004, **3**, 42.

[20] C. R. Gardner, C. T. Walsh and Ö. Almarsson, *Nat. Rev. Drug Discov.*, 2004, **3**, 926.

[21] G. R. Desiraju, *Acc. Chem. Res.*, 2002, **35**, 565.

[22] T. Steiner, *Acta Crystallogr. Sect. B: Struct. Sci. B*, 2001, **57**, 103.

[23] J. H. Clark, *Chem. Rev.*, 1980, **80**, 429.

[24] L. Leiserowitz and G. M. J. Schmidt, *J. Chem. Soc. (A)*, 1969, 2372.

[25] S. G. Fleischman, S. S. Kuduva, J. A. McMahon, B. Moulton, R. D. Bailey Walsh, N. Rodríguez-Hornedo and M. J. Zaworotko, *Cryst. Growth Des.*, 2003, **3**, 909.

[26] M. B. Hickey, M. L. Peterson, L. A. Scoppettuolo, S. L. Morrisette, A. Vetter, H. Guzman, J. F. Remenar, Z. Zhang, M. D. Tawa, S. Haley, M. J. Zaworotko and Ö. Almarsson, *Eur. J. Pharm. Biopharm.*, 2007, **67**, 112.

[27] F. Wöhler, *Annalen*, 1844, **51**, 153.

[28] A. R. Ling and J. L. Baker, *J. Chem. Soc. Trans.*, 1893, **63**, 1314.

[29] G. R. Desiraju, D. Y. Curtin and I. C. Paul, *Mol. Cryst. Liquid Cryst.*, 1979, **52**, 563.

[30] A. O. Patil, D. Y. Curtin and I. C. Paul, *J. Am. Chem. Soc.*, 1984, **106**, 4010.

[31] J. McIntosh, R. H. M. Robinson, F. R. Selbie, J. P. Reidy, H. Elliott Blake and L. Guttmann, *Lancet*, 1945, **246**, 97.

[32] M. R. Caira, *Mol. Pharmacol.*, 2007, **4**, 310.

[33] J. C. Krantz, Jr., J. M. Holbert, H. K. Iwamoto and C. J. Carr, *J. Am. Pharm. Assoc.*, 1947, **36**, 248.

[34] R. E. Meléndez and A. D. Hamilton, *Topics Curr. Chem.*, 1998, **198**, 97.

[35] M. C. Etter, *Acc. Chem. Res.*, 1990, **23**, 120.

[36] J. Maddox, *Nature*, 1988, **335**, 201.

[37] A. Gavezzotti, *Acc. Chem. Res.*, 1984, **27**, 309.

[38] J. D. Dunitz, *Chem. Commun.*, 2003, 545.

[39] M. A. Neumann, F. J. J. Leusen and J. Kendrick, *Angew. Chem. Int. Ed.*, 2008, **47**, 2427.

[40] S. L. Morissette, Ö. Almarsson, M. L. Peterson, J. F. Remenar, M. J. Read, A. V. Lemmo, S. Ellis, M. J. Cima and C. R. Gardner, *Adv. Drug Deliv. Rev.*, 2004, **56**, 275.

[41] G. G. Z. Zhang, R. F. Henry, T. B. Borchardt and X. Lou, *J. Pharm. Sci.*, 2007, **96**, 990.

[42] V. R. Peddireddi, W. Jones, A. P. Chorlton and R. Docherty, *Chem. Commun.*, 1996, 987.

[43] A. V. Trask, J. van de Streek, W. D. S. Motherwell and W. Jones, *Cryst. Growth Des.*, 2005, **5**, 2233.

[44] A. V. Trask, D. A. Haynes, W. D. S. Motherwell and W. Jones, *Chem. Commun.*, 2006, 51.

[45] T. Friščić, A. V. Trask, W. Jones and W. D. S. Motherwell, *Angew. Chem. Int. Ed.*, 2006, **45**, 7546.

[46] A. Jayasankar, A. Somwangthanaroj, Z. J. Shao and N. Rodríguez-Hornedo, *Pharm. Res.*, 2006, **23**, 2381.

[47] A. Jayasankar, D. J. Good and N. Rodríguez-Hornedo, *Mol. Pharm.*, 2007, **4**, 360.

[48] K. Chadwick, R. Davey and W. Cross, *Cryst. Eng. Commun.*, 2007, **9**, 732.

[49] N. Shan, F. Toda and W. Jones, *Chem. Commun.*, 2002, 2372.

[50] L. McCausland, *WO Patent, WO 2007/075793 A2*, July 5, 2007.

[51] G. Ruecroft; D. Hipkiss, T. Ly, N. Maxted and P. W. Cains, *Org. Proc. Res. Dev.*, 2005, **9**, 923.

[52] A. V. Trask, W. D. S. Motherwell and W. Jones, *Chem. Commun.*, 2004, 890.

[53] D. P. McNamara, S. L. Childs, J. Giordano, A. Iarriccio, J. Cassidy, M. S. Shet. R. Mannion, E. O'Donnell and A. Park, *Pharm. Res.*, 2006, **23**, 1888.

[54] G. Giuseppetti, C. Tadini, G. P. Bettinetti, F. Giordano and A. La Manna, *Il Farmaco. Ed. Sci.*, 1980, **35**, 138.

[55] H. Nakai, M. Takasuka and M. Shiro, *J. Chem. Soc., Perkin Trans.*, 1984, **2**, 1459.

[56] M. R. Caira, L. Nassimbeni, A. F. Wildervanck and J. Chem. Soc, *Perkin Trans.*, 1995, **2**, 2213.

[57] G. P. Stahly, *Cryst. Growth Des.*, 2007, **6**, 1007.

[58] G. Bettinetti, M. R. Caira, A. Callegari, M. Merli, M. Sorrenti and C. Tadini, *J. Pharm. Sci.*, 2000, **89**, 478.

[59] G. M. J. Schmidt, *Pure Appl. Chem.*, 1971, **27**, 647.

[60] Y. Matsuda, R. Akazawa, R. Teraoka and M. Otsuka, *J. Pharm. Pharmacol.*, 1994, **46**, 162.

[61] J. F. Remenar, M. L. Peterson, P. W. Stephens, Z. Zhang, Y. Zimenkov and M. B. Hickey, *Mol. Pharm.*, 2007, **4**, 386.

[62] W. Curatolo, *Pharm. Sci. Technol. Today*, 1998, **1**, 387.

[63] L.-F. Huang and W.-Q (T) Tong, *Adv. Drug Deliv. Rev.*, 2004, **56**, 321.

[64] A. Dokoumetzidis and P. Macheras, *Int. J. Pharm.*, 2006, **321**, 1.

[65] C. A. S. Bergström, C. M. Wassvik, K. Joahnsson and I. Hubatsch, *J. Med. Chem.*, 2007, **50**, 5858.

[66] J. F. Remenar, S. L. Morissette, M. L. Peterson, B. Moulton, J. M. MacPhee, H. R. Guzmán and Ö. Almarsson, *J. Am. Chem. Soc.*, 2003, **125**, 8456.

[67] S. L. Childs, G. P. Stahly and A. Park, *Mol. Pharm.*, 2007, **4**, 323.

[68] S. L. Childs, L. J. Chyall, J. T. Dunlap, V. N. Smolenskaya, B. C. Stahly and G. P. Stahly, *J. Am. Chem. Soc.*, 2004, **126**, 13335.

[69] T. Steiner, *Acta Crystallogr. Sect. B*, 1998, **54**, 456.

[70] P. K. Thallapally and A. Nangia, *Cryst. Eng. Commun.*, 2001, paper 27. Available at http://www.rsc.org/ej/ce/2001/B102780h/B102780h.pdf

[71] B. K. Saha, A. Nangia and M. Jaskólski, *Cryst. Eng. Commun.*, 2005, **7**, 355.

[72] V. Gobry, G. Bouchard, P.-A. Carrupt, B. Testa and H. Girault, *Helv. Chim. Acta*, 2000, **83**, 1465.

[73] H. S. Yathirajan, B. Nagaraj, P. Nagaraja and M. Bolte, *Acta Crystallogr. Sect. E: Struct. Rep. Online*, 2005, **E61**, o489.

[74] M. Zegarac, E. Mestrovic, A. Dumbovic, M. Devcic and P. Tudja, *WO Patent 2007/080362 A1*, July 19, 2007.

[75] J. A. McMahon, J. A. Bis, P. Vishweshwar, T. R. Shattock, O. L. McClaughlin and M. J. Zaworotko, *Z. Kristallogr.*, 2005, **220**, 340.

[76] S. Basavoju, D. Boström and S. P. Velaga, *Cryst. Growth Des.*, 2006, **6**, 2699.

[77] P. Revill, N. Serradell, J. Bolós and E. Rosa, *Drugs Future*, 2007, **32**, 788.

[78] M. Murphy, K. Dinehart, P. Hurter, P. Connelly and Y. Cui, *WO Patent 2005/123076 A2*, December 29, 2005.

[79] P. R. Connelly, *WO Patent 2007/098270 A2*, August 30th, 2007.

[80] G. L. Amidon, H. Lennernas, V. P. Shah and J. R. A. Crison, *Pharm. Res.*, 1995, **12**, 413.

[81] N. A. Kasim, M. Whitehouse, C. Ramachandran, M. Bermejo, H. Lennerna, A. S. Hussain, H. E. Junginger, S. A. Stavchansky, K. K. Midha, V. P. Shah and G. L. Amidon, *Mol. Pharm.*, 2004, **1**, 85.

[82] A. L. Grzesiak, M. Lang, K. Kim and A. J. Matzger, *J. Pharm. Sci.*, 2003, **92**, 2260.

[83] W. W. Porter III, S. C. Elie and A. J. Matzger, *Cryst. Growth Des.*, 2008, **8**, 14.

[84] V. I. Ilyin, J. D. Pomonis, G. T. Whiteside, J. E. Harrison, M. S. Pearson, L. Mark, P. I. Turchin, S. Gottshall, R. B. Carter, P. Nguyen, D. J. Hogenkamp, S. Olanrewaju, E. Benjamin and R. M. Woodward, *J. Pharmacol. Exp. Ther.*, 2006, **318**, 1083.

[85] N. Variankaval, R. Wenslow, J. Murry, R. Hartman, R. Helmy, E. Kwong, S.-D. Clas, C. Dalton and I. Santos, *Cryst. Growth Des.*, 2006, **6**, 690.

[86] E. M. Doherty, A. W. Bannon, Y. Bo, N. Chen, C. Dominguez, J. Falsey, C. Fotsch, N. R. Gavva, J. Katon, T. Nixey, V. I. Ognyanov, L. Pettus, R. Rzasa, M. Stec, S. Surapaneni, R. Tamir, J. Zhu, J. J. S. Treanor and M. H. Norman, *J. Med. Chem.*, 2007, **50**, 3515.

[87] A. Bak, A. Gore, E. Yanez, M. Stanton, S. Tufekcic, R. Syed, A. Akrami, M. Rose, S. Surapaneni, T. Bostick, A. King, S. Neervannan, D. Ostovic and A. Koparkar, *J. Pharm. Sci.*, 2008, **97**, 3942.

[88] M. R. Caira, T. G. Dekker and W. Liebenberg, *J. Chem. Crystallogr.*, 1998, **28**, 11.

[89] J. P. Jani, R. S. Finn, M. Campbell, K. G. Coleman, R. D. Connell, N. Currier, E. O. Emerson, E. Floyd, S. Harriman, J. C. Kath, J. Morris, J. D. Moyer, L. R. Pustilnik, K. Rafidi, S. Ralston, A. M. K. Rossi, S. J. Steyn, L. Wagner, S. M. Winter and S. K. Bhattacharya, *Cancer Res.*, 2007, **67**, 9887.

[90] D. H. Brown Ripin, D. E. Bourassa, T. Brandt, M. J. Castaldi, H. N. Frost, J. Hawkins, P. J. Johnson, S. S. Massett, K. Neumann, J. Phillips, J. W. Raggon, P. R. Rose, J. L. Rutherford, B. Sitter, A. M. Stewart, Jr., M. G. Vetelino and L. Wei, *Org. Proc. Res. Dev.*, 2005, **9**, 440.

[91] P. N. Munster, C. D. Britten, M. Mita, K. Gelmon, S. E. Minton, S. Moulder, D. J. Slamon, F. Guo, S. P. Letrent, L. Denis and A. W. Tolcher, *Clin. Cancer Res.*, 2007, **13**, 1238.

[92] Z. J. Li, Y. Abramov, J. Bordner, J. Leonard, A. Medek and A. V. Trask, *J. Am. Chem. Soc.*, 2006, **128**, 8199.

[93] T. Higuchi and I. A. Pitman, *J. Pharm. Sci.*, 1973, **62**, 55.

[94] K. S. Raghavan, S. Ranadive, K. S. Bembenek, L. Benkerrour, V. Trognon, R. G. Corrao and L. Esposito, *WO Patent* 2003/000175, January 3, 2003.

[95] A. M. Chen, M. E. Ellison, A. Peresypkin, R. M. Wenslow, N. Variankaval, C. G. Savarin, T. K. Natishan, D. J. Mathre, P. G. Dormer, D. H. Euler, R. G. Ball, Z. Ye, Y. Wang and I. Santos, *Chem. Commun.*, 2007, 419.

[96] R. Barrett and R. A. Ward, *WO Patent* 2005/037253 A1, April 28, 2005.

[97] M. Isaji, *Curr. Opin. Investig. Drugs*, 2007, **8**, 285.

[98] W. Meng, B. A. Ellsworth, A. A. Nirschl, P. J. McCann, M. Patel, R. N. Girotra, G. Wu, P. M. Sher, E. P. Morrison, S. A. Biller, R. Zahler, P. P. Deshpande, A. Pullockaran, D. L. Hagan, N. Morgan, J. R. Taylor, M. T. Obermeier, W. G. Humphreys, A. Khanna, L. Discenza, J. G. Robertson, A. Wang, S. Han, J. R. Wetterau, E. B. Janovitz, O. P. Flint, J. M. Whaley and W. N. Washburn, *J. Med. Chem.*, 2008, **51**, 1145.

[99] J. Z. Gougoutas, *WO Patent* 2002/083066 A2, October 24, 2002.

[100] M. Imamura, K. Nakanishi, R. Shiraki, K. Onda, D. Sasuga and M. Yuda, *WO Patent* 2007/114475 A1, October 11, 2007.

[101] P. P. Deshpande, B. A. Ellsworth, J. Singh, T. W. Denzel, C. Lai, G. Crispino, M. E. Randazzo and J. Z. Gougoutas, *US Patent* 2004/0138439 A1, July 15, 2004.

[102] A. V. Trask, W. D. S. Motherwell and W. Jones, *Cryst. Growth Des.*, 2005, **5**, 1013.

[103] S. Karki, T. Friščić, W. Jones and W. D. S. Motherwell, *Mol. Pharm.*, 2007, **4**, 347.

[104] D.-K. Bučar, R. F. Henry, X. Lou, T. B. Borchardt and G. G. Z. Zhang, *Chem. Commun.*, 2007, 525.

[105] D.-K. Bučar, R. F. Henry, X. Lou, R. W. Duerst, T. B. Borchardt, L. R. MacGillivray and G. G. Z. Zhang, *Mol. Pharm.*, 2007, **4**, 339.

[106] A. V. Trask, W. D. S. Motherwell and W. Jones, *Int. J. Pharm.*, 2006, **320**, 114.

[107] Z.-L. Wang and L.-H. Wei, *Acta Crystallogr. Sect. E: Struct. Rep. Online E*, 2007, **63**, o1681.

[108] G .J. Sependa and R. Storey, *WO Patent* 2007/132227 A1, November 22, 2007.

[109] M. R. Caira, L. R. Nassimbeni, J. L. Scott and A. F. Wildervanck, *J. Chem. Crystallogr.*, 1996, **26**, 117.

[110] G. B. Kauffman and R. D. Myers, *Chem. Ed.*, 1998, **3**, 1.

[111] F. Toda and K. Mori, *J. Org. Chem.*, 1988, **53**, 308.

[112] F. Toda, K. Tanaka and T. C. W. Mak, *Chem. Lett.*, 1984, 2085.

[113] F. Toda, K. Tanaka and S. Nagamatsu, *Tetrahedron Lett.*, 1984, **25**, 4929.

[114] J. Deng, Y. Chi, F. Fu. X. Cui, K. Yu. J. Zhu and Y. Jiang, *Tetrahedron Asymmetry*, 2000, **11**, 1729.

[115] T. H. Ha, W. J. Kim, H. S. Oh, S. H. Park, J. C. Lee, H. K. Kim and K.-H. Suh, *WO Patent* 2007/013743, February 1, 2007.

[116] F. Toda and K. Mori, *Chem. Commun.*, 1986, 1357.

[117] C.-Q. Kang, Y.-Q. Cheng, H.-Q. Guo, X.-P. Qiu and L.-X. Gao, *Tetrahedron Asymmetry*, 2005, **16**, 2141.

[118] M. Ratajczak-Sitarz, A. Katrusiak, K. Gawrońska and J. Gawroński, *Tetrahedron Asymmetry*, 2007, **18**, 765.

[119] Y. Wang, J. Sun and K. Ding, *Tetrahedron*, 2000, **56**, 4447.

[120] E. Tayama and H. Tanaka, *Tetrahedron Lett.*, 2007, **48**, 4183.

[121] H. Du, B. Ji, Y. Wang, J. Sun, J. Meng and K. Ding, *Tetrahedron Lett.*, 2002, **43**, 5273.

[122] M. Periasamy, A. S. B. Prasad, J. V. B. Kanth and C. K. Reddy, *Tetrahedron Asymmetry*, 1995, **6**, 341.

[123] F. Toda, K. Yoshizawa, S. Hyoda, S. Toyota, S. Chatziefthimiou and I. M. Mavridis, *Org. Biomol. Chem.*, 2004, **2**, 449.

[124] R. T. Owen, *Drugs Today*, 2007, **43**, 857.

[125] R.B. Silverman, *Angew. Chem. Int. Ed.*, 2008, **47**, 3500.

[126] M. S. Hoekstra, D. M. Sobieray, M. A. Schwindt, T. A. Mulhern, T. M. Grote, B. K. Huckabee, V. S. Hendrickson, L. C. Franklin, E. J. Granger and G. L. Karrick, *Org. Proc. Res. Dev.*, 1997, **1**, 26.

[127] B. Samas, W. Wang and D. B. Godrej, *Acta Crystallogr. Sect. E*, 2007, **63**, o3938.

[128] EAFUS: A Food Additive Database available at vm.cfsan.fda.gov/~dms/eafus.html

[129] G. S. Paulekuhn, J. B. Dressman and C. Saal, *J. Med. Chem.*, 2007, **50**, 6665.

[130] D. A. Haynes, W. Jones and W. D. S. Motherwell, *J. Pharm. Sci.*, 2005, **94**, 2111.

[131] N. J. Babu, L. S. Reddy and A. Nangia, *Mol. Pharm.*, 2007, **4**, 417.

[132] B. Puschner, R. H. Poppenga, L. J. Lowenstine, M. S. Filigenzi and P. A. Pesavento, *J. Vet. Diagn. Invest.*, 2007, **19**, 616.

[133] G. M. Whitesides, E. E. Simanek, J. P. Mathias, C. T. Seto, D. N. Chin, M. Mammen and D. M. Gordon, *Acc. Chem. Res.*, 1995, **28**, 37.

[134] A. Ranganathan, V. R. Pedireddi and C. N. R. Rao, *J. Am. Chem. Soc.*, 1999, **121**, 1752.

[135] C. B. Aakeröy, J. Desper and B. M. T. Scott, *Chem. Commun.*, 2006, **8**, 1445.

CHAPTER **23**

Pregnane X Receptor: Prediction and Attenuation of Human CYP3A4 Enzyme Induction and Drug–Drug Interactions

Michael W. Sinz

1. INTRODUCTION

Today, coadministration of multiple drugs is common in most drug treatment regimens, where a significant number of patients take five or more medications per week (prescription/over-the-counter drugs, vitamins/minerals, and herbal supplements) [1]. Metabolic drug–drug interactions are serious drug reactions and frequently occur when coadministered drugs interfere with the elimination of one another. A review of the top 200 marketed prescription drugs in 2002 indicated that ~50% of drugs are significantly eliminated by CYP3A4-mediated metabolism [2]. A more recent review of marketed drugs (2005–2006) indicates this trend continues, with 52% of new drugs either partially or exclusively eliminated by CYP3A4 [3,4]. Therefore, any increase or decrease in CYP3A4

Bristol-Myers Squibb Co., 5 Research Parkway, Wallingford, CT 06492, USA

Annual Reports in Medicinal Chemistry, Volume 43
ISSN 0065-7743, DOI 10.1016/S0065-7743(08)00023-7

enzyme activity will have a significant impact on the elimination of most marketed drugs and likely precipitate a drug–drug interaction.

Induction of the CYP3A4 enzyme increases the metabolic clearance of drugs significantly eliminated by CYP3A4, which can result in therapeutic failure, diminished efficacy, or increased dosage requirements. Rifampicin, a drug used commonly to treat tuberculosis, is often associated with drug–drug interactions due to its potent ability to induce CYP3A4 [5]. For example, rifampicin (**1**) (600 mg/day) when coadministered with a 40 mg dose of simvastatin (HMGCoA reductase inhibitor for the treatment of hypercholesterolemia) reduced the AUC of simvastatin and its active metabolite, simvastatin acid, by 87 and 93%, respectively, resulting in ineffective cholesterol lowering [6]. The interactions are not always mediated by combinations of prescribed drugs; often herbal supplements contain ingredients that affect CYP levels. St. John's wort is a medicinal plant which in clinical trials has demonstrated mood enhancement properties similar to prescribed antidepressants [7]. Unfortunately, St. John's wort contains a component known as hyperforin (**2**) which is a potent CYP3A4 enzyme inducer and has demonstrated multiple clinical drug–drug interactions when coadministered with CYP3A4 substrates, such as HIV protease inhibitors, immunosuppressants, and oral contraceptives [8]. Not all induction-mediated drug–drug interactions occur between two coadministered drugs. Autoinduction is defined as the ability of a drug to induce enzymes that enhance its own metabolism, resulting in lower (often sub-therapeutic) drug exposure. In some situations, a complete loss of efficacy can occur when drug exposures significantly decrease. For example, the clinical development of MKC-963, a potent inhibitor of platelet aggregation, was terminated when it was determined that the drug was predominantly eliminated by CYP3A4 and was also a potent inducer of CYP3A4. After multiple dosing (14 days, 120 mg, QD), the AUC and C_{max} of MKC-963 decreased by 69% and 77%, respectively [9].

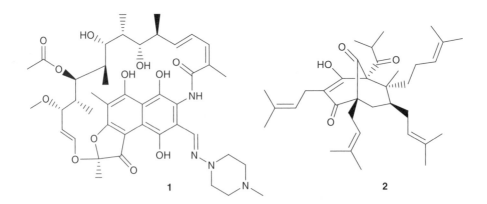

Induction of CYP enzymes is mediated predominately through nuclear hormone receptors or transcription factors. When drug binds to these receptors, they become activated and interact with specific DNA sequences found within the promoter region of target genes [10]. This drug–receptor–DNA interaction leads to increased gene transcription and subsequent translation of CYP enzyme. These receptor-mediated processes result in increased CYP enzyme activity which ultimately leads to increased metabolic elimination of drugs. Understanding the structural features of drugs and receptors that mediate or inhibit binding to these receptors will aid in the synthesis of new drug candidates which do not activate the receptor and enhance our ability to attenuate or eliminate adverse drug interactions due to enzyme induction.

2. PREGNANE X RECEPTOR

2.1 Function, structure, and species differences

The pregnane X receptor (PXR, also known as steroid and xenobiotic receptor, SXR) is a ligand (drug) activated nuclear hormone receptor found in liver and intestine and is responsible for transcriptional regulation of drug metabolizing enzymes and transporters [10]. PXR also interacts with endogenous components (pregnanes) and is involved in the homeostasis and regulation of cholesterol and bile acid metabolism. Figure 1 illustrates the major features of the PXR receptor. At the N-terminus is the activation function 1 (AF-1) region that regulates the receptor in a ligand-independent manner by making functional interactions with transcriptional machinery. The DNA binding domain (DBD) directs the receptor to its target genes by binding to specific DNA response elements found within the promoter region of target genes. Known target genes up-regulated by activated PXR include a variety of drug metabolizing enzymes and transporters, such as CYP2B6, CYP2C8/9/19, CYP3A4/5, CYP450 reductase, dehydrogenases, carboxylesterases, UGT1A1/3/6, SULTs, GSTs, MDR1, and MRP2 [11,12]. Near the C-terminus is the ligand binding domain (LBD) which contains the binding pocket and the ligand-dependent activation factor, AF-2 region. Binding of a ligand to the binding pocket initiates a series of corepressor displacements and recruitment of various coactivators, such as the steroid receptor coactivator, SRC-1, which interacts with the AF-2 region [11].

The DBD of PXR is fairly well conserved across multiple animal species with >90% sequence homology; however significant differences within the LBD are found between animals and humans [13]. For example, the known human CYP3A4 inducer (1) is a strong activator of human PXR, but a very weak rodent

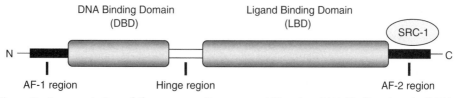

Figure 1 Representation of the pregnane X receptor: AF1 region, DNA binding domain (DBD), hinge region, ligand binding domain (LBD), and AF-2 region, along with SRC-1 coactivator.

PXR activator, whereas pregnenolone-16α-carbonitrile has been shown to be a weak activator of human PXR but a potent activator of rodent PXR [14]. Given the significant differences in agonist binding between species, human-based models must be employed to accurately assess PXR agonists and human enzyme induction potential. The most pronounced homology in ligand binding sequence to human PXR is the rhesus monkey which is 95% similar. It has been shown that monkey *in vitro* and *in vivo* models respond in a similar fashion to human *in vitro* and *in vivo* models with known human PXR agonists, such as (**1**) [15,16].

2.2 Biological models to assess PXR-mediated drug interactions

The assessment of CYP induction and drug–drug interaction potential is most often conducted with primary cultures of human hepatocytes. This cell-based model has the ability to respond to a variety of CYP inducers and isoforms, such as CYP1A, 2B, and 3A due to the presence and function of multiple nuclear hormone receptor systems and transcription factors. Although the primary human hepatocyte assay replicates a broad array of induction responses, it suffers from limited availability, random access, and low-moderate throughput. Newer cell-based models using immortalized hepatocytes have demonstrated the ability to respond to CYP-isoform induction similar to primary human hepatocytes. The most common immortalized hepatocyte cell lines reported to date are the Fa2N-4 and HepaRG cell lines [17,18]. A drawback of the immortalized hepatocyte model is the inability to replicate all CYP induction pathways similar to primary human hepatocytes. For example, the Fa2N-4 cell line appropriately reproduces CYP1A and 3A-mediated induction responses, but does not reproduce CYP2B-mediated induction responses due to very low expression of the nuclear hormone receptor CAR (constitutive androstane receptor) [17]. In both hepatocyte models, the ability of PXR ligands to bind and activate PXR are intact, as are the necessary coactivators, response elements and target genes (such as CYP3A4). However useful these cell based systems may be in assessing CYP-isoform induction, they are difficult to configure into routine and high throughput models and lack the ability to directly assess interactions between ligand and receptor. The human PXR protein has been the basis for several assays which directly measure agonist binding and activation, as well as predict CYP3A4 induction potential [19–23]. The most common human PXR models are binding and transactivation assays which are robust, reproducible, high throughput (384-well plate), and respond appropriately to known CYP3A4 enzyme inducers [24]. There are strong correlations between responses in the binding and transactivation assays, as well as between the PXR transactivation assay and primary human hepatocytes [24,25].

The ligand binding assays are competitive binding SPAs (scintillation proximity assays) which consist of expressed human PXR LBD in conjunction with the SRC-1 fragment incubated with test compound and a tight binding radiolabeled ligand, e.g., the potent PXR agonist SR12813 (**3**). Binding affinity is assessed by competition (IC_{50}) between test compound and displacement of (**3**) [24]. Cell-based PXR transactivation assays are constructed of: (1) an expression vector containing response elements from the CYP3A4 promoter region coupled to a reporter gene (typically luciferase) and (2) a human PXR expression vector. Both vectors are expressed in an appropriate cell line, such as HepG2, known to express the appropriate coactivators and corepressors. The potent PXR agonist (1) is generally run as a positive control and benchmark compound in PXR transactivation assays [24,26]. Agonist properties can be measured by determining the potency (EC_{50}) and efficacy (E_{max}) of a test compound and comparing these to (1).

In general, there is a good correlation between EC_{50} and IC_{50} values in PXR transactivation and binding assays, respectively [24]. However, there are examples where binding to the PXR receptor does not lead to transactivation or enzyme induction. This situation can be caused by PXR antagonists that bind to the receptor but do not elicit the appropriate conformational changes of the receptor, displacement of corepressors, or recruitment of coactivators. For example, ecteinascidin 743 (4) is an experimental anticancer drug that can bind/inhibit PXR ($IC_{50} = 3\,nM$) in a human PXR transactivation assay and does not induce CYP3A4 in primary human hepatocytes [27].

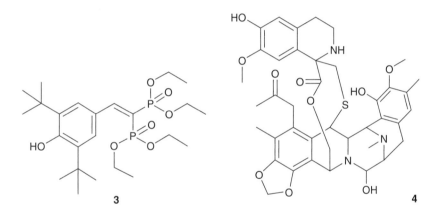

3   4

2.3 Human PXR ligands: agonist and antagonists

PXR is unlike most nuclear hormone receptors in that a wide variety of ligands can bind to the receptor as both agonists and antagonists. The most potent agonists are compounds (2) and (3) with submicromolar EC_{50}s in PXR transactivation assays: 0.04 and 0.22 µM, respectively [26]. Other common agonists of PXR include: (1), forskolin (5), ritonavir (6), nicardipine (7), bergamottin (8), paclitaxol (9), clotrimazole (10), reserpine (11), and mifepristone (12) [26]. These compounds are well characterized in primary human hepatocytes as inducing CYP3A4, as well as causing drug–drug interactions with coadministered CYP3A4 substrates [24,26]. However, several of the potent PXR agonists do not cause significant drug interactions in patients due to: (1) simultaneous CYP3A4 inhibition and induction (6, 8, 12), (2) route of administration (**10**-topical), or (3) low dose (exposure) required for efficacy which results in a significant margin between PXR activation and drug efficacy (**7, 9**) [26].

There is considerable structural diversity of PXR agonists which tend to be large (molecular weight ~225 to >800 Da) and hydrophobic (clogP >1 to 8) [11]. For the majority of PXR agonists, the molecular weight is ~500 Da and the clogP ~6 [24]. Additional observations made when describing PXR agonists are: they tend to have more hydrogen bond acceptor groups as compared to hydrogen bond donating groups and they often have halogenated aromatic rings which most likely increases the hydrophobicity of the ligand [28]. Both of these observations are illustrated throughout the agonists described herein.

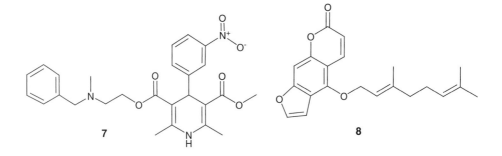

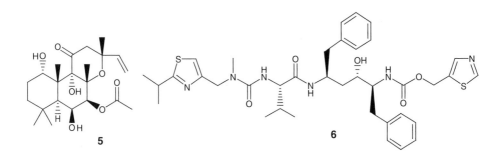

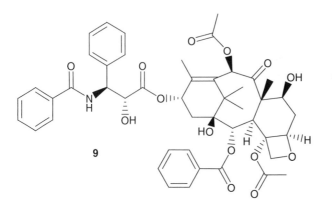

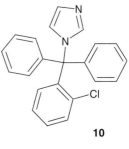

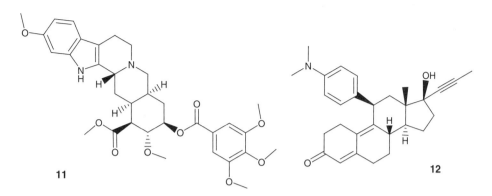

11   12

In addition to (4) several other PXR antagonists have been identified. The HIV protease inhibitor A-792611 (13) has recently been shown to be a PXR antagonist [29]. Interestingly the protease inhibitors demonstrate a wide range of interactions with PXR. Ritonavir (6) is a reasonably potent agonist of PXR while A-792611 is a good antagonist. Many of the other protease inhibitors generally show some interaction with PXR (albeit weak interactions) that do not result in significant induction or drug interactions, such as saquinavir or indinavir [26]. Several naturally occurring phytochemicals, such as sulforaphane (14) which is found in broccoli and coumestrol (15) are recent examples of PXR antagonists [30,31]. Other compounds, such as ketoconazole (16), inhibit PXR activation not through interactions at the LBD but through interactions at the AF-2 region by disrupting binding of the coactivator SRC-1 with PXR [32]. Therefore, ligands can inhibit or interfere with the activation of PXR through interactions in the LBD or allosteric interactions with the receptor. It has been suggested that an effective PXR antagonist may be of clinical value to reduce the induction of drug metabolizing enzymes during the therapeutic treatment of diseases, e.g., ketoconazole treatment in cancer therapy [33]. This approach to reduce the induction effect of known CYP3A4 inducers used in combination drug therapies could prove helpful if the PXR antagonist/inhibitor itself does not have any significant off target effects (e.g., CYP3A4 inhibition by ketoconazole). Apparently the continual antagonism or inactivation of PXR does not have any significant biological implications as the mouse PXR knockout model has been shown to be phenotypically normal [34,35].

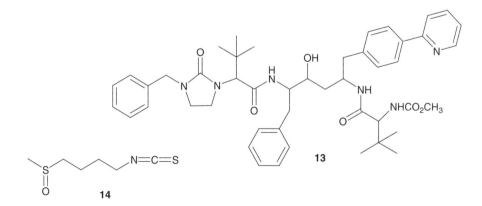

13

14

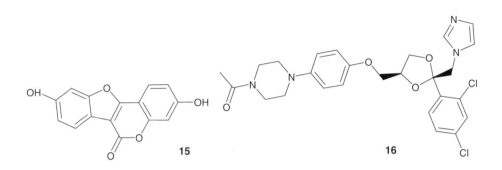

15   16

3. CRYSTAL STRUCTURE AND MOLECULAR MODELING OF PXR

Crystal structures of the human PXR LBD with the coactivator SRC-1 have been determined alone (apo-protein) and in complex with several ligands (**1**, **2**, **3**, estradiol) [36–40]. The ligand binding pocket of PXR was found to be large and flexible with ~30 amino acids forming an elliptically shaped pocket. The binding domain is predominately hydrophobic, however there are four polar and four charged residues distributed throughout the pocket with both hydrogen bond donating and accepting residues. The volume of the LBD of the apo-protein is 1,280 $\mathring{A}^3$, however the pocket has the ability to expand in order to accommodate larger ligands, such as (**2**) which has a volume of 1,544 $\mathring{A}^3$ [38,39]. Also, ligands are not required to fill the entire binding pocket. For example, estradiol binds in such a way that a significant portion of the binding pocket is unoccupied [40]. The first crystal structure of PXR with bound ligand indicated that (**3**) was able to bind in multiple orientations; however when the PXR LBD was co-expressed with the SRC-1 coactivator, (**3**) was found to bind in only one orientation [37]. These results confirm that SRC-1 is important in determining the proper orientation of ligands and has the ability to restrict orientation within the binding pocket.

In general terms, the ligand binding pocket has the ability to conform and modify its volume and shape depending on the ligand which is indicative of an 'induced fit' model. Given these properties of the PXR LBD and the variety of chemical structures that make up PXR agonists, it has been challenging to construct *in silico* models that mimic the initial binding and subsequent conformational changes within the binding pocket. Nonetheless, several key aspects of ligands and residues within the LBD have been established and some success in developing *in silico* models has been achieved [28,41]. By and large, hydrophobic and hydrogen bond acceptor features are important for PXR activators. Six residues within the LBD have been shown to consistently be involved in ligand binding interactions: hydrophobic interactions with Met243, Trp299, and Phe420; and hydrogen bond formation with Ser247, Gln285, and His407. In general, the pharmacophore models developed to date indicate that binding requires the ligand to have at least one hydrogen bond acceptor group and several hydrophobic groups that interact with residues in the binding pocket in order to obtain significant PXR binding. It has also been suggested that increasing the number of hydrophobic interactions leads to greater ligand affinity [42].

The following example illustrates how modeling can be used to assign key structural elements to residues within the ligand binding pocket. Lemaire et al., through the use of pharmacophore modeling and virtual screening, discovered a PXR agonist, C2BA-4 (17) which was more potent in a PXR transactivation assay than (3) (EC$_{50}$ = 49 vs. 137 nM, respectively) [43]. The pharmacophore model incorporated many of the aforementioned characteristics of PXR agonists, such as high molecular weight, high cLogP, multiple hydrophobic regions, and a hydrogen bond acceptor feature. When docking (17) into the model, the hydrophobic phenylethyl and 2-chlorobenzyl moieties were found to interact with Trp299 and Phe320 residues, respectively, while the sulfonamide hydrogen bonded to the Ser247 and His407 residues.

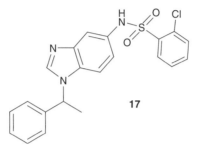

17

3.1 SAR examples to reduce PXR activation

Using the information describing hydrophobic and hydrogen binding features of agonists and residues within the ligand binding pocket, Gao et al., tested several hypotheses for potentially destabilizing PXR ligand affinity [44]. One such approach was to disrupt the hydrophobic nature of a ligand by introducing polar groups. Compounds (18–23) demonstrate how introducing more polar nitrogen heterocycles or a sulfamide to the terminal biphenyl position disrupts PXR transactivation. Compounds 18, 19, 20, 21, and 22 gave decreasing PXR transactivation responses from 30 down to 3, −7, −4, and −6%, respectively, when the polar groups were added. Interestingly, compound (23) gave a 48% response in the PXR transactivation assay despite introduction of polarity to the hydrophobic moiety, but this modification also provided a hydrogen bond acceptor feature in juxtaposition to Gln285 in the LBD that enhanced the PXR ligand interaction [44].

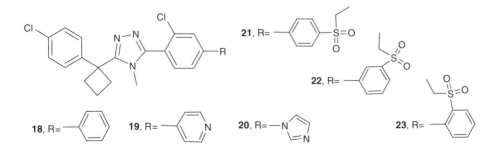

Another example of SAR was conducted by Harper et al., who reduced the PXR transactivation properties from a series of indole-*N*-acetamides being developed as potent allosteric inhibitors of HCV NS5B polymerase [45]. They determined that the PXR agonist properties were associated with the acetamide side chain attached at N1 of structures (24) and (25) and that by removing the sp^3 carbon link to the 6-membered ring they could reduce the PXR transactivation from 82% down to 14% between compounds (24) and (25), respectively. Further molecular modeling by Gao et al., indicated that the carboxylic acid group and the amide group interacted with amino acid residues Ser247 and His407, respectively [44]. They also hypothesized that removal of the carbon linker increased the rigidity of the molecule in such a way that the N-dimethyl group collides with the interior of the binding pocket which in turn moves the molecule out of a favorable binding position.

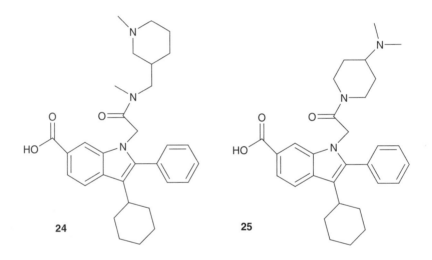

24   25

Sometimes structural changes need not be significant; such is the example of cortisone (26) and hydrocortisone (27). Compound (26) has been shown to weakly bind to PXR and does not activate PXR, while (27) does activate human PXR [20,46]. The two molecules differ only at the C-11 position with (26) having an oxo-group and (27) having a hydroxyl-group. Both the oxo- and hydroxyl-groups are hydrogen bond acceptors (necessary for all PXR ligands), however when Schuster and Langer fitted the two structures into their pharmacophore model, only (27) was found to have the correct orientation to interact within the binding pocket while (26) did not [42]. A similar situation exists with docetaxel (28) and paclitaxel (9), two structurally similar anticancer agents, in which only (9) results in significant PXR transactivation and induction of CYP3A4 in primary human hepatocytes [26,47]. Further studies indicated that (28) is a partial agonist of PXR and although it can bind, it is unable to activate PXR due to an inability to displace transcriptional corepressors from the transactivation domain of the receptor [27]. There are two structural differences between these compounds: (1) a terminal phenyl group on (**9**) vs. an O-*t*-butyl group on (**28**) and (2) an acetate group on (**9**) vs. a hydroxyl-group on (**28**) at position 10. Again, Schuster and Langer fitted the two molecules into their pharmacophore model and found that the terminal phenyl and O-*t*-butyl groups interacted with Tyr306 in the binding pocket of PXR. The researchers hypothesized that the phenyl group was able to participate in π–π stacking at this residue thereby enhancing the interaction between ligand and receptor, whereas the O-*t*-butyl group was not able to interact in this manner [42]. No mention of an interaction at position 10 of the compounds in the binding pocket was described.

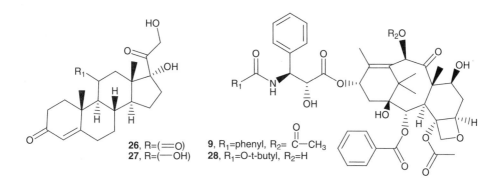

26, R=(=O)
27, R=(—OH)

9, R₁=phenyl, R₂= C—CH₃
28, R₁=O-t-butyl, R₂=H

Stereochemistry can also play a role in PXR ligand activation and species-specific activation. The enantiomers of a C-cyclopropylalkylamide, S20, were found to have differential potencies in a human PXR transactivation assay. The (+)-S20 (**29**) compound resulted in much greater activation of PXR ($EC_{50} = 400\,nM$) than the (−)-S20 enantiomer (**30**). In contrast, when evaluated in a mouse PXR transactivation assay, the reverse was observed: (**30**) was significantly more potent than (**29**) [48].

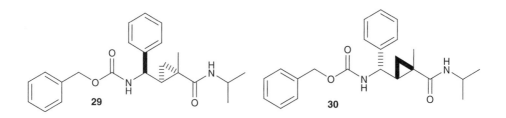

29   30

Based on the literature evidence to date, several key structural modifications are recommended to reduce the ligand binding affinity of PXR agonists: (1) reducing hydrophobicity by removing aromatic groups/halogens or attaching polar groups to hydrophobic residues thereby disrupting essential hydrophobic interactions, (2) structural modifications that make the molecule more rigid or conformational changes that move crucial hydrophobic or hydrogen bond accepting features in the molecule away from key residues in the binding pocket, and (3) removing the central hydrogen bond accepting feature from the molecule [44]. As shown with several examples, these modifications need not be dramatic if they are targeted toward key positions on the ligand.

4. SUMMARY

Drug-drug interactions are estimated to be responsible for 20–30% of all adverse drug reactions and the risk of significant drug interactions increases with the number of concomitantly administered medications [49]. CYP3A4-mediated drug interactions, due to enzyme induction, are predominately mediated through a PXR ligand activated mechanism. By using multiple PXR containing human *in vitro* and *in vivo* model systems, such as PXR binding/transactivation assays, immortalized and primary hepatocytes, and clinical drug interaction studies, we can assess the drug interaction potential of new chemical entities during drug discovery and development. Through the use of PXR binding and transactivation assays, we can probe the specific interactions of drugs (ligands) and their ability to initiate an induction response. PXR ligands are varied in structure yet tend to be large and hydrophobic while the ligand binding pocket of PXR is large, flexible, and interacts with hydrophobic and hydrogen bond acceptor groups on ligands. Based on recent pharmacophore modeling and structure–activity assessments, researchers have been able to ascertain those structural modifications that can reduce or eliminate PXR agonist activity. Most significant is the disruption of interactions between the ligand and LBD of PXR which involve hydrophobic and hydrogen bond acceptor interactions.

REFERENCES

[1] D. W. Kaufman, J. P. Kelly, L. Rosenberg, T. E. Anderson and A. A. Mitchell, *JAMA*, 2002, **287**, 337.
[2] J. A. Williams, R. Hyland, B. C. Jones, D. A. Smith, S. Hurst, T. C. Goosen, V. Peterkin, J. R. Koup and S. E. Ball, *Drug Metab. Dispos.*, 2004, **32**, 1201.
[3] S. Hedge and M. Schmidt, in *"Annual Reports in Medicinal Chemistry"* (ed. J. Macor), Vol. 42, Elsevier, New York, 2007.
[4] S. Hegde and M. Schmidt, in *"Annual Reports in Medicinal Chemistry"* (ed. A. Wood), Vol. 41, Elsevier, New York, 2006.
[5] M. Niemi, J. T. Backman, M. F. Fromm, P. J. Neuvonen and K. T. Kivisto, *Clin. Pharmacokinet.*, 2003, **42**, 819.
[6] C. Kyrklund, J. T. Backman, K. T. Kivisto, M. Neuvonen, J. Laitila and P. J. Neuvonen, *Clin. Pharmacol. Ther.*, 2000, **68**, 592.
[7] R. Madabushi, B. Frank, B. Drewelow, H. Derendorf and V. Butterweck, *Eur. J. Clin. Pharmacol.*, 2006, **62**, 225.
[8] M. Mannel, *Drug Saf.*, 2004, **27**, 773.
[9] T. Shimizu, K. Akimoto, T. Yoshimura, T. Niwa, K. Kobayashi, M. Tsunoo and K. Chiba, *Drug Metab. Dispos.*, 2006, **34**, 950.
[10] L. M. Tompkins and A. D. Wallace, *J. Biochem. Mol. Toxicol.*, 2007, **21**, 176.
[11] V. E. Carnahan and M. R. Redinbo, *Curr. Drug Metab.*, 2005, **6**, 357.

[12] S. Harmsen, I. Meijerman, J. H. Beijnen and J. H. Schellens, *Cancer Treat. Rev.*, 2007, **33**, 369.

[13] J. T. Moore, L. B. Moore, J. M. Maglich and S. A. Kliewer, *Biochim. Biophys. Acta*, 2003, **1619**, 235.

[14] F. J. Gonzalez and A. M. Yu, *Annu. Rev. Pharmacol. Toxicol.*, 2006, **46**, 41.

[15] M. Nishimura, A. Koeda, Y. Suganuma, E. Suzuki, T. Shimizu, M. Nakayama, T. Satoh, S. Narimatsu and S. Naito, *Drug Metab. Pharmacokinet.*, 2007, **22**, 178.

[16] T. Prueksaritanont, Y. Kuo, C. Tang, C. Li, Y. Qiu, B. Lu, K. Strong-Basalyga, K. Richards, B. Carr and J. H. Lin, *Drug Metab. Dispos.*, 2006, **34**, 1546.

[17] N. Hariparsad, B. A. Carr, R. Evers and X. Chu, *Drug Metab. Dispos.*, 2008, **36**, 1046.

[18] S. L. Ripp, J. B. Mills, O. A. Fahmi, K. A. Trevena, J. L. Liras, T. S. Maurer and S. M. de Morais, *Drug Metab. Dispos.*, 2006, **34**, 1742.

[19] B. Goodwin, E. Hodgson and C. Liddle, *Mol. Pharmacol.*, 1999, **56**, 1329.

[20] S. A. Jones, L. B. Moore, J. L. Shenk, G. B. Wisely, G. A. Hamilton, D. D. McKee, N. C. Tomkinson, E. L. LeCluyse, M. H. Lambert, T. M. Willson, S. A. Kliewer and J. T. Moore, *Mol. Endocrinol.*, 2000, **14**, 27.

[21] S. A. Kliewer, J. T. Moore, L. Wade, J. L. Staudinger, M. A. Watson, S. A. Jones, D. D. McKee, B. B. Oliver, T. M. Willson, R. H. Zetterstrom, T. Perlmann and J. M. Lehmann, *Cell*, 1998, **92**, 73.

[22] G. Luo, T. Guenthner, L. S. Gan and W. G. Humphreys, *Curr. Drug Metab.*, 2004, **5**, 483.

[23] J. Raucy, L. Warfe, M. F. Yueh and S. W. Allen, *J. Pharmacol. Exp. Ther.*, 2002, **303**, 412.

[24] Z. Zhu, S. Kim, T. Chen, J. H. Lin, A. Bell, J. Bryson, Y. Dubaquie, Y. Yan, J. Yanchunas, D. Xie, R. Stoffel, M. Sinz and K. Dickinson, *J. Biomol. Screen.*, 2004, **9**, 533.

[25] G. Luo, M. Cunningham, S. Kim, T. Burn, J. Lin, M. Sinz, G. Hamilton, C. Rizzo, S. Jolley, D. Gilbert, A. Downey, D. Mudra, R. Graham, K. Carroll, J. Xie, A. Madan, A. Parkinson, D. Christ, B. Selling, E. LeCluyse and L. S. Gan, *Drug Metab. Dispos.*, 2002, **30**, 795.

[26] M. Sinz, S. Kim, Z. Zhu, T. Chen, M. Anthony, K. Dickinson and A. D. Rodrigues, *Curr. Drug Metab.*, 2006, **7**, 375.

[27] T. W. Synold, I. Dussault and B. M. Forman, *Nat. Med.*, 2001, **7**, 584.

[28] C. Y. Ung, H. Li, C. W. Yap and Y. Z. Chen, *Mol. Pharmacol.*, 2007, **71**, 158.

[29] C. Healan-Greenberg, J. F. Waring, D. J. Kempf, E. A. Blomme, R. G. Tirona and R. B. Kim, *Drug Metab. Dispos.*, 2008, **36**, 500.

[30] H. Wang, H. Li, L. B. Moore, M. D. Johnson, J. M. Maglich, B. Goodwin, O. R. Ittoop, B. Wisely, K. Creech, D. J. Parks, J. L. Collins, T. M. Willson, G. V. Kalpana, M. Venkatesh, W. Xie, S. Y. Cho, J. Roboz, M. Redinbo, J. T. Moore and S. Mani, *Mol. Endocrinol.*, 2007, **22**, 838.

[31] C. Zhou, E. J. Poulton, F. Grun, T. K. Bammler, B. Blumberg, K. E. Thummel and D. L. Eaton, *Mol. Pharmacol.*, 2007, **71**, 220.

[32] H. Wang, H. Huang, H. Li, D. G. Teotico, M. Sinz, S. D. Baker, J. Staudinger, G. Kalpana, M. R. Redinbo and S. Mani, *Clin. Cancer Res.*, 2007, **13**, 2488.

[33] H. Huang, H. Wang, M. Sinz, M. Zoeckler, J. Staudinger, M. R. Redinbo, D. G. Teotico, J. Locker, G. V. Kalpana and S. Mani, *Oncogene*, 2007, **26**, 258.

[34] J. L. Staudinger, B. Goodwin, S. A. Jones, D. Hawkins-Brown, K. I. MacKenzie, A. LaTour, Y. Liu, C. D. Klaassen, K. K. Brown, J. Reinhard, T. M. Willson, B. H. Koller and S. A. Kliewer, *Proc. Natl. Acad. Sci. USA*, 2001, **98**, 3369.

[35] W. Xie, J. L. Barwick, M. Downes, B. Blumberg, C. M. Simon, M. C. Nelson, B. A. Neuschwander-Tetri, E. M. Brunt, P. S. Guzelian and R. M. Evans, *Nature*, 2000, **406**, 435.

[36] J. E. Chrencik, J. Orans, L. B. Moore, Y. Xue, L. Peng, J. L. Collins, G. B. Wisely, M. H. Lambert, S. A. Kliewer and M. R. Redinbo, *Mol. Endocrinol.*, 2005, **19**, 1125.

[37] R. E. Watkins, P. R. Davis-Searles, M. H. Lambert and M. R. Redinbo, *J. Mol. Biol.*, 2003, **331**, 815.

[38] R. E. Watkins, J. M. Maglich, L. B. Moore, G. B. Wisely, S. M. Noble, P. R. Davis-Searles, M. H. Lambert, S. A. Kliewer and M. R. Redinbo, *Biochemistry*, 2003, **42**, 1430.

[39] R. E. Watkins, G. B. Wisely, L. B. Moore, J. L. Collins, M. H. Lambert, S. P. Williams, T. M. Willson, S. A. Kliewer and M. R. Redinbo, *Science*, 2001, **292**, 2329.

[40] Y. Xue, L. B. Moore, J. Orans, L. Peng, S. Bencharit, S. A. Kliewer and M. R. Redinbo, *Mol. Endocrinol.*, 2007, **21**, 1028.

[41] S. Ekins, C. Chang, S. Mani, M. D. Krasowski, E. J. Reschly, M. Iyer, V. Kholodovych, N. Ai, W. J. Welsh, M. Sinz, P. W. Swaan, R. Patel and K. Bachmann, *Mol. Pharmacol.*, 2007, **72**, 592.

[42] D. Schuster and T. Langer, *J. Chem. Inf. Model*, 2005, **45**, 431.

[43] G. Lemaire, C. Benod, V. Nahoum, A. Pillon, A. M. Boussioux, J. F. Guichou, G. Subra, J. M. Pascussi, W. Bourguet, A. Chavanieu and P. Balaguer, *Mol. Pharmacol.*, 2007, **72**, 572.

[44] Y. D. Gao, S. H. Olson, J. M. Balkovec, Y. Zhu, I. Royo, J. Yabut, R. Evers, E. Y. Tan, W. Tang, D. P. Hartley and R. T. Mosley, *Xenobiotica*, 2007, **37**, 124.

[45] S. Harper, S. Avolio, B. Pacini, M. Di Filippo, S. Altamura, L. Tomei, G. Paonessa, S. Di Marco, A. Carfi, C. Giuliano, J. Padron, F. Bonelli, G. Migliaccio, R. De Francesco, R. Laufer, M. Rowley and F. Narjes, *J. Med. Chem.*, 2005, **48**, 4547.

[46] G. Bertilsson, J. Heidrich, K. Svensson, M. Asman, L. Jendeberg, M. Sydow-Backman, R. Ohlsson, H. Postlind, P. Blomquist and A. Berkenstam, *Proc. Natl. Acad. Sci. USA*, 1998, **95**, 12208.

[47] S. C. Nallani, B. Goodwin, A. R. Buckley, D. J. Buckley and P. B. Desai, *Cancer Chemother. Pharmacol.*, 2004, **54**, 219.

[48] Y. Mu, C. R. Stephenson, C. Kendall, S. P. Saini, D. Toma, S. Ren, H. Cai, S. C. Strom, B. W. Day, P. Wipf and W. Xie, *Mol. Pharmacol.*, 2005, **68**, 403.

[49] G. I. Kohler, S. M. Bode-Boger, R. Busse, M. Hoopmann, T. Welte and R. H. Boger, *Int. J. Clin. Pharmacol. Ther.*, 2000, **38**, 504.

CHAPTER **24**

Formulation in Drug Discovery

Robert G. Strickley

1. INTRODUCTION

The role of the formulation scientist within a drug discovery team, as outlined in three excellent publications [1–3], is to help design a new chemical entity that has the maximum chance of success from a pharmaceutical perspective. The long-term goal is a successful commercial product. The intermediate goal is a viable clinical formulation. The short-term goal is to support drug discovery by conducting preformulation studies on new chemical entities and to make the formulations for preclinical pharmacokinetic, efficacy, and toxicology studies.

Gilead Sciences, Foster City, CA, 94404, USA

Annual Reports in Medicinal Chemistry, Volume 43
ISSN 0065-7743, DOI 10.1016/S0065-7743(08)00024-9

Depending on the organizational structure, the formulation scientist in drug discovery research may or may not be the formulation scientist who develops the clinical and commercial formulation and manufacturing process. Early clinical studies may use non-commercial formulations to probe specific questions such as pharmacokinetics, drug–drug interactions, and efficacy for proof-of-concept. Later stage clinical studies should use the intended commercial formulation and manufacturing process. Preclinical studies in animals can use a much broader scope of formulations. This chapter is intended to help the medicinal chemist to understand issues confronted by the formulation scientist within drug discovery, and to provide detailed information and general references for use in discussing and designing preclinical studies.

2. EARLY CONSIDERATIONS OF A NEW CHEMICAL ENTITY

The intended route of administration (e.g., injectable vs. oral) is an essential initial consideration. If the goal is an injectable drug then chemical stability and solubility become the main criteria with little or no importance placed on permeability and absorption. If the goal is an oral dosage form then bioavailability is a critical issue, and the emphasis is placed on permeability, absorption, in vivo solubility, and dissolution rate. The term "developability" has been introduced to encompass the extent to which a compound exhibits such properties beyond the classical focus of potency and selectivity. Since the amount of a drug is limited at the early stage of drug discovery, in vitro assays must be conducted on a small or miniaturized scale, and the challenge for the formulation scientist is to acquire as much relevant information as possible with the small amount of material available within a relatively short period of time [4].

2.1 Maximum absorbable dose

Absorption is a complex process that involves permeability, solubility, and charge state as described in a very useful book by Avdeef [5]. Knowledge of the approximate human dose is very helpful, for the concept of the maximum absorbable dose (MAD) links the permeability of the drug with the aqueous solubility required. The MAD is quantitatively expressed by Equation 1 [3], and normally refers to a solid dosage form. However, the in vivo solubility can depend on fasted or fed conditions and also potential supersaturation upon dilution or dissolution.

$$MAD = S \times K_a \times SIWW \times SITT \tag{1}$$

where, S is the aqueous solubility in mg/mL, K_a the intestinal absorption rate constant (min^{-1}), SIWW the small intestinal water volume (mL) ~ 250 mL, and SITT the small intestine transit time (min) ~ 4.5 h.

2.2 Aqueous solubility

The choice of media for the measurement of solubility has been the subject of much discussion within the pharmaceutical community. Table 1 lists typical alternatives, ordered by increasing complexity. Chemical structures of the bile salts and lecithin that are constituents of simulated intestinal fluids are shown in Table 2. The compositions of various simulated intestinal fluids are listed in Table 3 [6,7], again ranked by complexity. For molecules in which the solubility is increased in simulated intestinal fluids relative to simple aqueous buffers, there is a strong possibility of an increased oral bioavailability in the fed state compared to the fasted state.

A typical initial screen in drug discovery is to measure the water solubility at pH 2 and pH 7. Common next steps are to measure the solubility in simulated intestinal fluids and to measure the water solubility over the entire pH range (2–12) thereby enabling the calculation of pK_a values (see section on solubility theory). The preferred experimental procedure to measure solubility is the "shake flask" or "equilibrium solubility" method, in which solid drug is added to the solvent of interest and mixed for a minimum of 2 h. The undissolved solids are then centrifuged and/or filtered out, and the supernatant is assayed by an analytical method such as HPLC or UV absorbance. By contrast, the measurement of "kinetic solubility" starts with the drug in a concentrated solution (e.g., 10–100 mg/mL in DMSO) and entails dilution into the solvent of interest. Assay may be performed in a manner similar to that described earlier. A 96-well format may be used for higher throughput. In the equilibrium solubility method the concentration of drug increases with time as it dissolves, whereas in the kinetic solubility method the concentration of drug decreases with time as it precipitates. The kinetic solubility method sometimes results in an overestimate of the equilibrium solubility. Sometimes the measured kinetic solubility results in an overestimate of the equilibrium (thermodynamic) solubility.

For oral drugs, what level of water solubility is required? Dressman suggested that "compounds with aqueous solubility lower than 100 µg/mL often present dissolution limitations to absorption" [7]. The required solubility is dependent upon the dose, as discussed in Section 2.1. As described by Lipinski [8,9], a low dose (0.1 mg/kg) of a drug with high, medium or low permeability requires solubilities of 1, 5, and 20 µg/mL, respectively. For a medium dose of 1.0 mg/kg the corresponding required solubilities are 10, 50, and 210 µg/mL, respectively, and for a high dose of 10 mg/kg they are 100, 520, and 2,100 µg/mL, respectively.

Table 1 Media for aqueous solubility measurements

Water
Water pH 2 (0.01 M HCl)
Water pH 2 (0.01 M sodium phosphate)
Water pH 7 (0.01 M sodium phosphate)
Simulated intestinal fluid — fasted
Simulated intestinal fluid — fed
Simulated intestinal bile lecithin mixture — fed SIBLM [6]

Table 2   Chemical structures of bile salts and lecithin

Bile salt	Chemical structure
Glycocholic acid	
Glycodeoxycholic acid	
Glycochenodeoxycholic acid	
Taurocholic acid	
Taurodeoxycholic acid	

Table 2 *(Continued)*

Bile salt	Chemical structure
Taurochenodeoxycholic acid	
Lecithin	

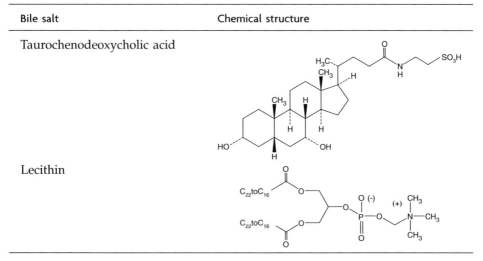

Table 3 Quantitative composition of aqueous-based simulated intestinal fluids

Simulated intestinal fluids	Bile salt		Lecithin (mM)	pH
	Type	Concentration (mM)		
SIF fasted [7]	Sodium taurocholate	3	0.75	6.5[a]
SIF fed [7]	Sodium taurocholate	15	3.75	5.0[b]
SIBLM	Sodium glycocholate	12		
(Simulated	Sodium glycodeoxycholate	6		
intestinal	Sodium glycochenodeoxycholate	12	11	6.5[a]
bile lecithin	Sodium taurodeoxycholate	4		
mixture) [6]	Sodium taurocholate	2		
	Sodium taurochenodeoxycholate	4		

[a]Water buffered with 10 mM sodium phosphate pH 6.5, ionic strength adjusted to 0.15 M with NaCl.
[b]Water buffered with 10 mM sodium phosphate pH 5.0, ionic strength adjusted to 0.15 M with NaCl.

Lin [10] suggests a minimum water solubility of 10 µg/mL for oral absorption, along with four other criteria:

(1) Permeability in Caco-2 cells $> 1 \times 10^{-7}$ cm/sec,
(2) Intestinal efflux Pgp (B to A)/(A to B) ratio < 2 (applied with caution since many drugs are Pgp substrates and have oral bioavailability $> 30\%$),
(3) Metabolism in liver microsomes (1 mg protein/mL) at 60 min $> 20\%$ remaining, and
(4) Polar surface area (sum of van der Waals surface area of oxygens and nitrogens) < 120 Å, and preferably < 60 Å.

2.3 Chemical stability

A successful commercial product requires a suitable shelf life typically at least 90% stability over two years of storage. The concept of solution and solid-state stabilization is beyond the scope of this chapter, but the formulation scientist and the medicinal chemist work together to eliminate any molecules that have obvious stability limitations. An excellent book is available that focuses on the topic of chemical stability of pharmaceuticals [11].

A common initial solution stability screen in drug discovery is to measure the concentration of the chemical in water at 40°C over time at pH 2 and pH 7. The next level of investigation would be to conduct the entire solution pH-rate profile to determine the pH of maximum chemical stability and any acid- or base-catalyzed decomposition pathways. Solid-state chemical stability can also be studied early in a discovery program, and the usual method is to expose the solid drug to 25°C/60% relative humidity (RH) and 40°C/75% RH. The drug may also be exposed to 25°C and 40°C while enclosed in containers with a desiccant to isolate the effect of temperature on stability from that of humidity. A stability-specific analytical method such as HPLC is preferred at this stage in order to accurately quantitate the loss of drug and simultaneously observe the growth of the degradation products.

Chemical instability, such as a half-life under 30 min upon storage at pH 2–7 and 40°C, can result in development issues. Instability at low pH can result in chemical degradation in the stomach, and instability at pH 4–7 can result in chemical degradation in the intestine. Molecules that are unstable in acid can potentially be successfully formulated using enteric coating with a carboxylic acid-based polymer that dissolves in the intestine where the pH is above 4. Instability at neutral pH can be more challenging not only for the performance of *in vitro* biological assays, but also because the *in vivo* transit time in the small intestine is up to 4–5 h; such a limitation is difficult to overcome through formulation approaches. In the early stages of drug discovery, chemical stability is measured in order to rank new chemical entities and also to allow the formulation scientist to properly prepare formulations for preclinical *in vivo* studies.

2.4 Pro-drug potential

A pro-drug is a molecule that is converted *in vivo* to the active drug molecule. The pro-drug approach in drug discovery can be quite rewarding, but has unique challenges as described in two well-written recent books [12,13] and a book chapter [14]. While more detailed discussion is beyond the scope of this chapter, the formulation scientist with the proper training can assist the medicinal chemist in identifying structural analogs as potential pro-drugs.

A pro-drug approach can provide improvements in solubility (usually in water, but occasionally in lipids), permeability, oral bioavailability, or stability. Obviously the active drug must contain a structural motif that enables the synthesis of pro-drugs; thus, given the choice, the medicinal chemist is best advised to incorporate functional groups that can be converted into a pro-drug. Examples include a carboxylic acid or alcohol for conversion to an ester (enalapril), an alcohol to a phosphate ester (fosamprenavir), a phosphonic acid to a phosphonate ester (tenofovir disoproxil), a sulfonamide to an acylsulfonamide (parecoxib sodium), an amine to an amide or carbamate, and a hydantoin to a phosphonooxymethyl derivative (fosphenytoin).

2.5 Final form

The solid-state form of the drug can be either amorphous or crystalline. An amorphous solid is a higher energy solid than a crystalline solid and thus will have a higher solubility. An amorphous solid does not exhibit a melting point but rather displays a glass transition temperature, at which it is seen to collapse and lose volume or even become an oil. The glass transition temperature of an amorphous solid is usually lower than the melting point of a crystalline form of that substance. Crystalline solids can exist in different polymorphic forms, and the crystalline form with the highest melting point is often but not always the most thermodynamically stable form at room temperature.

In the drug discovery setting new chemical entities are normally isolated as solids by crystallization, precipitation, solvent evaporation, or lyophilization (freeze-drying). The solid can be a free base, free acid, or a salt form depending on the presence or absence of a suitable salt-forming counterion during isolation; for example, where the material has been purified by reverse-phase chromatography with trifluoroacetic acid in the mobile phase it is common to isolate weak bases as the trifluoroacetate salt. If crystallized, the polymorph formed may or may not be the most thermodynamically stable. If the solid is obtained by either solvent evaporation or (particularly) lyophilization it will likely be amorphous.

In the early stages of a drug discovery project while still screening many new chemical entities, an amorphous solid is acceptable. The further along the discovery process the greater is the need to identify and use the most thermodynamically stable form. An exception can be for conducting toxicology studies, where the greater solubility of an amorphous solid can be helpful in achieving the high exposures desired. In general, the role of the medicinal chemist in exploring crystalline forms of drugs is somewhat unclear; the task of identifying the final form of the drug to be entered into human clinical trials is commonly the purview of fellow chemists in process chemistry (large scale synthesis).

2.6 Pharmaceutical salts

Common counterions for the formation of pharmaceutically acceptable salts of both acids and bases are listed in Table 4 [15].

Table 4   Common pharmaceutical salt forming anions and cations [15]

Anions	Cations
HCl	Sodium
HBr	Potassium
Sulfate	Calcium
Nitrate	Magnesium
Phosphate	Triethyamine
Mesylate	Ethanolamine
Esylate	Triethanolamine
Isethionate	Meglumine
Tosylate	Choline
Napsylate	Arginine
Besylate	Lysine
Acetate	
Maleate	
Fumarate	
Citrate	
Succinate	

3. SOLUBILITY THEORY

Solubility involves the free energy of the solid and the free energy of the molecules in solution [16]. The free energy of a specific solid is fixed (i.e., a property of that solid), but the free energy of the molecules in solution is a function of the solvent and the solution concentration. When the solution free energy is less than the solid free energy molecules will dissolve from the solid until the free energy of the molecules in solution equals the free energy of the solid. An increase in solubility at constant temperature and pressure can occur by increasing the free energy of the solid, either by chemical means such as varying the salt form, or by physical means such as creating an amorphous solid, polymorphs, or particle size reduction (i.e., micronization or nanoparticle formation). The more practical effect of particle size reduction is to increase the rate of dissolution. The increase in equilibrium solubility via solid-state alteration is only maintained if the solid phase at equilibrium remains the same as the initial solid. Thus, solubility manipulation via solid-state properties is inherently more difficult to achieve and to reproducibly control than is alteration of the solution properties.

A more common and controlled means to increase solubility by the appropriate choice of solubilizing excipient(s). Chemical potential is the incremental increase in the free energy of the solution per incremental increase in the number of molecules in solution (Equation (2)). Excipients that solubilize a molecule via bulk solution properties provide an environment that requires a higher solution concentration (i.e., solubility) to reach a solution chemical potential that matches the chemical potential of the solid (Equation (3)).

$$\mu_{solution} = \left(\frac{dG_{solution}}{dN_{solution}}\right) T, P \tag{2}$$

$$\mu_{solid} = \mu_{solution} \tag{3}$$

If the molecule is ionizable, then *pH adjustment* can be utilized to increase water solubility since the ionized molecular species has higher water solubility than its neutral counterpart. Equations (4) and (5) show that the total solubility, S_T, is a function of the intrinsic solubility of the neutral species, S_o, and the difference between the molecule's pK_a and the solution pH. For every pH unit away from the pK_a, the weak acid/base solubility increases 10-fold. Thus, solubility enhancements more than 1,000 times the intrinsic solubility can be achieved as long as the formulation pH is at least 3 units away from the pK_a.

$$\text{For a weak acid} \quad S_T = So \, (1 + 10^{pH-pKa}) \tag{4}$$

$$\text{For a weak base} \quad S_T = So \, (1 + 10^{pKa-pH}) \tag{5}$$

Cosolvents are solutions of organic solvent(s) and water and are often used to solubilize water-insoluble drugs, particularly those with no ionizable groups. Solubility in a cosolvent mixture typically increases logarithmically with a linear increase in the fraction of organic solvent(s) as shown in Equation (6)

$$\log S_m = \log S_w + f_\sigma \tag{6}$$

where S_m is the total solubility in the cosolvent mixture, S_w the solubility in water, f the fraction of organic solvent in the cosolvent mixture, and the parameter σ the slope of the plot of log S_m versus f, and can be used as a measure of the solubilization potential of a given cosolvent.

Complexation between a ligand and a complexing agent can increase the ligand's solubility if both the ligand and complexing agent have the proper size, lipophilicity, and charge that allow for favorable solubility-enhancing non-covalent interactions. If the ligand and complexing agent combine to form a 1:1 complex, the total ligand solubility, S_T, can be described by Equation (7):

$$S_T = S_w + K_{11}S_oL_T/(1 + K_{11}S_o) \tag{7}$$

where S_w is the ligand solubility in water, K_{11} the formation constant of the 1:1 complex, and L_T the total concentration of the complexing agent. Thus the total ligand solubility is a linear function of the concentration of the complexing agent. Using cyclodextrins as complexing agents, solubility enhancements as high as 10^4–10^5 can be achieved.

Emulsions are a mixture of water, oil, surfactant, and other excipients. If a poorly water-soluble molecule is soluble in oil, then it can be solubilized in an emulsion where it partitions into the oil phase. If the solubility in water is S_w and the solubility in oil is S_{oil}, the total solubility in an emulsion, S_{Te}, is

$$S_{Te} = (1 - f_{oil})S_w + (S_{oil})(f_{oil}) \tag{8}$$

where f_{oil} is the fraction of the emulsion composed of oil.

4. FORMULATION

4.1 Preclinical pharmacokinetic screening

Several formulation options exist for exploratory pharmacokinetic studies. Table 5 lists typical designs for preclinical studies in different species.

In such studies the dose of a drug is usually administered on a mg/kg basis (mg of drug per kg-body weight of the animal). Thus to accommodate different body weights, formulations for which the dose is readily adjusted, such as solutions or suspensions, are preferred. Selections of specific formulations are based on initial data on aqueous solubility and chemical stability. If the drug is water-soluble and chemically stable at a suitable pH, then the formulation can be aqueous-based. However, many drugs have limited water solubility and require extra means to generate a solution. Pharmaceutically acceptable excipients, as listed in Table 6, are commonly used for this purpose. Table 7 lists the chemical structure of various solubilizing solvents. Flow charts of the suggested order of solubilization approaches for injectable formulations are outlined in Table 8.

4.2 Intravenous formulations

An intravenous formulation for practical purposes must be a solution. The ideal intravenous formulation is an aqueous solution that is isotonic with plasma. Typical intravenous fluids include saline, which is water with 0.9% sodium chloride (9 mg/mL), and dextrose 5% water (D5W), which is water with 50 mg/mL dextrose. The range of pH for intravenous solutions is 3–11, and the aim is to minimize buffer capacity, especially at the extremes of pH. The ideal is phosphate-buffered saline (PBS), which is water with 0.01 M sodium phosphate, pH 7.4 and made isotonic with sodium chloride ($\sim$0.14 M). Sterility is achieved in preclinical injectable formulations by filtration through a sterile 0.2 μm filter

Table 5 Preclinical pharmacokinetic study design

Species	Mass (kg)	Route of administration	Maximum volume	Dose (mg/kg)
Dog	10	IV bolus	1 mL/kg	1
		IV infusion	2 mL/kg	1
		Oral	5 mL/kg	5
Monkey	3–5	IV bolus	1 mL/kg	1
		IV infusion	2 mL/kg	1
		Oral	5 mL/kg	5
Rat	0.3–0.5	IV bolus	2 mL/kg	1
		IV infusion	5 mL/kg	1
		Oral	10 mL/kg	10
Mouse	$\sim$0.05	IV bolus	0.2 mL	1
		Intraperitoneal	0.3 mL	2
		Oral	0.5 mL	10

Table 6 Solubilizing excipients used in commercially available solubilized oral and injectable formulations (adapted from Refs [16] and [17])

Water-soluble	Water-insoluble	Surfactants
Solvents	Beeswax	Polyoxyl 35 castor oil (Cremophor EL)
Dimethylacetamide (DMA)	Corn oil mono-, di-, tri-glycerides	Polyoxyl 40 hydrogenated castor oil (Cremophor RH40)
Dimethylsulfoxide (DMSO)	Glyceryl monolinoleate (Maisine™ 35-1)	Polyoxyl 60 hydrogenated castor oil (Cremophor RH60)
Ethanol	Glyceryl monooleate (Peceol™)	Polysorbate 20 (Tween 20)
Glycerin	Glycerol esters of fatty acids (Gelucire® 39/01)	Polysorbate 80 (Tween 80)
N-methyl-2-pyrrolidone (NMP)	Medium chain monoglycerides (Capmul MCM)	d-α-tocopheryl polyethylene glycol 1000 succinate (TPGS)
PEG 300	Medium chain diglycerides	Solutol HS-15
PEG 400	Oleic acid	Sorbitan monooleate (Span 20)
Poloxamer 407 (solid)	PEG 300 caprylic/capric propylene glycol	Sorbitan monooleate (Span 80)
Propylene glycol	diesters (Captex 200)	PEG 300 caprylic/capric glycerides (Softigen® 767)
	Polyglyceryl oleate (Plurol® Oleique)	PEG 400 caprylic/capric glycerides (Labrasol®)
Cyclodextrin	Propylene glycol monolaurate (Lauroglycol™	PEG 300 oleic glycerides (Labrafil® M-1944CS)
Hydroxypropyl-β-cyclodextrin	FCC)	PEG 300 linoleic glycerides (Labrafil® M-2125CS)
Sulfobutylether-β-cyclodextrin	Propylene glycol dicaprylocaprate	PEG 1500 lauric glycerides (Gelucire® 44/14)
(Captisol®)	(Labrafac™ PG)	PEG 1500 stearic glycerides (Gelucire® 50/13)
α-Cyclodextrin	Soy fatty acids	Polyoxyl 8 stearate (PEG 400 monostearate)
	d-α-tocopherol (Vitamin E)	Polyoxyl 40 stearate (PEG 1750 monostearate)
Phospholipids		
Hydrogenated soy	*Long-chain triglycerides*	
phosphatidylcholine (HSPC)	Castor Oil, Corn oil	
Distearoylphosphatidyl-	Cottonseed oil, Olive oil	
glycerol (DSPG)	Peanut oil, Peppermint oil	
L-α-dimyristoylphos-phatidylcholine	Safflower oil, Sesame oil	
(DMPC)	Soybean oil, Hydrogenated soybean oil	
L-α-dimyristoylphos-phatidylglycerol	Hydrogenated vegetable oils	
(DMPG)		
	Medium-chain triglycerides	
	Caprylic/capric triglycerides derived from	
	coconut oil or palm seed oil	

Table 7 Chemical structures of selected solubilizing excipients [16]

Excipient name or common name(s)	Chemical structure
Sorbitan monooleate, Span 80, MW = 428	
Polyoxyethylene 20 sorbitan monooleate, Polysorbate 80, Tween 80, MW = 1310	
d-α tocopheryl polyethylene glycol 1000 succinate, Vitamin E TPGS, MW ~1513	
Solutol® HS 15 (polyethyleneglycol 660 12 hydroxystearate)[a]	

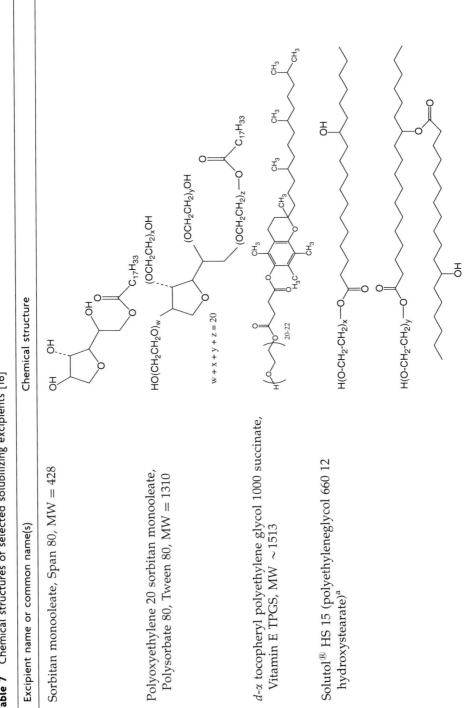

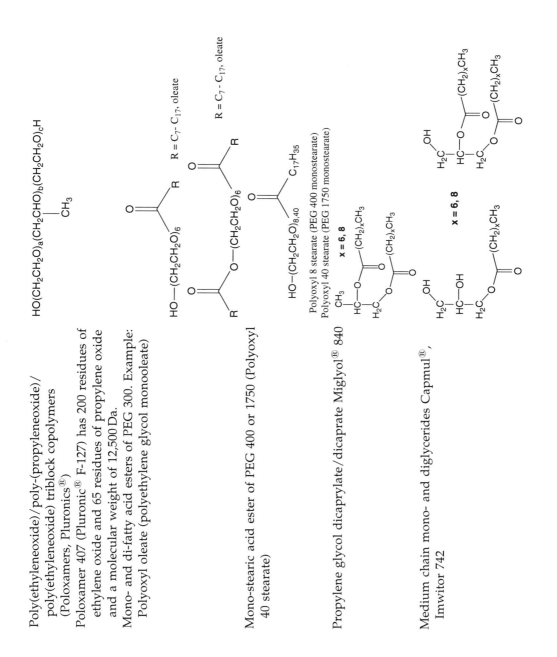

Poly(ethyleneoxide)/poly-(propyleneoxide)/
poly(ethyleneoxide) triblock copolymers
(Poloxamers, Pluronics®)

$HO(CH_2CH_2O)_a(CH_2CHO)_b(CH_2CH_2O)_cH$
$|$
CH_3

Poloxamer 407 (Pluronic® F-127) has 200 residues of
ethylene oxide and 65 residues of propylene oxide
and a molecular weight of 12,500 Da.

Mono- and di-fatty acid esters of PEG 300. Example:
Polyoxyl oleate (polyethylene glycol monooleate)

$R = C_7 - C_{17}$, oleate

$R = C_7 - C_{17}$, oleate

Mono-stearic acid ester of PEG 400 or 1750 (Polyoxyl
40 stearate)

$HO-(CH_2CH_2O)_{8,40}$ $C_{17}H_{35}$

Polyoxyl 8 stearate (PEG 400 monostearate)
Polyoxyl 40 stearate (PEG 1750 monostearate)

Propylene glycol dicaprylate/dicaprate Miglyol® 840

x = 6, 8

Medium chain mono- and diglycerides Capmul®,
Imwitor 742

x = 6, 8

Table 7 *(Continued)*

Excipient name or common name(s)	Chemical structure
Medium chain triglycerides (Caprylic and capric triglycerides) Labrafac® Miglyol® 810, 812 Crodamol GTCC-PN Softison 378 (x = 10)	
Long chain monoglycerides Glyceryl monooleate (Peceol®) Glyceryl monolinoleate (Maisine®)	
Polyoxyethylene castor oil derivatives, Cremophor EL and Cremophor RH40[b] Complex mixture of 75%–83% relatively hydrophobic molecules, and 17%–25% relatively hydrophilic molecules (polyethylene glycol and glycerol ethoxylates)	

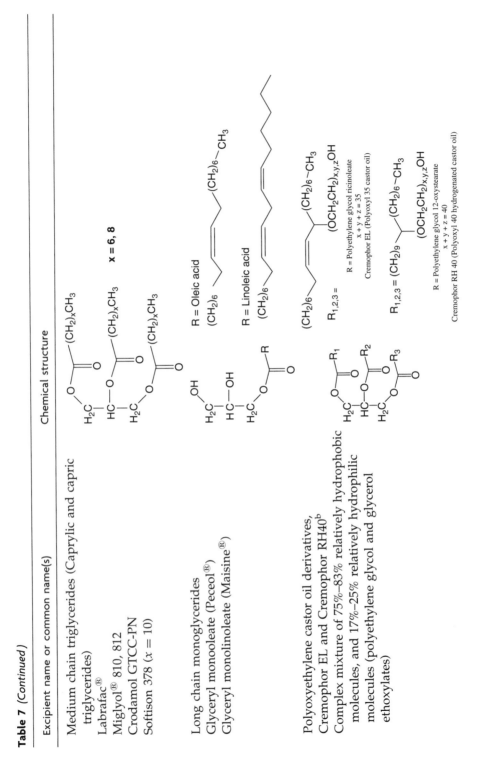

Mono-, di- and triglycerides and mono- and di-fatty acid esters of PEG (also contains glycerol and PEG)

Example	x	R
Softigen® 767	6	C_8, C_{10}
Labrasol®[c]	8	C_8, C_{10}
Labrafil® M-1944CS[d]	6	$C_{18:1}$
Labrafil® M-2125CS[e]	6	$C_{18:2}$
Gelucire® 44/14[f]	32	C_{12}, C_{14}

Phospholipids

Example	R	R'
DSPG	C_{18}	Glycerol
DMPC	C_{14}	Choline
DMPG	C_{14}	Glycerol

Hydroxypropyl-β-cyclodextrin[g] average degree of substitution:
4 (Encapsin™)
8 (Molecusol®)

$R = H$ or CH_2CHCH_3

Table 7 *(Continued)*

Excipient name or common name(s)	Chemical structure
Sulfobutylether-β-cyclodextrin[g], average degree of substitution: 6.5 MW = 2163, Captisol®	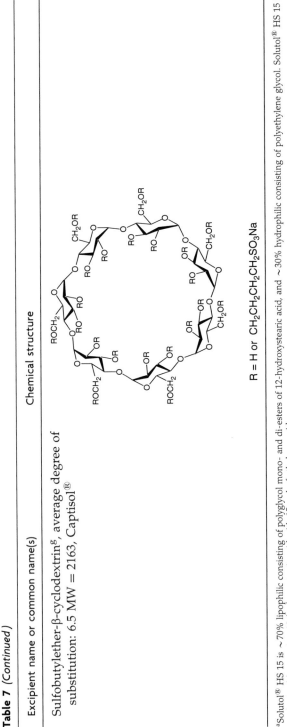 R = H or $CH_2CH_2CH_2CH_2SO_3Na$

[a]Solutol® HS 15 is ~70% lipophilic consisting of polyglycol mono- and di-esters of 12-hydroxystearic acid, and ~30% hydrophilic consisting of polyethylene glycol. Solutol® HS 15 is synthesized by reacting 12-hydroxystearic acid with 15 mol of ethylene oxide.

[b]Cremophors are complex mixtures of various hydrophobic and hydrophilic components. Cremophor EL is obtained by reacting 35 mol of ethyleneoxide with 1 mol of castor oil, and comprises about 83% hydrophobic constituents of which the main component is glycerol polyethylene glycol ricinoleate. Cremophor RH 40 is obtained by reacting 40 mol of ethyleneoxide with 1 mol of hydrogenated castor oil, and comprises about 75% hydrophobic constituents of which the main component is glycerol polyethylene glycol 12-hydroxystearate.

[c]Labrasol® is a mixture of mono-, di-, and triglycerides and mono- and di-fatty acid esters of PEG 400. Labrasol® is synthesized by an alcoholysis/esterification reaction using medium chain triglycerides from coconut oil and PEG 400, and the main fatty acid is caprylic/capric acids.

[d]Labrafil® M-1944 CS is a mixture of mono-, di-, and triglycerides and mono- and di-fatty acid esters of PEG 300. Labrafil® M-1944 CS is synthesized by an alcoholysis/esterification reaction using apricot kernel oil and PEG 300, and the main fatty acid is oleic acid (58–80%).

[e]Labrafil® M-2125 CS is a mixture of mono-, di-, and triglycerides and mono- and di-fatty acid esters of PEG 300. Labrafil® M-2125 CS is synthesized by an alcoholysis/esterification reaction using corn oil and PEG 300, and the main fatty acid is linoleic acid (50–65%).

[f]Gelucire® 44/14 is a mixture of mono-, di-, and triglycerides and mono- and di-fatty acid esters of PEG 1500. Gelucire® 44/14 is synthesized by an alcoholysis/esterification reaction using palm kernel oil and PEG 1500, and the main fatty acid is lauric acid.

[g]It is unclear on which of the three hydroxyls (primary 6, or the secondary 2' or 3') substitution occurs.

Table 8 Flow charts of suggested order of solubilization approaches for injectable formulations [16]

Intravenous	Subcutaneous	Intramuscular
Aqueous isotonic solution	Aqueous isotonic solution	Aqueous isotonic solution
↓	↓	↓
Aqueous, pH 3–11	Aqueous, pH 3–11	Aqueous, pH 3–11
↓	↓	↓
Cosolvent (organic solvent ≤ 60%)	Cosolvent (organic solvent ≤ 15%)	Cosolvent (organic solvent ≤ 100%)
↓	↓	↓
Cosolvent and pH 3–11	Cosolvent and pH 3–9	Cosolvent and pH 3–11
↓	↓	↓
Aqueous with complexation	Aqueous with complexation	Aqueous with complexation
↓	↓	↓
		Organic solvent (100%)
		↓
Organic solvent/ surfactant for dilution with an aqueous diluent	Organic solvent/ surfactant for dilution with an aqueous diluent	Organic solvent/surfactant for dilution with an aqueous diluent
↓	↓	↓
Oil-in-water emulsion	Oil-in-water emulsion	Oil-in-water emulsion
↓	↓	↓
Liposome	Organic solvent up to 100% but minimize volume	Oil (extended release)

using syringes and glassware that have been sterilized by autoclaving (heating to 120°C for 30 min).

Intravenous formulations that have proven particularly useful in the author's experience for formulation of exploratory compounds with limited aqueous solubility are listed in Table 9.

4.2.1 pH adjustment

An ionizable functional group in a drug is very useful for the formulation scientist and allows for pH adjustment to increase water solubility (Section 3). Tables 10 and 11 list such weakly acidic and weakly basic groups, and Table 12 illustrates examples of their combination in zwitterionic drugs that are commercially available as intravenous solutions [19]. Weak acids can be formulated in their anionic form at pH values above their respective pK_a, and weak bases can be formulated in their cationic form at pH values below their respective pK_a. Zwitterions can be formulated in either their cationic or anionic form. Table 13 lists the buffers that are used in commercially available intravenous formulations [19].

Table 9 Intravenous formulations for preclinical studies of poorly water-soluble molecules

Formulation	Species	Route	Volume (mL/kg)
60% PEG 400, 40% water pH 3–11 [13]	Rat	IV$_{bolus}$	2
	Rat	IV$_{infusion}$	5
	Dog	IV$_{bolus}$	1
	Dog	IV$_{infusion}$	2
80% PEG 400, 20% water pH 3–11	Dog	IV$_{infusion}$	0.5
50% PEG 400, 50% DMA	Dog	IV$_{infusion}$	0.1
40% PG, 60% Water pH 3–11 (50% PG	Rat	IV$_{bolus}$	2
causes hemolysis)	Rat	IV$_{infusion}$	5
	Dog	IV$_{bolus}$	1
	Dog	IV$_{infusion}$	2
5% Ethanol, 45% PEG 400, 50% water pH	Rat	IV$_{bolus}$	2
3–11 (ethanol < 10% due to risk of	Rat	IV$_{infusion}$	5
intoxication)	Dog	IV$_{bolus}$	1
	Dog	IV$_{infusion}$	2
5% Ethanol, 40% PG, 55% water pH 3–11	Rat	IV$_{bolus}$	2
(ethanol < 10% due to risk of	Rat	IV$_{infusion}$	5
intoxication)	Dog	IV$_{bolus}$	1
	Dog	IV$_{infusion}$	2
20% Ethanol, 60% PG, 20% water [21]	Dog	IV$_{bolus}$	0.5
5% Ethanol, 20% PG, 75% PEG 400	Rat	IV$_{bolus}$	1.0
5% Ethanol, 20% PG, 40% PEG 400, 35% water, pH 3	Dog	IV$_{infusion}$	2.5
5% Solutol HS-15, 20% DMSO, 75% water pH 3–11 [20]	Rat	IV$_{bolus}$	2.0
12% Captisol®, 87% water, pH 3–11	Rat	IV$_{bolus}$	2
	Rat	IV$_{infusion}$	5
	Dog	IV$_{bolus}$	1
	Dog	IV$_{infusion}$	2
5% Ethanol, 7% Captisol®, 45% PEG 400, 43% water, pH 3	Dog	IV$_{infusion}$	1.0

In practice, when formulating for a preclinical pharmacokinetic study using a low pH it is typical to use hydrochloric acid and at high pH sodium hydroxide. For example, 1 equivalent of hydrochloric acid is added to lower the pH to ∼2 units below the pK_a of a weakly basic drug, with more added to lower the pH to as low as ∼3 if necessary. For a weakly acidic compound the upper limit of pH is ∼11. Such examples of in-situ salt formation have minimal buffer capacity, which minimizes pain upon injection. Also, the slower the injection the more biocompatible the result. Methanesulfonic acid is another commonly used acid for in-situ salt formation.

Table 10 Examples of weak acid chemical functional groups, their approximate pK$_a$'s and formulation pH's [19]

Functional group name	Functional group structure	Functional group pK$_a$	Formulation pH	Selected examples
Sulfonic acid		<1	Neutral	Aztreonam
Phosphate ester		2	Neutral	Fosphenytoin Bethamethasone Dexamethasone Fludarapine
Carboxylic acid		2.5–5	5–8	Penicillin Ketorolac
4-Hydroxy coumarin		~8	8.3	Warfarin
Uracil		~8	9.2	Flurouracil
Sulfonamide		7–9	9–11.6	Acetazolamide Clorothiazide Diazoxide

Table 10 (*Continued*)

Functional group name	Functional group structure	Functional group pKa	Formulation pH	Selected examples
Barbituric acid		7–9	9.5–11	Methohexital Pentobarbital Phenobarbital Secobarbital
Guanine		2.2 (basic), 9.4 (acidic)	11	Acyclovir Gancyclovir
Hydantoin		~10	10–12	Phenytoin
Phenol		8–10	10.5 emulsion organic organic	Liothyronine Propofol Etoposide Teniposide

Table 11 Examples of weak base chemical functional groups, their approximate pK_a's and formulation pH's [19]

Functional group name	Functional group structure	Functional group pK_a	Formulation pH	Selected examples
1H-Imidazole	(chemical structure)	~4–7	< 6	Miconazole Ondansetron
Pyridine	(chemical structure)	~5	2–4	Amronone Milrinone Papaverine Pyridoxine
Aniline	(chemical structure)	~5	2–6	Metoclopramide Minocycline (Procaine Procainamide also have a tertiary amine)
4,5-Imidazoline	(chemical structure)	~6	3–4	Tolazoline
Amine	(chemical structure)	7–10	3–7	Atenolol Codeine Daunorubicin Morphine Verapamil

Table 11 (*Continued*)

Functional group name	Functional group structure	Functional group pK_a	Formulation pH	Selected examples
N-Alky morpholine		7.4	< 5	Doxapram
Imidazole		~7	3–6.5	Cimetidine Dacarbazine Phentolamine
Amidine		~9–11	< 8	Pentamidine

Table 12 Examples of zwitterionic drugs, approximate pK_a's and formulation pH's [19]

Selected example	Chemical structure	Acidic functional group name and pK_a	Basic functional group name and pK_a	Formulation pH (ionic state)
Ciprofloxacin		Carboxylic acid ~4	Aniline ~4 Amine ~9	3–4 (Cationic)
Sufentanil		Carboxylic acid ~4	Amine ~8	3.5–6 (Cationic)
Ampicillin		Carboxylic acid ~4	Amine ~8	8–10 (Anionic)
Cephapirin		Carboxylic acid ~3	Pyridine ~5	6–8 (Anionic)

Table 13   List of buffers used in commercial parenteral formulations [19]

pH	Buffer (pK_a's)	Concentration in formulation, molarity	Concentration administered, molarity	Route of administration
2.5–4.0	Tartaric acid (2.9, 4.2)	0.04	0.04	IM, IV
3	Maleic acid (1.9, 6.2)	0.14	0.14	IM
3	Glycine (2.3, 9.6)	0.2	0.05	IV infusion
3.0–4.5	Sodium lactate/ Lactic acid (3.8)	0.17 0.02	0.17 0.085 0.02	IV IV infusion, SC
3–5	Ascorbic acid (4.2, 11.6)	0.02	0.01 0.02	IM IV
3.0–7	Sodium citrates/ Citric acid (3.1, 4.8, 6.4)	0.6 0.1	0.6 0.1	IM, IV, IV infusion, SC
4–6	Sodium acetate/Acetic acid (4.75)	0.01	0.01	IV, SC
4–6.5	Sodium bicarbonate/ Carbonic acid (6.3, 10.3)	0.08	0.08 0.001	IM, IV, IV Infusion
4.2–6	Sodium succinate/ Succinic acid (4.2, 5.6)	0.04	0.005 0.04	IV infusion, SC
6	Histidine (1.8, 6.0, 9.2)	0.005 0.05	0.0005 0.05	IV infusion IM
6–7	Sodium benzoate/ Benzoic acid (4.2)	0.5	0.5	IV
3–8	Sodium phosphates (2.2, 7.2, 12.4)	0.08	0.08	IV, IV infusion IM
7.4–9.0	Tris(hydroxy-methyl)amino-methane (8.3)	0.01	0.01	IM, IV infusion Intra-arterially, Intrathecal
8.7–11	Sodium bicarbonate/ Sodium carbonate (6.3, 10.3)	0.01	0.01	IV, IV infusion, Intravitreal (Fomiversen)

Note: IM, intramuscular; IV, intravenous; SC, subcutaneous.

4.2.2 Cosolvents

In using an organic solvent, considerations include the nature and proportion of the specific solvent and the total amount administered. The philosophy and practice is to minimize the amount of organic solvent used to achieve one's goal. Pharmaceutically acceptable solvents include ethanol, propylene glycol (PG), PEG 400, DMSO, DMA, and NMP [17]. In intravenous formulations intended for bolus injection at 1–2 mL/kg, some general guidelines are that the amount of water should be ≥ 40%, the amount of ethanol ≤ 10%, the amount of PG ≤ 40%, the amount of PEG 400 ≤ 60%, and the amount of DMSO ≤ 20%. PEG 400 is more biocompatible than PG, which can cause hemolysis at the site of injection (particularly if the proportion of PG is ≥ 50%). Ethanol is commonly combined with PEG 400 (or PG) both for increased solubilization capacity and to reduce the viscosity.

Formulations that are entirely organic can have strong solubilizing capacity and can be administered by intravenous bolus injection if the dose volume is reduced to 0.5–1 mL/kg. The concern is precipitation upon injection, and thus a slow bolus is used in practice.

Intravenous infusions are normally administered over 15–30 min in volumes that are larger than intravenous bolus injections. Typical volumes for intravenous infusions are 2–5 mL/kg. Any intravenous infusion formulation can be administered by intravenous bolus since the volume of intravenous infusions is much higher than a bolus injection (<2 mL/kg). The % of organic solvent in an intravenous bolus can be higher than an intravenous infusion formulation, but the limitation is dose volume and the total amount of excipient administered intravenously.

A very powerful formulation approach is to generate a supersaturated solution that stays in solution long enough for handling and dose administration. The surfactant Solutol HS-15 was specifically designed for this purpose as an aid for intravenous administration. Thus, for a formulation of 5% Solutol HS-15, 20% DMSO, and 75% water (pH 3–11), the drug is first dissolved in DMSO at a five-fold higher concentration than ultimately required. This solution is then added slowly with stirring (minimizing local high concentrations to minimize the risk of precipitation) to an aqueous Solutol HS-15 solution, followed by sterile filtration [20]. The length of time to precipitation is dependent upon on the drug, but may be many hours.

4.2.3 Complexation

A very effective and safe complexing agent is the sulfobutylether-β-cyclodextrin Captisol® (see Table 7). Captisol® is used in at least two commercially available injectable formulations: Vfend® (intravenous infusion) and Geodeon® (intramuscular injection). Captisol® forms a non-covalent complex with molecules that fit inside the cyclodextrin cavity. It contains negatively charged sulfates and is highly water-soluble. Water with 12% Captisol® is isotonic and can be administered by intravenous bolus injection at 1 mL/kg or by infusion at 2–5 mL/kg. The pH of the formulation can be adjusted to 3–11 to take advantage of any pH-dependent complexation that the drug may exhibit.

4.3 Oral formulations

Poor oral bioavailability may be due to many factors, including chemical stability (pH 2–7), dissolution rate, solubility, permeability, efflux, and metabolism. Formulation modifications can affect oral bioavailability if the limitations are chemical stability, dissolution, or solubility. They can also overcome efflux limitations if the efflux transport process can be saturated at higher drug concentrations and the modification allows for *in vivo* solubilization or supersaturation. However, formulation modifications cannot affect oral bioavailability if the limitation is passive permeability or metabolism (assuming the formulation does not contain excipients that perturb these factors), and these two factors are therefore critical to the successful design of effective oral agents.

The typical means of oral administration of a solution or suspension is by gavage, which uses a blunt-tip syringe. Depending on the project needs, an oral formulation can be a bulk solution, suspension, liquid-filled capsule, solid in a capsule, or a tablet. The choice between a solution or a suspension is context dependent and is a balance between the desire to maximize the probability of absorption and the need for relevance to a (solid) commercial dosage form.

4.3.1 Oral solution

Table 14 contains a flow chart of suggested solubilization approaches of increasing complexity for oral solutions [17]. The typical volume of an oral solution is 1–4 mL/kg in non-rodent species, and 2–10 mL/kg in rodents. If the oral solution is entirely organic the dose volume should be minimized to ≤ 2 mL/kg. In general any intravenous solution can also be administered orally. More complex alternatives include microemulsions, which are thermodynamically stable solutions of water, oil, surfactant, and solvent; or self-emulsifying drug delivery systems (SEDDS) which are microemulsion preconcentrates composed of an oil, surfactant, cosurfactant and solvent(s).

4.3.2 Liquid-filled capsule

Table 14 also contains a flow chart of the suggested order of solubilization approaches for liquid-filled capsules [17]. While beyond the scope of this chapter, oral lipid-based formulations can be extremely useful in formulating and developing poorly water-soluble molecules and have been the topic of many publications [18–21] and an excellent recent book [22]. Capsules are made of either gelatin or hydroxypropyl methylcellulose and are available in a two-piece format for hand filling on a small scale [23]. Volumes that can be filled into a typical capsule are ~1 mL, but larger alternatives with capacities of 3–5 mL are available for administration to dogs. Acceptable solvents are included in Table 6. One concern is the physical compatibility of the solvent with the capsule shell; in general, water-insoluble solvents are compatible, but water-soluble solvents that are liquid at room temperature lead to softening of the shell over a period of time. However in the drug discovery setting, the time required for physical

Table 14 Flow chart of suggested order of solubilization approaches for oral liquid formulations: simple to complex [17]

Capsule	Oral solution
Water-soluble organic solvent	Aqueous, pH 2–10
↓	↓
Long-chain triglyceride	Cosolvent (aqueous/organic solvent), pH 2–10
↓	↓
Medium-chain triglyceride	Organic solvent(s) (100%)
↓	↓
Water-insoluble organic solvent	Aqueous with complexation, pH 2–10
↓	↓
Organic solvents and surfactant	Oil-in-water emulsion
↓	↓
↓	Microemulsion
Triglyceride and surfactant	↓
↓	SEDDS
Microemulsion	↓
↓	SMEDDS
SEDDS (self-emulsifying drug delivery system)	↓
↓	S-SEDDS
SMEDDS (self-microemulsifying drug delivery system)	
↓	
S-SEDDS (supersaturated self-emulsifying drug delivery system)	

compatibility is short, and if necessary the capsules can be filled just prior to administration.

4.3.3 Oral suspension

Oral suspensions need not be aqueous-based and can include any oral solution that has been saturated with excess solid. The typical oral aqueous suspension is composed of water with a surfactant such as 0.1% v/v TWEEN 80 and a suspending agent such as 0.5% w/v carboxymethylcellulose or hydroxypropyl-methylcellulose. The pH may be adjusted with a buffer or with strong acid or base. The challenge in preparing a suspension is to disperse the solid evenly, either gently by means such as manual mixing with a stirrer bar or swirling, or more energetically through the use of a small scale hand-held low-shear homogenizer. Practical concerns with suspensions include ease of transfer by syringe, agglomeration, particle settling, and avoiding/minimizing foam or bubbles. When a very high dose of a drug is desired (such as in toxicology studies), a suspension may be the only choice.

4.3.4 Oral solid

Although the typical ultimate goal is to commercialize a solid oral dosage form, there is commonly not enough drug available at the early stages of drug discovery to make a tablet. Therefore, a typical means to dose a solid in drug discovery uses a "powder in a capsule", in which the solid is hand-mixed with pregelatinized starch in a 1:1 mass ratio and filled into a two-piece capsule. This form of administration is convenient for dogs and monkeys, but suspensions are preferred for rodents.

4.4 Drug safety (Toxicology)

One of the most challenging aspects of preclinical formulation science is to enable the administration of poorly water-soluble drugs for drug safety studies at very high doses so as to achieve maximal plasma exposure. For example, in rodents administration of 10 mL/kg of a 200 mg/mL formulation equates to a dose of 2,000 mg/kg. In non-rodents (e.g., dogs) the highest dose is commonly 100 mg/kg, which can be achieved by administering 5 mL/kg of a 20 mg/mL formulation. To minimize the dose volume, a regimen of 2 mL/kg of a 50 mg/mL formulation is also often used. The challenge in designing the drug safety study is to balance dose volume with biocompatibility and the required concentration of the drug. In longer-term studies such as those lasting six months, the issue of formulation biocompatibility is critical and is one of the practical limitations in study design. An excellent recent review article lists many formulations used in preclinical toxicology studies in various species [24]. Selected examples of these are listed in Table 15. Again, oral suspensions need not be solely aqueous-based but can include any oral solution that has been saturated with excess solid. This approach can be quite successful in achieving the desired dose level and resulting plasma concentration.

5. CONCLUSION

The medicinal chemist and the formulation scientist work together within drug discovery research to identify a drug that has the best chance of clinical and eventual commercial success. The partnership produces structural modifications to improve solubility, stability, permeability, oral bioavailability, and allows for pro-drug strategies to be implemented where necessary. The classical role of the formulation scientist is recognized as preparing formulations for preclinical *in vivo* studies, and the identification of the final solid-state form of the drug, but collaboration at earlier stages of drug discovery increases the speed of preclinical evaluation and reduces attrition in development.

Table 15 Oral formulations used in toxicological and clinical studies

Drug (Company)	Formulation	Species	Route/high dose	Volume administered	Dose frequency	NDA application no.
Amprenavir (GlaxoSmith-Kline)	20/80 Water/PEG 400 (200 mg/mL)	Monkey	Oral gavage	10 mL/kg	Single dose	21007
	Water and PEG 400	Cyno monkey	Oral gavage/200 mg/kg	1–2 mL/kg	b.i.d. 6 h apart for 1 month	
	PEG 400/TPGS/PG	Dog	Oral gavage/350 mg/kg	1–2 mL/kg	12 months	
	100% PEG 400 (100 mg/mL)	Sprague–Dawley rat	Oral gavage	5 and 10 mL/kg	Single dose and 7 days of dosing	21007
	20/80 Water/PEG 400, pH 2 (200 mg/mL)	Sprague–Dawley rat	Oral gavage/1,000 mg/kg	5 mL/kg	b.i.d. 6 h apart for 1 month 32 days	21007, 21039
	PEG 400/TPGS/PG	Sprague–Dawley rat	Oral gavage/375 mg/kg	2–5 mL/kg	b.i.d. for 6 months	21007, 21039
Lopinavir (Abbott)	5/95 Ethanol/PG	Rat	Oral gavage	4 mL/kg/day	Single dose	21226
	5/95 Ethanol/PG	Mice	Intravenous	0.5 mL	Single dose	21226
	5/95 Ethanol/PG 1/12/13/74	Sprague–Dawley rat	Oral gavage	2 ml/kg/day	14 days of dosing	21226
	Water/Cremophor EL/Ethanol/Oleic acid	Beagle dog	Oral gavage	1.25 mL/kg/day	14 and 90 days of dosing	21226
Ritonavir (Abbott)	5/95 Ethanol /PG	Rat	Oral gavage	2 mL/kg/day	1 month	20–659
	5/95 Ethanol /PG	Dog	Oral gavage	2 mL/kg/day	1 month	20–659
	10/27/31/32 PG/Glycerin/Ethanol/others	Juvenile Sprague–Dawley rat	Oral gavage/100/50 mg/kg	2 mL/kg coformulation	4 weeks	21–226, 21–251
		Sprague–Dawley rat		LPV: 1 mL/kg		

Table 15 (*Continued*)

Drug (Company)	Formulation	Species	Route/high dose	Volume administered	Dose frequency	NDA application no.
	LPV: 5/95 Ethanol / PG RTV: 5/50/45 Ethanol /PG/Water Ph 2		Oral gavage/150/ 75 mg/kg	RTV: 1 mL/kg	3 months + 1 month recovery	
	Semisolid: Oleic acid, ethanol, Cremophor EL, BHT	Beagle dog	Oral capsule/60/ 20 mg/kg and 45/15 mg/kg	–	6 months	
Lopinavir (LPV) and Ritonavir (RTV) (Abbott)	Semisolid: 65.5/12.5/12.5/9/0.13 Oleic acid/ethanol/ Cremophor EL/PG/ BHT Semisolid: 73.5/13.2/12.2/1.1/ 0.01	Beagle dog	Oral capsule/30/ 15 mg/kg	1.25 mL/kg	3 months	
	Oleic acid/ ethanol/ Cremophor EL/water/ BHT	Beagle dog	Oral capsule/30/ 15 mg/kg	1.25 mL/kg	3 months	

Compound	Formulation	Species	Route/dose	Dose volume	Schedule/duration	References
	35.6/16.7/15.3/5.9/3.1/1/6.9 Ethanol/corn syrup/PG/Glycerin/Povidone/Cremophor RH40/Water	Beagle dog	Oral capsule/50/25 mg/kg	–	3 months	
Tipranavir (TPV)/and Ritonavir (RTV) (Boeheringer Ingelheim)	TPV: Aqueous solution pH 10.5 RTV: PG	Sprague–Dawley rat	Oral gavage/200 mg/kg	TPV: 2 mL/kg RTV: 2 mL/kg	b.i.d. 8 h apart, 26 weeks+13 week recovery	21–814
	TPV: Aqueous solution, pH 10.5 RTV: PG	Beagle dog	Oral gavage/75/25 mg/kg and 150/40 mg/kg	TPV: 2 mL/kg RTV: 2 mL/kg	RTV followed immediately by TPV then cherry syrup wash 12–14 weeks	21–814
BILN 2061 (Boeheringer Ingelheim)	20/80 Ethanol/PEG 400	Human	Oral gavage	10 mL	b.i.d. for 2 days	[25]

Note: b.i.d., twice-a-day; BHT, butylated hydroxytoluene (anti-oxidant).

REFERENCES

[1] S. Venkatesh and R. A. Lipper, Role of the Development Scientist in Compound Lead Selection and Optimization, *J. Pharm. Sci.*, 2000, **89**, 145.

[2] E. F. G. Fiese, General Pharmaceutics — The New Physical Pharmacy, *J. Pharm. Sci.*, 2003, **92**, 1331.

[3] W. Curatolo, Physical Chemical Properties of Oral Drug Candidates in the Discovery and Exploratory Development Settings, *Pharm. Sci. Technol. Today*, 1998, **1**, 387.

[4] S. Balbach and C. Korn, Pharmaceutical Evaluation of Early Development Candidates "The 100 mg-Approach", *Intl. J. Pharm.*, 2004, **275**, 1.

[5] A. Avdeef, *Absorption and Drug Development: Solubility, Permeability and Charge State*, Wiley, 2003.

[6] V. J. Stella, S. Martodihardjo, K. Terada and V. M. Rao, Some Relationships between the Physical Properties of Various 3-Acyloxymethyl Prodrugs of Phenytoin to Structure: Potential In Vivo Performance Implications, *J. Pharm. Sci.*, 1998, **87**, 1235.

[7] D. Hörter and J. B. Dressman, Influence of Physicochemical Properties on Dissolution of Drugs in the Gastrointestinal Tract, *Adv. Drug Deliv. Rev.*, 2001, **46**, 75.

[8] C. A. Lipinski, Solubility in Water and DMSO: Issues and Potential Solutions, in *"Pharmaceutical Profiling in Pharmaceutical Profiling Drug Discovery for Lead Selection"* (eds R. T. Borchardt, E. H. Kerns, C. A. Lipinski, D. R. Thakker, and B. Wang), AAPS Press, 2004 pp. 93–126; Volume I in Biotechnology: Pharmaceutical Aspects (series eds R.T. Borchardt and C.R. Middaugh).

[9] C. A. Lipinski, F. Lombardo, B. W. Dominy and P. J. Feeney, Experimental and Computational Approaches to Estimate Solubility and Permeability in Drug Discovery and Development Settings, *Adv. Drug Deliv. Rev.*, 1997, **23**, 3.

[10] J. H. Lin, Challenges in Drug Discovery: Lead Optimization and Prediction of Human Pharmacokinetics, in *Pharmaceutical Profiling Drug Discovery for Lead Selection* (eds R. T. Borchardt, E. H. Kerns, C. A. Lipinski, D. R. Thakker, and B. Wang), AAPS Press, 2004 pp. 293–325; Volume I in Biotechnology: Pharmaceutical Aspects (series eds R.T. Borchardt and C.R. Middaugh).

[11] S. Yoshioka and V. J. Stella, *Stability of Drugs and Dosage Forms*, Kluwer Academic, 2000.

[12] V. J. Stella, R. T. Borchardt, M. J. Hageman, R. Oliyai, H. Maag and J. W. Tilley (eds), *Prodrugs: Challenges and Rewards*, Part 1, Volume V in Biotechnology: Pharmaceutical Aspects, 2007 (series eds R. T. Borchardt and C. R. Middaugh).

[13] V. J. Stella, R. T. Borchardt, M. J. Hageman, R. Oliyai, H. Maag and J. W. Tilley (eds), *Prodrugs: Challenges and Rewards*, Part 2, Volume V in Biotechnology: Pharmaceutical Aspects, 2007 (series eds R. T. Borchardt and C. R. Middaugh).

[14] V. J. Stella, Prodrug Strategies for Improving Drug-Like Properties, in *"Optimizing the "Drug-Like" Properties of Leads in Drug Discovery"* (eds R. T. Borchardt, E. H. Kerns, M. J. Hageman, D. R. Thakker, and J. L. Stevens), AAPS Press, 2006 pp. 221–242; Volume IV in Biotechnology: Pharmaceutical Aspects (series eds R.T. Borchardt and C.R. Middaugh).

[15] P. H. Stahl and C. G. Wermuth (eds), *Handbook of Pharmaceutical Salts, Properties, Selection, and Use*, Verlag Helvetica Chimica Acts, Zurich, 2002.

[16] R. G. Strickley, Solubilizing Excipients in Oral and Injectable Formulations, *Pharm. Res.*, 2004, **21**(2), 201–230.

[17] R. G. Strickley and R. Oliyai, Solubilizing Vehicles for Oral Formulation Development in Solvent Systems and Their Selection, in *"Pharmaceutics and Biopharmaceutics"* (eds P. Augustijns and M. E. Brewster), AAPS Press, UK, 2007 pp. 257–308; Volume VI in Biotechnology: Pharmaceutical Aspects (series eds R.T. Borchardt and C.R. Middaugh).

[18] D. L. Pole, Physical and Biological Considerations for the Use of Nonaqueous Solvents in Oral Bioavailability Enhancement, *J. Pharm. Sci.*, 2008, **97**, 1071.

[19] R. G. Strickley, Parenteral Formulations of Small Molecules Therapeutics Marketed in the United States (1999) — Part I., *PDA J. Pharm. Sci. Technol.*, 1999, **53**(6), 324–349.

[20] R. G. Strickley, L. Liu and P. Lapresca, Preclinical Parenteral and Oral Formulations of Water Insoluble Molecules, ISSX Meeting, Nashville, TN (USA), Abstract # 291, 1999.

[21] B. J. Aungst, N. H. Nguyen, N. J. Taylor and D. S. Bindra, Formulation and Food Effect on the Oral Absorption of a Poorly Water Soluble, Highly Permeable Antiretroviral Agent, *J. Pharm. Sci.*, 2002, **91**, 1391–1395.

[22] D. J. Hauss (ed.), *Oral Lipid-Based Formulations, Enhancing the Bioavailability of Poorly Water-Soluble Drugs*, Informa Healthcare, 2007.

[23] F. Podczeck and B. E. Jones (eds), *Pharmaceutical Capsules*, Pharmaceutical Press, 2004.

[24] S. C. Gad, C. D. Cassidy, N. Aubert, B. Spainhour and H. Robbe, Nonclinical Vehicle Use in Studies by Multiple Routes in Multiple Species, *International Journal of Toxicology*, 2006, **25**, 499–512.

[25] D. Lamarre, An NS3 Protease Inhibitor with Antiviral Effects in Humams Infected with Hepatitis C Virus, *Nature*, 2003, **426**, 186–189.

PART VIII:
Trends and Perspectives

CHAPTER 25

To Market, To Market — 2007

Shridhar Hegde and **Michelle Schmidt**

1. INTRODUCTION

Compared to previous years, the pharmaceutical market experienced a slight reduction in the introduction of new molecular entities (NMEs) in 2007. Out of the 20 NMEs, including 1 biologic agent, three represented novel mechanisms of

Pfizer Global Research & Development, St. Louis, Missouri 63017

Annual Reports in Medicinal Chemistry, Volume 43
ISSN 0065-7743, DOI 10.1016/S0065-7743(08)00025-0

action with two targeting the same disease [1–2]. Reflecting an ongoing trend, investment in the anticancer and anti-infective areas produced nearly half of the newcomers to the market. While the United States continued to lead as the most active arena for new product launches, Europe and Asia followed at a distant second with three apiece, and Brazil welcomed one. Novartis enjoyed three entries while GSK was accredited with two, including one vaccine. Of the remaining major pharmaceutical companies, Pfizer, BMS, Merck, J&J, and Wyeth each contributed one. While new combinations of existing drugs continued to be popular in providing enhanced patient benefit, extensive coverage of these launches is beyond the scope of this chapter. Likewise, line extensions involving new formulations and new indications will not be fully addressed in this venue although this group comprised a significant percentage of new products in 2007.

In addition to ushering in five NMEs, the anticancer sector has garnered competition for Merck's Gardasil® vaccine against human papillomavirus (HPV) with the introduction of GlaxoSmithKline's Cervarix® for the prevention of HPV-induced cervical cancer and infections. While Gardasil® is approved for use in females aged 9–26 and is prophylactically effective against four strains of HPV (6, 11, 16, and 18), Cervarix® is recommended for a wider age range of 10–45 years, but only targets the two strains (16 and 18) that are responsible for approximately 70% of cervical cancer cases. GSK has committed to a head-to-head comparative trial of the two vaccines evaluating immune response and protection in more than 1,000 US women. Among the new oncolytic drugs, receptor tyrosine kinases (RTK) continue to be attractive targets. Tasigna® (nilotinib) is Novartis's second generation BCR-ABL tyrosine kinase inhibitor for the treatment of chronic myeloid leukemia that is resistant or intolerant to Gleevec® (imatinib), the leading therapy. Replacement of the N-methylpiperazine portion of imatinib increases the affinity for the inactive conformation of wild-type BCR-ABL by 20–30-fold and confers nilotinib with activity against several imatinib-resistant BCR-ABL mutants. Nilotinib will also have competition from the previous year's Sprycel™ (dasatanib). While both drugs have demonstrated success against multiple imatinib-resistant mutants, the T315I mutant remains refractory to these new treatment alternatives. Tykerb® (lapatinib) is another RTK inhibitor having dual affinity for epidermal growth factor receptor (EGFR) and human epidermal receptor type 2 (HER2). This adenosine triphosphate (ATP)-competitive inhibitor is the first targeted, oral treatment option for patients with advanced or metastatic HER2-positive breast cancer with prior drug experience, including anthracycline, a taxane, and Herceptin® (trastuzumab), and is indicated as a combination therapy with Xeloda® (capecitabine). The refractory breast cancer population has also received hope with the introduction of Ixempra™ (ixabepilone). This novel cytotoxic macrolide binds to and stabilizes microtubules, resulting in mitotic arrest and apoptosis, and is approved as monotherapy for metastatic or locally advanced breast cancer in patients with prior resistance to anthracyclines, taxanes, and capecitabine or as combination therapy with capecitabine. For the treatment of renal cell carcinoma, Torisel™ (temsirolimus) is the first marketed drug to specifically inhibit mTOR (mammalian target of rapamycin), a serine/threonine kinase. It possesses orphan drug status in the United States and joins the previously marketed RTK inhibitors Sutent® (sunitinib) and Nexavar® (sorafenib) for this indication. While the final anticancer drug exhibits wide-spectrum activity against a variety of human tumors, Yondelis® (trabectedin) has been initially launched for the treatment of advanced soft tissue sarcoma (STS) after failure of first-line therapy with anthracyclines or ifosfamide or in patients who are unsuited to receive these agents. Its antineoplastic action derives from the ability to bind to the minor groove of DNA via N2 alkylation of guanine, ultimately disrupting the binding of transcription factors involved in cell proliferation. In addition to the NMEs, Evista® (raloxifene), a selective estrogen receptor modulator (SERM) marketed since 1998 for the post-menopausal treatment of osteoporosis, has recently been approved for reducing the risk of invasive breast cancer in post-menopausal women, with and without osteoporosis, who are considered high risk.

The anti-infective arena contributed four NMEs, a mixed composition drug, and three vaccines. In the battle against human immunodeficiency virus (HIV), two new weapons were added to the arsenal. Both drugs are heralded as exploiting novel mechanisms of action to combat a disease plagued with cross-resistance and side effect issues. Selzentry™ (maraviroc) targets CCR5, one of the co-receptors required for viral entry into the host cell. Isentress™ (raltegravir) inhibits HIV-1 integrase, the enzyme responsible for integration of the viral DNA into the host genome, effectively halting viral replication. Both drugs received accelerated approval from the FDA for combination therapy with traditional antiretrovirals (reverse transcriptase and protease inhibitors (PIs)). Also reaching the market was Geninax® (garenoxacin), a quinolone antimicrobial with broad-spectrum activity against both Gram-negative and Gram-positive bacteria. It was initially launched in Japan for the treatment of respiratory tract and otorhinolaryngological infections. Revisiting the pleuromutilin class of agents led to the development and launch of Altabax™ (retapamulin). As the first in its class to be approved for human use, this tricyclic diterpenoid is a treatment for impetigo, a highly contagious skin infection caused by bacteria such as *S. aureus* and/or *S. pyogenes*. While not covered in extensive detail in this review, Veregen™ entered the market in December for the treatment of external genital and perianal warts in adult, immunocompetent patients. The drug is actually a defined mixture of eight catechins extracted from green tea leaves. Although the exact mechanism of action is unknown, the composition exhibits antioxidant activity *in vitro*. Finally, two influenza vaccines and one small pox vaccine debuted. Afluria® is an inactivated vaccine against subtypes A and B and is recommended for patients aged 18 and older. The other influenza vaccine protects against the H5N1 avian strain. It was licensed by the FDA from sanofi pasteur in collaboration with the National Institute of Health. While this vaccine will not be sold commercially, it will be stockpiled for distribution by health officials to high-risk individuals. ACAM-2000™ has been developed and launched for protection against small pox. This live virus will likewise be stockpiled for emergency use and for the protection of military personnel.

The CNS field was represented by the entry of three new drugs and new indications for existing drugs. Joining other marketed amphetamine and methylphenidate formulations for the treatment of attention-deficit/hyperactivity disorder (ADHD), Vyvanse™ (lisdexamfetamine) is a prodrug of D-amphetamine that boasts reduced abuse potential with an extended-release benefit. Schizophrenia patients have a new extended-release option in Invega™ (paliperidone). This dual 5-HT$_2$ and dopamine D$_2$ receptor antagonist, which is the active metabolite of Johnson and Johnson's antipsychotic agent risperidone, employs an osmotic-controlled release oral delivery system (OROS™) for sustained plasma levels over a 24-h period. Inovelon® (rufinamide) was launched primarily as adjunctive therapy for Lennox–Gastaut Syndrome, a debilitating childhood-onset epilepsy. Two existing drugs established extended market value. To improve patient compliance and maintain more consistent plasma levels, a transdermal form of rivastigmine, marketed as Exelon® Patch, was launched as an option for Alzheimer's disease. For the 3–6 million Americans afflicted with fibromyalgia, a disease characterized by widespread, chronic musculoskeletal pain of unknown etiology, an option finally exists for alleviation of symptoms. With demonstrated efficacy against neuropathic pain, Lyrica® (pregabalin) was approved for treatment of fibromyalgia, which also encompasses fatigue, interrupted sleep patterns, cognitive issues, and psychological suffering.

The cardiovascular domain welcomed two NMEs along with numerous combinations of existing drugs. Representing a novel approach to treating hypertension, Tekturna® (aliskiren) is the first marketed direct inhibitor of renin, part of the renin–angiotensin–aldosterone system (RAAS). Previously marketed modulators of this system include angiotensin-converting enzyme (ACE) inhibitors and angiotensin II receptor blockers (ARBs). Aliskeren is approved as monotherapy or in combination with other antihypertensive medications; however, an added benefit was observed with combination therapy. Letairis™ (ambrisentan) is the third endothelin antagonist to reach the market for the treatment of pulmonary arterial hypertension, following Tracleer® and Thelin®. Owing to concerns about potential hepatic toxicity and birth defects, ambrisentan is only available through a restricted distribution program. With the loss of exclusivity of Pfizer's Norvasc® (amlodipine) in 2007, the calcium channel blocker has been incorporated into two combination products with angiotensin receptor blockers. Exforge® couples amlodipine with valsartan while Azor™ utilizes olmesartan as the ARB component. A third antihypertensive combination, Zanipress®, joins the calcium channel blocker lercanidipine with the ACE inhibitor enalapril. These combination products strive to improve patient compliance with a reduced pill burden. Keeping with this theme, Torent Pharma launched the CVpill, a kit packaging a capsule containing the statin atorvastatin, the ACE inhibitor ramipril, and the antiplatelet agent enteric-coated aspirin with a tablet comprising the β blocker metoprolol. The kit conveniently provides a once-daily regimen for patients with multiple cardiovascular risk factors.

In the respiratory area, two new drugs emerged. Veramyst™ (fluticasone furoate) is a nasal spray launched for the treatment of both seasonal and year-round allergy symptoms in patients 2 years and older. Veramyst represents another new agent with long duration of action and low systemic exposure. Brovana™ (arformoterol) is a nebulized β_2-adrenoceptor agonist approved for relieving the symptoms of bronchoconstriction in patients with chronic obstructive pulmonary disease (COPD), including emphysema and chronic bronchitis. The corticosteroid ciclesonide, previously launched as an asthma treatment, experienced two line extensions. As an intranasal formulation marketed as *Osonase*, the drug is indicated for allergic rhinitis. For once-daily, inhaled treatment of asthma, the drug is marketed as *Osovair*, a combination product with the β_2-adrenoceptor agonist formoterol.

Concerning endocrine diseases, diabetes patients received two new options. Galvus® (vildagliptin) is the second dipeptidyl peptidase IV (DPP-4) inhibitor to achieve marketing status. As one of the evolving antidiabetic targets, inhibition of DPP-4 modulates the activity of incretins, endogenous peptides involved in glucose homeostasis. The first marketed DPP-4 inhibitor sitagliptin was united with the glucose-lowering agent metformin to provide the combination drug Janumet™. The product is approved for use in adult patients with type 2 diabetes to improve blood sugar control, particularly when inadequate results are obtained with the individual components alone or to improve patient compliance in individuals already receiving both drugs. The remaining endocrine disease launches targeted the female population. For women desiring birth control without a monthly menstruation, Lybrel™, a combination of levonorgestrel and ethinyl estradiol, may be the contraceptive of choice; it is taken 365 days a year without a placebo or pill-free hiatus. For fertility therapy, Pergoveris™ (follitropin alfa/lutropin alfa) stimulates follicular development in women with severe luteinizing hormone and follicle-stimulating hormone deficiencies. This is the first combination of these two hormones for subcutaneous injection. Follitrope® (recombinant human follicle-stimulating hormone) is another fertility treatment that induces ovulation. As a recombinant product, it possesses a similar safety profile to endogenous FSH. Intrinsa® (testosterone transdermal patch) is indicated as a replacement for natural testosterone levels in women experiencing early menopause following hysterectomy or ovary removal. Treatment should alleviate low sexual desire that is typical in this patient population. While the progestogen dienogest has been a component in oral contraceptive and hormone replacement therapy formulations for years, dienogest was approved as a monotherapy for endometriosis in 2007, sold as Dinagest®.

The metabolic section saw the introduction of the only new biologic product of the year. Soliris™ (eculizumab) is a humanized monoclonal antibody targeting the C5 protein of the complement system whose cleavage is instrumental to the destructive inflammatory cascade culminating in hemolytic anemia. Interference of the complement system by this antibody led to the first approved therapy for paroxysmal nocturnal hemoglobinuria (PNH), a rare and life-threatening form of hemolytic anemia. For the treatment of anemia associated with chronic kidney disease, Roche launched Mircera® (methoxy polyethylene glycol-epoetin beta). This PEGylated erythropoiesis-stimulating agent is delivered by either subcutaneous or intravenous injection once every 2 weeks to correct the condition.

Muscarinic antagonists continue to be appealing targets for the treatment of overactive bladder (OAB). The latest introduction is the M_1/M_3 antagonist imidafenacin that is co-marketed by Kyorin and Ono under the trade names Uritos® and Staybla®, respectively. Another muscarinic M3 antagonist tolterodine was combined with the α_1-adrenoceptor antagonist tamsulosin in a single, once-daily, extended-release capsule. The product is Roliflo® OD, and the indication is the management of concomitant OAB and bladder outlet obstruction, a condition prevalent in men with benign prostate hyperplasia. The final renal-urologic is another combination therapy incorporating calcium acetate and magnesium carbonate to afford the phosphate-binding agent OsvaRen®. For patients on dialysis, OsvaRen can assist in controlling phosphate blood levels and prevent excess phosphate-driven bone and blood vessel damage.

The gastrointestinal category contained one NME in addition to new combinations, formulations, and line extensions. *Levovir* (clevudine) is a recent option for hepatitis B virus (HBV). As a fluorinated β-L-nucleoside analog, it is the fifth of the marketed HBV drugs to act by interference with hepatitis B viral replication. A breakthrough in the treatment of *Helicobacter pylori* surfaced with the three-in-one capsule Pylera™, which is a mixture of metronidazole, tetracycline, and bismuth biskalcitrate. The result is a significant reduction in pill burden for the eradication of this bacteria. Another approval directed at the improvement of dosing regimen is Lialda™ (mesalamine), which was reformulated as a once-daily treatment for ulcerative colitis. With the previous formulation, some patients were required to take between 6 and 16 pills daily. Following its approval for rheumatoid arthritis, psoriatic arthritis, and ankylosing spondylitis, the use of Humira® (adalimumab), a recombinant anti-tissue necrosis factor (TNF)-α monoclonal antibody, has been extended to the battle against Crohn's disease.

While the remaining therapeutic areas did not generate NMEs in 2007, a number of valuable contributions to the pharmaceutical market were gained through line extensions, improved formulations, and novel combinations. Focusing on hematologic agents, ATryn®, a recombinant form of human antithrombin III, emerged for the prophylaxis of post-surgical venous thromboembolism in patients with congenital antithrombin deficiency. As an alternative to the bovine protein-based thrombin used to control bleeding during surgical procedures, Omrix Biopharmaceuticals launched Evithrom™, a novel thrombin product derived from human plasma. On the dermatology front, the oral contraceptive Yaz® (drospirenone/ethinyl estradiol) was approved for concomitant treatment of moderate acne. From the ophthalmic section, AzaSite™ was created as an improved therapy for bacterial conjunctivitis by exploiting the DuraSite® system, a polymer-based formulation that stabilizes antibiotics in an aqueous matrix. The analgesic and anesthetic segment provided Zingo™ (lidocaine), a fast-acting, transdermal treatment to reduce the pain associated with venous access procedures. Compared to other local anesthetics with an onset of approximately 20 min, Zingo relinquishes relief in only 3 min. Two existing drugs achieved new indications in the therapy of musculoskeletal and connective tissue diseases. Lupus is a connective tissue disorder that can affect multiple organs and emanates from loss of immune tolerance to self. Lupus nephritis is the manifestation of the disease attacking the kidneys. Prograf® (tacrolimus) was launched to combat lupus and is the seventh indication for this immunosuppressive agent. Tracleer® (bosentan) was approved for the new indication of treating digital ulcers in patients with systemic sclerosis and continuing digital ulcer disease. Finally, while chondroitin and glucosamine have been known to promote joint health, Zylera coupled the two ingredients to yield Relamine™, a prescription treatment for osteoarthritis.

Since diagnostic agents are not considered therapeutic pharmaceuticals, they are not discussed in detail in this review. They do, however, deserve mention for their critical role in disease assessment and elucidation of treatment options. Although approved in 2006, the ultrasound contrast agent Sonazoid® (perflubutane) launched for the first time in Japan in 2007. This diagnostic agent is utilized in the visualization of liver lesions and in the evaluation of post-treatment effectiveness. Interestingly, the contrast agent is subsequently cleared via exhalation.

2. ALISKIREN (ANTIHYPERTENSIVE) [3–8]

Country of origin:	Switzerland
Originator:	Novartis
First introduction:	US
Introduced by:	Novartis
Trade name:	Tekturna
CAS registry no:	173334-57-1
Molecular weight:	551.76

Aliskiren is a first-in-class antihypertensive drug that acts by direct inhibition of renin. It is indicated for oral administration either as monotherapy or in combination with other antihypertensive agents. Renin catalyzes the cleavage of angiotensin to form angiotensin I, the first and the rate-limiting step of the RAAS. The inhibition of renin by aliskiren results in reduced levels of angiotensin I, angiotensin II, and aldosterone, all of which contribute to the antihypertensive effect. RAAS modulators are among some of the most commonly prescribed antihypertensive agents to date. Previously marketed drugs with this general mechanism include ACE inhibitors and ARBs, both of which act on steps downstream from renin. All inhibitors of RAAS, including aliskiren, can cause a compensatory rise in plasma renin concentration by suppressing the negative feedback loop. When this rise occurs during treatment with ACE inhibitors and ARBs, the result is increased levels of plasma renin activity (PRA) and incomplete inhibition of RAAS. Elevated PRA levels are thought to increase cardiovascular risk among hypertensive patients. During treatment with aliskiren, however, the effect of increased renin levels is blocked, so that PRA is reduced, whether aliskiren is used as monotherapy or in combination with other antihypertensive agents. Aliskiren is a highly potent inhibitor of human renin *in vitro* ($IC_{50} = 0.6$ nM). In sodium-depleted marmosets, oral administration of aliskiren leads to a marked, dose-dependent (1–10 mg/kg) and long-lasting (up to 24 h) reduction in mean arterial pressure and complete inhibition of PRA. Oral aliskiren is poorly absorbed, with an absolute bioavailability of approximately 2.5%. The peak plasma concentration (C_{max}) is reached in 2.5–3.0 h after administration of a single 150- or 300-mg dose of the drug. Increases in C_{max} and AUC are dose-proportionate over the range of 75–600 mg, and the steady-state plasma concentrations are reached within 7–8 days. Aliskiren has moderate protein binding (50%) and a half-life of approximately 24 h. Co-administration with a high-fat meal reduces the C_{max} and AUC of aliskiren by 85% and 71%, respectively. Aliskiren concentrations are greater in the kidneys than in plasma. The high concentration of aliskiren at the site of renin release could account for the profound effect of the drug on PRA. Aliskiren is detectable in the kidneys up to 3 weeks after discontinuation of the drug, whereas plasma levels of aliskiren are undetectable at this time point. Much of the orally administered aliskiren is excreted as unabsorbed drug in the feces. The absorbed portion is eliminated via the hepatobiliary route, and to some extent through oxidative metabolism. The recommended starting dose of aliskiren is 150 mg once daily, which may be increased to 300 mg once daily after 2 weeks if the blood pressure (BP) goal is not achieved. The efficacy and safety of aliskiren, alone and in combination with other commonly used antihypertensive drugs, have been evaluated in several clinical trials. For example, in one trial involving 672 patients, oral aliskiren at 150 and 300 mg once daily for 8 weeks reduced mean sitting systolic/diastolic BP by 13.0/10.3 and 14.7/11.1 mm Hg, respectively, compared with 3.8/4.9 mm Hg with placebo. The antihypertensive effect of aliskiren was sustained throughout the 24-h dosing interval as assessed by ambulatory BP monitoring. In clinical trials that included an active comparator, 150-mg once-daily aliskiren provided similar BP reduction as other antihypertensive agents, such as the diuretic hydrochlorothiazide or the ARBs irbesartan, losartan, and valsartan. The combinations of aliskiren with either hydrochlorothiazide or any of the ARBs provided a significantly greater BP reduction as compared to the respective monotherapy components. The most common side effect of aliskiren is diarrhea, particularly with doses higher than 300 mg daily. Unlike ACE inhibition, direct inhibition of renin does not increase bradykinin levels. Elevated bradykinin levels are thought to be responsible for the angioderma and cough that occur commonly during treatment with ACE inhibitors. Overall rates of angioderma and cough in aliskiren clinical trials were 0.06% and 1.1%, respectively. The key differentiators for aliskiren as compared to the existing antihypertensive drugs appear to be its long duration of action and a favorable side effect profile.

Aliskiren has not yet been evaluated in patients with significantly impaired renal function. As with ACE inhibitors, renin inhibitors should not be used in pregnancy, specifically the second and third trimesters, during which they will interfere with fetal kidney development. The chemical synthesis of aliskiren is achieved in a multistep sequence with the key step involving a Grignard reaction of (2R)-2-isopropyl-3-(3-methoxypropoxy-4-methoxy)phenylpropyl-magnesium chloride with (2S,5S)-2-isopropylbutyrolactone-5-carboxylic acid chloride to produce a keto-lactone intermediate. Reduction of the keto-lactone with sodium borohydride gives a diastereomeric mixture of the corresponding hydroxy-lactones, from which the desired isomer is isolated by chromatography. Subsequent conversion of the hydroxyl group to an azido group via nucleophilic displacement of the corresponding mesylate with sodium azide, opening of the lactone ring with 4-amino-3,3-dimethylbutyramide, and reduction of the azido group to an amino group by catalytic hydrogenation lead to aliskiren.

3. AMBRISENTAN (PULMONARY ARTERIAL HYPERTENSION) [9–13]

Country of origin:	US	
Originator:	Abbott	
First introduction:	US	
Introduced by:	Gilead	
Trade name:	Letairis	
CAS registry no:	177036-94-1	
Molecular weight:	378.42	

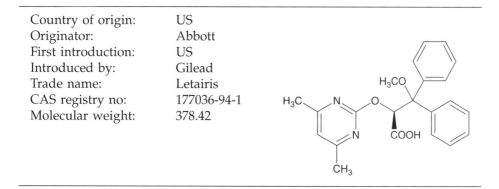

Ambrisentan is a selective endothelin-A (ET_A) receptor antagonist introduced last year for the oral treatment of patients with pulmonary arterial hypertension (PAH), to improve exercise capacity and delay clinical worsening. It is the third ET-receptor antagonist to be marketed for this indication behind bosentan and sitaxsentan. PAH is a rare disease of the small pulmonary arteries characterized by vascular proliferation and remodeling, resulting in a progressive increase in pulmonary vascular resistance and pulmonary arterial pressure, and ultimately, right ventricular failure and premature death. Early symptoms of PAH include gradual onset of shortness of breath, fatigue, palpitation, edema, and fainting. Endothelin-1 (ET-1), a potent vasoconstrictor and smooth muscle mitogen, is a key contributor to the acceleration of the disease, and its effects are mediated through activation of ET_A and ET_B receptors. ET_A receptors are found primarily on smooth muscle cells and, when activated, induce vasoconstriction and cellular proliferation. ET_B receptors are expressed in both pulmonary vascular endothelial cells and smooth muscle cells. Activation of ET_B receptors promotes clearance of ET and increases nitric oxide and prostacyclin release, which induces vasodilation and antiproliferative effects. Ambrisentan shows approximately 200-fold selectivity for ET_A over ET_B receptors (ET_A $K_i = 1\,nM$, ET_B $K_i = 195\,nM$). By comparison, sitaxsentan is significantly more selective for ET_A (approximately 7,000-fold), and bosentan is less selective (approximately 20-fold). Although all three drugs are efficacious in the treatment of PAH, it is hypothesized that the ET_A receptor selectivity of sitaxsentan and ambrisentan may confer a greater clinical benefit by inhibiting the deleterious ET_A-mediated vasoconstriction while preserving the beneficial vasodilator and clearance functions of ET_B receptors. Oral ambrisentan has a T_{max} of approximately 2 h. The oral bioavailability is about 80%, and it is not affected by food. Ambrisentan is highly bound to plasma proteins (99%). The elimination of ambrisentan is predominantly by non-renal pathways, and the effective half-life is about 9 h. *In vitro* studies indicate that ambrisentan is a substrate of P-glycoprotein (Pgp) efflux transporter; hence, drug–drug interactions are possible during concomitant therapy with strong inhibitors of Pgp. The recommended initial dosing regimen of ambrisentan is 5 mg once daily with or without food. The dose can then be increased to 10 mg once daily if the 5 mg dose is tolerated. In two separate 12-week Phase III clinical trials (ARIES-1 and ARIES-2) involving 393 PAH patients, treatment with ambrisentan resulted in significant improvement in the primary endpoint of exercise capacity as assessed by a 6-min walk distance. In ARIES-1, the mean increases from baseline in 6-min walk distance were 23 and 44 m with 5 and 10 mg doses of ambrisentan, respectively, compared to a decrease of 8 m with placebo. In ARIES 2, the mean increases from baseline in 6-min walk distance were 22 and 49 m with 2.5 and 5 mg doses, respectively, compared to a decrease of 10 m with placebo. Ambrisentan also significantly delayed time to clinical worsening of PAH. In an open-label extension study ($n = 383$), 95% of patients who continued to take the drug were alive for 1 year. The most common side effects of ambrisentan included swelling of legs and ankles, nasal congestion, sinusitis, flushing, and palpitations. Treatment with ET-receptor antagonists has been associated with dose-dependent hepatic injury, manifested primarily by serum aminotransferase (ALT and AST) elevations but sometimes accompanied by abnormal liver function (bilirubin elevations). Serum aminotransferase levels must be measured before initiation of treatment and at least monthly during treatment. Ambrisentan is chemically derived in five steps starting with a Darzen condensation reaction of benzophenone with methyl 2-chloroacetate in the presence of sodium methoxide to give 3,3-diphenyloxirane-2-carboxylic acid methyl ester. Treatment of this ester with boron trifluoride etherate and methanol, followed by optical resolution, yields the pure enantiomer 2(S)-hydroxy-3-methoxy-3,3-diphenylpropionic acid methyl ester. Subsequent condensation with 4,6-dimethyl-2-(methylsulfonyl)-pyrimidine using potassium carbonate followed by hydrolysis of the ester group with potassium hydroxide gives ambrisentan.

4. ARFORMOTEROL (COPD) [14–17]

Country of origin: US
Originator: Sepracor
First introduction: US
Introduced by: Sepracor
Trade name: Brovana
CAS registry no: 067346-49-0
Molecular weight: 344.40

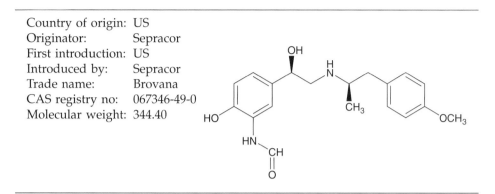

Arformoterol, a selective long-acting β_2-agonist, was launched last year as an inhalation solution for treatment of bronchoconstriction associated with COPD. It is the active (R,R)-enantiomer of the previously marketed β_2-agonist formoterol, which is registered as a racemic mixture. Similar to other β_2-agonists, the mechanism of action of formoterol and arformoterol involves activation of adenyl cyclase, leading to increased production of cyclic adenosine monophosphate (cAMP) via ATP. Increased levels of intracellular cAMP result in relaxation of the bronchial smooth muscle, and also prevent mast cells from releasing inflammatory mediators. Arformoterol has high binding affinity for the human β_2 receptor ($K_d = 2.9\,\mathrm{nM}$) and approximately 39-fold selectivity over the β_1 receptor ($K_d = 113\,\mathrm{nM}$). It is $>1,000$ times more potent β_2 ligand than the corresponding (S,S)-enantiomer ($K_d = 3100\,\mathrm{nM}$), and twice as potent as racemic formoterol ($K_d = 5.2\,\mathrm{nM}$). Additionally, arformoterol acts as a full or nearly full agonist of the β_2 receptor, whereas the (S,S)-enantiomer acts as an inverse agonist. Therefore, the (R,R)-enantiomer may account exclusively for the activation of β_2 receptors by the racemic mixture. The recommended dosage of arformoterol is 15 µg administered twice daily (morning and evening) via a nebulizer. Following a 15-µg dose, onset of bronchodilation occurs within 7 min when defined as a 15% increase in forced expiratory volume in 1 s (FEV_1), and 20 min when defined as an increase in FEV_1 of 12% and 200 mL. The peak bronchodilation effect occurs in 1–3 h. A substantial portion of systemic exposure of arformoterol is due to pulmonary absorption, with a peak plasma concentration occurring 0.25–1 h after dosing and the plasma levels increasing linearly with dosage. Arformoterol primarily undergoes hepatic metabolism through glucuronidation and also through O-demethylation by CYP2D6 and CYP2C19. The elimination half-life is approximately 26 h. Arformoterol is excreted in the urine (67%) and feces (22%) as metabolites. The clinical efficacy of arformoterol has been evaluated in two 12-week trials in COPD patients ($n = 1456$). Both trials compared three doses of arformoterol (15 µg BID, 25 µg BID, and 50 µg QD) to placebo and to salmeterol (42 µg BID). In both studies, mean percent change from baseline FEV_1 over 12 weeks was significantly higher with 15 µg BID arformoterol (+16.9% and +15.7%)

than placebo (+6% and +5.3%). Higher doses of arformoterol did not produce significantly greater improvement. The improvement in salmeterol-treated subjects was +17.4%. In both studies, patients in the arformoterol groups also demonstrated improvements in peak expiratory flow rate and decreases in the use of rescue albuterol and supplemental ipratropium relative to patients in the placebo group. Some tolerance to the bronchodilator effect of arformoterol was observed upon chronic dosing. The FEV_1 improvement at the end of the 12-h dosing interval decreased by approximately one-third over the course of the study (22.1% mean improvement after the first dose compared with 14.6% at week 12). In these trials, the most commonly reported adverse events with arformoterol were similar to those seen with other β_2 agonists, and included pain, chest pain, back pain, diarrhea, sinusitis, leg cramps, dyspnea, and rash. Arformoterol is chemically derived via the condensation of two chiral intermediates, 4-benzyloxy-3-formamido-(R)-phenyloxirane and N-benzyl-1-(4-methoxyphenyl)-(R)-2-propanamine, and subsequent removal of benzyl protecting groups via catalytic hydrogenolysis. The chiral phenyloxirane intermediate is obtained by enantioselective reduction of the corresponding phenacyl bromide to the bromohydrin with a chiral oxazaborolidine reagent, followed by cyclization with potassium carbonate in methanol. The chiral 2-propanamine intermediate is obtained by reductive amination of (4-methoxyphenyl)acetone with benzylamine followed by chiral resolution with (S)-mandelic acid.

5. CLEVUDINE (HEPATITIS B) [18–21]

Country of origin:	US
Originator:	University of Georgia/ Yale University
First introduction:	South Korea
Introduced by:	Bukwang
Trade name:	Levovir
CAS registry no:	163252-36-6
Molecular weight:	260.22

Clevudine is a fluorinated β-L-nucleoside analog launched last year for the oral treatment of chronic HBV infection. It is the fifth nucleoside or nucleotide analog to be marketed for this indication. The previous drugs from this class include lamivudine, adefovir, entecavir, and telbivudine. In HBV-expressing human hepatoma cell line 2.2.15, clevudine inhibits HBV DNA synthesis with an EC_{50} of $0.1\,\mu M$, and does not show cytotoxicity up to $200\,\mu M$. It is phosphorylated by cellular kinases to the active triphosphate derivative, which subsequently inhibits HBV DNA polymerase and HBV replication.

Clevudine-5'-triphosphate has an intracellular half-life of 16.5 h. Interestingly, it is a non-competitive inhibitor of viral polymerase, and inhibits HBV replication without being incorporated into the DNA. This mechanism of action is different from previously marketed nucleoside inhibitors, which act as alternate substrates and are incorporated into the viral DNA leading to DNA chain termination. Furthermore, clevudine-5'-triphosphate is neither an inhibitor nor a substrate for human DNA polymerases α, β, γ, and ε. Clevudine is also active against Epstein-Barr virus (EBV; $EC_{50} = 1\,\mu M$), but lacks activity against HSV (herpes simplex virus) or HIV. In pharmacokinetic studies in healthy volunteers, single oral doses of clevudine ranging from 150 to 1,200 mg gave dose-proportionate increases in C_{max} and AUC, and the oral bioavailability was not influenced by food. Clevudine has a relatively long half-life of 7.9–13.5 h. The recovery of unchanged clevudine from urine ranges from 21% to 31%. In a dose-escalating Phase I/II study, clevudine 10 mg ($n = 4$), 50 mg ($n = 10$), and 100 mg ($n = 10$) was administered once daily to HBV-infected patients over a 28-day period. The pharmacokinetic profile of clevudine was linear with a plasma half-life of approximately 60 h. Clevudine was undetectable in plasma after 4 weeks following the cessation of dosing. The clinical efficacy of clevudine has been demonstrated in three different Phase III trials. In a 48-week study involving 55 treatment-naïve patients, 30 mg/day clevudine for 24 weeks followed by a maintenance dose of 10 mg/day for a further 24 weeks resulted in 68% of hepatitis B e-antigen (HBeAg)-positive patients and 100% of HBeAg-negative patients achieving HBV DNA levels below the lower limit of detection (300 copies/mL) by PCR assay, and 89% and 100% of patients, respectively, showing normal ALT levels. In two additional trials in HBeAg-positive and HBeAg-negative patients ($n = 248$ and 89, respectively), clevudine 30 mg/day for 24 weeks resulted in 59% of the HBeAg-positive and 92% of the HBeAg-negative patients showing below detectable levels of HBV DNA at the end of treatment period. The most common adverse events reported with clevudine treatment include infection, asthenia, dyspepsia, abdominal pain, headache, and diarrhea. As with other nucleoside analogs, long-term treatment with clevudine has the potential to result in drug resistance. In animal models, resistance occurred in the B domain of the polymerase gene within 12 months of treatment. *In vitro* studies suggest the possibility of cross-resistance with lamivudine-resistant HBV mutants. A unique characteristic of clevudine is the slow rebound of viremia after cessation of treatment. In a placebo-controlled Phase II trial, HbeAg-positive patients were treated with 30- or 50-mg once-daily clevudine for 12 weeks and subsequently monitored for a further 24 weeks off therapy. Median serum HBV DNA reductions from baseline at week 12 were 0.20, 4.49, and 4.45 $\log_{10}$ copies/mL in the placebo, 30-mg clevudine, and 50-mg clevudine groups, respectively. By comparison, HBV DNA $\log_{10}$ reductions were 3.32 and 2.99 at week 12 off therapy and 2.28 and 1.40 at week 24 off therapy in the 30- and 50-mg clevudine groups, respectively. Clevudine is chemically derived from L-ribose by first incorporating acyl protective groups to produce 1-*O*-acetyl-2,3,5-tri-*O*-benzoyl-β-L-ribofuranose intermediate, which is then converted to 1,3,5-tri-*O*-benzoyl-α-L-ribofuranose in two steps by treating with hydrogen chloride and subsequent hydrolysis and acyl migration. The remaining steps leading to clevudine include conversion of the C2-hydroxy group to C2-fluoro group with triethylamine trihydrofluoride, formation of the corresponding ribofuranosyl bromide intermediate with hydrogen bromide and acetic acid, condensation with silylated thymine, and removal of benzoyl protective groups with methanolic ammonia.

6. ECULIZUMAB (HEMOGLOBINURIA) [22–24]

Country of origin:	US	Class:		Humanized IgG2/4κ
Originator:	Alexion			Monoclonal antibody
First introduction:	US	Type:		Anti-C5
Introduced by:	Alexion	Molecular weight:		148 kDa
Trade name:	Soliris	Expression system:		Murine myeloma cell culture
CAS registry no:	219685-50-4	Manufacturer:		Alexion

Eculizumab, a fully humanized anti-C5 monoclonal antibody, was introduced last year for treating patients with PNH to reduce hemolysis. It is the first therapy to be approved for this rare and life-threatening form of hemolytic anemia. PNH is a clonal hematopoietic stem-cell disorder that is characterized by the production of abnormal red blood cells (RBCs) with a deficiency of surface proteins that protect the cells against attack by the body's complement system. Complement-mediated destruction of the susceptible RBCs results in intravascular hemolysis, the primary clinical manifestation in all PNH patients. Previously, patients with PNH have mainly been managed supportively, with red cell transfusions as required, and treatments such as folate and iron supplementation, anticoagulation for thrombotic disease, and the occasional use of steroids during hemolytic crises. Allogenic stem cell transplantation is currently the only curative option for PNH; however, it is associated with significant morbidity and mortality. Eculizumab therapy is aimed at preventing red cell lysis through blockade of complement activation process and the production of the membrane attack complex. Eculizumab specifically binds to the human complement protein C5 with high affinity ($IC_{50} = 2$ nM) and inhibits its cleavage to C5a and C5b, which is a key step in the pathway leading to the membrane attack complex C5b-C9. Eculizumab has been granted orphan drug status from both the FDA and European regulatory agencies. Eculizumab is given as a 35-min intravenous infusion in a dosage regimen of 600 mg every 7 days for the first 4 weeks, followed by 900 mg for the fifth dose 7 days later, then 900 mg every 14 days thereafter. It has an average clearance of 22 mL/h and an average volume of distribution of 7.7 L. The mean half-life is 272 h. The efficacy of eculizumab was assessed in a randomized, double-blind, placebo-controlled, 26-week study involving 87 PNH patients with hemolysis. Patients receiving eculizumab showed significantly reduced hemolysis (p < 0.001), resulting in improvements in anemia, as indicated by an increase in hemoglobin stabilization in 49% of the patients in the eculizumab group as compared with 0% in the placebo group. Patients also had a reduced need for RBC transfusions (transfusion avoidance: eculizumab = 51%; placebo = 0%). Less fatigue and improved health-related quality of life were also reported by the patients receiving eculizumab compared to those receiving placebo. Efficacy was maintained with eculizumab therapy for up to 54 months in an extension study. The most serious adverse reaction associated with eculizumab therapy is meningococcal infections. Eculizumab is contraindicated in patients who are not vaccinated against *Neisseria meningitidis* or who have *N. meningitidis* infections. The most common adverse reactions with eculizumab include headache (44%), nasopharyngitis (23%), back pain (19%), and nausea (16%).

7. FLUTICASONE FUROATE (ANTIALLERGY) [25–27]

Country of origin:	US
Originator:	GSK
First introduction:	US
Introduced by:	GSK
Trade name:	Veramyst
CAS registry no:	397864-44-7
Molecular weight:	538.58

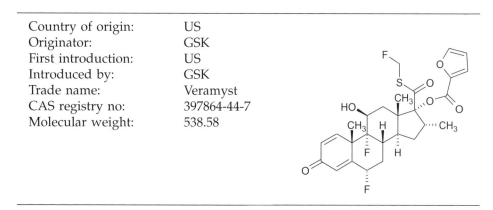

Fluticasone furoate is a new corticosteroid derivative launched last year as a nasal spray for the treatment of seasonal and perennial allergic rhinitis in adults and in children aged ⩾2 years. It is structurally closely related to the previously marketed intranasal corticosteroid fluticasone propionate (FP). Both of these steroids contain the unusual S-fluoromethyl carbothioate group, which confers high lipophilicity and hence enhanced binding and retention of the drug by the nasal tissue. Additionally, the carbothioate group rapidly undergoes first-pass metabolism by CYP3A4, thus minimizing systemic exposure of the parent drug. Fluticasone furoate is a potent ligand for the glucocorticoid receptor (GR), with a relative receptor affinity (RRA) of 2,989 with reference to dexamethasone RRA of 100. As seen with other inhaled corticosteroids marketed in recent years (e.g. ciclesonide), fluticasone furoate exhibits a long duration of action and has minimal systemic exposure. Its binding kinetics are characterized by a fast on-rate and a slow off-rate. It has an equilibrium dissociation constant (K_d) of 0.3 nM, as compared to dexamethasone K_d value of 8.8 nM. The recommended starting dose of fluticasone furoate is 110 µg (two sprays per nostril) once daily for patients ⩾12 years of age, and 55 µg (one spray per nostril) once daily for children 2–11 years of age. Children not adequately responding to 55 µg may use 110 µg once daily. Following intranasal administration, most of the dose is eventually swallowed and undergoes incomplete absorption and extensive first-pass metabolism in the liver and gut, resulting in negligible systemic exposure. At the highest recommended dosage of 110 µg once daily for up to 12 months in adults and up to 12 weeks in children, plasma concentrations of fluticasone furoate are typically below the lower limit of quantification of 10 pg/mL. Its average systemic bioavailability following intranasal administration is estimated at 0.5%. Pharmacokinetic studies using oral solution dosing and intravenous dosing show that at least 30% of fluticasone furoate is absorbed and then rapidly cleared from plasma. Binding of fluticasone furoate to human plasma proteins is >99%. Following intravenous dose, the mean volume of distribution at steady state is 608 L. Fluticasone furoate is cleared from systemic circulation principally by hepatic metabolism via CYP3A4. The principal route of metabolism is hydrolysis of the S-fluoromethyl carbothioate function to form the inactive 17β-carboxylic acid metabolite. In vivo studies show no evidence of cleavage of the furoate moiety to form fluticasone. Fluticasone furoate and its metabolites are eliminated primarily in the feces, with urinary excretion accounting for <2% of an orally or intravenously administered dose. The elimination phase half-life is about 15 h. The most common adverse reactions (>1% incidence) included headache, epistatix, sinus and throat pain, nasal ulceration, back pain, pyrexia, and cough. The clinical efficacy of fluticasone furoate in patients ⩾12 years old with symptoms of seasonal or perennial allergic rhinitis has been evaluated in five randomized, double-blind, placebo-controlled studies of 2–4 weeks duration ($n = 1,829$). Assessment of efficacy was based on total nasal symptom score. Secondary efficacy variables included the total ocular symptom score (TOSS), and rhinoconjunctivitis quality of life questionnaire (RQLQ). In each study, fluticasone furoate was shown to be more effective than placebo in relieving overall nasal allergy symptoms. In addition, three seasonal allergy studies showed that fluticasone furoate was significantly better than placebo in relieving ocular symptoms and in providing a significant and clinically meaningful improvement in overall RQLQ. However, these effects were not seen in the perennial allergic rhinitis trials. In all studies, fluticasone furoate sustained its effectiveness for a full 24 h. In clinical trials involving children 2–11 years old with seasonal and perennial allergic rhinitis, fluticasone furoate was effective in treating overall nasal symptoms. However, changes in TOSS in the seasonal allergic rhinitis trial were not statistically significant, and TOSS was not assessed in the perennial allergic rhinitis trial. Fluticasone furoate is chemically derived starting from a readily available corticosteroid, 6α,9α-difluoro-11β,17α-dihydroxy-16α-methyl-3-oxoandrosta-1,4-diene-17β-carbo-xylic acid, by first converting the carboxylic acid group to the corresponding carbothioic acid via activation with carbonyl diimidazole followed by reaction with hydrogen sulfide gas. Subsequently, selective acylation of the 17α-hydroxyl group with 2-furoyl chloride and alkylation of the 17β-carbothioic acid group with bromofluoromethane under basic conditions provides fluticasone furoate.

8. GARENOXACIN (ANTI-INFECTIVE) [28–32]

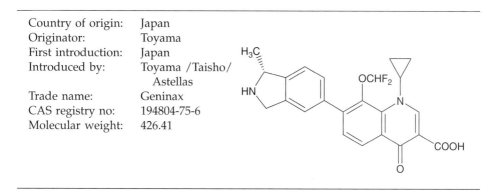

Country of origin:	Japan
Originator:	Toyama
First introduction:	Japan
Introduced by:	Toyama /Taisho/ Astellas
Trade name:	Geninax
CAS registry no:	194804-75-6
Molecular weight:	426.41

Garenoxacin is a new quinolone antimicrobial agent that exhibits a broad spectrum of activity against both Gram-negative and Gram-positive organisms, including the important community-acquired respiratory pathogens *S. pneumoniae, Haemophilus influenzae,* and *Moraxella catarrhalis.* In addition, it has potent activity against several resistant strains such as multidrug-resistant *S. pneumoniae,* methicillin-resistant *S. aureus* (MRSA), and vancomycin-resistant enterococci (VRE). It was launched last year in Japan as an oral treatment for respiratory tract and otorhinolaryngological infections. The recommended adult dosage of garenoxacin is 400 mg once daily. Garenoxacin differs from the conventional quinolone antibiotics by its lack of the characteristic 6-position fluorine substituent. As with other quinolone antibiotics marketed in recent years, the mechanism of action of garenoxacin involves dual inhibition of two essential bacterial enzymes, DNA gyrase and topoisomerase IV. The lack of 6-fluoro substituent does not adversely affect its potency of inhibiting DNA gyrase (*Escherichia coli*, $IC_{50} = 0.17\,\mu g/mL$) or topoisomerase IV (*S. aureus*, $IC_{50} = 2.19\,\mu g/$ mL). *In vitro,* the minimum inhibitory concentration for 90% growth reduction (MIC_{90}) values of garenoxacin against *S. pneumoniae* ($0.12\,\mu g/mL$), *H. influenzae* ($0.03\,\mu g/mL$), and *M. catarrhalis* ($0.03\,\mu g/mL$) are generally similar to those of gemifloxacin and moxifloxacin, both of which are also marketed for treating respiratory infections. However, as compared to other quinolones, garenoxacin exhibits improved PK/PD parameters, such as the high ratios between unbound fraction of 24 h AUC ($f\,AUC_{0-24}$) and the corresponding MICs for various pathogens. A high $f\,AUC_{0-24}:MIC_{90}$ ratio is generally correlated to good clinical success rate and low probability of resistance development. Against *S. pneumoniae,* the $f\,AUC_{0-24}:MIC_{90}$ ratio for garenoxacin (400 mg QD) is 130, compared to gemifloxacin (320 mg QD) and moxifloxacin (400 mg QD) ratios of 65 and 88, respectively. In human pharmacokinetic studies, oral garenoxacin results in dose-proportionate increases of plasma C_{max} and AUC values at doses ranging from 50 to 1,200 mg. Its oral bioavailability is about 95%, and the mean terminal half-life is 15.4 h. Approximately 87% of the drug is protein-bound. About 30–40% of the dose is excreted unchanged in the urine, suggesting more than one route of elimination. It does not undergo CYP450-mediated metabolism to a clinically important extent. In multiple clinical studies, oral garenoxacin for 5 or 10 days was shown to be as effective as 10-day treatment with amoxicillin/clavulanic acid in patients with acute bacterial exacerbations of chronic bronchitis (AECB) or acute bacterial sinusitis, with clinical cure rates among garenoxacin recipients ranging between 84–87% and 88–91%, respectively. In a Phase III trial comparing 5-day garenoxacin (400 QD) to 10-day amoxicillin (1 g TID) for community-acquired pneumonia (CAP), clinical response rate was 91% for garenoxacin and 87% for amoxicillin, and the overall bacterial eradication rate was 88% for garenoxacin as compared with 91% for amoxicillin. Garenoxacin had a higher eradication rate for *S. pneumoniae* and *H. influenzae* whereas amoxicillin was more effective against *Haemophilus* spp. The most common adverse events associated with garenoxacin were rash, dizziness, nausea, headache, and pruritus. Garenoxacin is chemically derived in a sequence of 14 steps, 12 of which entail the construction of a key quinolone intermediate, 7-bromo-1-cyclopropyl-8-(difluoromethoxy)-1,4-dihydro-4-oxoquinoline-3-carboxylic acid ethyl ester. Subsequently, Suzuki cross-coupling reaction of this intermediate with (R)-[1-methyl-2-(trityl)isoindolin-5-yl]boronic acid, followed by deprotection of the trityl group and ester hydrolysis with hydrochloric acid gives garenoxacin. The chiral isoindoline reagent is obtained from racemic 5-bromo-2-methylisoindoline via coupling with N-CBZ-L-phenylalanine, separation of diastereomers by chromatography, cleavage of the chiral auxiliary, N-tritylation, halogen-metal exchange with butyllithium, and boronation with triisopropyl borate.

9. IMIDAFENACIN (OVERACTIVE BLADDER) [33–37]

Country of origin:	Japan	
Originator:	Kyorin	
First introduction:	Japan	
Introduced by:	Kyorin/Ono	
Trade name:	Staybla	
CAS registry no:	170105-16-5	
Molecular weight:	319.40	

Imidafenacin, an M3/M1 muscarinic receptor antagonist, was introduced in Japan last year for the oral treatment of OAB. The majority of OAB symptoms are thought to result from overactivity of the detrusor muscle, which is primarily mediated by acetylcholine-induced stimulation of muscarinic M_3 receptors in the bladder. Previously marketed muscarinic antagonists for OAB include propiverine, tolterodine, oxybutynin, trospium, darifenacin, and solifenacin. *In vitro*, imidafenacin is equally active against M1 and M3 receptors ($K_b = 0.32$ and 0.55 nM, respectively), and approximately 10-fold less active against M2 receptors ($K_b = 4.13$ nM). By comparison, darifenacin and solifenacin show higher selectivity for M3 versus M1 and M2 receptors, whereas propiverine, tolterodine, oxybutynin, and trospium are approximately equipotent against all three receptors. In addition to receptor selectivity, the organ selectivities of antimuscarinics play an important role in their therapeutic potential and adverse effect profile. For example, blockade of M3 receptors in the salivary gland contributes to dry mouth, a frequently reported adverse event in antimuscarinic therapy. *In vivo*, in rat models, imidafenacin is more potent and more bladder selective than oxybutynin (isovolumetric bladder contraction $ED_{30} = 0.11$ versus 2.1 mg/kg, decreased bladder capacity $ED_{50} = 0.074$ versus 1.1 mg/kg, salivary secretion $ED_{50} = 1.1$ versus 3.6 mg/kg). Imidafenacin is chemically synthesized in three steps starting with alkylation of diphenylacetonitrile with dibromoethane, followed by condensation with 2-methylimidazole, and hydrolysis of the cyano group to a carboxamide group with 70% sulfuric acid. The recommended dosage of oral imidafenacin is 0.1 mg twice daily. In healthy volunteers, imidafenacin is rapidly absorbed with an absolute bioavailability of approximately 58% and maximal plasma concentrations (C_{max}) achieved in 1–3 h. Imidafenacin plasma concentrations increase dose dependently, and the elimination half-life is approximately 3 h. Approximately 15% of the imidafenacin dose is excreted unchanged in the urine, and the remainder as metabolites in the urine and the feces. Imidafenacin is primarily metabolized by CYP3A4 and UGT1A4. N-glucuronide conjugate and products derived from oxidation of the imidazole ring are the major metabolites identified in human plasma after oral administration. Accumulation of imidafenacin in the plasma and urine is shown to be insignificant during multiple oral administration of 0.25 mg doses twice daily. The clinical efficacy and safety of imidafenacin have been evaluated in a randomized, double-blind, placebo-controlled Phase III study in OAB patients ($n = 750$). The enrollment criteria included ⩾20-year-old patients who experienced five or more urinary incontinence episodes per week and had a mean number of micturition per day of ⩾8. Although the study was completed in 2007, the clinical outcome data was not available in the public domain at the time of this writing.

10. IXABEPILONE (ANTICANCER) [38–42]

Country of origin:	US
Originator:	BMS
First introduction:	US
Introduced by:	BMS
Trade name:	Ixempra
CAS registry no:	219989-84-1
Molecular weight:	506.70

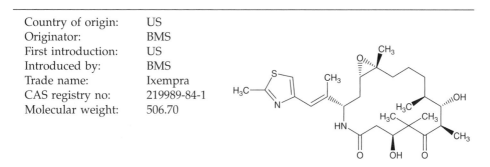

Ixabepilone, a semisynthetic analog of epothilone B, was launched last year for the treatment of metastatic or locally advanced breast cancer. It is indicated for use in combination with capecitabine in patients who have previously failed treatment with an anthracycline such as doxorubicin and a taxane such as paclitaxel. It is also approved as monotherapy for the treatment of metastatic or locally advanced breast cancer in patients whose tumors are resistant or refractory to anthracyclines, taxanes, and capecitabine. Ixabepilone is the first member of the epothilone family of anticancer agents to be approved. Epothilones are novel cytotoxic macrolides derived from bacterial fermentation. Like the taxanes, their mechanism of action involves binding to and stabilizing microtubules, which results in mitotic arrest and apoptosis. However, the tubulin-binding properties of taxanes and epothilones are not identical and show distinct contact sites, drug-induced resistance mechanisms, and sensitivity to β-tubulin mutations. Epothilones A and B exhibit potent *in vitro* anticancer activity, including activity against taxane-resistant cell lines, but their *in vivo* activity is modest, owing to poor metabolic stability and unfavorable pharmacokinetics. Ixabepilone is a semisynthetic analog of epothilone B, wherein the naturally existing lactone moiety is replaced by a lactam. It is synthesized from epothilone B in three steps via cleavage of the lactone group with sodium azide and tetrakis(triphenylphosphine) palladium to an azido carboxylic acid intermediate, which is subsequently reduced to the corresponding amino acid with hydrogen over platinum oxide and cyclized to the lactam by means of diphenylphosphoryl azide. *In vitro*, ixabepilone demonstrates a similar spectrum of cytotoxic activity as epothilone B against 21 different cancer cell lines ($IC_{50} = 1.4–34.5\,nM$), including three different cell lines with paclitaxel resistance. It is as potent as epothilone B in a microtubule stabilization assay and as effective in its ability to arrest proliferating tumor cells in mitosis. Ixabepilone is a parenterally administered drug. The recommended dosage is $40\,mg/m^2$ administered via intravenous infusion over 3 h every 3 weeks. Plasma concentrations typically peak toward the end of a 3-h infusion. Approximately 67–77% of the administered dose is bound to plasma proteins. Ixabepilone is extensively metabolized by CYP3A4 in the liver. Dose adjustments may be necessary during concomitant administration of ixabepilone with CYP3A4 inhibitors and inducers. Ixabepilone is primarily eliminated as metabolites in the feces (65%) and urine (21%). The terminal elimination half-life of ixabepilone is approximately 52 h. The efficacy of ixabepilone in combination with capecitabine for treating anthracycline- and taxane-resistant breast cancer was evaluated in a randomized, open-label clinical trial ($n = 752$). Patients received either ixabelipone ($40\,mg/m^2$ i.v. infusion once every 3 weeks) plus capecitabine ($1,000\,mg/m^2$ twice daily for 2 weeks followed by 1 week rest) for a median of five cycles, or capecitabine alone ($1,250\,mg/m^2$ twice daily for 2 weeks followed by 1 week rest) for a median of four cycles. The combination arm prolonged progression-free survival (PFS) compared to capecitabine alone (5.8 versus 4.2 months; $p = 0.0003$). The objective tumor response rate was also greater in the combination arm (35% versus 14%; $p < 0.0001$). The efficacy of ixabepilone as monotherapy (40 mg/m^2 i.v. infusion once every 3 weeks) was assessed in a single-arm trial involving 126 patients with tumors that had recurred or progressed after $\geqslant 2$ chemotherapy regimens including an anthracycline, a taxane, and capecitabine. Patients received a median of four treatment cycles. Independent radiologic review demonstrated an objective tumor response rate of 12.4% (95% CI, 6.9–19.9%) and a median duration of response of 6 months (95% CI, 5.0–7.6%). The most common adverse reactions ($\geqslant 20\%$) associated with ixabelipone as monotherapy or in combination with capecitabine were peripheral sensory neuropathy, fatigue/asthenia, myalgia/arthralgia, alopecia, nausea, vomiting, stomatitis/mucositis, diarrhea, and musculoskeletal pain. The most common hematologic abnormalities ($> 40\%$) include neutropenia, leukopenia, anemia, and thrombocytopenia. Ixabelipone in combination with capecitabine is contraindicated in patients with AST or ALT $> 2.5 \times$ ULN (upper limit of normal) or bilirubin $> 1 \times$ ULN due to increased risk of toxicity and neutropenia-related death.

11. LAPATINIB (ANTICANCER) [43–47]

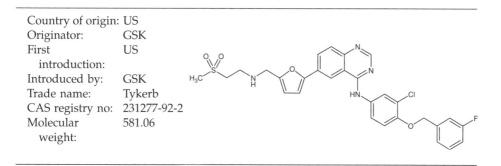

Country of origin: US
Originator: GSK
First   US
  introduction:
Introduced by: GSK
Trade name: Tykerb
CAS registry no: 231277-92-2
Molecular   581.06
  weight:

Lapatinib, a new member of the 4-anilinoquinazoline class of RTK inhibitors (RTKIs), was launched last year as an oral treatment for breast cancer. Lapatinib has dual affinity for EGFR and HER2 tyrosine kinases. It is indicated in combination with capecitabine for treating patients with advanced or metastatic breast cancer whose tumors overexpress HER2 and who have received prior therapy including an anthracycline, a taxane, and trastuzumab. Previously marketed drugs from the 4-anilinoquinazoline class include erlotinib (Tarceva®) and gefitinib (Iressa™), both of which are indicated for treating non-small-cell lung cancer (NSCLC). As with erlotinib and gefitinib, lapatinib is an ATP-competitive kinase inhibitor. It inhibits the tyrosine kinase activity EGFR and HER-2 with apparent K_i values of 3 and 13 nM, respectively, and has slow off-rate kinetics ($t_{1/2} \geqslant 300$ min). Oral absorption of lapatinib is incomplete and variable. Peak plasma concentrations (C_{max}) are achieved approximately 4 h after administration, and steady state is achieved within 6–7 days of daily dosing. Systemic exposure to lapatinib increases when administered with food. In addition, dividing the daily dose of lapatinib results in approximately 2-fold higher exposure at steady state compared to the same total dose administered once daily. The terminal half-life following a single dose of lapatinib is 14.2 h, and accumulation with repeated dosing indicates an effective half-life of 24 h. Lapatinib has high protein binding (>99%). *In vitro* studies indicate that lapatinib is a substrate for the efflux transporters breast cancer resistance protein and Pgp. Elimination of lapatinib is primarily through metabolism, with approximately 27% of the oral dose excreted unchanged in feces and <2% in urine. The metabolism of lapatinib is mediated mainly by CYP3A4 and CYP3A5, with minor contributions from CYP2C19 and CYP2C8, to a variety of oxidation products. The recommended dose of lapatinib is 1,250 mg (five tablets) given orally once daily on days 1–21 continuously in combination with capecitabine 2,000 mg/m^2/day (taken orally in 2 doses approximately 12 h apart) on days 1–14 in a repeating 21-day cycle. The clinical efficacy of lapatinib in combination with capecitabine was demonstrated in a Phase III trial in 399 patients with locally advanced or metastatic breast cancer refractory to anthracyclines, taxanes, and trastuzumab. Subjects were randomized to receive either lapatinib 1,250 mg/day plus capecitabine, or capecitabine alone. The primary endpoint was time to progression (TTP) defined as time from randomization to tumor progression or death related to breast cancer. The median TTP was 23.9 weeks for the combination treatment versus 18.3 weeks for capecitabine alone, with a response rate of 31.8% versus 17.4%, respectively. The most common adverse events associated with lapatinib treatment include diarrhea, hand–foot syndrome, nausea, rash, vomiting, and fatigue. Lapatinib has been reported to decrease left ventricular ejection fraction (LVEF), a measure of the strength of the heart's pumping capacity. LVEF should be assessed in all patients before and during treatment. QT prolongation also has been reported in patients treated with lapatinib. Caution should be exercised in administering this agent to patients who have or who may develop prolongation of QTc. In addition, dose adjustments may be necessary when administered concomitantly with drugs that are strong inhibitors or inducers of CYP3A4. The chemical synthesis of lapatinib entails the condensation of 4-chloro-6-iodoquinazoline and 3-chloro-4-(3-fluorobenzyloxy)aniline to produce a diaryl amine intermediate followed by Stille coupling of the iodo group with 5-dioxolanyl-2-(tributylstannyl)furan and subsequent acid hydrolysis of the cyclic ketal to the corresponding aldehyde. Finally, reductive amination of the aldehyde intermediate with 2-(methanesulfonyl) ethylamine in the presence of sodium triacetoxyborohydride produces lapatinib.

12. LISDEXAMFETAMINE (ADHD) [48–51]

Country of origin:	US
Originator:	New River Pharmaceuticals
First introduction:	US
Introduced by:	Shire
Trade name:	Vyvanse
CAS registry no:	608137-33-3
Molecular weight:	455.59

ADHD is a neurobehavioral disorder characterized by varying degrees of inattention, hyperactivity, and impulsivity. ADHD is typically diagnosed in childhood and affects 7–12% of the pediatric population in the United States with the condition often enduring into adulthood. Treatment options include psychostimulant drugs, such as methylphenidate and amphetamines, and the only non-stimulant atomoxetine. While the psychostimulants have been effective, they are controlled substances due to their potential for abuse. Lisdexamfetamine, the recent introduction into the psychostimulant class, has been designed to minimize the risk of abuse potential. This L-lysine prodrug of D-amphetamine is hydrolyzed to the active drug following oral absorption; the possibility of amphetamine reaching blood levels which elicit euphoria if inhaled or injected is, therefore, diminished. In addition, the prodrug effectively establishes an extended-release benefit. While the precise mechanism of action of lisdexamfetamine in treating ADHD is not known, amphetamines are believed to inhibit the reuptake of the neurotransmitters dopamine and noradrenaline (norepinephrine), thereby increasing their presynaptic availability and release into extraneuronal space. The prodrug is constructed by the condensation of D-amphetamine with the activated ester (N-hydroxysuccinimide) of bis-*tert*-butoxycarbonyl-protected L-lysine. Lisdexamfetamine is ultimately generated by treatment with hydrochloric acid in dioxane. The formulation for this once-a-day oral medication utilizes the dimesylate salt of lisdexamfetamine in capsules containing 30, 50, or 70 mg. The recommended starting dose in children aged 6–12 years is 30 mg/day. If necessary, the dose may be increased by 20 mg/day on a weekly basis with 70 mg being the maximum recommended dose. Using the 70 mg dose, an open-label study conducted in healthy volunteers evaluated the pharmacokinetics of lisdexamfetamine for 7 consecutive days. Steady-state levels of D-amphetamine were achieved by day 5. The mean C_{max} of 90.1 ng/mL was reached with a mean T_{max} of 3.7 h. The AUC_{0-24h} was 1,113 ng·h/mL, and the mean plasma elimination half-life was 10.1 h. Following the final dose on day 7, lisdexamfetamine was completely eliminated by 6 h while D-amphetamine persisted for 48 h. After oral administration of radiolabeled lisdexamfetamine, the prodrug was rapidly absorbed from the gastrointestinal tract and converted to L-lysine and the active D-amphetamine by first-pass intestinal and/or hepatic metabolism. The cytochrome P450 enzymes are not involved in the metabolism, and in combination with data from *in vitro* inhibition of the CYP isozymes, lisdexamfetamine should possess a low potential for drug–drug interactions. Approximately 96% of the radioactivity was recovered in the urine with only 0.3% found in the feces. Regarding the efficacy of lisdexamfetamine, a double-blind, randomized, placebo-controlled study was conducted in 290 children aged 6–12 years who met DSM-IV criteria for ADHD. Over the course of 4 weeks, patients were randomized to a fixed dose of lisdexamfetamine (30, 50, or 70 mg) or placebo once daily in the morning. Employing the ADHD rating scale, significant improvements in patient behavior were observed at endpoint for all doses compared to placebo. While the mean effects were similar for all doses, the 70-mg dose was superior to the lower two doses. Furthermore, as assessed by the Connor's Parent rating scale, the effects were maintained throughout the day with culmination at the 6 p.m. evaluation. The most common adverse events, comparable to other amphetamine formulations, were decreased appetite, insomnia, upper abdominal pain, and irritability. Lisdexamfetamine is contra-indicated in patients with advanced arteriosclerosis, symptomatic cardiovascular disease, moderate-to-severe hypertension, hyperthyroidism, known hypersensi-tivity to the sympathomimetic amines, glaucoma, a predisposition to agitated states, and a history of drug abuse. In addition, the drug should not be administered during or within 14 days of treatment with monoamine oxidase inhibitors. It has also been noted that psychostimulants may exacerbate symptoms of pre-existing psychotic disorders, so caution and close observation are recommended in this patient population.

13. MARAVIROC (ANTI-INFECTIVE-HIV) [52–55]

Country of origin:	US
Originator:	Pfizer
First introduction:	US
Introduced by:	Pfizer
Trade name:	Selzentry
CAS registry no:	376348-65-1
Molecular weight:	513.67

Maraviroc is the first CCR5 receptor antagonist that has been developed and launched for the treatment of HIV-1. Maraviroc binds in a slowly reversible, allosteric manner to CCR5, which is one of two principle chemokine co-receptors for viral entry into the host cell, the other being CXCR4. Binding of maraviroc to CCR5 induces conformational changes within the chemokine receptor, thereby preventing CCR5 binding to the viral gp120 protein and the ultimate CCR5-mediated virus-cell fusion that is a prerequisite for HIV invasion. Maraviroc, with its unique mechanism of action as a fusion inhibitor, joins the greater than 20 marketed antiretrovirals, including nucleotide reverse transcriptase inhibitors (NRTIs), non-nucleoside reverse transcriptase inhibitors (NNRTIs), and PIs. It is approved for use in combination with these other antiretroviral drugs in adult patients with R5-tropic HIV-1 infection (but not X4 or dual/mixed tropic HIV-1). With an IC_{50} of 11 nM for inhibition of gp120 binding to CCR5 and enhanced solubility compared to most HIV drugs, maraviroc has a lower pill burden (one pill BID). The synthesis of maraviroc involves the convergent connection of a triazole-substituted tropane moiety, a phenylpropyl fragment with a benzylic chiral center, and a 4,4-difluorocyclohexyl unit. In the presence of aqueous HCl and sodium acetate, 2,5-dimethoxytetrahydrofuran is cyclized with benzylamine and 2-oxomalonic acid to afford 8-benzyl-8-azabicyclo[3.2.1]octan-3-one. The ketone is converted to an amine via reduction of an intermediate oxime. Carbodiimide-mediated coupling of this amine with isobutyric acid yields the isobutyramide that is subsequently cyclized to the 1,2,4-triazole with acetic hydrazide. The benzyl-protected amine is then liberated by transfer hydrogenation (ammonium formate and palladium hydroxide) and subjected to reductive amination with 3(*S*)-(*tert*-butoxycarbonylamino)-3-phenylpropionaldehyde by means of sodium triacetoxyborohydride. Removal of the BOC-protecting group establishes the handle for the final amide coupling with 4,4-difluorocyclohexane carboxylic acid to provide maraviroc. The drug is administered orally (23–33% absolute bioavailability, higher for higher dose), and its dose is dependent on concomitant therapy (150, 300, or 600 mg twice daily). The pharmacokinetic properties were evaluated in a Phase I study in healthy male volunteers for 12 days. Following a 100-mg BID dose, a C_{max} of 131–155 ng/mL was achieved with a T_{max} of 2.4–3.3 h. The AUC ranged from 487 to 619 ng·h/mL with a terminal half-life of 16.1–17 h and a clearance of 10.5 ± 1.3 mL/min/kg. Approximately 75% of maraviroc was plasma protein-bound while the volume of distribution was estimated to be approximately 2.8 L/kg. In a dose range of 30–300 mg, a non-linear increase in dose-normalized exposure, evidenced by increases in both AUC and C_{max}, was observed accompanied by an increase in the rate of absorption. Saturation of Pgp-mediated efflux at higher doses was most likely responsible for this observation. CYP3A4 is the primary cytochrome P450 responsible for the 65% metabolism of maraviroc. Of the unchanged drug, 26.4% was eliminated via the feces and 8.3% in the urine. While maraviroc concentrations were reduced when co-administered with food, no food restrictions were deemed necessary since the change in plasma concentration did not adversely affect viral load. The safety and efficacy of maraviroc was gleaned from the 24-week analyses of two ongoing, double-blind, randomized, placebo-controlled studies involving more than 600 R5 HIV-1-infected patients with previous antiretroviral treatment experience. The prerequisite for patient entry was an HIV-1 RNA of >5,000 copies/mL despite prior therapy of at least 6 months with more than one antiretroviral from three classes (NRTIs, NNRTIs, and PIs (≥ 2 for this class)). Excluding low-dose ritonavir, all patients received an optimized background regimen (OBR) of three to six antiretrovirals derived from a historical assessment of each patient's baseline genotypic and phenotypic viral resistance measurements. Once the OBR was established, patients were randomized to maraviroc or placebo. The proportion of patients with HIV-1 RNA <400 copies/mL after 24 weeks of treatment was 61% for the maraviroc group compared to 28% for placebo. From baseline to week 24, the mean changes in plasma HIV-1 RNA was $-1.96 \log_{10}$ copies/mL following maraviroc therapy and $-.99 \log_{10}$ copies/mL for placebo treatment. Overall, maraviroc was well tolerated with the most common adverse events being cough, fever, colds, rash, muscle and joint pain, stomach pain, and dizziness. While some patients did experience liver enzyme elevation, these events did not appear to be dose-related. Since hepatotoxicity did occur in one patient with prior liver function abnormalities, maraviroc's label warns of a potentially increased risk of hepatoxicity with treatment. Postural hypotension was also observed in a dose-dependent manner; however, no patients discontinued therapy as a result. As a substrate for CYP3A4, the dose of maraviroc should be reduced by 50% in the presence of strong CYP3A4 inhibitors. Conversely, concomitant use of strong CYP3A4 inducers requires a 50% increase in maraviroc dose. The recommended dose of maraviroc is 150 mg BID in patients receiving strong CYP3A4 inhibitors including PIs and delaviridine with or without CYP3A4 inducers; with concomitant CYP3A4 inducers without any CYP3A4 inhibitors, the recommended dose is 600 mg BID. While there are no contraindications, maraviroc should be used with caution in patients with liver dysfunction, high risk of cardiovascular events, and pre-existing postural hypotension.

14. NILOTINIB (ANTICANCER — CML) [56–59]

Country of origin:	Switzerland
Originator:	Novartis
First introduction:	Switzerland
Introduced by:	Novartis
Trade name:	Tasigna
CAS registry no:	641571-10-0
Molecular weight:	529.52

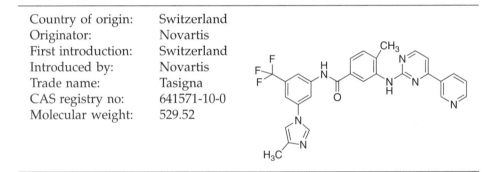

Chronic myeloid leukemia (CML), a hematological stem-cell disorder, is definitively diagnosed by the detection of the Philadelphia chromosome, a truncated version of chromosome 22 resulting from the reciprocal translocation of chromosomes 9 and 22 induced by a single mutagenic event. The consequence is the juxtaposition of two genes creating a fusion gene BCR-ABL. This gene leads to the translation of a fusion protein with increased tyrosine kinase activity that contributes to the pathogenesis of CML. Targeting the BCR-ABL protein has led to the successful intervention of the disease. Now established as first-line therapy for CML, imatinib was the first selective tyrosine kinase inhibitor of BCR-ABL. Since imatinib only binds to an inactive conformation of the ABL kinase portion, the conformational restrictions contribute to its selectivity. While imatinib has produced clinical remissions in most patients, encounters with subsequent drug resistance are increasing. The culprit is the emergence of mutations in the kinase domain of BCR-ABL that interfere with drug binding. Last year, dasatinib was launched with less stringent conformational constraints for ABL binding and demonstrated activity against all imatinib-resistant mutants except the T315I mutant. Nilotinib is the newest option for imatinib-resistant CML. Nilotinib and imatinib are phenylaminopyrimidinyl pyridine derivatives that differ in the substituent attached to the central phenyl ring. Replacement of the N-methylpiperazine portion of imatinib with the imidazophenyl amide in nilotinib increases the affinity for the inactive conformation of wild-type BCR-ABL by 20–30-fold; nilotinib reduced cellular BCR-ABL autophosphorylation in the CML cell lines K-562 and KU812F and the pro-B-cell line Ba/F3 expressing wild-type BCR-ABL with $IC_{50} = 42$, 60, and 23 nM, respectively, compared to 470, 399, and 231, respectively, for imatinib. Regarding proliferation of the same cell lines, nilotinib again displayed enhanced potency with $IC_{50} = 11$, 8, and 23 nM, respectively, compared to imatinib ($IC_{50} = 272$, 80, and 643 nM, respectively). Against the imatinib-resistant BCR-ABL mutants M351, F317L, and E255V, nilotinib displayed IC_{50} values for inhibition of autophosphorylation of 3, 43, and 245 nM, respectively; however, nilotinib was ineffective against the T315I mutant. Similar to imatinib ($IC_{50} = 74$ and 96 nM, respectively), nilotinib inhibited the cellular tyrosine kinase activity of platelet-derived growth factor receptor β ($IC_{50} = 85$ nM) and c-Kit ($IC_{50} = 192$ nM). The first step in the synthesis of nilotinib involves the nucleophilic aromatic substitution of 3-fluoro-5-(trifluoromethyl)benzonitrile with 2-methylimidazole. The nitrile is then hydrolyzed with sodium hydroxide in aqueous dioxane. A Curtius rearrangement employing diphenylphosphoryl azide in tert-butanol affords the tert-butyl carbamate. Deprotection of the Boc group provides the 3-(4-methylimidazol-1-yl)-5-(trifluoromethyl)aniline piece for the convergent synthesis. Construction of the other half begins with the condensation of 3-amino-4-methylbenzoic acid methyl ester with cyanamide in refluxing ethanolic HCl to generate the 3-guanidinobenzoate. An enamino ketone, prepared by a Claisen condensation of 3-acetylpyridine with ethyl formate in the presence of sodium metal in hot toluene, is then cyclized with the guanidine to yield the pyridylpyrimidine. Following saponification of the ethyl ester, the resultant 4-methyl-3-[4-(3-pyridyl)pyrimidin-2-ylamino]benzoic acid is finally coupled with the aniline utilizing diethyl cyanophosphate to provide nilotinib. The drug is formulated in 100-mg hard capsules, and the recommended dose of nilotinib is 400 mg orally BID, approximately 12 h apart. Food should not be consumed 2 h before and 1 h after dosing; an 82% increase in AUC was observed when the dose was given 30 min after a high-fat meal. Following 400 mg/BID dosing in patients with imatinib-resistant, Ph+ CML, the drug is absorbed rapidly with a T_{max} of 3 h. The mean peak steady-state serum concentration was 3.6 µmol/L, and the mean trough serum concentration, at steady state, was 1.7 µmol/L. Steady-state conditions were achieved by day 8. While the volume of distribution has not been reported, the blood-to-serum ratio was 0.68. The apparent elimination half-life was approximately 17 h. Using radioactive nilotinib, over 75% of dose recovered was intact parent; the major metabolite (approximately 7%) was the carboxylic acid, which did not contribute significantly to the pharmacological activity of nilotinib.

The predominant route of excretion was via the feces. Since nilotinib is metabolized by CYP3A4, strong inhibitors and inducers of this CYP isoform can significantly affect nilotinib concentrations; therefore, concomitant treatment should be avoided. Nilotinib is also an inhibitor of CYP3A4, CYP2C8, CYP2C9, CYP2D6, UGT1A1, and Pgp, so caution should be exercised with co-administration of substrates of these enzymes. The efficacy of nilotinib was evaluated in several Phase II clinical studies; however, this discussion will focus on the results in patients with chronic phase CML ($n = 316$) with imatinib resistance or intolerance. Following an oral dosing regimen of nilotinib at 400 mg/BID, permitting dose escalation to 600 mg/BID in patients with inadequate response, the complete hematologic response rate was 74% after a median follow-up of 6 months. Furthermore, major cytogenetic responses were recorded in 52% of patients, of which 34% were complete. From this data, the estimated 1-year survival rate was 95%. The most common adverse events included rash, nausea, pruritus, headache, fatigue, constipation, diarrhea, and vomiting. Regarding the serious adverse reactions, QT prolongation and sudden deaths are at the top of the list and are captured in a black box warning in the prescribing information. The drug is contraindicated in patients with long QT interval and in patients with hypokalemia and hypomagnesemia since these conditions promote QT prolongation. Ventricular repolarization abnormalities have been implicated in the sudden death situations. Myelosuppression, associated with neutropenia, thrombocytopenia, and anemia, may also occur, but it is reversible upon dose withholding. Serum lipase and liver enzyme (billirubin, AST/ALT, and alkaline phosphatase) elevations are a concern, so these enzyme levels should be monitored periodically. In addition, nilotinib may cause electrolyte abnormalities (hypophosphatemia, hypokalemia, hyperkalemia, hypocalcemia, and hyponatremia), and verified acceptable levels are required before commencing treatment. Finally, women are advised not to become pregnant while taking nilotinib since fetal harm may occur.

15. PALIPERIDONE (ANTIPSYCHOTIC) [60–63]

Country of origin:	US
Originator:	Johnson & Johnson
First introduction:	US
Introduced by:	Johnson & Johnson
Trade name:	Invega
CAS registry no:	144598-75-4
Molecular weight:	426.4840

Schizophrenia is diagnosed by the presence of positive psychotic symptoms, such as delusional thoughts, auditory hallucinations, and irrational fears. In addition to these symptoms, many patients also experience the negative symptoms of social alienation, difficulty articulating, apathy, and lack of energy.

The first generation drugs, known as typical antipsychotics, antagonize only the dopamine D_2 receptors. Chlorpromazine falls into this category, and while it demonstrates efficacy against psychosis, it is ineffective against the negative symptoms, and dose-limiting extrapyrimidal symptoms (EPS) are a concern. The second generation, or atypical antipsychotics, antagonizes both the mesolimbic pathway dopamine D_2 receptors and the serotonin 5-HT_{2A} receptors in the prefrontal cortex. This dual antagonism retains the antipsychotic activity with an improvement in the negative symptoms and a reduction in EPS. Paliperidone, the C-9 hydroxylated active metabolite of the antipsychotic agent risperidone, is the newest atypical antipsychotic to join the market following the introductions of olanzapine (Zyprexa^{TM}), risperidone (Risperdal^{TM}), quetiapine(Seroquel^{TM}), and ziprasidone (Geodon^{TM}). Compared to its parent, paliperidone has improved PK properties and a reduced potential for drug interactions. In terms of receptor affinity, the two drugs are equipotent; the K_i values for binding to D_2 and 5HT_{2A} for risperidone and paliperidone are 5.9 and 4.8 nM, and 0.16 and 0.25 nM, respectively. The synthesis of paliperidone involves the reaction of 2-acetyl-γ-butyrolactone with 2-amino-3-benzyloxypyridine in the presence of phosphoryl chloride. The benzyl protecting group of the intermediate 9-benzyloxy-3-(2-chloroethyl)-2-methylpyrido[1,2-a]pyrimidin-4-one is then removed by hydrogenolysis over Pd/C followed by the nucleophilic displacement of the chlorine with 6-fluoro-3-(4-piperidinyl)-1,2-benzisoxazole to provide racemic paliperidone. While paliperidone can be resolved, both enantiomers are equipotent and interconvert *in vivo* obviating the need for separation. The formulation of paliperidone utilizes an OROS^{TM} for sustained plasma levels over a 24-h period. This extended release formulation permits once-daily administration, and by reducing fluctuations in peak-to-trough concentrations, initial dose titration is not necessary. Fluctuation-related adverse effects should also be minimized. The pharmacokinetic properties of paliperidone were evaluated in healthy volunteers. Compared to risperidone, paliperidone experienced a greater than 50% reduction in oral bioavailability (28% versus 70% for risperidone). Within a dose range of 3–12 mg, a linear dose-response in pharmacokinetic properties was observed. Following a single dose of 6 mg, a mean C_{max} of 11.7 ng/mL was achieved after a T_{max} of 25.1 h; this resulted in an AUC of 302 ng·h/mL. Steady-state levels were reached within 4–5 days. At 22 h, a median D_2 receptor occupancy of 64% was realized dropping to 53% at 46 h. Based on this data, the estimated efficacious dose was predicted to be >3 mg. The apparent volume of distribution was 487 L, but there were differences between the two interconverting enantiomers; the median volume of distribution was 192 L for the (−) enantiomer compared to 70.6 L for the (+) enantiomer. The enantiomers also had significantly different clearances — 8.15 L/h for the (−) enantiomer versus 1.41 L/h for the (+) enantiomer. The interconversion clearance was 7.9 L/h in both directions. Racemic paliperidone was approximately 74% plasma protein-bound, and the terminal half-life was approximately 23 h. While ^{14}C-labeled paliperidone was excreted predominantly via the renal route (79.6%), the majority of the dose (approximately 60%) was eliminated unchanged. The CYP isoenzymes CYP2D6 and, to a lesser extent, 3A4 were responsible for the minor metabolites formed from dealkylation, hydroxylation, dehydrogenation, and benzisoxazole scission. Preliminary data showed that co-administration of a potent CYP2D6 inhibitor, paroxetine, slightly increased the C_{max} and AUC of paliperidone, but the increase was not considered statistically significant. The safety and tolerability of paliperidone ER were evaluated in a double-blind study in elderly schizophrenic patients ($n = 114$). Over 6 weeks, patients were randomized to receive a daily oral dose of paliperidone ER (3–12 mg with a 3 mg incremental option from day 7) or placebo. Discontinuation due to adverse effects was similar to placebo (7% and 8%, respectively). Overall, paliperidone was well tolerated with hypertonia and tremor (3% each) surfacing only in the treatment group. Also for a 6-week duration, the antipsychotic efficacy of paliperidone ER was evaluated in three similar double-blind, parallel group, dose-finding studies in young adults ($n = 243$ with a mean age of 22.5 years). With a baseline mean positive and negative syndrome scale (PANSS) score of 95, patients were randomized to receive paliperidone ER at doses of 3, 6, 9, 12, or 15 mg once daily, placebo, or olanzapine 10 mg once a day (for study validation). PANSS scores increased in a dose-dependent manner from −16 (3 mg) to −23 (15 mg) compared to −4 for placebo. The personal and social performance (PSP) scale also improved from +7.2 to +15.8 in this dose range compared to −1.2 for placebo. The frequency of adverse events was similar for the treatment groups (64–76%) and placebo (70%). The results were similar for olanzapine which had a mean improvement of −20.4 in PANSS score, a mean improvement of +11 in PSP, and a 78% incidence of adverse events. The most common adverse events included tachycardia, QTc prolongation, headache, anxiety, dizziness, tremors, and insomnia along with the dose-related events of somnolence, orthostatic hypotension, salivary hypersecretion, and extrapyramidal disorder. Weight gain was also observed in 6–9% of patients which may be attributable to paliperidone's lower affinity for H_1-histaminergic and α1- and α2-adrenergic receptors. Patients with renal impairment require dose adjustments since elimination of paliperidone is altered. Paliperidone is contraindicated in patients with a hypersensitivity to risperidone. Concomitant use of class III antiarrhythmic agents should be avoided since this may result in additive QT interval prolongation. Also, loss of levodopa efficacy is expected with this D_2 antagonist. Finally, paliperidone is marketed with a black box warning of increased mortality in elderly patients with dementia-related psychosis.

16. RALTEGRAVIR (ANTI-INFECTIVE-HIV) [64–67]

Country of origin:	US
Originator:	Merck
First introduction:	Canada
Introduced by:	Merck
Trade name:	Isentress
CAS registry no:	518048-05-0 (free base)
CAS registry no:	871038-72-1 (monopotassium salt)
Molecular weight:	446.43 (free base)
Molecular weight:	482.51 (monopotassium salt)

Joining maraviroc as a unique approach to battling HIV-1, raltegravir, an inhibitor of HIV-1 integrase, represents the first in its class to be developed and launched as a combination treatment with other antiretroviral agents (NRTIs, NNRTIs, and PIs). HIV-1 integrase is essential for replication of the virus as a virally encoded enzyme that integrates the viral DNA into the genome of the host cell. Inhibition of HIV-1 integrase prevents the two-step process of endonucleolytic removal of the terminal dinucleotide from each 3′ end of the viral DNA followed by the covalent integration of the viral DNA, at these modified 3′ ends, into the host DNA, thereby representing a viable intervention in the viral life cycle. *In vitro*, raltegravir inhibited the strand transfer activity of HIV-1 integrase with an IC_{50} of 2–7 nM with $> 1,000$-fold selectivity over other phosphoryltransferases. In addition, its *in vitro* IC_{95} for HIV-1 in 10% fetal bovine serum and 50% human serum was 19 and 33 nM, respectively. As with its diketo acid structural predecessors, raltegravir coordinates to divalent metal ions in the integrase active site. The synthesis of raltegravir begins with the treatment of acetone cyanohydrin with liquid ammonia in a pressure vessel. The resulting aminonitrile is protected as the benzyl carbamate before reaction of the nitrile moiety with hydroxylamine to afford the amidoxime. The pyrimidone ring is then constructed by condensation with dimethyl acetylenedicarboxylate and subsequent cyclization in hot xylene. Methylation of the pyrimidone is performed next with iodomethane and magnesium methoxide in dimethylsulfoxide followed by conversion of the methyl ester to an amide with 4-fluorobenzylamine. The amine, liberated from hydrogenolytic removal of the carbobenzyloxy-protecting group, is acylated with oxadiazolecarbonyl chloride, prepared in three steps from 5-methyltetrazole, to afford raltegravir. The drug is formulated as its crystalline potassium salt in 400-mg tablets. The recommended daily dose is 400 mg orally BID. While the absolute bioavailability of raltegravir has not been determined, the drug is absorbed with a T_{max} of approximately 3 h, and the AUC and C_{max} increase dose proportionally over a range of 100–1,600 mg. Following a 400-mg dose BID, a geometric mean AUC_{0-12h} of 14.3 µM.h and C_{12h} of 143 nM are realized. Within 2 days of multiple dosing, steady state is achieved. The pharmacokinetic properties are not affected by food intake, and the protein binding of raltegravir is approximately 83%. Elimination occurs in a biphasic manner with a terminal half-life of 7–12 h. Using ^{14}C-labeled drug, most of the dose is recovered within 24 h in urine and feces (32% versus 51%, respectively). Raltegravir is metabolized primarily by glucuronidation by the enzyme UDP-glucuronosyltransferase 1A1 (UGT1A1). Since it is not a substrate, nor an inhibitor or inducer, of the CYP450 isoenzymes, it is not expected to have drug interactions with other antiretrovirals or concomitantly administered drugs metabolized by the CYP450 pathway. Conversely, dose adjustments may be necessary with co-administration of strong inducers and inhibitors of UGT1A1. For example, co-dosing with the UGT1A1 inducer rifampin resulted in a 40% decrease in raltegravir AUC. Similar to the clinical studies with maraviroc, the efficacy and safety of raltegravir were determined from analyses of 24-week data from two ongoing, randomized, double-blind, placebo-controlled trials with patients ($n = 436$) on an optimized background therapy (OBT). Patients were randomized to receive either raltegravir (400 mg BID orally) or placebo in combination with OBT. For inclusion, patients had to possess a documented resistance to at least one drug in each of the three traditional antiretroviral classes (NRTIs, NNRTIs, and PIs). At the 24-week mark, 76% of patients in the raltegravir arm displayed HIV-1 RNA < 400 copies/mL compared to 39% in the placebo arm. Furthermore, the proportion of patients with HIV-1 RNA < 50 copies/mL was 63% with raltegravir treatment versus 33% with placebo. Following raltegravir treatment, the mean change in plasma HIV-1 RNA from baseline was greater than placebo ($-1.85 \log_{10}$ copies/mL for raltegravir and $-0.84 \log_{10}$ for placebo). In accordance with the reduction in viral load, $CD4^+$ cell counts also increased from baseline for the raltegravir arm (89 cells/mm^3) versus placebo (33 cells/mm^3). Overall, raltegravir was well tolerated with no dose-related toxicities and a safety profile comparable to placebo. The most common clinical adverse events were diarrhea, nausea, vomiting, fatigue, headache, flushing, pruritus, and injection-site reactions.

17. RETAPAMULIN (ANTI-INFECTIVE) [68–71]

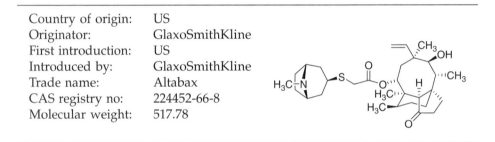

Country of origin:	US
Originator:	GlaxoSmithKline
First introduction:	US
Introduced by:	GlaxoSmithKline
Trade name:	Altabax
CAS registry no:	224452-66-8
Molecular weight:	517.78

Typical treatments of impetigo, a highly contagious skin infection caused by bacteria such as *S. aureus* and/or *S. pyogenes*, include the topical antibiotics mupirocin, fusidic acid, and bacitracin. The emergence of multidrug-resistant pathogens, however, has necessitated the development of novel antibiotics with alternative mechanisms of action. Retapamulin has surfaced from a reinvestigation of the pleuromutilin class of agents. Pleuromutilin is a tricyclic diterpenoid that was first isolated in 1951 from the edible mushroom *Pleurotus mutilus*. The first semisynthetic analogs tiamulin and valnemulin, developed for veterinary use, have been shown to interact uniquely with bacterial ribosomes by high affinity binding to a site on the 50S subunit. Binding to this site interferes with ribosomal peptidyl transferase activity, blocks P-site interactions, and prevents the evolution of active 50S ribosomal subunits. Retapamulin, the first pleuromutilin approved for human use, behaves similarly to selectively inhibit bacterial protein synthesis. This novel mechanism of action has been implicated in the lack of *in vitro* target-specific cross-resistance with other classes of antibiotics. Against *S. aureus* and *S. pyogenes* isolates, retapamulin exhibited MIC_{90} values of 0.12 and 0.03 µg/mL, respectively. The synthesis of retapamulin begins with generation of the mesylate of pleuromutilin, isolated through fermentation of *Clitopilus passeckerianus*, followed by nucleophilic substitution with *exo*-8-methyl-8-azabicyclo[3.2.1]octan-3-thiol under basic conditions (potassium-*tert*-butoxide in ethanol or tetrabutylammonium hydrogen sulfate in dichloromethane/water and sodium hydroxide at pH 12.5). The azabicyclic thiol derivative may be prepared via a Mitsunobu reaction between tropine and thioacetic acid. While retapamulin displayed initial evidence of antimicrobial activity, its poor pharmacokinetic properties (poor oral bioavailability and short half-life) precluded its development as an oral antibiotic. The drug was, therefore, formulated as a 1% ointment for topical administration, supplied in 5, 10, and 15 g tubes. The recommended dose is the application of a thin layer to the affected area (2% total body surface area in pediatric patients (9 months and older) or up to $100 \, cm^2$ in total area in adults) twice daily for 5 days. The pharmacokinetic properties of retapamulin were evaluated in healthy adults receiving once-daily application to intact and abraded skin surfaces for up to 7 days. While only 3% of blood samples had quantifiable levels of retapamulin at day 1, 82% of blood samples of the intact skin group elicited measurable levels on day 7 with a C_{max} of 3.5 ng/mL. In contrast, the abraded skin group reached detectable levels in 97% of blood samples on day 1 reaching 100% by day 7. In this situation, the median C_{max} was 9.0 ng/mL on day 7. Overall, the systemic exposure was low. Retapamulin that does reach the plasma is 94% protein-bound while the apparent volume of distribution has not been determined. Through *in vitro* analysis in human liver microsomes, it was determined that CYP3A4 is responsible for the majority of the metabolites of retapamulin, resulting from predominantly mono-oxygenation and *N*-demethylation. Elimination of intact retapamulin and its metabolites, however, has not been investigated in humans due to low systemic exposure. While co-administration of a known CYP3A4 inhibitor resulted in an increased serum concentration of retapamulin in humans, the overall systemic levels were still low enough to obviate the need for dose adjustments. The efficacy of retapamulin was evaluated in a double-blind, randomized, placebo-controlled study. Although the majority of patients enrolled (164 out of 210) were under the age of 13, the study included adult and pediatric patients (aged 9 months or older) with impetigo not exceeding $100 \, cm^2$ in total area (up to 10 lesions) or a total surface area not exceeding 2%. If underlying skin disease or secondary skin or systemic infection were detected, the patient was excluded. Successful endpoints were defined as the absence of treated lesions, remaining lesions were dry without crusts with or without erythema compared to baseline, or had improved to the extent that no further antimicrobial therapy was required. The data were grouped according to four populations; the intent-to-treat clinical (ITTC) population consisted of all randomized patients that received at least one dose of retapamulin, the clinical per protocol (PPC) population satisfied the inclusion/exclusion criteria and subsequently adhered to the protocol, the intent-to-treat bacteriological (ITTB) contained all randomized patients who obtained one dose of medication and had an identified pathogen at the initiation of the study, and the bacteriological per protocol (PPB) population included all ITTB patients who satisfied the inclusion/exclusion criteria and subsequently adhered to the protocol. At the completion of therapy, population success rates versus placebo were as follows: ITTC — 85.6% versus 52.1%, PPC — 89.5% versus 53.2%, ITTB — 88.6% versus 49.1%, and PPB — 89.7% versus 50.0%.

At 9 days post-treatment, the margins were further improved versus placebo: ITTC — 75.5% versus 39.4%, PPC — 82.4% versus 43.1%, ITTB — 79.8% versus 33.3%, and PPB — 84.3% versus 37.5%. The safety of retapamulin was also evaluated in over 2,000 adult and pediatric patients. The most frequent adverse event was application site irritation, but other side effects, occurring in <2% of patients, included headache, diarrhea, nausea, and nasopharyngitis. While there are no contraindications, it is recommended that pregnant women only use retapamulin when the potential benefits outweigh the potential risks since animal reproductive studies are not always predictive of human response. Likewise, nursing mothers are cautioned about the unknown possibility of retapamulin excretion in breast milk.

18. RUFINAMIDE (ANTICONVULSANT) [72–75]

Country of origin:	US	
Originator:	Novartis	
First introduction:	Germany	
Introduced by:	Eisai	
Trade name:	Inovelon	
CAS registry no:	106308-44-5	
Molecular weight:	238.20	

Approximately 2.5 million people worldwide are afflicted with epilepsy, a devastating neurological disorder diagnosed by the tendency toward recurrent, unprovoked seizures, often of unknown etiology. Despite an existing arsenal of antiepileptic drugs (phenobarbital, phenytoin, carbamazepine, valproate, lamorrigine, topiramate, zonisamide, oxcarbazepine, and levetiracetam), refractory forms persist, particularly in Lennox–Gastaut syndrome (LGS), a debilitating childhood-onset epilepsy. Rufinamide has been launched primarily as adjunctive therapy of LGS. Its proposed mechanism of action involves the limitation of firing of sodium-dependent action potentials. The ultimate result is membrane stabilization. Since it does not exhibit measurable binding to monoamine, acetylcholine, histamine, glycine, AMPA/kainate, NMDA, or GABA receptors or systems, these receptor-mediated pathways are not anticipated to be involved in the exertion of rufinamide's effects. Rufinamide displayed efficacy in several electrical and chemical animal seizure models. In the maximal electroshock (MES)-induced tonic–clonic seizure model in rodents, rufinamide inhibited seizures with ED_{50} values of 5–17 mg/kg p.o. While rufinamide may be prepared by several related routes, the preferred starting material is 2,6-difluorobenzyl azide. Reaction with either 2-propynoic acid or 2-chloroacrylonitrile affords 1-(2,6-difluorobenzyl)-1H-1,2,3-triazole-4-carboxylic acid or 1-(2,6-difluorobenzyl)-1H-1,2,3-triazole-4-carbonitrile, respectively. For the carboxylic acid intermediate, treatment with thionyl chloride followed by concentrated aqueous ammonium hydroxide or conversion to its methyl ester (methanol/sulfuric acid) with subsequent ammonolyis provides rufinamide. From the carbonitrile intermediate, sodium hydroxide treatment in hot toluene/water yields rufinamide. Available in a coated tablet or suspension form, the dose of rufinamide, delivered BID, is dependent on weight and concomitant drug usage. Rufinamide has no solubility in water and is poorly dissolved in gastric and intestinal fluids. Absorption is significantly improved with food intake, reaching 70%. In the fed state, AUC increases (57.2–81.7 mcg.h/mL) as does C_{max} (2.19–4.29 mcg.h/mL) while T_{max} decreases from 8 to 6 h. Peak plasma concentrations are reached in 5–6 h. Rufinamide is 34% protein-bound and is extensively metabolized by non-CYP450 enzymes; only 2% is excreted intact in the urine and feces. The main metabolite is derived from hydrolysis of the amide and is pharmacologically inactive. While the mean half-life is 9.5 h, the clearance is proportional to body surface area; therefore, children display a lower plasma clearance compared to adolescents and adults. The efficacy of rufinamide has been evaluated in multiple clinical trials; however, only the results involving LGS patients will be discussed here. The randomized, placebo-controlled add-on to a polytherapy trial enrolled 138 patients (aged 4–37 years) with a well-documented history of LGS with the appropriate seizure types. Before entry, all had >90 seizures in the prior month, and the number of seizure types was comparable between the placebo and treatment arms. Patients were randomized to receive twice daily doses of rufinamide (titrated to 45 mg/kg/day, achieving a median dose of 1,800 mg/day) or placebo. Following a 28-day baseline phase, a 14-day titration phase and 70 days of maintenance ensued. The primary endpoints of the study were overall decrease in seizure frequency, decrease in the frequency of tonic–atonic seizures, and decreased severity of seizures. Improvement was observed in all three parameters. Regarding the median percent reduction in seizure frequency per 28 days relative to the baseline phase, the rufinamide group reported a 32.7% reduction versus 11.7% for the placebo group. Examination of the tonic–atonic seizures revealed a 42.5% median reduction in the treatment arm compared to a 1.4% increase in the placebo arm. Finally, an improvement in seizure severity was noted in 53.4% of rufinamide subjects versus 30.6% with placebo. While no patient achieved long-term seizure freedom in either group, a subsequent, open-label observation period concluded that the benefits endured. Overall, rufinamide was well tolerated with the most common adverse events including fatigue, somnolence, tremors, mild-to-moderate dizziness, nausea, headache, and diplopia. Since rufinamide is not metabolized by the CYP450 system, it is anticipated to have a low potential for interaction with drugs metabolized by the CYP isozymes. Rufinamide, however, does exhibit clinically relevant interactions with other antiepileptic drugs; concomitant treatment with valproate results in a reduction in rufinamide clearance while concomitant treatment with phenytoin, primidone, phenobarbital, carbamazepine, or vigabatrin causes an increase in rufinamide clearance. In these situations, rufinamide dosage adjustment may be required.

19. TEMSIROLIMUS (ANTICANCER) [76–80]

Country of origin:	US
Originator:	Wyeth
First introduction:	US
Introduced by:	Wyeth
Trade name:	Torisel
CAS registry no:	162635-04-3
Molecular weight:	1,030.30

While renal cell carcinoma (RCC) accounts for only 2–3% of all cancers, the 5-year survival rate for advanced RCC disease is only 5–10%, with approximately 13,000 deaths occurring annually (US statistics only). Immunotherapeutic cytokine options, such as IFN-α and IL-2, have traditionally been frontline treatments, but these agents are not efficacious in all patients and can cause serious side effects. Sunitinib and sorafenib, which are recent introductions to the market for RCC, target vascular endothelial growth factor (VEGF), among other tyrosine kinases, to block blood vessel growth and cancer cell proliferation. In addition, bevacizumab, a monoclonal antibody against VEGF, has also demonstrated prolongation of PFS. The newest entry for this indication focuses on targets that are downstream from VEGF. Temsirolimus is an inhibitor of the serine/threonine kinase mTOR, which is the mammalian target of rapamycin. mTOR has been implicated in cell replication through control of the cell cycle translation of specific mRNAs. Inhibition of mTOR prevents phosphorylation of the 4E binding protein-1 and the 40S ribosomal protein S6 kinase that are responsible for cell cycle protein translation initiation; cell cycle arrest occurs as the result of termination of cell division from the G_1 to the S phase. Disruption of mTOR signaling also has antiangiogenic effects that could be deemed essential in combating RCC, which is driven by unregulated angiogenesis. Temsirolimus is the 2,2-bis(hydroxymethyl)propionate ester of rapamycin (sirolimus), a macrolide fungicide isolated from the bacteria *Streptomyces hygroscopicus*. Similar to its parent sirolimus, temsirolimus interacts with mTOR through its complex with FK-506 binding protein 12. Unlike its predecessor, the ester side chain imparts significantly improved aqueous solubility. The most efficient and selective synthesis involves a lipase-catalyzed acylation of the sterically less hindered 42-OH position of sirolimus with a ketal-protected vinyl ester intermediate. Subsequent acid-catalyzed deprotection provides temsirolimus. This enzyme-catalyzed route obviates the need for multiple protection/deprotection steps. Temsirolimus is administered as an infusion of a recommended 25-mg dose over a 30–60-min period once a week, typically with prior antihistamine administration. The pharmacokinetic analysis of temsirolimus was performed at a single 25-mg dose in cancer patients. A mean AUC of $1,627\,ng \cdot h/mL$ was achieved with a C_{max} of $585\,ng/mL$, typically reached at the end of infusion. Exposure increased less than proportionally with dose in the 1–25 mg range. The steady-state volume of distribution was $172\,L$, and the mean half-life was $17.3\,h$. The predominant route of excretion was via the feces with a mean systemic clearance of $16.2\,L/h$. The major metabolite, formed by CYP3A4, is the result of ester cleavage to generate sirolimus. The concentration of this metabolite increases with concomitant usage of strong inducers of CYP3A4 (dexamethasone, rifampin) and decreases with administration of CYP3A4 inhibitors (ketoconazole). Dose adjustments, should therefore, be considered if alternative treatment is not an option. In a Phase II trial, patients were randomized to receive one of three doses (25, 75, $250\,mg/m^2$). While there did not appear to be a clear dose-response for efficacy or toxicity, 17% of patients maintained stable disease for >6 months. Furthermore, responses were observed in very high-risk patients, and survival doubled compared to historical controls. A Phase III clinical trial evaluated the use of temsirolimus alone, in comparison to IFN-α, and in combination with IFN-α. Previously untreated patients (625) with advanced RCC were randomized to receive IFN-α alone (207 patients), 25-mg weekly infusions of temsirolimus, or 15-mg weekly infusions of temsirolimus in combination with IFN-α. The endpoints of the study included comparison of overall survival (OS), progression-free survival (PFS), and objective response rate (ORR). In comparison to IFN-α alone, temsirolimus treatment resulted in a statistically significant improvement in OS and PFS from randomization to death (median OS = 10.9 months and median PFS = 5.5 months for temsirolimus compared to median OS = 7.3 months and median PFS = 3.1 months for IFN-α). The combination, however, did not fare any better than IFN-α alone. The most common adverse events reported with temsirolimus treatment include maculopapular rash, mucositis, asthenia, anemia, nausea/vomiting, diarrhea, dyspnea, edema, and disturbance of lipid levels. In addition to monitoring serum cholesterol and triglycerides, glucose levels should also be tested as hyperglycemia is a possibility with temsirolimus treatment. Concomitant use of hypoglycemic and/or lipid-lowering drugs with temsirolimus may, therefore, be necessary. Infrequent but severe complications include interstitial lung disease, bowel perforation, and renal failure, so implicating symptoms should be reported promptly. Temsirolimus is contraindicated postoperatively since it may cause immunosuppression leaving the patient open to opportunistic infection. Likewise, the use of live vaccines and close contact to individuals receiving live vaccines should be avoided. Finally, pregnant women should not use temsirolimus since the drug has demonstrated teratogenic effects in animal models. Both women and men should utilize reliable contraception methods during and for 3 months following the final dose of temsirolimus.

20. TRABECTEDIN (ANTICANCER) [81–86]

Country of origin:	US
Originator:	University of Illinois
First introduction:	UK, Germany
Introduced by:	Pharma Mar/J&J
Trade name:	Yondelis
CAS registry no:	114899-77-3
Molecular weight:	761.84

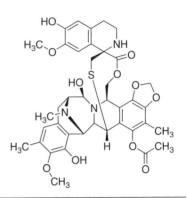

Trabectedin is a marine natural product derived from the tunicate *Ecteinascidia turbinate*. With its demonstrated *in vitro* and *in vivo* activity against a range of solid tumor cell lines, human xenografts and tumor explants, this antineoplastic agent has been developed and launched for the treatment of advanced STS after failure of first-line therapy with anthracyclines or ifosfamide or in patients who are unsuited to receive these agents. Its proposed mechanism of action involves binding to the N2 position of guanine in the minor groove demonstrating a preference for sequences containing 5′-PuGC and 5′-PyGG motifs. Subsequent alkylation of DNA, via an iminium intermediate generated from an intra-molecular acid-catalyzed activation and dehydration of the carbinolamine, induces a curvature of the DNA toward the major groove that ultimately disrupts the binding of transcription factors involved in cell proliferation. Evaluation of trabectedin against the National Cancer Institute's human *in vitro* cell line panel, including melanoma, non-small-cell lung, ovarian, renal, prostate, and breast cancer, demonstrated potencies ranging from 1 pM to 10 nM. While the challenging total synthesis of trabectedin has been accomplished by a few research groups, the commercial preparation begins from readily available cyanosafracin B in 21 steps. Following the Boc and MOM protection of the amine and phenol functionalities, respectively, the methoxy-*p*-quinone was hydrolyzed with sodium hydroxide in methanol. The resulting quinine was reduced (hydrogen over Pd/C) to give an unstable hydroquinone that was subsequently alkylated with bromochloromethane and allyl bromide. The MOM and Boc groups were then removed followed by cleavage of the amide by Edman degradation (through formation of the thiourea with phenyl isothiocyanate and treatment with HCl in 1,4-dioxane). At this point, the amine was protected as the TROC carbamate before reprotecting the phenol as the MOM ether and then liberating the amine with zinc in acetic acid. The amine was converted to an alcohol moiety with sodium nitrite in acetic acid, and this handle was acylated with (*S*)-*N*-[(trichloroethoxy)carbonyl]-*S*-(9-fluorenylmethyl)cysteine. Removal of the allyl-protecting group followed by oxidation provided an alpha-hydroxy ketone intermediate. Dehydration and deprotection of the cysteine established the Michael addition of the thiol to the *o*-quinone methide with concomitant trapping with acetic anhydride. The remaining steps involved removal of protecting groups, installation of the tetrahydroisoquinoline ring via a Pictet–Spengler reaction, and conversion of the cyano to the alcohol of the carbinolamine with silver nitrate in acetonitrile and water. The drug is formulated as a powder (1 mg/vial) for reconstitution (1 mL) for infusion. The pharmacokinetic properties were evaluated following 24-h intravenous infusion every 3 weeks in patients with advanced or metastatic STS. Linear pharmacokinetic properties were observed over a dose range of 0.05–1.8 mg/m^2. After the first round of treatment in patients receiving 1.5 mg/m^2, the recommended dose, the mean C_{max} was 1.04–1.34 ng/mL, and the mean AUC was 39.9–45.5 ng·h/mL. It should be noted, however, that inter-patient variability resulted in a coefficient of variation of >39% for the C_{max} and AUC values. The mean clearance was 35.6–63.2 L/h/m^2, the mean terminal half-life was 89 h, the mean volume of distribution was 3,900 L, and the drug was 94–98% protein-bound. Trabectedin was extensively metabolized in the liver, predominately via CYP3A4 with minor contributions from a number of other CYP isozymes. Excretion was mainly through the fecal route. The Phase II clinical trials focused on the efficacy of trabectedin in treating STS. The first study enrolled 54 heavily pretreated patients; 52% had received three or more drugs. Of the patient population, 41% had leiomyosarcoma. With a treatment regimen of a 24-h infusion of trabectedin at a dose of 1.5 mg/m^2 every 3 weeks, an overall response rate of 4% was observed with two partial responses. In addition, four minor responses and nine disease stabilizations occurred, lasting for more than 6 months. At 6 months, 24% of patients were free from progression. At 2 years, there was a 30% survival rate; however, two treatment-related deaths also occurred. A second clinical study enrolled 36 patients, and a response rate of 8% was achieved with one complete response and two partial responses. The clinical benefit rate was determined to be 14%. The 1-year OS was 53%, and prolonged responses (up to 20 months) were also observed. This response was considered remarkable in this patient population. The most common adverse events included nausea, fatigue, vomiting, anorexia, neutropenia, and increases in aspartate aminotransferase and alanine aminotransferase liver enzymes. Hyperbillirubinemia (grades 1 and 2) also occurred in 23% of patients, but it typically resolved within 2 weeks of drug termination. To combat the nausea and vomiting, all patients must receive 20 mg of dexamethasone intravenously before the trabectedin infusion. Not only does this pretreatment have an antiemetic effect, it also appears to offer a hepatoprotective benefit. Concomitant administration of potent inhibitors and inducers of CYP3A4 should be avoided since plasma levels of this CYP3A4-metabolized drug will be affected. Trabectedin is contraindicated in patients with a known hypersensitivity to trabectedin, concurrent serious or uncontrolled infection, or breast-feeding. Trabectedin should also not be administered in combination with yellow fever vaccine. Finally, there are strict criteria regarding absolute neutrophil count, platelet count, and renal and hepatic enzyme levels for permission to initiate or continue treatment with trabectedin.

21. VILDAGLIPTIN (ANTIDIABETIC) [87–92]

Country of origin:	US
Originator:	Novartis
First introduction:	Brazil
Introduced by:	Novartis
Trade name:	Galvus
CAS registry no:	274901-16-5
Molecular weight:	303.40

Vildagliptin, a DPP-4 inhibitor, was launched last year for the oral treatment of type 2 diabetes. It is approved as an add-on therapy at 50-mg once-daily dose in combination with a sulfonylurea as well as 50-mg twice-daily combination with either metformin or a thiazolidinedione. Vildagliptin is the second DPP-4 inhibitor to reach the market behind sitagliptin, which was introduced in 2006. DPP-4 inhibitors act by slowing the inactivation of incretins, which are endogenous peptides involved in the physiologic regulation of glucose homeostasis. Incretin hormones, including glucagon-like peptide-1 (GLP-1) and glucose-dependent insulinotropic polypeptide (GIP), are released by the intestine throughout the day, and levels are increased in response to a meal. When blood glucose concentrations are normal or elevated, GLP-1 and GIP increase the synthesis and release of insulin from pancreatic β cells via intracellular signaling pathways involving cAMP. GLP-1 also lowers glucagon secretion from pancreatic α cells, which leads to reduced hepatic glucose production. However, although GLP-1 and GIP effectively lower blood glucose, they are short-lived as a result of rapid inactivation by the ubiquitous serine protease DPP-4. By inhibiting DPP-4, vildagliptin increases the concentration and duration of active incretin levels, which in turn results in increased insulin release and decreased glucagon levels in a glucose-dependent manner. Both vildagliptin and sitagliptin are potent, competitive, reversible inhibitors of DPP-4 ($IC_{50} = 3.5$ and 18 nM, respectively), and they both show slow, tight-binding inhibition kinetics. Vildagliptin is chemically derived in three steps starting from L-prolinamide via acylation with chloroacetyl chloride to produce 1-(chloroacetyl)-L-prolinamide, subsequent dehydration of the carboxamide group to the nitrile with trifluoroacetic anhydride, and condensation of the 1-(chloroacetyl)-L-prolinenitrile intermediate with 3-hydroxyadamantan-1-amine. Oral vildagliptin is rapidly absorbed with the maximum plasma concentration (C_{max}) achieved within 1–2 h of administration. Vildgliptin has an absolute oral bioavailability of 85%, and its pharmacokinetics is not affected by food. The mean elimination half-life is 1.68 h with 100-mg once-daily dosage, and 2.54 h with 100-mg twice-daily dosage. Vildagliptin-DPP4 complex exhibits a slow dissociation half-life of 55 min. Maximum inhibition of DPP-4 activity is seen 30 min after a vildagliptin dose, and $\geq 50\%$ inhibition of DPP-4 continues for 10 h or longer. Vildaglaptin has low protein binding (>10%), and a volume of distribution at steady state of 70.5 L. In humans, the metabolism of vildagliptin occurs primarily via hydrolysis of the cyano moiety to the corresponding carboxylic acid, accounting for approximately 55% of the circulating drug-related material following an oral dose. The carboxylic acid metabolite is not pharmacologically active. Vildagliptin does not inhibit or induce cytochrome P450. Excretion of vildagliptin occurs mainly via urine (85%). Approximately 18–22% of the amount excreted is unmetabolized drug. The clinical efficacy of vildagliptin as monotherapy as well as combination therapy has been established in numerous studies. In two 12-week studies in drug-naive patients, vildagliptin monotherapy at 25- to 50-mg BID doses demonstrated a reduction in HbA_{1c} of 0.6%, with a greater response of 1.2% occurring in those patients with an $HbA_{1c} > 8\%$. In a 24-week study in drug-naive patients, comparing monotherapy with vildagliptin 50 mg BID versus rosiglitazone 8 mg QD showed a similar significant HbA_{1c} reduction (1.1% versus 1.3%). In combination therapy studies in patients inadequately controlled on metformin, vildagliptin as an add-on therapy at a dose of 50 or 100 mg resulted in HbA_{1c} reduction by 0.7% and 1.1%, respectively. Additionally, in patients inadequately controlled by prior pioglitazone 45-mg monotherapy, vildagliptin was used as add-on therapy at doses of 50 and 100 mg, resulting in reductions in HbA_{1c} of 0.8% and 1.0%, respectively. The most common adverse events reported in patients receiving vildagliptin included headache, nasopharyngitis, cough, constipation, dizziness, and increased sweating. Vildagliptin is not recommended for patients with liver impairment, and liver monitoring should be conducted at the start of treatment, every 3 months for the first year, and periodically thereafter. The pooled safety data from clinical studies suggests an increased frequency of liver enzyme elevations depending on the dose. The results showed 0.86% of vildagliptin patients taking the 100-mg once-daily dose, 0.34% of those taking the 50-mg twice-daily dose, and 0.21% of those taking the 50-mg once-daily dose had AST and ALT elevations of greater than three times the ULN. At a 50-mg daily dosage, the incidence rate was comparable to the 0.20% rate in the pooled comparator group of patients taking metformin, a thiazolidinedione, a sulfonylurea, or a placebo. Vildagliptin is currently on the market in Brazil and Mexico and approved for marketing in Europe. Approval in the US is still pending. In February 2007, FDA issued an approvable letter requesting additional data on the safety of the drug in patients with renal impairment.

REFERENCES

[1] The collection of new therapeutic entities first launched in 2007 originated from the following sources: (a) CIPSLINE, Prous database; (b) Iddb, Current Drugs database; (c) IMS R&D Focus; (d) Adis Business Intelligence R&D Insight; (e) Pharmaprojects.

[2] A. I. Graul, J. R. Prous, M. Barrionuevo, J. Bozzo, R. Castañer, E. Cruces, L. Revel, E. Rosa, N. Serradell and L. A. Sorbera, *Drug News Perspect.*, 2008, **21**, 7.

[3] J. E. Frampton and M. P. Curran, *Drugs*, 2007, **67**, 1767.

[4] H. M. Siragy, S. Kar and P. Kirkpatrick, *Nat. Rev. Drug Discov.*, 2007, **6**, 779.

[5]   N. E. Mealy, J. Castañer and R. M. Castañer, *Drugs Future*, 2001, **26**, 1139.
[6]   E. O'Brien, *Expert Opin. Investig. Drugs*, 2006, **15**, 1269.
[7]   K. K. Sureshkumar, S. Vasudevan, R. J. Marcus, S. M. Hussain and R. L. McGill, *Expert Opin. Pharmacother.*, 2008, **9**, 825.
[8]   H. Rüeger, S. Stutz, R. Göschke, F. Spindler and J. Maibaum, *Tetrahedron Lett.*, 2000, **41**, 10085.
[9]   L. A. Sorbera and J. Castañer, *Drugs Future*, 2005, **30**, 765.
[10] J. H. Newman, S. Kar and P. Kirkpatrick, *Nat. Rev. Drug Discov.*, 2007, **6**, 697.
[11] D. J. Cada, T. Levien and D. E. Baker, *Hosp. Pharm.*, 2007, **42**, 1145.
[12] H. Reichers, H.-P. Albrecht, W. Amberg, E. Baumann, H. Bernard, H.-J. Bohm, D. Klinge, A. Kling, S. Muller, M. Rashack, L. Unger, N. Walker and W. Wernet, *J. Med. Chem.*, 1996, **39**, 2123.
[13] H. Vatter and V. Seifert, *Cardiovasc. Drug Rev.*, 2006, **24**, 63.
[14] P. Revill, N. Serradell, J. Bolós and M. Bayés, *Drugs Future*, 2006, **31**, 944.
[15] R. Hett, Q. K. Fang, Y. Gao, S. A. Wald and C. H. Senanayake, *Org. Process Res. Dev.*, 1998, **2**, 96.
[16] J. P. Harahan, N. A. Hanna, W. J. Calhoun, S. A. Shan, K. Scarps and R. A. Baumgartner, *COPD*, 2008, **5**, 25.
[17] M. G. Matera and M. Mazola, *Drugs*, 2007, **67**, 503.
[18] B. E. Korba, P. A. Furman and M. J. Otto, *Expert Rev. Anti Infect. Ther.*, 2006, **4**, 549.
[19] C.-K. Hui and G. K. K. Lau, *Expert Opin. Investig. Drugs*, 2005, **14**, 1277.
[20] J. Du, Y. Choi, K. Lee, B. K. Chun, J. H. Hong and C. K. Chu, *Nucleosides Nucleotides Nucleic Acids*, 1999, **18**, 187.
[21] C. K. Chu, J. H. Hong, Y. Choi, J. Du, K. Lee, B. K. Chun, F. D. Boudinot, S. F. Peek, B. E. Korba, B. C. Tennant and Y.-C. Cheng, *Drugs Future*, 1998, **23**, 821.
[22] K. M. Zareba, *Drugs Today*, 2007, **43**, 539.
[23] J. Harding, *Drugs Future*, 2004, **29**, 673.
[24] C. J. Parker, S. Kar and P. Kirkpatrick, *Nat. Rev. Drug Discov.*, 2007, **6**, 515.
[25] P. L. McCormack and L. J. Scott, *Drugs*, 2007, **67**, 1905.
[26] L. A. Sorbera, N. Serradell and J. Bolós, *Drugs Future*, 2007, **32**, 12.
[27] M. B. Berry, M. J. Hughes, D. Parry-Jones and S. J. Skittrall, *WO Patent* 07144363, 2007.
[28] A. Graul, X. Rabasseda and J. Castañer, *Drugs Future*, 1999, **24**, 1324.
[29] M. I. Andersson and A. P. MacGowan, *J. Antimicrob. Chemother.*, 2003, **51**(Suppl. S1), 1.
[30] A. Dalhoff and F. J. Schmitz, *Eur. J. Clin. Microbiol. Infect. Dis.*, 2003, **22**, 203.
[31] K. Hayashi, M. Takahata, Y. Kawamura and Y. Todo, *Arzneim.-Forsch./Drug Res.*, 2002, **52**, 903.
[32] F. J. Boswell, J. M. Andrews and R. Wise, *J. Antimicrob. Chemother.*, 2000, **48**, 446.
[33] H. Miyachi, H. Kiyota, H. Uchiki and M. Segawa, *Bioorg. Med. Chem.*, 1999, **7**, 1151.
[34] S. Ohmori, M. Miura, C. Toriumi, Y. Satoh and T. Ooie, *Drug Metab. Dispos.*, 2007, **35**, 1624.
[35] F. Kobayashi, Y. Yageta, M. Segawa and S. Matsuzawa, *Arzneim.-Forsch./Drug Res.*, 2007, **57**, 92.
[36] F. Kobayashi, Y. Yageta, T. Yamazaki, E. Wakabayashi, M. Inoue, M. Segawa and S. Matsuzawa, *Arzneim.-Forsch./Drug Res.*, 2007, **57**, 147.
[37] T. Ohno, S. Nakade, K. Nakayama, J. Kitagawa, S. Ueda, H. Miyabe, Y. Masuda and Y. Miyata, *Br. J. Clin. Pharmacol.*, 2007, **65**, 197.
[38] A. Conlin, M. Fornier, C. Hudis, S. Kar and P. Kirkpatrick, *Nat. Rev. Drug Discov.*, 2007, **6**, 953.
[39] J. Cortes and J. Baselga, *Oncologist*, 2007, **12**, 271.
[40] F. Y. F. Lee, R. Borzilleri, C. R. Fairchild, S-H. Kim, B. H. Long, C. Reventos-Suarez, G. D. Vite, W. C. Rose and R. A. Kramer, *Clin. Cancer Res.*, 2001, **7**, 1429.
[41] N. Lin, K. Brakora and M. Seiden, *Curr. Opin. Invest. Drugs*, 2003, **4**, 746.
[42] R. M. Borzilleri, X. Zheng, R. J. Schmidt, J. A. Johnson, S-H. Kim, J. D. DiMarco, C. R. Fairchild, J. Z. Gougoutas, F. Y. F. Lee and G. D. Vite, *J. Am. Chem. Soc.*, 2000, **122**, 8890.
[43] S. Dhillon and A. J. Wagstaff, *Drugs*, 2007, **67**, 2101.
[44] B. Boyd, J. Bozzo and J. Castañer, *Drugs Future*, 2005, **30**, 1225.
[45] A. Mukherjee, A. S. Dhadda, M. Shehata and S. Chan, *Expert Opin. Pharmacother.*, 2007, **8**, 2189.
[46] B. Moy, P. Kirkpatrick, S. Kar and P. Goss, *Nat. Rev. Drug Discov.*, 2007, **6**, 431.

[47] F. Montemurro, G. Valabrega and M. Aglietta, *Expert Opin. Biol. Ther.*, 2007, **7**, 257.

[48] L. A. Sorbera, N. Serradell, E. Rosa and J. Bolós, *Drugs Future*, 2007, **32**, 223.

[49] S. Krishnan and Y. Zhang, *J. Clin. Pharmacol.*, 2008, **48**, 293.

[50] S. K. A. Blick and G. M. Keating, *Pediatr. Drugs*, 2007, **9**, 129.

[51] J. Elia, C. Easley and P. Kirkpatrick, *Nat. Rev. Drug Discov.*, 2007, **6**, 343.

[52] J. F. Girotto, *Formulary*, 2007, **42**, 601.

[53] J. Barretina Ginesta, J. Castañer, J. Bozzo and M. Bayés, *Drugs Future*, 2005, **30**, 469.

[54] D. Kuritzkes, S. Kar and P. Kirkpatrick, *Nat. Rev. Drug Discov.*, 2008, **7**, 15.

[55] H. Fadel and Z. Temesgen, *Drugs Today*, 2007, **43**, 749.

[56] G. L. Plosker and D. M. Robinson, *Drugs*, 2008, **68**, 459.

[57] S. L. Davies, J. Bolós, N. Serradell and M. Bayés, *Drugs Future*, 2008, **32**, 17.

[58] A. Quintás-Cardama and J. Cortes, *Drugs Today*, 2007, **43**, 691.

[59] A. Quintás-Cardama, H. Kantarjian and J. Cortes, *Nat. Rev. Drug Discov.*, 2007, **6**, 834.

[60] P. Revill, N. Serradell and J. Bolós, *Drugs Future*, 2006, **31**, 579.

[61] F. H. L. Awouters and P. J. Lewi, *Antipsychotic Drug Res.*, 2007, **57**, 625.

[62] E. Spina and R. Cavallaro, *Expert Opin. Drug Saf.*, 2007, **6**, 651.

[63] R. T. Owen, *Drugs Today*, 2007, **43**, 249.

[64] M. Anker and R. B. Corales, *Expert Opin. Investig Drugs*, 2008, **17**, 97.

[65] Y. Wang, N. Serrandell, J. Bolós and E. Rosa, *Drugs Future*, 2007, **32**, 118.

[66] S. G. Deeks, S. Kar, S. I. Gubernick and P. Kirkpatrick, *Nat. Rev. Drug Discov.*, 2008, **7**, 117.

[67] T. H. Evering and M. Markowitz, *Drugs Today*, 2007, **43**, 865.

[68] R. S. Daum, S. Kar and P. Kirkpatrick, *Nat. Rev. Drug Discov.*, 2007, **6**, 865.

[69] B. Boyd and J. Castañer, *Drugs Future*, 2006, **31**, 107.

[70] M. R. Jacobs, *Future Microbiol.*, 2007, **2**, 591.

[71] O. A. Phillips and L. H. Sharaf, *Expert Opin. Ther. Patents*, 2007, **17**, 429.

[72] E. D. Deeks and L. J. Scott, *CNS Drugs*, 2006, **20**, 751.

[73] D. Heaney and M. C. Walker, *Drugs Today*, 2007, **43**, 455.

[74] S. Hakimian, A. Cheng-Hakimian, G. D. Anderson and J. W. Miller, *Expert Opin. Pharmacother.*, 2007, **8**, 1931.

[75] L. A. Sorbera, P. A. Leeson, X. Rabasseda and J. Castañer, *Drugs Future*, 2000, **25**, 1145.

[76] B. Rini, S. Kar and P. Kirkpatrick, *Nat. Rev. Drug Discov.*, 2007, **6**, 599.

[77] W. W. Ma and A. Jimeno, *Drugs Today*, 2007, **43**, 659.

[78] W. E. Samlowski and N. J. Vogelzang, *Expert Opin. Emerg. Drugs*, 2007, **12**, 605.

[79] M. L. Harrison, A. Montes and M. E. Gore, *Expert Rev. Anticancer Ther.*, 2007, **7**, 57.

[80] J. Gu, M. E. Ruppen and P. Cai, *Org. Lett.*, 2005, **7**, 3945.

[81] C. h. Van Kesteren, M. M. M. de Vooght, L. López-Lázaro, R. A. A. Mathôt, J. H. M. Schellens, J. M. Jimeno and J. H. Beijnen, *Anticancer Drugs*, 2003, **14**, 487.

[82] M. D'Incalci and J. Jimeno, *Expert Opin. Investig. Drugs*, 2003, **12**, 1843.

[83] C. Cuevas, M. Pérez, M. J. Martin, J. L. Chicharro, C. Fernández-Rivas, M. Flores, A. Francesch, P. Gallego, M. Zarzuelo, F. de la Calle, J. Garcia, C. Polanco, I. Rodriquez and I. Manzanares, *Org. Lett.*, 2000, **2**, 2545.

[84] N. J. Carter and S. J. Keam, *Drugs*, 2007, **67**, 2257.

[85] Anonymous., *J. Clin. Drugs R&D*, 2003, **4**, 75.

[86] J. Fayette, I. Ray Coquard, L. Alberti, H. Boyle, P. Méeus, A.-V. Decouvelaere, P. Thiesse, M.-P. Sunyach, D. Ranchère and J.-Y. Blay, *Curr. Opin. Oncol.*, 2006, **18**, 347.

[87] E. L. Kleppinger and K. Helms, *Ann. Pharmacother.*, 2007, **41**, 824.

[88] S. Henness and S. J. Keam, *Drugs*, 2006, **66**, 1989.

[89] J. A. McIntyre and J. Castañer, *Drugs Future*, 2004, **29**, 887.

[90] S. Ristic and P. C. Bates, *Drugs Today*, 2006, **42**, 519.

[91] E. B. Villhauer, J. A. Brinkman, G. B. Naderi, B. F. Burkey, B. E. Dunning, K. Prasad, B. L. Mangold, M. E. Russell and T. E. Hughes, *J. Med. Chem.*, 2003, **46**, 2774.

[92] A. J. Garber and M. D. Sharma, *Expert. Opin. Investig. Drugs*, 2008, **17**, 105.

CUMULATIVE NCE INTRODUCTION INDEX, 1983–2007

GENERAL NAME	INDICATION	YEAR INTRO.	ARMC VOL., PAGE
abacavir sulfate	antiviral	1999	35, 333
abarelix	anticancer	2004	40, 446
abatacept	rheumatoid arthritis	2006	42, 509
acarbose	antidiabetic	1990	26, 297
aceclofenac	antiinflammatory	1992	28, 325
acemannan	wound healing agent	2001	37, 259
acetohydroxamic acid	hypoammonuric	1983	19, 313
acetorphan	antidiarrheal	1993	29, 332
acipimox	hypolipidemic	1985	21, 323
acitretin	antipsoriatic	1989	25, 309
acrivastine	antihistamine	1988	24, 295
actarit	antirheumatic	1994	30, 296
adalimumab	rheumatoid arthritis	2003	39, 267
adamantanium bromide	antiseptic	1984	20, 315
adefovir dipivoxil	antiviral	2002	38, 348
adrafinil	psychostimulant	1986	22, 315
AF-2259	antiinflammatory	1987	23, 325
afloqualone	muscle relaxant	1983	19, 313
agalsidase alfa	fabry's disease	2001	37, 259
alacepril	antihypertensive	1988	24, 296
alclometasone dipropionate	topical antiinflammatory	1985	21, 323
alefacept	plaque psoriasis	2003	39, 267
alemtuzumab	anticancer	2001	37, 260
alendronate sodium	osteoporosis	1993	29, 332
alfentanil HCl	analgesic	1983	19, 314
alfuzosin HCl	antihypertensive	1988	24, 296
alglucerase	enzyme	1991	27, 321
alglucosidase alfa	Pompe disease	2006	42, 511
aliskiren	antihypertensive	2007	43, 461
alitretinoin	anticancer	1999	35, 333
alminoprofen	analgesic	1983	19, 314
almotriptan	antimigraine	2000	36, 295
anakinra	antiarthritic	2001	37, 261
anidulafungin	antifungal	2006	42, 512
alosetron hydrochloride	irritable bowel syndrome	2000	36, 295
alpha-1 antitrypsin	protease inhibitor	1988	24, 297
alpidem	anxiolytic	1991	27, 322
alpiropride	antimigraine	1988	24, 296
alteplase	thrombolytic	1987	23, 326

GENERAL NAME	INDICATION	YEAR INTRO.	ARMC VOL., PAGE
ambrisentan	pulmonary arterial hypertension	2007	43, 463
amfenac sodium	antiinflammatory	1986	22, 315
amifostine	cytoprotective	1995	31, 338
aminoprofen	topical antiinflammatory	1990	26, 298
amisulpride	antipsychotic	1986	22, 316
amlexanox	antiasthmatic	1987	23, 327
amlodipine besylate	antihypertensive	1990	26, 298
amorolfine HCl	topical antifungal	1991	27, 322
amosulalol	antihypertensive	1988	24, 297
ampiroxicam	antiinflammatory	1994	30, 296
amprenavir	antiviral	1999	35, 334
amrinone	cardiotonic	1983	19, 314
amrubicin HCl	antineoplastic	2002	38, 349
amsacrine	antineoplastic	1987	23, 327
amtolmetin guacil	antiinflammatory	1993	29, 332
anagrelide HCl	hematological	1997	33, 328
anastrozole	antineoplastic	1995	31, 338
angiotensin II	anticancer adjuvant	1994	30, 296
aniracetam	cognition enhancer	1993	29, 333
anti-digoxin polyclonal antibody	antidote	2002	38, 350
APD	calcium regulator	1987	23, 326
apraclonidine HCl	antiglaucoma	1988	24, 297
aprepitant	antiemetic	2003	39, 268
APSAC	thrombolytic	1987	23, 326
aranidipine	antihypertensive	1996	32, 306
arbekacin	antibiotic	1990	26, 298
arformoterol	chronic obstructive pulmonary disease	2007	43, 465
argatroban	antithromobotic	1990	26, 299
arglabin	anticancer	1999	35, 335
aripiprazole	neuroleptic	2002	38, 350
arotinolol HCl	antihypertensive	1986	22, 316
arteether	antimalarial	2000	36, 296
artemisinin	antimalarial	1987	23, 327
aspoxicillin	antibiotic	1987	23, 328
astemizole	antihistamine	1983	19, 314
astromycin sulfate	antibiotic	1985	21, 324
atazanavir	antiviral	2003	39, 269
atomoxetine	attention deficit hyperactivity disorder	2003	39, 270
atorvastatin calcium	dyslipidemia	1997	33, 328
atosiban	preterm labor	2000	36, 297
atovaquone	antiparasitic	1992	28, 326
auranofin	chrysotherapeutic	1983	19, 143
azacitidine	anticancer	2004	40, 447

GENERAL NAME	INDICATION	YEAR INTRO.	ARMC VOL., PAGE
azelnidipine	antihypertensive	2003	39, 270
azelaic acid	antiacne	1989	25, 310
azelastine HCl	antihistamine	1986	22, 316
azithromycin	antibiotic	1988	24, 298
azosemide	diuretic	1986	22, 316
aztreonam	antibiotic	1984	20, 315
balofloxacin	antibacterial	2002	38, 351
balsalazide disodium	ulcerative colitis	1997	33, 329
bambuterol	bronchodilator	1990	26, 299
barnidipine HCl	antihypertensive	1992	28, 326
beclobrate	hypolipidemic	1986	22, 317
befunolol HCl	antiglaucoma	1983	19, 315
belotecan	anticancer	2004	40, 449
benazepril HCl	antihypertensive	1990	26, 299
benexate HCl	antiulcer	1987	23, 328
benidipine HCl	antihypertensive	1991	27, 322
beraprost sodium	platelet aggreg. inhibitor	1992	28, 326
betamethasone butyrate prospinate	topical antiinflammatory	1994	30, 297
betaxolol HCl	antihypertensive	1983	19, 315
betotastine besilate	antiallergic	2000	36, 297
bevacizumab	anticancer	2004	40, 450
bevantolol HCl	antihypertensive	1987	23, 328
bexarotene	anticancer	2000	36, 298
biapenem	antibacterial	2002	38, 351
bicalutamide	antineoplastic	1995	31, 338
bifemelane HCl	nootropic	1987	23, 329
bimatoprost	antiglaucoma	2001	37, 261
binfonazole	hypnotic	1983	19, 315
binifibrate	hypolipidemic	1986	22, 317
bisantrene HCl	antineoplastic	1990	26, 300
bisoprolol fumarate	antihypertensive	1986	22, 317
bivalirudin	antithrombotic	2000	36, 298
bopindolol	antihypertensive	1985	21, 324
bortezomib	anticancer	2003	39, 271
bosentan	antihypertensive	2001	37, 262
brimonidine	antiglaucoma	1996	32, 306
brinzolamide	antiglaucoma	1998	34, 318
brodimoprin	antibiotic	1993	29, 333
bromfenac sodium	NSAID	1997	33, 329
brotizolam	hypnotic	1983	19, 315
brovincamine fumarate	cerebral vasodilator	1986	22, 317
bucillamine	immunomodulator	1987	23, 329
bucladesine sodium	cardiostimulant	1984	20, 316
budipine	antiParkinsonian	1997	33, 330
budralazine	antihypertensive	1983	19, 315
bulaquine	antimalarial	2000	36, 299

GENERAL NAME	INDICATION	YEAR INTRO.	ARMC VOL., PAGE
bunazosin HCl	antihypertensive	1985	21, 324
bupropion HCl	antidepressant	1989	25, 310
buserelin acetate	hormone	1984	20, 316
buspirone HCl	anxiolytic	1985	21, 324
butenafine HCl	topical antifungal	1992	28, 327
butibufen	antiinflammatory	1992	28, 327
butoconazole	topical antifungal	1986	22, 318
butoctamide	hypnotic	1984	20, 316
butyl flufenamate	topical antiinflammatory	1983	19, 316
cabergoline	antiprolactin	1993	29, 334
cadexomer iodine	wound healing agent	1983	19, 316
cadralazine	hypertensive	1988	24, 298
calcipotriol	antipsoriatic	1991	27, 323
camostat mesylate	antineoplastic	1985	21, 325
candesartan cilexetil	antihypertension	1997	33, 330
capecitabine	antineoplastic	1998	34, 319
captopril	antihypertensive agent	1982	13, 086
carboplatin	antibiotic	1986	22, 318
carperitide	congestive heart failure	1995	31, 339
carumonam	antibiotic	1988	24, 298
carvedilol	antihypertensive	1991	27, 323
caspofungin acetate	antifungal	2001	37, 263
cefbuperazone sodium	antibiotic	1985	21, 325
cefcapene pivoxil	antibiotic	1997	33, 330
cefdinir	antibiotic	1991	27, 323
cefditoren pivoxil	oral cephalosporin	1994	30, 297
cefepime	antibiotic	1993	29, 334
cefetamet pivoxil HCl	antibiotic	1992	28, 327
cefixime	antibiotic	1987	23, 329
cefmenoxime HCl	antibiotic	1983	19, 316
cefminox sodium	antibiotic	1987	23, 330
cefodizime sodium	antibiotic	1990	26, 300
cefonicid sodium	antibiotic	1984	20, 316
ceforanide	antibiotic	1984	20, 317
cefoselis	antibiotic	1998	34, 319
cefotetan disodium	antibiotic	1984	20, 317
cefotiam hexetil HCl	antibiotic	1991	27, 324
cefozopran HCl	injectable cephalosporin	1995	31, 339
cefpimizole	antibiotic	1987	23, 330
cefpiramide sodium	antibiotic	1985	21, 325
cefpirome sulfate	antibiotic	1992	28, 328
cefpodoxime proxetil	antibiotic	1989	25, 310
cefprozil	antibiotic	1992	28, 328
ceftazidime	antibiotic	1983	19, 316
cefteram pivoxil	antibiotic	1987	23, 330
ceftibuten	antibiotic	1992	28, 329
cefuroxime axetil	antibiotic	1987	23, 331

GENERAL NAME	INDICATION	YEAR INTRO.	ARMC VOL., PAGE
cefuzonam sodium	antibiotic	1987	23, 331
celecoxib	antiarthritic	1999	35, 335
celiprolol HCl	antihypertensive	1983	19, 317
centchroman	antiestrogen	1991	27, 324
centoxin	immunomodulator	1991	27, 325
cerivastatin	dyslipidemia	1997	33, 331
cetirizine HCl	antihistamine	1987	23, 331
cetrorelix	female infertility	1999	35, 336
cetuximab	anticancer	2003	39, 272
cevimeline hydrochloride	anti-xerostomia	2000	36, 299
chenodiol	anticholelithogenic	1983	19, 317
CHF-1301	antiparkinsonian	1999	35, 336
choline alfoscerate	nootropic	1990	26, 300
cibenzoline	antiarrhythmic	1985	21, 325
ciclesonide	asthma, COPD	2005	41, 443
cicletanine	antihypertensive	1988	24, 299
cidofovir	antiviral	1996	32, 306
cilazapril	antihypertensive	1990	26, 301
cilostazol	antithrombotic	1988	24, 299
cimetropium bromide	antispasmodic	1985	21, 326
cinacalcet	hyperparathyroidism	2004	40, 451
cinildipine	antihypertensive	1995	31, 339
cinitapride	gastroprokinetic	1990	26, 301
cinolazepam	hypnotic	1993	29, 334
ciprofibrate	hypolipidemic	1985	21, 326
ciprofloxacin	antibacterial	1986	22, 318
cisapride	gastroprokinetic	1988	24, 299
cisatracurium besilate	muscle relaxant	1995	31, 340
citalopram	antidepressant	1989	25, 311
cladribine	antineoplastic	1993	29, 335
clarithromycin	antibiotic	1990	26, 302
clevudine	hepatitis B	2007	43, 466
clobenoside	vasoprotective	1988	24, 300
cloconazole HCl	topical antifungal	1986	22, 318
clodronate disodium	calcium regulator	1986	22, 319
clofarabine	anticancer	2005	41, 444
clopidogrel hydrogensulfate	antithrombotic	1998	34, 320
cloricromen	antithrombotic	1991	27, 325
clospipramine HCl	neuroleptic	1991	27, 325
colesevelam hydrochloride	hypolipidemic	2000	36, 300
colestimide	hypolipidaemic	1999	35, 337
colforsin daropate HCl	cardiotonic	1999	35, 337
conivaptan	hyponatremia	2006	42, 514
crotelidae polyvalent immune fab	antidote	2001	37, 263
cyclosporine	immunosuppressant	1983	19, 317
cytarabine ocfosfate	antineoplastic	1993	29, 335

GENERAL NAME	INDICATION	YEAR INTRO.	ARMC VOL., PAGE
dalfopristin	antibiotic	1999	35, 338
dapiprazole HCl	antiglaucoma	1987	23, 332
daptomycin	antibiotic	2003	39, 272
darifenacin	urinary incontinence	2005	41, 445
darunavir	HIV	2006	42, 515
dasatinib	anticancer	2006	42, 517
decitabine	myelodysplastic syndromes	2006	42, 519
defeiprone	iron chelator	1995	31, 340
deferasirox	chronic iron overload	2005	41, 446
defibrotide	antithrombotic	1986	22, 319
deflazacort	antiinflammatory	1986	22, 319
delapril	antihypertensive	1989	25, 311
delavirdine mesylate	antiviral	1997	33, 331
denileukin diftitox	anticancer	1999	35, 338
denopamine	cardiostimulant	1988	24, 300
deprodone propionate	topical antiinflammatory	1992	28, 329
desflurane	anesthetic	1992	28, 329
desloratadine	antihistamine	2001	37, 264
dexfenfluramine	antiobesity	1997	33, 332
dexibuprofen	antiinflammatory	1994	30, 298
dexmedetomidine hydrochloride	sedative	2000	36, 301
dexmethylphenidate HCl	psychostimulant	2002	38, 352
dexrazoxane	cardioprotective	1992	28, 330
dezocine	analgesic	1991	27, 326
diacerein	antirheumatic	1985	21, 326
didanosine	antiviral	1991	27, 326
dilevalol	antihypertensive	1989	25, 311
dirithromycin	antibiotic	1993	29, 336
disodium pamidronate	calcium regulator	1989	25, 312
divistyramine	hypocholesterolemic	1984	20, 317
docarpamine	cardiostimulant	1994	30, 298
docetaxel	antineoplastic	1995	31, 341
dofetilide	antiarrhythmic	2000	36, 301
dolasetron mesylate	antiemetic	1998	34, 321
donepezil HCl	anti-Alzheimer	1997	33, 332
dopexamine	cardiostimulant	1989	25, 312
doripenem	antibiotic	2005	41, 448
dornase alfa	cystic fibrosis	1994	30, 298
dorzolamide HCL	antiglaucoma	1995	31, 341
dosmalfate	antiulcer	2000	36, 302
doxacurium chloride	muscle relaxant	1991	27, 326
doxazosin mesylate	antihypertensive	1988	24, 300
doxefazepam	hypnotic	1985	21, 326
doxercalciferol	vitamin D prohormone	1999	35, 339
doxifluridine	antineoplastic	1987	23, 332
doxofylline	bronchodilator	1985	21, 327

GENERAL NAME	INDICATION	YEAR INTRO.	ARMC VOL., PAGE
dronabinol	antinauseant	1986	22, 319
drospirenone	contraceptive	2000	36, 302
drotrecogin alfa	antisepsis	2001	37, 265
droxicam	antiinflammatory	1990	26, 302
droxidopa	antiparkinsonian	1989	25, 312
duloxetine	antidepressant	2004	40, 452
dutasteride	5a reductase inhibitor	2002	38, 353
duteplase	anticougulant	1995	31, 342
eberconazole	antifungal	2005	41, 449
ebastine	antihistamine	1990	26 302
ebrotidine	antiulcer	1997	33, 333
ecabet sodium	antiulcerative	1993	29, 336
eculizumab	hemoglobinuria	2007	43, 468
edaravone	neuroprotective	2001	37, 265
efalizumab	psoriasis	2003	39, 274
efavirenz	antiviral	1998	34, 321
efonidipine	antihypertensive	1994	30, 299
egualen sodium	antiulcer	2000	36, 303
eletriptan	antimigraine	2001	37, 266
emedastine difumarate	antiallergic/antiasthmatic	1993	29, 336
emorfazone	analgesic	1984	20, 317
emtricitabine	antiviral	2003	39, 274
enalapril maleate	antihypertensive	1984	20, 317
enalaprilat	antihypertensive	1987	23, 332
encainide HCl	antiarrhythmic	1987	23, 333
enfuvirtide	antiviral	2003	39, 275
enocitabine	antineoplastic	1983	19, 318
enoxacin	antibacterial	1986	22, 320
enoxaparin	antithrombotic	1987	23, 333
enoximone	cardiostimulant	1988	24, 301
enprostil	antiulcer	1985	21, 327
entacapone	antiparkinsonian	1998	34, 322
entecavir	antiviral	2005	41, 450
epalrestat	antidiabetic	1992	28, 330
eperisone HCl	muscle relaxant	1983	19, 318
epidermal growth factor	wound healing agent	1987	23, 333
epinastine	antiallergic	1994	30, 299
epirubicin HCl	antineoplastic	1984	20, 318
eplerenone	antihypertensive	2003	39, 276
epoprostenol sodium	platelet aggreg. inhib.	1983	19, 318
eprosartan	antihypertensive	1997	33, 333
eptazocine HBr	analgesic	1987	23, 334
eptilfibatide	antithrombotic	1999	35, 340
erdosteine	expectorant	1995	31, 342
erlotinib	anticancer	2004	40, 454
ertapenem sodium	antibacterial	2002	38, 353
erythromycin acistrate	antibiotic	1988	24, 301

GENERAL NAME	INDICATION	YEAR INTRO.	ARMC VOL., PAGE
erythropoietin	hematopoetic	1988	24, 301
escitalopram oxolate	antidepressant	2002	38, 354
esmolol HCl	antiarrhythmic	1987	23, 334
esomeprazole magnesium	gastric antisecretory	2000	36, 303
eszopiclone	hypnotic	2005	41, 451
ethyl icosapentate	antithrombotic	1990	26, 303
etizolam	anxiolytic	1984	20, 318
etodolac	antiinflammatory	1985	21, 327
etoricoxibe	antiarthritic/analgesic	2002	38, 355
everolimus	immunosuppressant	2004	40, 455
exemestane	anticancer	2000	36, 304
exenatide	anti-diabetic	2005	41, 452
exifone	nootropic	1988	24, 302
ezetimibe	hypolipidemic	2002	38, 355
factor VIIa	haemophilia	1996	32, 307
factor VIII	hemostatic	1992	28, 330
fadrozole HCl	antineoplastic	1995	31, 342
falecalcitriol	vitamin D	2001	37, 266
famciclovir	antiviral	1994	30, 300
famotidine	antiulcer	1985	21, 327
fasudil HCl	neuroprotective	1995	31, 343
felbamate	antiepileptic	1993	29, 337
felbinac	topical antiinflammatory	1986	22, 320
felodipine	antihypertensive	1988	24, 302
fenbuprol	choleretic	1983	19, 318
fenoldopam mesylate	antihypertensive	1998	34, 322
fenticonazole nitrate	antifungal	1987	23, 334
fexofenadine	antiallergic	1996	32, 307
filgrastim	immunostimulant	1991	27, 327
finasteride	5a-reductase inhibitor	1992	28, 331
fisalamine	intestinal antiinflammatory	1984	20, 318
fleroxacin	antibacterial	1992	28, 331
flomoxef sodium	antibiotic	1988	24, 302
flosequinan	cardiostimulant	1992	28, 331
fluconazole	antifungal	1988	24, 303
fludarabine phosphate	antineoplastic	1991	27, 327
flumazenil	benzodiazepine antag.	1987	23, 335
flunoxaprofen	antiinflammatory	1987	23, 335
fluoxetine HCl	antidepressant	1986	22, 320
flupirtine maleate	analgesic	1985	21, 328
flurithromycin ethylsuccinate	antibiotic	1997	33, 333
flutamide	antineoplastic	1983	19, 318
flutazolam	anxiolytic	1984	20, 318
fluticasone furoate	anti-allergy	2007	43, 469
fluticasone propionate	antiinflammatory	1990	26, 303
flutoprazepam	anxiolytic	1986	22, 320

GENERAL NAME	INDICATION	YEAR INTRO.	ARMC VOL., PAGE
flutrimazole	topical antifungal	1995	31, 343
flutropium bromide	antitussive	1988	24, 303
fluvastatin	hypolipaemic	1994	30, 300
fluvoxamine maleate	antidepressant	1983	19, 319
follitropin alfa	fertility enhancer	1996	32, 307
follitropin beta	fertility enhancer	1996	32, 308
fomepizole	antidote	1998	34, 323
fomivirsen sodium	antiviral	1998	34, 323
fondaparinux sodium	antithrombotic	2002	38, 356
formestane	antineoplastic	1993	29, 337
formoterol fumarate	bronchodilator	1986	22, 321
fosamprenavir	antiviral	2003	39, 277
foscarnet sodium	antiviral	1989	25, 313
fosfosal	analgesic	1984	20, 319
fosfluconazole	antifungal	2004	40, 457
fosinopril sodium	antihypertensive	1991	27, 328
fosphenytoin sodium	antiepileptic	1996	32, 308
fotemustine	antineoplastic	1989	25, 313
fropenam	antibiotic	1997	33, 334
frovatriptan	antimigraine	2002	38, 357
fudosteine	expectorant	2001	37, 267
fulveristrant	anticancer	2002	38, 357
gabapentin	antiepileptic	1993	29, 338
gadoversetamide	MRI contrast agent	2000	36, 304
gallium nitrate	calcium regulator	1991	27, 328
gallopamil HCl	antianginal	1983	19, 319
galsulfase	mucopolysaccharidosis VI	2005	41, 453
ganciclovir	antiviral	1988	24, 303
ganirelix acetate	female infertility	2000	36, 305
garenoxacin	anti-infective	2007	43, 471
gatilfloxacin	antibiotic	1999	35, 340
gefitinib	antineoplastic	2002	38, 358
gemcitabine HCl	antineoplastic	1995	31, 344
gemeprost	abortifacient	1983	19, 319
gemifloxacin	antibacterial	2004	40, 458
gemtuzumab ozogamicin	anticancer	2000	36, 306
gestodene	progestogen	1987	23, 335
gestrinone	antiprogestogen	1986	22, 321
glatiramer acetate	Multiple Sclerosis	1997	33, 334
glimepiride	antidiabetic	1995	31, 344
glucagon, rDNA	hypoglycemia	1993	29, 338
GMDP	immunostimulant	1996	32, 308
goserelin	hormone	1987	23, 336
granisetron HCl	antiemetic	1991	27, 329
guanadrel sulfate	antihypertensive	1983	19, 319
gusperimus	immunosuppressant	1994	30, 300

GENERAL NAME	INDICATION	YEAR INTRO.	ARMC VOL., PAGE
halobetasol propionate	topical antiinflammatory	1991	27, 329
halofantrine	antimalarial	1988	24, 304
halometasone	topical antiinflammatory	1983	19, 320
histrelin	precocious puberty	1993	29, 338
hydrocortisone aceponate	topical antiinflammatory	1988	24, 304
hydrocortisone butyrate	topical antiinflammatory	1983	19, 320
ibandronic acid	osteoporosis	1996	32, 309
ibopamine HCl	cardiostimulant	1984	20, 319
ibudilast	antiasthmatic	1989	25, 313
ibutilide fumarate	antiarrhythmic	1996	32, 309
ibritunomab tiuxetan	anticancer	2002	38, 359
idarubicin HCl	antineoplastic	1990	26, 303
idebenone	nootropic	1986	22, 321
idursulfase	mucopolysaccharidosis II (Hunter syndrome)	2006	42, 520
iloprost	platelet aggreg. inhibitor	1992	28, 332
imatinib mesylate	antineoplastic	2001	37, 267
imidafenacin	overactive bladder	2007	43, 472
imidapril HCl	antihypertensive	1993	29, 339
imiglucerase	Gaucher's disease	1994	30, 301
imipenem/cilastatin	antibiotic	1985	21, 328
imiquimod	antiviral	1997	33, 335
incadronic acid	osteoporosis	1997	33, 335
indalpine	antidepressant	1983	19, 320
indeloxazine HCl	nootropic	1988	24, 304
indinavir sulfate	antiviral	1996	32, 310
indisetron	antiemetic	2004	40, 459
indobufen	antithrombotic	1984	20, 319
influenza virus (live)	antiviral vaccine	2003	39, 277
insulin lispro	antidiabetic	1996	32, 310
interferon alfacon-1	antiviral	1997	33, 336
interferon gamma-1b	immunostimulant	1991	27, 329
interferon, gamma	antiinflammatory	1989	25, 314
interferon, gamma-1a	antineoplastic	1992	28, 332
interferon, b-1a	multiple sclerosis	1996	32, 311
interferon, b-1b	multiple sclerosis	1993	29, 339
interleukin-2	antineoplastic	1989	25, 314
ioflupane	diagnosis CNS	2000	36, 306
ipriflavone	calcium regulator	1989	25, 314
irbesartan	antihypertensive	1997	33, 336
irinotecan	antineoplastic	1994	30, 301
irsogladine	antiulcer	1989	25, 315
isepamicin	antibiotic	1988	24, 305
isofezolac	antiinflammatory	1984	20, 319
isoxicam	antiinflammatory	1983	19, 320
isradipine	antihypertensive	1989	25, 315
itopride HCl	gastroprokinetic	1995	31, 344

GENERAL NAME	INDICATION	YEAR INTRO.	ARMC VOL., PAGE
itraconazole	antifungal	1988	24, 305
ivabradine	angina	2006	42, 522
ivermectin	antiparasitic	1987	23, 336
ixabepilone	anticancer	2007	43, 473
ketanserin	antihypertensive	1985	21, 328
ketorolac tromethamine	analgesic	1990	26, 304
kinetin	skin photodamage/ dermatologic	1999	35, 341
lacidipine	antihypertensive	1991	27, 330
lafutidine	gastric antisecretory	2000	36, 307
lamivudine	antiviral	1995	31, 345
lamotrigine	anticonvulsant	1990	26, 304
landiolol	antiarrhythmic	2002	38, 360
lanoconazole	antifungal	1994	30, 302
lanreotide acetate	acromegaly	1995	31, 345
lansoprazole	antiulcer	1992	28, 332
lapatinib	anticancer	2007	43, 475
laronidase	mucopolysaccaridosis I	2003	39, 278
latanoprost	antiglaucoma	1996	32, 311
lefunomide	antiarthritic	1998	34, 324
lenalidomide	myelodysplastic syndromes, multiple myeloma	2006	42, 523
lenampicillin HCl	antibiotic	1987	23, 336
lentinan	immunostimulant	1986	22, 322
lepirudin	anticoagulant	1997	33, 336
lercanidipine	antihyperintensive	1997	33, 337
letrazole	anticancer	1996	32, 311
leuprolide acetate	hormone	1984	20, 319
levacecarnine HCl	nootropic	1986	22, 322
levalbuterol HCl	antiasthmatic	1999	35, 341
levetiracetam	antiepileptic	2000	36, 307
levobunolol HCl	antiglaucoma	1985	21, 328
levobupivacaine hydrochloride	local anesthetic	2000	36, 308
levocabastine HCl	antihistamine	1991	27, 330
levocetirizine	antihistamine	2001	37, 268
levodropropizine	antitussive	1988	24, 305
levofloxacin	antibiotic	1993	29, 340
levosimendan	heart failure	2000	36, 308
lidamidine HCl	antiperistaltic	1984	20, 320
limaprost	antithrombotic	1988	24, 306
linezolid	antibiotic	2000	36, 309
liranaftate	topical antifungal	2000	36, 309
lisdexamfetamine	ADHD	2007	43, 477
lisinopril	antihypertensive	1987	23, 337
lobenzarit sodium	antiinflammatory	1986	22, 322
lodoxamide tromethamine	antiallergic ophthalmic	1992	28, 333

GENERAL NAME	INDICATION	YEAR INTRO.	ARMC VOL., PAGE
lomefloxacin	antibiotic	1989	25, 315
lomerizine HCl	antimigraine	1999	35, 342
lonidamine	antineoplastic	1987	23, 337
lopinavir	antiviral	2000	36, 310
loprazolam mesylate	hypnotic	1983	19, 321
loprinone HCl	cardiostimulant	1996	32, 312
loracarbef	antibiotic	1992	28, 333
loratadine	antihistamine	1988	24, 306
lornoxicam	NSAID	1997	33, 337
losartan	antihypertensive	1994	30, 302
loteprednol etabonate	antiallergic ophthalmic	1998	34, 324
lovastatin	hypocholesterolemic	1987	23, 337
loxoprofen sodium	antiinflammatory	1986	22, 322
lulbiprostone	chronic idiopathic constipation	2006	42, 525
luliconazole	antifungal	2005	41, 454
lumiracoxib	anti-inflammatory	2005	41, 455
Lyme disease	vaccine	1999	35, 342
mabuterol HCl	bronchodilator	1986	22, 323
malotilate	hepatoprotective	1985	21, 329
manidipine HCl	antihypertensive	1990	26, 304
maraviroc	anti-infective – HIV	2007	43, 478
masoprocol	topical antineoplastic	1992	28, 333
maxacalcitol	vitamin D	2000	36, 310
mebefradil HCl	antihypertensive	1997	33, 338
medifoxamine fumarate	antidepressant	1986	22, 323
mefloquine HCl	antimalarial	1985	21, 329
meglutol	hypolipidemic	1983	19, 321
melinamide	hypocholesterolemic	1984	20, 320
meloxicam	antiarthritic	1996	32, 312
mepixanox	analeptic	1984	20, 320
meptazinol HCl	analgesic	1983	19, 321
meropenem	carbapenem antibiotic	1994	30, 303
metaclazepam	anxiolytic	1987	23, 338
metapramine	antidepressant	1984	20, 320
mexazolam	anxiolytic	1984	20, 321
micafungin	antifungal	2002	38, 360
mifepristone	abortifacient	1988	24, 306
miglitol	antidiabetic	1998	34, 325
miglustat	gaucher's disease	2003	39, 279
milnacipran	antidepressant	1997	33, 338
milrinone	cardiostimulant	1989	25, 316
miltefosine	topical antineoplastic	1993	29, 340
miokamycin	antibiotic	1985	21, 329
mirtazapine	antidepressant	1994	30, 303
misoprostol	antiulcer	1985	21, 329
mitiglinide	antidiabetic	2004	40, 460

GENERAL NAME	INDICATION	YEAR INTRO.	ARMC VOL., PAGE
mitoxantrone HCl	antineoplastic	1984	20, 321
mivacurium chloride	muscle relaxant	1992	28, 334
mivotilate	hepatoprotectant	1999	35, 343
mizolastine	antihistamine	1998	34, 325
mizoribine	immunosuppressant	1984	20, 321
moclobemide	antidepressant	1990	26, 305
modafinil	idiopathic hypersomnia	1994	30, 303
moexipril HCl	antihypertensive	1995	31, 346
mofezolac	analgesic	1994	30, 304
mometasone furoate	topical antiinflammatory	1987	23, 338
montelukast sodium	antiasthma	1998	34, 326
moricizine HCl	antiarrhythmic	1990	26, 305
mosapride citrate	gastroprokinetic	1998	34, 326
moxifloxacin HCL	antibiotic	1999	35, 343
moxonidine	antihypertensive	1991	27, 330
mozavaptan	hyponatremia	2006	42, 527
mupirocin	topical antibiotic	1985	21, 330
muromonab-CD3	immunosuppressant	1986	22, 323
muzolimine	diuretic	1983	19, 321
mycophenolate mofetil	immunosuppressant	1995	31, 346
mycophenolate sodium	immunosuppressant	2003	39, 279
nabumetone	antiinflammatory	1985	21, 330
nadifloxacin	topical antibiotic	1993	29, 340
nafamostat mesylate	protease inhibitor	1986	22, 323
nafarelin acetate	hormone	1990	26, 306
naftifine HCl	antifungal	1984	20, 321
naftopidil	dysuria	1999	35, 344
nalmefene HCl	dependence treatment	1995	31, 347
naltrexone HCl	narcotic antagonist	1984	20, 322
naratriptan HCl	antimigraine	1997	33, 339
nartograstim	leukopenia	1994	30, 304
natalizumab	multiple sclerosis	2004	40, 462
nateglinide	antidiabetic	1999	35, 344
nazasetron	antiemetic	1994	30, 305
nebivolol	antihypertensive	1997	33, 339
nedaplatin	antineoplastic	1995	31, 347
nedocromil sodium	antiallergic	1986	22, 324
nefazodone	antidepressant	1994	30, 305
nelarabine	anticancer	2006	42, 528
nelfinavir mesylate	antiviral	1997	33, 340
neltenexine	cystic fibrosis	1993	29, 341
nemonapride	neuroleptic	1991	27, 331
nepafenac	anti-inflammatory	2005	41, 456
neridronic acid	calcium regulator	2002	38, 361
nesiritide	congestive heart failure	2001	37, 269
neticonazole HCl	topical antifungal	1993	29, 341
nevirapine	antiviral	1996	32, 313

GENERAL NAME	INDICATION	YEAR INTRO.	ARMC VOL., PAGE
nicorandil	coronary vasodilator	1984	20, 322
nifekalant HCl	antiarrythmic	1999	35, 344
nilotinib	anticancer – CML	2007	43, 480
nilutamide	antineoplastic	1987	23, 338
nilvadipine	antihypertensive	1989	25, 316
nimesulide	antiinflammatory	1985	21, 330
nimodipine	cerebral vasodilator	1985	21, 330
nimotuzumab	anticancer	2006	42, 529
nipradilol	antihypertensive	1988	24, 307
nisoldipine	antihypertensive	1990	26, 306
nitisinone	antityrosinaemia	2002	38, 361
nitrefazole	alcohol deterrent	1983	19, 322
nitrendipine	hypertensive	1985	21, 331
nizatidine	antiulcer	1987	23, 339
nizofenzone fumarate	nootropic	1988	24, 307
nomegestrol acetate	progestogen	1986	22, 324
norelgestromin	contraceptive	2002	38, 362
norfloxacin	antibacterial	1983	19, 322
norgestimate	progestogen	1986	22, 324
OCT-43	anticancer	1999	35, 345
octreotide	antisecretory	1988	24, 307
ofloxacin	antibacterial	1985	21, 331
olanzapine	neuroleptic	1996	32, 313
olimesartan Medoxomil	antihypertensive	2002	38, 363
olopatadine HCl	antiallergic	1997	33, 340
omalizumab	allergic asthma	2003	39, 280
omeprazole	antiulcer	1988	24, 308
ondansetron HCl	antiemetic	1990	26, 306
OP-1	osteoinductor	2001	37, 269
orlistat	antiobesity	1998	34, 327
ornoprostil	antiulcer	1987	23, 339
osalazine sodium	intestinal antinflamm.	1986	22, 324
oseltamivir phosphate	antiviral	1999	35, 346
oxaliplatin	anticancer	1996	32, 313
oxaprozin	antiinflammatory	1983	19, 322
oxcarbazepine	anticonvulsant	1990	26, 307
oxiconazole nitrate	antifungal	1983	19, 322
oxiracetam	nootropic	1987	23, 339
oxitropium bromide	bronchodilator	1983	19, 323
ozagrel sodium	antithrombotic	1988	24, 308
paclitaxal	antineoplastic	1993	29, 342
palifermin	mucositis	2005	41, 461
paliperidone	antipsychotic	2007	43, 482
palonosetron	antiemetic	2003	39, 281
panipenem/betamipron	carbapenem antibiotic	1994	30, 305
panitumumab	anticancer	2006	42, 531
pantoprazole sodium	antiulcer	1995	30, 306

GENERAL NAME	INDICATION	YEAR INTRO.	ARMC VOL., PAGE
parecoxib sodium	analgesic	2002	38, 364
paricalcitol	vitamin D	1998	34, 327
parnaparin sodium	anticoagulant	1993	29, 342
paroxetine	antidepressant	1991	27, 331
pazufloxacin	antibacterial	2002	38, 364
pefloxacin mesylate	antibacterial	1985	21, 331
pegademase bovine	immunostimulant	1990	26, 307
pegaptanib	age-related macular degeneration	2005	41, 458
pegaspargase	antineoplastic	1994	30, 306
pegvisomant	acromegaly	2003	39, 281
pemetrexed	anticancer	2004	40, 463
pemirolast potassium	antiasthmatic	1991	27, 331
penciclovir	antiviral	1996	32, 314
pentostatin	antineoplastic	1992	28, 334
pergolide mesylate	antiparkinsonian	1988	24, 308
perindopril	antihypertensive	1988	24, 309
perospirone HCL	neuroleptic	2001	37, 270
picotamide	antithrombotic	1987	23, 340
pidotimod	immunostimulant	1993	29, 343
piketoprofen	topical antiinflammatory	1984	20, 322
pilsicainide HCl	antiarrhythmic	1991	27, 332
pimaprofen	topical antiinflammatory	1984	20, 322
pimecrolimus	immunosuppressant	2002	38, 365
pimobendan	heart failure	1994	30, 307
pinacidil	antihypertensive	1987	23, 340
pioglitazone HCL	antidiabetic	1999	35, 346
pirarubicin	antineoplastic	1988	24, 309
pirmenol	antiarrhythmic	1994	30, 307
piroxicam cinnamate	antiinflammatory	1988	24, 309
pitavastatin	hypocholesterolemic	2003	39, 282
pivagabine	antidepressant	1997	33, 341
plaunotol	antiulcer	1987	23, 340
polaprezinc	antiulcer	1994	30, 307
porfimer sodium	antineoplastic adjuvant	1993	29, 343
posaconazole	antifungal	2006	42, 532
pramipexole HCl	antiParkinsonian	1997	33, 341
pramiracetam H_2SO_4	cognition enhancer	1993	29, 343
pramlintide	anti-diabetic	2005	41, 460
pranlukast	antiasthmatic	1995	31, 347
pravastatin	antilipidemic	1989	25, 316
prednicarbate	topical antiinflammatory	1986	22, 325
pregabalin	antiepileptic	2004	40, 464
prezatide copper acetate	vulnery	1996	32, 314
progabide	anticonvulsant	1985	21, 331
promegestrone	progestogen	1983	19, 323
propacetamol HCl	analgesic	1986	22, 325

GENERAL NAME	INDICATION	YEAR INTRO.	ARMC VOL., PAGE
propagermanium	antiviral	1994	30, 308
propentofylline propionate	cerebral vasodilator	1988	24, 310
propiverine HCl	urologic	1992	28, 335
propofol	anesthetic	1986	22, 325
prulifloxacin	antibacterial	2002	38, 366
pumactant	lung surfactant	1994	30, 308
quazepam	hypnotic	1985	21, 332
quetiapine fumarate	neuroleptic	1997	33, 341
quinagolide	hyperprolactinemia	1994	30, 309
quinapril	antihypertensive	1989	25, 317
quinfamide	amebicide	1984	20, 322
quinupristin	antibiotic	1999	35, 338
rabeprazole sodium	gastric antisecretory	1998	34, 328
raloxifene HCl	osteoporosis	1998	34, 328
raltegravir	anti-infective – HIV	2007	43, 484
raltitrexed	anticancer	1996	32, 315
ramatroban	antiallergic	2000	36, 311
ramelteon	insomnia	2005	41, 462
ramipril	antihypertensive	1989	25, 317
ramosetron	antiemetic	1996	32, 315
ranibizumab	age-related macular degeneration	2006	42, 534
ranimustine	antineoplastic	1987	23, 341
ranitidine bismuth citrate	antiulcer	1995	31, 348
ranolazine	angina	2006	42, 535
rapacuronium bromide	muscle relaxant	1999	35, 347
rasagiline	parkinson's disease	2005	41, 464
rebamipide	antiulcer	1990	26, 308
reboxetine	antidepressant	1997	33, 342
remifentanil HCl	analgesic	1996	32, 316
remoxipride HCl	antipsychotic	1990	26, 308
repaglinide	antidiabetic	1998	34, 329
repirinast	antiallergic	1987	23, 341
retapamulin	anti-infective	2007	43, 486
reteplase	fibrinolytic	1996	32, 316
reviparin sodium	anticoagulant	1993	29, 344
rifabutin	antibacterial	1992	28, 335
rifapentine	antibacterial	1988	24, 310
rifaximin	antibiotic	1985	21, 332
rifaximin	antibiotic	1987	23, 341
rilmazafone	hypnotic	1989	25, 317
rilmenidine	antihypertensive	1988	24, 310
riluzole	neuroprotective	1996	32, 316
rimantadine HCl	antiviral	1987	23, 342
rimexolone	antiinflammatory	1995	31, 348
rimonabant	anti-obesity	2006	42, 537
risedronate sodium	osteoporosis	1998	34, 330

GENERAL NAME	INDICATION	YEAR INTRO.	ARMC VOL., PAGE
risperidone	neuroleptic	1993	29, 344
ritonavir	antiviral	1996	32, 317
rivastigmin	anti-Alzheimer	1997	33, 342
rizatriptan benzoate	antimigraine	1998	34, 330
rocuronium bromide	neuromuscular blocker	1994	30, 309
rofecoxib	antiarthritic	1999	35, 347
rokitamycin	antibiotic	1986	22, 325
romurtide	immunostimulant	1991	27, 332
ronafibrate	hypolipidemic	1986	22, 326
ropinirole HCl	antiParkinsonian	1996	32, 317
ropivacaine	anesthetic	1996	32, 318
rosaprostol	antiulcer	1985	21, 332
rosiglitazone maleate	antidiabetic	1999	35, 348
rosuvastatin	hypocholesterolemic	2003	39, 283
rotigotine	parkinson's disease	2006	42, 538
roxatidine acetate HCl	antiulcer	1986	22, 326
roxithromycin	antiulcer	1987	23, 342
rufinamide	anticonvulsant	2007	43, 488
rufloxacin HCl	antibacterial	1992	28, 335
rupatadine fumarate	antiallergic	2003	39, 284
RV-11	antibiotic	1989	25, 318
salmeterol hydroxynaphthoate	bronchodilator	1990	26, 308
sapropterin HCl	hyperphenylalaninemia	1992	8, 336
saquinavir mesvlate	antiviral	1995	31 349
sargramostim	immunostimulant	1991	27, 332
sarpogrelate HCl	platelet antiaggregant	1993	29, 344
schizophyllan	immunostimulant	1985	22, 326
seratrodast	antiasthmatic	1995	31, 349
sertaconazole nitrate	topical antifungal	1992	28, 336
sertindole	neuroleptic	1996	32, 318
setastine HCl	antihistamine	1987	23, 342
setiptiline	antidepressant	1989	25, 318
setraline HCl	antidepressant	1990	26, 309
sevoflurane	anesthetic	1990	26, 309
sibutramine	antiobesity	1998	34, 331
sildenafil citrate	male sexual dysfunction	1998	34, 331
silodosin	dysuria	2006	42, 540
simvastatin	hypocholesterolemic	1988	24, 311
sitagliptin	antidiabetic	2006	42, 541
sitaxsentan	pulmonary hypertension	2006	42, 543
sivelestat	anti-inflammatory	2002	38, 366
SKI-2053R	anticancer	1999	35, 348
sobuzoxane	antineoplastic	1994	30, 310
sodium cellulose PO4	hypocalciuric	1983	19, 323
sofalcone	antiulcer	1984	20, 323
solifenacin	pollakiuria	2004	40, 466

GENERAL NAME	INDICATION	YEAR INTRO.	ARMC VOL., PAGE
somatomedin-1	growth hormone insensitivity	1994	30, 310
somatotropin	growth hormone	1994	30, 310
somatropin	hormone	1987	23, 343
sorafenib	anticancer	2005	41, 466
sorivudine	antiviral	1993	29, 345
sparfloxacin	antibiotic	1993	29, 345
spirapril HCl	antihypertensive	1995	31, 349
spizofurone	antiulcer	1987	23, 343
stavudine	antiviral	1994	30, 311
strontium ranelate	osteoporosis	2004	40, 466
succimer	chelator	1991	27, 333
sufentanil	analgesic	1983	19, 323
sulbactam sodium	b-lactamase inhibitor	1986	22, 326
sulconizole nitrate	topical antifungal	1985	21, 332
sultamycillin tosylate	antibiotic	1987	23, 343
sumatriptan succinate	antimigraine	1991	27, 333
sunitinib	anticancer	2006	42, 544
suplatast tosilate	antiallergic	1995	31, 350
suprofen	analgesic	1983	19, 324
surfactant TA	respiratory surfactant	1987	23, 344
tacalcitol	topical antipsoriatic	1993	29, 346
tacrine HCl	Alzheimer's disease	1993	29, 346
tacrolimus	immunosuppressant	1993	29, 347
tadalafil	male sexual dysfunction	2003	39, 284
talaporfin sodium	anticancer	2004	40, 469
talipexole	antiParkinsonian	1996	32, 318
taltirelin	CNS stimulant	2000	36, 311
tamibarotene	anticancer	2005	41, 467
tamsulosin HCl	antiprostatic hypertrophy	1993	29, 347
tandospirone	anxiolytic	1996	32, 319
tasonermin	anticancer	1999	35, 349
tazanolast	antiallergic	1990	26, 309
tazarotene	antipsoriasis	1997	33, 343
tazobactam sodium	b-lactamase inhibitor	1992	28, 336
tegaserod maleate	irritable bowel syndrome	2001	37, 270
teicoplanin	antibacterial	1988	24, 311
telbivudine	hepatitis B	2006	42, 546
telithromycin	antibiotic	2001	37, 271
telmesteine	mucolytic	1992	28, 337
telmisartan	antihypertensive	1999	35, 349
temafloxacin HCl	antibacterial	1991	27, 334
temocapril	antihypertensive	1994	30, 311
temocillin disodium	antibiotic	1984	20, 323
temoporphin	antineoplastic/ photosensitizer	2002	38, 367
temozolomide	anticancer	1999	35, 349

GENERAL NAME	INDICATION	YEAR INTRO.	ARMC VOL., PAGE
temsirolimus	anticancer	2007	43, 490
tenofovir disoproxil fumarate	antiviral	2001	37, 271
tenoxicam	antiinflammatory	1987	23, 344
teprenone	antiulcer	1984	20, 323
terazosin HCl	antihypertensive	1984	20, 323
terbinafine HCl	antifungal	1991	27, 334
terconazole	antifungal	1983	19, 324
tertatolol HCl	antihypertensive	1987	23, 344
thymopentin	immunomodulator	1985	21, 333
tiagabine	antiepileptic	1996	32, 319
tiamenidine HCl	antihypertensive	1988	24, 311
tianeptine sodium	antidepressant	1983	19, 324
tibolone	anabolic	1988	24, 312
tigecycline	antibiotic	2005	41, 468
tilisolol HCl	antihypertensive	1992	28, 337
tiludronate disodium	Paget's disease	1995	31, 350
timiperone	neuroleptic	1984	20, 323
tinazoline	nasal decongestant	1988	24, 312
tioconazole	antifungal	1983	19, 324
tiopronin	urolithiasis	1989	25, 318
tiotropium bromide	bronchodilator	2002	38, 368
tipranavir	HIV	2005	41, 470
tiquizium bromide	antispasmodic	1984	20, 324
tiracizine HCl	antiarrhythmic	1990	26, 310
tirilazad mesylate	subarachnoid hemorrhage	1995	31, 351
tirofiban HCl	antithrombotic	1998	34, 332
tiropramide HCl	antispasmodic	1983	19, 324
tizanidine	muscle relaxant	1984	20, 324
tolcapone	antiParkinsonian	1997	33, 343
toloxatone	antidepressant	1984	20, 324
tolrestat	antidiabetic	1989	25, 319
topiramate	antiepileptic	1995	31, 351
topotecan HCl	anticancer	1996	32, 320
torasemide	diuretic	1993	29, 348
toremifene	antineoplastic	1989	25, 319
tositumomab	anticancer	2003	39, 285
tosufloxacin tosylate	antibacterial	1990	26, 310
trabectedin	anticancer	2007	43, 492
trandolapril	antihypertensive	1993	29, 348
travoprost	antiglaucoma	2001	37, 272
treprostinil sodium	antihypertensive	2002	38, 368
tretinoin tocoferil	antiulcer	1993	29, 348
trientine HCl	chelator	1986	22, 327
trimazosin HCl	antihypertensive	1985	21, 333
trimegestone	progestogen	2001	37, 273

GENERAL NAME	INDICATION	YEAR INTRO.	ARMC VOL., PAGE
trimetrexate glucuronate	Pneumocystis carinii pneumonia	1994	30, 312
troglitazone	antidiabetic	1997	33, 344
tropisetron	antiemetic	1992	28, 337
trovafloxacin mesylate	antibiotic	1998	34, 332
troxipide	antiulcer	1986	22, 327
ubenimex	immunostimulant	1987	23, 345
udenafil	erectile dysfunction	2005	41, 472
unoprostone isopropyl ester	antiglaucoma	1994	30, 312
valaciclovir HCl	antiviral	1995	31, 352
vadecoxib	antiarthritic	2002	38, 369
vaglancirclovir HCL	antiviral	2001	37, 273
valrubicin	anticancer	1999	35, 350
valsartan	antihypertensive	1996	32, 320
vardenafil	male sexual dysfunction	2003	39, 286
varenicline	nicotine-dependence	2006	42, 547
venlafaxine	antidepressant	1994	30, 312
verteporfin	photosensitizer	2000	36, 312
vesnarinone	cardiostimulant	1990	26, 310
vigabatrin	anticonvulsant	1989	25, 319
vildagliptin	antidiabetic	2007	43, 494
vinorelbine	antineoplastic	1989	25, 320
voglibose	antidiabetic	1994	30, 313
voriconazole	antifungal	2002	38, 370
vorinostat	anticancer	2006	42, 549
xamoterol fumarate	cardiotonic	1988	24, 312
ximelagatran	anticoagulant	2004	40, 470
zafirlukast	antiasthma	1996	32, 321
zalcitabine	antiviral	1992	28, 338
zaleplon	hypnotic	1999	35, 351
zaltoprofen	antiinflammatory	1993	29, 349
zanamivir	antiviral	1999	35, 352
ziconotide	severe chronic pain	2005	41, 473
zidovudine	antiviral	1987	23, 345
zileuton	antiasthma	1997	33, 344
zinostatin stimalamer	antineoplastic	1994	30, 313
ziprasidone hydrochloride	neuroleptic	2000	36, 312
zofenopril calcium	antihypertensive	2000	36, 313
zoledronate disodium	hypercalcemia	2000	36, 314
zolpidem hemitartrate	hypnotic	1988	24, 313
zomitriptan	antimigraine	1997	33, 345
zonisamide	anticonvulsant	1989	25, 320
zopiclone	hypnotic	1986	22, 327
zuclopenthixol acetate	antipsychotic	1987	23, 345

CUMULATIVE NCE INTRODUCTION INDEX, 1983–2007 (BY INDICATION)

GENERIC NAME	INDICATION	YEAR INTRO.	ARMC VOL., (PAGE)
gemeprost	ABORTIFACIENT	1983	19 (319)
mifepristone		1988	24 (306)
lanreotide acetate	ACROMEGALY	1995	31 (345)
pegvisomant		2003	39 (281)
lisdexamfetamine	ADHD	2007	43 (477)
pegaptanib	AGE-RELATED MACULAR DEGENERATION	2005	41 (458)
ranibizumab		2006	42 (534)
nitrefazole	ALCOHOL DETERRENT	1983	19 (322)
omalizumab	ALLERGIC ASTHMA	2003	39 (280)
tacrine HCl	ALZHEIMER'S DISEASE	1993	29 (346)
quinfamide	AMEBICIDE	1984	20 (322)
tibolone	ANABOLIC	1988	24 (312)
mepixanox	ANALEPTIC	1984	20 (320)
alfentanil HCl	ANALGESIC	1983	19 (314)
alminoprofen		1983	19 (314)
dezocine		1991	27 (326)
emorfazone		1984	20 (317)
eptazocine HBr		1987	23 (334)
etoricoxib		2002	38 (355)
flupirtine maleate		1985	21 (328)
fosfosal		1984	20 (319)
ketorolac tromethamine		1990	26 (304)
meptazinol HCl		1983	19 (321)
mofezolac		1994	30 (304)
parecoxib sodium		2002	38 (364)
propacetamol HCl		1986	22 (325)
remifentanil HCl		1996	32 (316)
sufentanil		1983	19 (323)
suprofen		1983	19 (324)
desflurane	ANESTHETIC	1992	28 (329)
propofol		1986	22 (325)
ropivacaine		1996	32 (318)
sevoflurane		1990	26 (309)
levobupivacaine hydrochloride	ANESTHETIC, LOCAL	2000	36 (308)
ivabradine	ANGINA	2006	42 (522)
ranolazine		2006	42 (535)
azelaic acid	ANTIACNE	1989	25 (310)
betotastine besilate	ANTIALLERGIC	2000	36 (297)
emedastine difumarate		1993	29 (336)
epinastine		1994	30 (299)
fexofenadine		1996	32 (307)

GENERIC NAME	INDICATION	YEAR INTRO.	ARMC VOL., (PAGE)
nedocromil sodium		1986	22 (324)
olopatadine hydrochloride		1997	33 (340)
ramatroban		2000	36 (311)
repirinast		1987	23 (341)
suplatast tosilate		1995	31 (350)
tazanolast		1990	26 (309)
lodoxamide tromethamine	ANTIALLERGIC	1992	28 (333)
rupatadine fumarate		2003	39 (284)
fluticasone furoate	ANTI-ALLERGY	2007	43 (469)
loteprednol etabonate	OPHTHALMIC	1998	34 (324)
donepezil hydrochloride	ANTI-ALZHEIMERS	1997	33 (332)
rivastigmin		1997	33 (342)
gallopamil HCl	ANTIANGINAL	1983	19 (319)
cibenzoline	ANTIARRHYTHMIC	1985	21 (325)
dofetilide		2000	36 (301)
encainide HCl		1987	23 (333)
esmolol HCl		1987	23 (334)
ibutilide fumarate		1996	32 (309)
landiolol		2002	38 (360)
moricizine hydrochloride		1990	26 (305)
nifekalant HCl		1999	35 (344)
pilsicainide hydrochloride		1991	27 (332)
pirmenol		1994	30 (307)
tiracizine hydrochloride		1990	26 (310)
anakinra	ANTIARTHRITIC	2001	37 (261)
celecoxib		1999	35 (335)
etoricoxib		2002	38 (355)
meloxicam		1996	32 (312)
leflunomide		1998	34 (324)
rofecoxib		1999	35 (347)
valdecoxib		2002	38 (369)
amlexanox	ANTIASTHMATIC	1987	23 (327)
emedastine difumarate		1993	29 (336)
ibudilast		1989	25 (313)
levalbuterol HCl		1999	35 (341)
montelukast sodium		1998	34 (326)
pemirolast potassium		1991	27 (331)
seratrodast		1995	31 (349)
zafirlukast		1996	32 (321)
zileuton		1997	33 (344)
balofloxacin	ANTIBACTERIAL	2002	38 (351)
biapenem		2002	38 (351)
ciprofloxacin		1986	22 (318)
enoxacin		1986	22 (320)
ertapenem sodium		2002	38 (353)
fleroxacin		1992	28 (331)
gemifloxacin		2004	40 (458)
norfloxacin		1983	19 (322)

GENERIC NAME	INDICATION	YEAR INTRO.	ARMC VOL., (PAGE)
ofloxacin		1985	21 (331)
pazufloxacin		2002	38 (364)
pefloxacin mesylate		1985	21 (331)
pranlukast		1995	31 (347)
prulifloxacin		2002	38 (366)
rifabutin		1992	28 (335)
rifapentine		1988	24 (310)
rufloxacin hydrochloride		1992	28 (335)
teicoplanin		1988	24 (311)
temafloxacin hydrochloride		1991	27 (334)
tosufloxacin tosylate		1990	26 (310)
arbekacin	ANTIBIOTIC	1990	26 (298)
aspoxicillin		1987	23 (328)
astromycin sulfate		1985	21 (324)
azithromycin		1988	24 (298)
aztreonam		1984	20 (315)
brodimoprin		1993	29 (333)
carboplatin		1986	22 (318)
carumonam		1988	24 (298)
cefbuperazone sodium		1985	21 (325)
cefcapene pivoxil		1997	33 (330)
cefdinir		1991	27 (323)
cefepime		1993	29 (334)
cefetamet pivoxil hydrochloride		1992	28 (327)
cefixime		1987	23 (329)
cefmenoxime HCl		1983	19 (316)
cefminox sodium		1987	23 (330)
cefodizime sodium		1990	26 (300)
cefonicid sodium		1984	20 (316)
ceforanide		1984	20 (317)
cefoselis		1998	34 (319)
cefotetan disodium		1984	20 (317)
cefotiam hexetil hydrochloride		1991	27 (324)
cefpimizole		1987	23 (330)
cefpiramide sodium		1985	21 (325)
cefpirome sulfate		1992	28 (328)
cefpodoxime proxetil		1989	25 (310)
cefprozil		1992	28 (328)
ceftazidime		1983	19 (316)
cefteram pivoxil		1987	23 (330)
ceftibuten		1992	28 (329)
cefuroxime axetil		1987	23 (331)
cefuzonam sodium		1987	23 (331)
clarithromycin		1990	26 (302)
dalfopristin		1999	35 (338)
dirithromycin		1993	29 (336)

GENERIC NAME	INDICATION	YEAR INTRO.	ARMC VOL., (PAGE)
doripenem		2005	41 (448)
erythromycin acistrate		1988	24 (301)
flomoxef sodium		1988	24 (302)
flurithromycin ethylsuccinate		1997	33 (333)
fropenam		1997	33 (334)
gatifloxacin		1999	35 (340)
imipenem/cilastatin		1985	21 (328)
isepamicin		1988	24 (305)
lenampicillin HCl		1987	23 (336)
levofloxacin		1993	29 (340)
linezolid		2000	36 (309)
lomefloxacin		1989	25 (315)
loracarbef		1992	28 (333)
miokamycin		1985	21 (329)
moxifloxacin HCl		1999	35 (343)
quinupristin		1999	35 (338)
rifaximin		1985	21 (332)
rifaximin		1987	23 (341)
rokitamycin		1986	22 (325)
RV-11		1989	25 (318)
sparfloxacin		1993	29 (345)
sultamycillin tosylate		1987	23 (343)
telithromycin		2001	37 (271)
temocillin disodium		1984	20 (323)
tigecycline		2005	41 (468)
trovafloxacin mesylate		1998	34 (332)
meropenem	ANTIBIOTIC,	1994	30 (303)
panipenem/betamipron	CARBAPENEM	1994	30 (305)
mupirocin	ANTIBIOTIC, TOPICAL	1985	21 (330)
nadifloxacin		1993	29 (340)
abarelix	ANTICANCER	2004	40 (446)
alemtuzumab		2001	37 (260)
alitretinoin		1999	35 (333)
arglabin		1999	35 (335)
azacitidine		2004	40 (447)
belotecan		2004	40 (449)
bevacizumab		2004	40 (450)
bexarotene		2000	36 (298)
bortezomib		2003	39 (271)
cetuximab		2003	39 (272)
clofarabine		2005	41 (444)
dasatinib		2006	42 (517)
denileukin diftitox		1999	35 (338)
erlotinib		2004	40 (454)
exemestane		2000	36 (304)
fulvestrant		2002	38 (357)
gemtuzumab ozogamicin		2000	36 (306)

GENERIC NAME	INDICATION	YEAR INTRO.	ARMC VOL., (PAGE)
ibritumomab tiuxetan		2002	38 (359)
ixabepilone		2007	43 (473)
lapatinib		2007	43 (475)
letrazole		1996	32 (311)
nelarabine		2006	42 (528)
nimotuzumab		2006	42 (529)
OCT-43		1999	35 (345)
oxaliplatin		1996	32 (313)
panitumumab		2006	42 (531)
pemetrexed		2004	40 (463)
raltitrexed		1996	32 (315)
SKI-2053R		1999	35 (348)
sorafenib		2005	41 (466)
sunitinib		2006	42 (544)
talaporfin sodium		2004	40 (469)
tamibarotene		2005	41 (467)
tasonermin		1999	35 (349)
temozolomide		1999	35 (350)
temsirolimus		2007	43 (490)
topotecan HCl		1996	32 (320)
tositumomab		2003	39 (285)
trabectedin		2007	43 (492)
valrubicin		1999	35 (350)
vorinostat		2006	42 (549)
angiotensin II	ANTICANCER ADJUVANT	1994	30 (296)
nilotinib	ANTICANCER – CML	2007	43 (480)
chenodiol	ANTICHOLELITHOGENIC	1983	19 (317)
duteplase	ANTICOAGULANT	1995	31 (342)
lepirudin		1997	33 (336)
parnaparin sodium		1993	29 (342)
reviparin sodium		1993	29 (344)
ximelagatran		2004	40 (470)
lamotrigine	ANTICONVULSANT	1990	26 (304)
oxcarbazepine		1990	26 (307)
progabide		1985	21 (331)
rufinamide		2007	43 (488)
vigabatrin		1989	25 (319)
zonisamide		1989	25 (320)
bupropion HCl	ANTIDEPRESSANT	1989	25 (310)
citalopram		1989	25 (311)
duloxetine		2004	40 (452)
escitalopram oxalate		2002	38 (354)
fluoxetine HCl		1986	22 (320)
fluvoxamine maleate		1983	19 (319)
indalpine		1983	19 (320)
medifoxamine fumarate		1986	22 (323)
metapramine		1984	20 (320)
milnacipran		1997	33 (338)
mirtazapine		1994	30 (303)

GENERIC NAME	INDICATION	YEAR INTRO.	ARMC VOL., (PAGE)
moclobemide		1990	26 (305)
nefazodone		1994	30 (305)
paroxetine		1991	27 (331)
pivagabine		1997	33 (341)
reboxetine		1997	33 (342)
setiptiline		1989	25 (318)
sertraline hydrochloride		1990	26 (309)
tianeptine sodium		1983	19 (324)
toloxatone		1984	20 (324)
venlafaxine		1994	30 (312)
acarbose	ANTIDIABETIC	1990	26 (297)
epalrestat		1992	28 (330)
exenatide		2005	41 (452)
glimepiride		1995	31 (344)
insulin lispro		1996	32 (310)
miglitol		1998	34 (325)
mitiglinide		2004	40 (460)
nateglinide		1999	35 (344)
pioglitazone HCl		1999	35 (346)
pramlintide		2005	41 (460)
repaglinide		1998	34 (329)
rosiglitazone maleate		1999	35 (347)
sitagliptin		2006	42 (541)
tolrestat		1989	25 (319)
troglitazone		1997	33 (344)
vildagliptin		2007	43 (494)
voglibose		1994	30 (313)
acetorphan	ANTIDIARRHEAL	1993	29 (332)
anti-digoxin polyclonal antibody	ANTIDOTE	2002	38 (350)
crotelidae polyvalent immune fab		2001	37 (263)
fomepizole		1998	34 (323)
aprepitant	ANTIEMETIC	2003	39 (268)
dolasetron mesylate		1998	34 (321)
granisetron hydrochloride		1991	27 (329)
indisetron		2004	40 (459)
ondansetron hydrochloride		1990	26 (306)
nazasetron		1994	30 (305)
palonosetron		2003	39 (281)
ramosetron		1996	32 (315)
tropisetron		1992	28 (337)
felbamate	ANTIEPILEPTIC	1993	29 (337)
fosphenytoin sodium		1996	32 (308)
gabapentin		1993	29 (338)
levetiracetam		2000	36 (307)

GENERIC NAME	INDICATION	YEAR INTRO.	ARMC VOL., (PAGE)
pregabalin		2004	40 (464)
tiagabine		1996	32 (320)
topiramate		1995	31 (351)
centchroman	ANTIESTROGEN	1991	27 (324)
anidulafungin	ANTIFUNGAL	2006	42 (512)
caspofungin acetate		2001	37 (263)
eberconazole		2005	41 (449)
fenticonazole nitrate		1987	23 (334)
fluconazole		1988	24 (303)
fosfluconazole		2004	40 (457)
itraconazole		1988	24 (305)
lanoconazole		1994	30 (302)
luliconazole		2005	41 (454)
micafungin		2002	38 (360)
naftifine HCl		1984	20 (321)
oxiconazole nitrate		1983	19 (322)
posaconazole		2006	42 (532)
terbinafine hydrochloride		1991	27 (334)
terconazole		1983	19 (324)
tioconazole		1983	19 (324)
voriconazole		2002	38 (370)
amorolfine hydrochloride	ANTIFUNGAL, TOPICAL	1991	27 (322)
butenafine hydrochloride		1992	28 (327)
butoconazole		1986	22 (318)
cloconazole HCl		1986	22 (318)
liranaftate		2000	36 (309)
flutrimazole		1995	31 (343)
neticonazole HCl		1993	29 (341)
sertaconazole nitrate		1992	28 (336)
sulconizole nitrate		1985	21 (332)
apraclonidine HCl	ANTIGLAUCOMA	1988	24 (297)
befunolol HCl		1983	19 (315)
bimatroprost		2001	37 (261)
brimonidine		1996	32 (306)
brinzolamide		1998	34 (318)
dapiprazole HCl		1987	23 (332)
dorzolamide HCl		1995	31 (341)
latanoprost		1996	32 (311)
levobunolol HCl		1985	21 (328)
travoprost		2001	37 (272)
unoprostone isopropyl ester		1994	30 (312)
acrivastine	ANTIHISTAMINE	1988	24 (295)
astemizole		1983	19 (314)
azelastine HCl		1986	22 (316)
cetirizine HCl		1987	23 (331)
desloratadine		2001	37 (264)
ebastine		1990	26 (302)
levocabastine hydrochloride		1991	27 (330)
levocetirizine		2001	37 (268)

GENERIC NAME	INDICATION	YEAR INTRO.	ARMC VOL., (PAGE)
loratadine		1988	24 (306)
mizolastine		1998	34 (325)
setastine HCl		1987	23 (342)
alacepril	ANTIHYPERTENSIVE	1988	24 (296)
alfuzosin HCl		1988	24 (296)
aliskiren		2007	43 (461)
amlodipine besylate		1990	26 (298)
amosulalol		1988	24 (297)
aranidipine		1996	32 (306)
arotinolol HCl		1986	22 (316)
azelnidipine		2003	39 (270)
barnidipine hydrochloride		1992	28 (326)
benazepril hydrochloride		1990	26 (299)
benidipine hydrochloride		1991	27 (322)
betaxolol HCl		1983	19 (315)
bevantolol HCl		1987	23 (328)
bisoprolol fumarate		1986	22 (317)
bopindolol		1985	21 (324)
bosentan		2001	37 (262)
budralazine		1983	19 (315)
bunazosin HCl		1985	21 (324)
candesartan cilexetil		1997	33 (330)
carvedilol		1991	27 (323)
celiprolol HCl		1983	19 (317)
cicletanine		1988	24 (299)
cilazapril		1990	26 (301)
cinildipine		1995	31 (339)
delapril		1989	25 (311)
dilevalol		1989	25 (311)
doxazosin mesylate		1988	24 (300)
efonidipine		1994	30 (299)
enalapril maleate		1984	20 (317)
enalaprilat		1987	23 (332)
eplerenone		2003	39 (276)
eprosartan		1997	33 (333)
felodipine		1988	24 (302)
fenoldopam mesylate		1998	34 (322)
fosinopril sodium		1991	27 (328)
guanadrel sulfate		1983	19 (319)
imidapril HCl		1993	29 (339)
irbesartan		1997	33 (336)
isradipine		1989	25 (315)
ketanserin		1985	21 (328)
lacidipine		1991	27 (330)
lercanidipine		1997	33 (337)
lisinopril		1987	23 (337)
losartan		1994	30 (302)

GENERIC NAME	INDICATION	YEAR INTRO.	ARMC VOL., (PAGE)
manidipine hydrochloride		1990	26 (304)
mebefradil hydrochloride		1997	33 (338)
moexipril HCl		1995	31 (346)
moxonidine		1991	27 (330)
nebivolol		1997	33 (339)
nilvadipine		1989	25 (316)
nipradilol		1988	24 (307)
nisoldipine		1990	26 (306)
olmesartan medoxomil		2002	38 (363)
perindopril		1988	24 (309)
pinacidil		1987	23 (340)
quinapril		1989	25 (317)
ramipril		1989	25 (317)
rilmenidine		1988	24 (310)
spirapril HCl		1995	31 (349)
telmisartan		1999	35 (349)
temocapril		1994	30 (311)
terazosin HCl		1984	20 (323)
tertatolol HCl		1987	23 (344)
tiamenidine HCl		1988	24 (311)
tilisolol hydrochloride		1992	28 (337)
trandolapril		1993	29 (348)
treprostinil sodium		2002	38 (368)
trimazosin HCl		1985	21 (333)
valsartan		1996	32 (320)
zofenopril calcium		2000	36 (313)
captopril	ANTIHYPERTENSIVE AGENT	1982	13 (086)
daptomycin	ANTI INFECTIVE	2003	39 (272)
garenoxacin		2007	43 (471)
retapamulin		2007	43 (486)
maraviroc	ANTI-INFECTIVE – HIV	2007	43 (478)
raltegravir		2007	43 (484)
aceclofenac	ANTIINFLAMMATORY	1992	28 (325)
AF-2259		1987	23 (325)
amfenac sodium		1986	22 (315)
ampiroxicam		1994	30 (296)
amtolmetin guacil		1993	29 (332)
butibufen		1992	28 (327)
deflazacort		1986	22 (319)
dexibuprofen		1994	30 (298)
droxicam		1990	26 (302)
etodolac		1985	21 (327)
flunoxaprofen		1987	23 (335)
fluticasone propionate		1990	26 (303)
interferon, gamma		1989	25 (314)
isofezolac		1984	20 (319)

GENERIC NAME	INDICATION	YEAR INTRO.	ARMC VOL., (PAGE)
isoxicam		1983	19 (320)
lobenzarit sodium		1986	22 (322)
loxoprofen sodium		1986	22 (322)
lumiracoxib		2005	41 (455)
nabumetone		1985	21 (330)
nepafenac		2005	41 (456)
nimesulide		1985	21 (330)
oxaprozin		1983	19 (322)
piroxicam cinnamate		1988	24 (309)
rimexolone		1995	31 (348)
sivelestat		2002	38 (366)
tenoxicam		1987	23 (344)
zaltoprofen		1993	29 (349)
fisalamine	ANTIINFLAMMATORY,	1984	20 (318)
osalazine sodium	INTESTINAL	1986	22 (324)
alclometasone dipropionate	ANTIINFLAMMATORY,	1985	21 (323)
aminoprofen	TOPICAL	1990	26 (298)
betamethasone butyrate propionate		1994	30 (297)
butyl flufenamate		1983	19 (316)
deprodone propionate		1992	28 (329)
felbinac		1986	22 (320)
halobetasol propionate		1991	27 (329)
halometasone		1983	19 (320)
hydrocortisone aceponate		1988	24 (304)
hydrocortisone butyrate propionate		1983	19 (320)
mometasone furoate		1987	23 (338)
piketoprofen		1984	20 (322)
pimaprofen		1984	20 (322)
prednicarbate		1986	22 (325)
pravastatin	ANTILIPIDEMIC	1989	25 (316)
arteether	ANTIMALARIAL	2000	36 (296)
artemisinin		1987	23 (327)
bulaquine		2000	36 (299)
halofantrine		1988	24 (304)
mefloquine HCl		1985	21 (329)
almotriptan	ANTIMIGRAINE	2000	36 (295)
alpiropride		1988	24 (296)
eletriptan		2001	37 (266)
frovatriptan		2002	38 (357)
lomerizine HCl		1999	35 (342)
naratriptan hydrochloride		1997	33 (339)
rizatriptan benzoate		1998	34 (330)
sumatriptan succinate		1991	27 (333)

GENERIC NAME	INDICATION	YEAR INTRO.	ARMC VOL., (PAGE)
zolmitriptan		1997	33 (345)
dronabinol	ANTINAUSEANT	1986	22 (319)
amrubicin HCl	ANTINEOPLASTIC	2002	38 (349)
amsacrine		1987	23 (327)
anastrozole		1995	31 (338)
bicalutamide		1995	31 (338)
bisantrene hydrochloride		1990	26 (300)
camostat mesylate		1985	21 (325)
capecitabine		1998	34 (319)
cladribine		1993	29 (335)
cytarabine ocfosfate		1993	29 (335)
docetaxel		1995	31 (341)
doxifluridine		1987	23 (332)
enocitabine		1983	19 (318)
epirubicin HCl		1984	20 (318)
fadrozole HCl		1995	31 (342)
fludarabine phosphate		1991	27 (327)
flutamide		1983	19 (318)
formestane		1993	29 (337)
fotemustine		1989	25 (313)
geftimib		2002	38 (358)
gemcitabine HCl		1995	31 (344)
idarubicin hydrochloride		1990	26 (303)
imatinib mesylate		2001	37 (267)
interferon gamma-1a		1992	28 (332)
interleukin-2		1989	25 (314)
irinotecan		1994	30 (301)
lonidamine		1987	23 (337)
mitoxantrone HCl		1984	20 (321)
nedaplatin		1995	31 (347)
nilutamide		1987	23 (338)
paclitaxal		1993	29 (342)
pegaspargase		1994	30 (306)
pentostatin		1992	28 (334)
pirarubicin		1988	24 (309)
ranimustine		1987	23 (341)
sobuzoxane		1994	30 (310)
temoporphin		2002	38 (367)
toremifene		1989	25 (319)
vinorelbine		1989	25 (320)
zinostatin stimalamer		1994	30 (313)
porfimer sodium	ANTINEOPLASTIC ADJUVANT	1993	29 (343)
masoprocol	ANTINEOPLASTIC, TOPICAL	1992	28 (333)
miltefosine		1993	29 (340)
dexfenfluramine	ANTIOBESITY	1997	33 (332)

GENERIC NAME	INDICATION	YEAR INTRO.	ARMC VOL., (PAGE)
rimonabant		2006	42 (537)
orlistat		1998	34 (327)
sibutramine		1998	34 (331)
atovaquone	ANTIPARASITIC	1992	28 (326)
ivermectin		1987	23 (336)
budipine	ANTIPARKINSONIAN	1997	33 (330)
CHF-1301		1999	35 (336)
droxidopa		1989	25 (312)
entacapone		1998	34 (322)
pergolide mesylate		1988	24 (308)
pramipexole hydrochloride		1997	33 (341)
ropinirole HCl		1996	32 (317)
talipexole		1996	32 (318)
tolcapone		1997	33 (343)
lidamidine HCl	ANTIPERISTALTIC	1984	20 (320)
gestrinone	ANTIPROGESTOGEN	1986	22 (321)
cabergoline	ANTIPROLACTIN	1993	29 (334)
tamsulosin HCl	ANTIPROSTATIC HYPERTROPHY	1993	29 (347)
acitretin	ANTIPSORIATIC	1989	25 (309)
calcipotriol		1991	27 (323)
tazarotene		1997	33 (343)
tacalcitol	ANTIPSORIATIC, TOPICAL	1993	29 (346)
amisulpride	ANTIPSYCHOTIC	1986	22 (316)
paliperidone		2007	43 (482)
remoxipride hydrochloride		1990	26 (308)
zuclopenthixol acetate		1987	23 (345)
actarit	ANTIRHEUMATIC	1994	30 (296)
diacerein		1985	21 (326)
octreotide	ANTISECRETORY	1988	24 (307)
adamantanium bromide	ANTISEPTIC	1984	20 (315)
drotecogin alfa	ANTISEPSIS	2001	37 (265)
cimetropium bromide	ANTISPASMODIC	1985	21 (326)
tiquizium bromide		1984	20 (324)
tiropramide HCl		1983	19 (324)
argatroban	ANTITHROMBOTIC	1990	26 (299)
bivalirudin		2000	36 (298)
defibrotide		1986	22 (319)
cilostazol		1988	24 (299)
clopidogrel hydrogensulfate		1998	34 (320)
cloricromen		1991	27 (325)
enoxaparin		1987	23 (333)
eptifibatide		1999	35 (340)
ethyl icosapentate		1990	26 (303)
fondaparinux sodium		2002	38 (356)
indobufen		1984	20 (319)

GENERIC NAME	INDICATION	YEAR INTRO.	ARMC VOL., (PAGE)
limaprost		1988	24 (306)
ozagrel sodium		1988	24 (308)
picotamide		1987	23 (340)
tirofiban hydrochloride		1998	34 (332)
flutropium bromide	ANTITUSSIVE	1988	24 (303)
levodropropizine		1988	24 (305)
nitisinone	ANTITYROSINAEMIA	2002	38 (361)
benexate HCl	ANTIULCER	1987	23 (328)
dosmalfate		2000	36 (302)
ebrotidine		1997	33 (333)
ecabet sodium		1993	29 (336)
egualen sodium		2000	36 (303)
enprostil		1985	21 (327)
famotidine		1985	21 (327)
irsogladine		1989	25 (315)
lansoprazole		1992	28 (332)
misoprostol		1985	21 (329)
nizatidine		1987	23 (339)
omeprazole		1988	24 (308)
ornoprostil		1987	23 (339)
pantoprazole sodium		1994	30 (306)
plaunotol		1987	23 (340)
polaprezinc		1994	30 (307)
ranitidine bismuth citrate		1995	31 (348)
rebamipide		1990	26 (308)
rosaprostol		1985	21 (332)
roxatidine acetate HCl		1986	22 (326)
roxithromycin		1987	23 (342)
sofalcone		1984	20 (323)
spizofurone		1987	23 (343)
teprenone		1984	20 (323)
tretinoin tocoferil		1993	29 (348)
troxipide		1986	22 (327)
abacavir sulfate	ANTIVIRAL	1999	35 (333)
adefovir dipivoxil		2002	38 (348)
amprenavir		1999	35 (334)
atazanavir		2003	39 (269)
cidofovir		1996	32 (306)
delavirdine mesylate		1997	33 (331)
didanosine		1991	27 (326)
efavirenz		1998	34 (321)
emtricitabine		2003	39 (274)
enfuvirtide		2003	39 (275)
entecavir		2005	41 (450)
famciclovir		1994	30 (300)
fomivirsen sodium		1998	34 (323)
fosamprenavir		2003	39 (277)
foscarnet sodium		1989	25 (313)

GENERIC NAME	INDICATION	YEAR INTRO.	ARMC VOL., (PAGE)
ganciclovir		1988	24 (303)
imiquimod		1997	33 (335)
indinavir sulfate		1996	32 (310)
interferon alfacon-1		1997	33 (336)
lamivudine		1995	31 (345)
lopinavir		2000	36 (310)
nelfinavir mesylate		1997	33 (340)
nevirapine		1996	32 (313)
oseltamivir phosphate		1999	35 (346)
penciclovir		1996	32 (314)
propagermanium		1994	30 (308)
rimantadine HCl		1987	23 (342)
ritonavir		1996	32 (317)
saquinavir mesylate		1995	31 (349)
sorivudine		1993	29 (345)
stavudine		1994	30 (311)
tenofovir disoproxil fumarate		2001	37 (271)
valaciclovir HCl		1995	31 (352)
zalcitabine		1992	28 (338)
zanamivir		1999	35 (352)
zidovudine		1987	23 (345)
influenza virus live	ANTIVIRAL VACCINE	2003	39 (277)
cevimeline hydrochloride	ANTI-XEROSTOMIA	2000	36 (299)
alpidem	ANXIOLYTIC	1991	27 (322)
buspirone HCl		1985	21 (324)
etizolam		1984	20 (318)
flutazolam		1984	20 (318)
flutoprazepam		1986	22 (320)
metaclazepam		1987	23 (338)
mexazolam		1984	20 (321)
tandospirone		1996	32 (319)
ciclesonide	ASTHMA, COPD	2005	41 (443)
atomoxetine	ATTENTION DEFICIT HYPERACTIVITY DISORDER	2003	39 (270)
flumazenil	BENZODIAZEPINE ANTAG.	1987	23 (335)
bambuterol	BRONCHODILATOR	1990	26 (299)
doxofylline		1985	21 (327)
formoterol fumarate		1986	22 (321)
mabuterol HCl		1986	22 (323)
oxitropium bromide		1983	19 (323)
salmeterol hydro-xynaphthoate		1990	26 (308)
tiotropium bromide		2002	38 (368)
APD	CALCIUM REGULATOR	1987	23 (326)
clodronate disodium		1986	22 (319)
disodium pamidronate		1989	25 (312)

GENERIC NAME	INDICATION	YEAR INTRO.	ARMC VOL., (PAGE)
gallium nitrate		1991	27 (328)
ipriflavone		1989	25 (314)
neridronic acid		2002	38 (361)
dexrazoxane	CARDIOPROTECTIVE	1992	28 (330)
bucladesine sodium	CARDIOSTIMULANT	1984	20 (316)
denopamine		1988	24 (300)
docarpamine		1994	30 (298)
dopexamine		1989	25 (312)
enoximone		1988	24 (301)
flosequinan		1992	28 (331)
ibopamine HCl		1984	20 (319)
loprinone hydrochloride		1996	32 (312)
milrinone		1989	25 (316)
vesnarinone		1990	26 (310)
amrinone	CARDIOTONIC	1983	19 (314)
colforsin daropate HCL		1999	35 (337)
xamoterol fumarate		1988	24 (312)
cefozopran HCL	CEPHALOSPORIN, INJECTABLE	1995	31 (339)
cefditoren pivoxil	CEPHALOSPORIN, ORAL	1994	30 (297)
brovincamine fumarate	CEREBRAL VASODILATOR	1986	22 (317)
nimodipine		1985	21 (330)
propentofylline		1988	24 (310)
succimer	CHELATOR	1991	27 (333)
trientine HCl		1986	22 (327)
fenbuprol	CHOLERETIC	1983	19 (318)
lulbiprostone	CHRONIC IDIOPATHIC CONSTIPATION	2006	42 (525)
deferasirox	CHRONIC IRON OVERLOAD	2005	41 (446)
arformoterol	CHRONIC OBSTRUCTIVE PULMONARY DISEASE	2007	43 (465)
auranofin	CHRYSOTHERAPEUTIC	1983	19 (314)
taltirelin	CNS STIMULANT	2000	36 (311)
aniracetam	COGNITION ENHANCER	1993	29 (333)
pramiracetam H_2SO_4		1993	29 (343)
carperitide	CONGESTIVE HEART	1995	31 (339)
nesiritide	FAILURE	2001	37 (269)
drospirenone	CONTRACEPTIVE	2000	36 (302)
norelgestromin		2002	38 (362)
nicorandil	CORONARY VASODILATOR	1984	20 (322)
dornase alfa	CYSTIC FIBROSIS	1994	30 (298)
neltenexine		1993	29 (341)
amifostine	CYTOPROTECTIVE	1995	31 (338)
nalmefene HCL	DEPENDENCE TREATMENT	1995	31 (347)
ioflupane	DIAGNOSIS CNS	2000	36 (306)
azosemide	DIURETIC	1986	22 (316)
muzolimine		1983	19 (321)
torasemide		1993	29 (348)
atorvastatin calcium	DYSLIPIDEMIA	1997	33 (328)
cerivastatin		1997	33 (331)

GENERIC NAME	INDICATION	YEAR INTRO.	ARMC VOL., (PAGE)
naftopidil	DYSURIA	1999	35 (343)
silodosin		2006	42 (540)
alglucerase	ENZYME	1991	27 (321)
udenafil	ERECTILE DYSFUNCTION	2005	41 (472)
erdosteine	EXPECTORANT	1995	31 (342)
fudosteine		2001	37 (267)
agalsidase alfa	FABRY'S DISEASE	2001	37 (259)
cetrorelix	FEMALE INFERTILITY	1999	35 (336)
ganirelix acetate		2000	36 (305)
follitropin alfa	FERTILITY ENHANCER	1996	32 (307)
follitropin beta		1996	32 (308)
reteplase	FIBRINOLYTIC	1996	32 (316)
esomeprazole magnesium	GASTRIC ANTISECRETORY	2000	36 (303)
lafutidine		2000	36 (307)
rabeprazole sodium		1998	34 (328)
cinitapride	GASTROPROKINETIC	1990	26 (301)
cisapride		1988	24 (299)
itopride HCL		1995	31 (344)
mosapride citrate		1998	34 (326)
imiglucerase	GAUCHER'S DISEASE	1994	30 (301)
miglustat		2003	39 (279)
somatotropin	GROWTH HORMONE	1994	30 (310)
somatomedin-1	GROWTH HORMONE INSENSITIVITY	1994	30 (310)
factor VIIa	HAEMOPHILIA	1996	32 (307)
levosimendan	HEART FAILURE	2000	36 (308)
pimobendan		1994	30 (307)
anagrelide hydrochloride	HEMATOLOGIC	1997	33 (328)
erythropoietin	HEMATOPOETIC	1988	24 (301)
eculizumab	HEMOGLOBINURIA	2007	43 (468)
factor VIII	HEMOSTATIC	1992	28 (330)
telbivudine	HEPATITIS B	2006	42 (546)
clevudine		2007	43 (466)
malotilate	HEPATOPROTECTIVE	1985	21 (329)
mivotilate		1999	35 (343)
darunavir	HIV	2006	42 (515)
tipranavir		2005	41 (470)
buserelin acetate	HORMONE	1984	20 (316)
goserelin		1987	23 (336)
leuprolide acetate		1984	20 (319)
nafarelin acetate		1990	26 (306)
somatropin		1987	23 (343)
zoledronate disodium	HYPERCALCEMIA	2000	36 (314)
cinacalcet	HYPERPARATHYROIDISM	2004	40 (451)
sapropterin hydrochloride	HYPERPHENYL-ALANINEMIA	1992	28 (336)
quinagolide	HYPERPROLACTINEMIA	1994	30 (309)
cadralazine	HYPERTENSIVE	1988	24 (298)
nitrendipine		1985	21 (331)
binfonazole	HYPNOTIC	1983	19 (315)
brotizolam		1983	19 (315)

GENERIC NAME	INDICATION	YEAR INTRO.	ARMC VOL., (PAGE)
butoctamide		1984	20 (316)
cinolazepam		1993	29 (334)
doxefazepam		1985	21 (326)
eszopiclone		2005	41 (451)
loprazolam mesylate		1983	19 (321)
quazepam		1985	21 (332)
rilmazafone		1989	25 (317)
zaleplon		1999	35 (351)
zolpidem hemitartrate		1988	24 (313)
zopiclone		1986	22 (327)
acetohydroxamic acid	HYPOAMMONURIC	1983	19 (313)
sodium cellulose PO4	HYPOCALCIURIC	1983	19 (323)
divistyramine	HYPOCHOLESTEROLEMIC	1984	20 (317)
lovastatin		1987	23 (337)
melinamide		1984	20 (320)
pitavastatin		2003	39 (282)
rosuvastatin		2003	39 (283)
simvastatin		1988	24 (311)
glucagon, rDNA	HYPOGLYCEMIA	1993	29 (338)
acipimox	HYPOLIPIDEMIC	1985	21 (323)
beclobrate		1986	22 (317)
binifibrate		1986	22 (317)
ciprofibrate		1985	21 (326)
colesevelam hydrochloride		2000	36 (300)
colestimide		1999	35 (337)
ezetimibe		2002	38 (355)
fluvastatin		1994	30 (300)
meglutol		1983	19 (321)
ronafibrate		1986	22 (326)
conivaptan	HYPONATREMIA	2006	42 (514)
mozavaptan		2006	42 (527)
modafinil	IDIOPATHIC HYPERSOMNIA	1994	30 (303)
bucillamine	IMMUNOMODULATOR	1987	23 (329)
centoxin		1991	27 (325)
thymopentin		1985	21 (333)
filgrastim	IMMUNOSTIMULANT	1991	27 (327)
GMDP		1996	32 (308)
interferon gamma-1b		1991	27 (329)
lentinan		1986	22 (322)
pegademase bovine		1990	26 (307)
pidotimod		1993	29 (343)
romurtide		1991	27 (332)
sargramostim		1991	27 (332)
schizophyllan		1985	22 (326)

GENERIC NAME	INDICATION	YEAR INTRO.	ARMC VOL., (PAGE)
ubenimex		1987	23 (345)
cyclosporine	IMMUNOSUPPRESSANT	1983	19 (317)
everolimus		2004	40 (455)
gusperimus		1994	30 (300)
mizoribine		1984	20 (321)
muromonab-CD3		1986	22 (323)
mycophenolate sodium		2003	39 (279)
mycophenolate mofetil		1995	31 (346)
pimecrolimus		2002	38 (365)
tacrolimus		1993	29 (347)
ramelteon	INSOMNIA	2005	41 (462)
defeiprone	IRON CHELATOR	1995	31 (340)
alosetron hydrochloride	IRRITABLE BOWEL SYNDROME	2000	36 (295)
tegasedor maleate		2001	37 (270)
sulbactam sodium	b-LACTAMASE INHIBITOR	1986	22 (326)
tazobactam sodium		1992	28 (336)
nartograstim	LEUKOPENIA	1994	30 (304)
pumactant	LUNG SURFACTANT	1994	30 (308)
sildenafil citrate	MALE SEXUAL DYSFUNCTION	1998	34 (331)
gadoversetamide	MRI CONTRAST AGENT	2000	36 (304)
telmesteine	MUCOLYTIC	1992	28 (337)
laronidase	MUCOPOLYSACCARIDOSIS	2003	39 (278)
galsulfase	MUCOPOLYSACCHARIDOSIS VI	2005	41 (453)
idursulfase	MUCOPOLYSACCHARIDOSIS II (HUNTER SYNDROME)	2006	42 (520)
palifermin	MUCOSITIS	2005	41 (461)
interferon X-1a	MULTIPLE SCLEROSIS	1996	32 (311)
interferon X-1b		1993	29 (339)
glatiramer acetate		1997	33 (334)
natalizumab		2004	40 (462)
afloqualone	MUSCLE RELAXANT	1983	19 (313)
cisatracurium besilate		1995	31 (340)
doxacurium chloride		1991	27 (326)
eperisone HCl		1983	19 (318)
mivacurium chloride		1992	28 (334)
rapacuronium bromide		1999	35 (347)
tizanidine		1984	20 (324)
decitabine	MYELODYSPLASTIC SYNDROMES	2006	42 (519)
lenalidomide	MYELODYSPLASTIC SYNDROMES, MULTIPLE MYELOMA	2006	42 (523)
naltrexone HCl	NARCOTIC ANTAGONIST	1984	20 (322)
tinazoline	NASAL DECONGESTANT	1988	24 (312)
aripiprazole	NEUROLEPTIC	2002	38 (350)
clospipramine hydrochloride		1991	27 (325)

GENERIC NAME	INDICATION	YEAR INTRO.	ARMC VOL., (PAGE)
nemonapride		1991	27 (331)
olanzapine		1996	32 (313)
perospirone hydrochloride		2001	37 (270)
quetiapine fumarate		1997	33 (341)
risperidone		1993	29 (344)
sertindole		1996	32 (318)
timiperone		1984	20 (323)
ziprasidone hydrochloride		2000	36 (312)
rocuronium bromide	NEUROMUSCULAR BLOCKER	1994	30 (309)
edaravone	NEUROPROTECTIVE	1995	37 (265)
fasudil HCL		1995	31 (343)
riluzole		1996	32 (317)
varenicline	NICOTINE-DEPENDENCE	2006	42 (547)
bifemelane HCl	NOOTROPIC	1987	23 (329)
choline alfoscerate		1990	26 (300)
exifone		1988	24 (302)
idebenone		1986	22 (321)
indeloxazine HCl		1988	24 (304)
levacecarnine HCl		1986	22 (322)
nizofenzone fumarate		1988	24 (307)
oxiracetam		1987	23 (339)
bromfenac sodium	NSAID	1997	33 (329)
lornoxicam		1997	33 (337)
OP-1	OSTEOINDUCTOR	2001	37 (269)
alendronate sodium	OSTEOPOROSIS	1993	29 (332)
ibandronic acid		1996	32 (309)
incadronic acid		1997	33 (335)
raloxifene hydrochloride		1998	34 (328)
risedronate sodium		1998	34 (330)
strontium ranelate		2004	40 (467)
imidafenacin	OVERACTIVE BLADDER	2007	43 (472)
tiludronate disodium	PAGET'S DISEASE	1995	31 (350)
rasagiline	PARKINSON'S DISEASE	2005	41 (464)
rotigotine		2006	42 (538)
tadalafil	PDE5 INHIBITOR	2003	39 (284)
vardenafil		2003	39 (286)
temoporphin	PHOTOSENSITIZER	2002	38 (367)
verteporfin		2000	36 (312)
alefacept	PLAQUE PSORIASIS	2003	39 (267)
beraprost sodium	PLATELET AGGREG. INHIBITOR	1992	28 (326)
epoprostenol sodium		1983	19 (318)
iloprost		1992	28 (332)

GENERIC NAME	INDICATION	YEAR INTRO.	ARMC VOL., (PAGE)
sarpogrelate HCl	PLATELET ANTIAGGREGANT	1993	29 (344)
trimetrexate glucuronate	PNEUMOCYSTIS CARINII PNEUMONIA	1994	30 (312)
solifenacin	POLLAKIURIA	2004	40 (466)
alglucosidase alfa	POMPE DISEASE	2006	42 (511)
histrelin	PRECOCIOUS PUBERTY	1993	29 (338)
atosiban	PRETERM LABOR	2000	36 (297)
gestodene	PROGESTOGEN	1987	23 (335)
nomegestrol acetate		1986	22 (324)
norgestimate		1986	22 (324)
promegestrone		1983	19 (323)
trimegestone		2001	37 (273)
alpha-1 antitrypsin	PROTEASE INHIBITOR	1988	24 (297)
nafamostat mesylate		1986	22 (323)
adrafinil	PSYCHOSTIMULANT	1986	22 (315)
dexmethylphenidate HCl		2002	38 (352)
dutasteride		2002	38 (353)
efalizumab	PSORIASIS	2003	39 (274)
ambrisentan	PULMONARY ARTERIAL HYPERTENSION	2007	43 (463)
sitaxsentan	PULMONARY HYPERTENSION	2006	42 (543)
finasteride	5a-REDUCTASE INHIBITOR	1992	28 (331)
surfactant TA	RESPIRATORY SURFACTANT	1987	23 (344)
abatacept	RHEUMATOID ARTHRITIS	2006	42 (509)
Adalimumab		2003	39 (267)
dexmedetomidine hydrochloride	SEDATIVE	2000	36 (301)
ziconotide	SEVERE CHRONIC PAIN	2005	41 (473)
kinetin	SKIN PHOTODAMAGE/ DERMATOLOGIC	1999	35 (341)
tirilazad mesylate	SUBARACHNOID HEMORRHAGE	1995	31 (351)
APSAC	THROMBOLYTIC	1987	23 (326)
alteplase		1987	23 (326)
balsalazide disodium	ULCERATIVE COLITIS	1997	33 (329)
darifenacin	URINARY INCONTINENCE	2005	41 (445)
tiopronin	UROLITHIASIS	1989	25 (318)
propiverine hydrochloride	UROLOGIC	1992	28 (335)
Lyme disease	VACCINE	1999	35 (342)
clobenoside	VASOPROTECTIVE	1988	24 (300)
falecalcitriol	VITAMIN D	2001	37 (266)
maxacalcitol		2000	36 (310)
paricalcitol		1998	34 (327)
doxercalciferol	VITAMIN D PROHORMONE	1999	35 (339)
prezatide copper acetate	VULNERARY	1996	32 (314)
acemannan	WOUND HEALING AGENT	2001	37 (257)
cadexomer iodine		1983	19 (316)
epidermal growth factor		1987	23 (333)

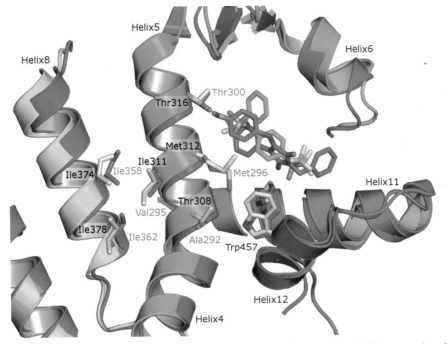

Plate 7.1 Different conformations of a key methionine residue in mLXRα [27] (magenta) and hLXRβ [30] (coloured from blue to red along the sequence).

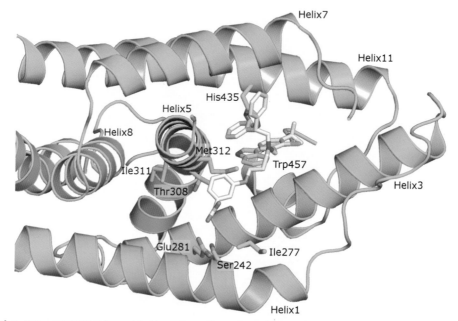

Plate 7.2 GW3965 **2** bound in Trp/His pocket of LXRβ extending the phenylacetic acid group into the solvent-exposed area.

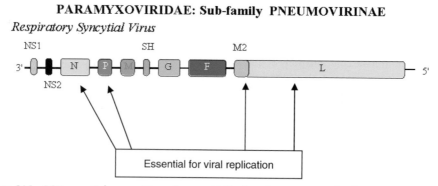

PARAMYXOVIRIDAE: Sub-family PNEUMOVIRINAE

Respiratory Syncytial Virus

Plate 14.1 RSV essential genes. N, nucleocapsid, P, phosphoprotein, and L, long protein including polymerase. (Acknowledgement, Andrew Easton, University of Warwick.)

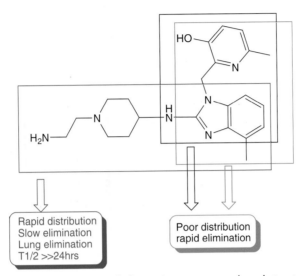

Plate 14.2 JNJ-2408086 distribution and elimination assessment by substructure.

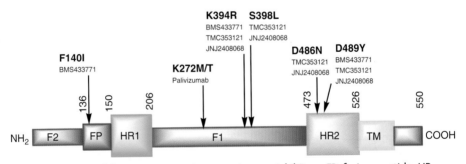

Plate 14.3 Reported fusion protein point mutations to inhibitors. FP, fusion peptide, HR, heptad repeat and TM, transmembrane.

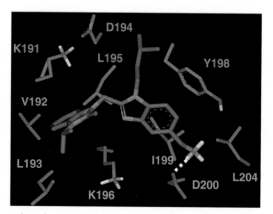

Plate 14.4   Binding pocket showing 5-aminomethylene interaction with an aspartic acid (D200).

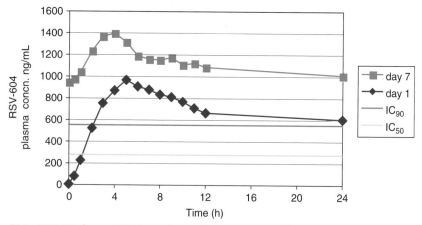

Plate 14.5   RSV-604 human PK data, where volunteers received 600 mg on day 1 followed by 450 mg on days 2–7.